Mycology: Current and Future Developments

(Volume 3)

Myconanotechnology: Green Chemistry for Sustainable Development

Edited by

Savita
Department of Botany, Hindu College
University of Delhi, India

Anju Srivastava
Department of Chemistry, Hindu College
University of Delhi, India

Reena Jain
Department of Chemistry, Hindu College
University of Delhi, India

&

Pratap Kumar Pati
Department of Biotechnology
Guru Nanak Dev University, Amritsar, Punjab, India

Mycology: Current and Future Developments

Volume # 3

Myconanotechnology: Green Chemistry for Sustainable Development

Editors: Savita, Anju Srivastava, Reena Jain & Pratap Kumar Pati

ISSN (Online): 2452-0780

ISSN (Print): 2452-0772

ISBN (Online): 978-981-5051-36-0

ISBN (Print): 978-981-5051-37-7

ISBN (Paperback): 978-981-5051-38-4

need for a court order if at any point you breach any terms of this License Agreement. In no event will any delay or failure by Bentham Science Publishers in enforcing your compliance with this License Agreement constitute a waiver of any of its rights.

3. You acknowledge that you have read this License Agreement, and agree to be bound by its terms and conditions. To the extent that any other terms and conditions presented on any website of Bentham Science Publishers conflict with, or are inconsistent with, the terms and conditions set out in this License Agreement, you acknowledge that the terms and conditions set out in this License Agreement shall prevail.

Bentham Science Publishers Pte. Ltd.
80 Robinson Road #02-00
Singapore 068898
Singapore
Email: subscriptions@benthamscience.net

BENTHAM SCIENCE

CONTENTS

FOREWORD

Nanotechnology is a fast-growing field of science that involves the synthesis and development of various nanomaterials, production, manipulation and use of materials ranging in size from less than a micron to that of individual atoms. The most commonly employed procedures are based on physical and chemical techniques therefore it is reasonable to develop alternative biological methods with an important ecological impact.

Fungal systems have attracted much attention in this regard, as they are eco-friendly, biodegradable, easy to culture, nontoxic, low-cost and scale up, with a high wall binding capacity, thus heralding great promise for their application in biotechnology and industry. Formation of nanoparticles employing fungi and their application in medicine, agriculture and other areas is known as myconanotechnology. Nowadays, this kind of green synthesized nanoparticles have attracted great attention due to their outstanding biocompatibility and eco-friendly nature leading to explore new commercial uses and benefits of fungi.

The present book is organised into two sections, A and B. Section A will update readers on several cutting-edge aspects of the synthesis and characterization of nanoparticles through the use of fungi; then section B will describe some applications of myconanotechnology for management of bacterial and fungal diseases, pest control, as well as for application in medicine and agriculture making this a fairly new and exciting area of research with considerable potential for further development.

Annalisa De Girolamo
Senior Scientist
Institute of Sciences of Food Production (ISPA)
National Research Council of Italy (CNR)
Via G. Amendola 122/O, 70126 Bari (Italy)

PREFACE

Mycology is the study of fungi which has become a valuable science in the last 100 years as it has wide applications in agriculture, food and beverages industry, biofuel production, biotechnology, medicine, *etc*. Recently nanotechnology has emerged as a potential candidate due to its immense applications in other fields of science such as chemistry, biology, physics, materials science, engineering and biomedicine. Nanotechnology is the art of science of manipulating matter at nano scale design (1-100 nm). Myconanotechnology is an interface between mycology and nanotechnology which can be defined as the green synthesis of nanoparticles using fungi which is recently gaining attention due to their cost-effective, sustainable, resource efficient, simplicity and ecofriendly nature. While the synthesis of nanoparticles through conventional chemical and physical methods requires high energy, they are toxic and expensive too. Myconanotechnology has a wide range of applications like in agriculture, biomedical, electronics, textiles, cell and molecular biology, nanodevices, and many more. In the proposed book entitled Myconanotechnology, the editors have included the recent advanced approaches for studying mycology for the benefit of humankind. The book will cover the history and scope of myconanotechnology, green synthesis of nanoparticles by using fungi, the mechanism involved in such biosynthesis, applications of green nanomaterials in agriculture, industries, cosmetics, wood industry, biofuel production, therapeutics, diagnostics, *etc*. This book will be immensely useful to researchers, scientists of fungal biology, teachers and students, microbiologists, nanotechnologists, technocrats, and to those who are interested in myconanotechnology. The present book is intended to provide both students and researchers with a broad background to some of the fastest developing areas in current myconanotechnology. The chapters in the book are loosely grouped into two sections (Synthesis of nanoparticles and their applications in environment, industry, medicine and agriculture) in order to reflect the wider 'customers' or context of the particular myconanotechnological areas or activities. Each chapter is contributed by internationally recognized researchers/scientists, so the reader is given up-to-date and detailed information about myconanotechnology and various applications of fungi. It is to be noted that the authors of individual chapters are solely responsible for any scientific queries. We are indebted to the scientists and active researchers who have contributed their research work in the form of chapters and made this compilation a unique collection of recent studies in the field of myconanotechnology.

Reena Jain
Department of Chemistry
Hindu College
University of Delhi, India

Savita
Department of Botany
Hindu College
University of Delhi, India

Pratap Kumar Pati
Professor
Department of Biotechnology
Guru Nanak Dev University
Amritsar, Punjab, India

Anju Srivastava
Department of Chemistry
Hindu College
University of Delhi, India

List of Contributors

Ana Elsi Ulloa Pérez   Instituto Politécnico Nacional, CIIDIR Unidad Sinaloa, Depto. de Medio Ambiente, Blvd. Juan de Dios Bátiz Paredes 250 Guasave, Sinaloa, CP 81101, Mexico

Anamika Das   Department of Paramedical Sciences, Guru Kashi University, Talwandi Sabo, Bathinda, Punjab-151302, India

Anita Puyam   Department of Plant Pathology, College of Agriculture, Rani Lakshmi Bai Central Agricultural University, Jhansi, Uttar Pradesh-284003, India

Anitha Roy   Department of Pharmacology, Saveetha Dental College and Hospitals, Saveetha Institute Medical and Technical Sciences, Chennai-600077, India

Anjana K. Vala   Department of Life Sciences, Maharaja Krishnakumarsinhji Bhavnagar University, Bhavnagar, Gujarat, India

Ankita Verma   Division of Agricultural Chemicals, ICAR-Indian Agricultural Research Institute, New Delhi-110 012, India

Ankur Maheshwari   Department of Zoology, Zakir Husain Dehli College, University of Dehli, Dehli, India

Apurva Mishra   Topological Molecular Biology Laboratory, Department of Microbiology and Molecular Genetics, University of California, Davis, California 95616, USA

Babita Singh   ICAR-National Research Centre for Integrated Pest Management, New Delhi-110012, India

Bharti P. Dave   Department of Biosciences, School of Sciences, Indrashi lUniversity, Rajpur-Kadi, Mehasana, Gujarat, India

Conor F. McGee   Department of Agriculture, Food and the Marine, Backweston Laboratories, Celbridge, Co. Kildare, Ireland

Eduardo Fidel Héctor Ardisana   Instituto de Posgrado, Universidad Técnica de Manabí, Portoviejo, Ecuador, C. P. EC130105, Mexico

Evelyn M. Doyle   School of Biology and Environmental Science, University College Dublin, Belfield, D4, Co. Dublin, Ireland

Eygenia Stamatelopoulou   Department of Food Science and Technology, University of the Peloponnese, Antikalamos, 24100 Kalamata, Greece

Garima Deswal   Department of Biotechnology, Deenbandhu Chhotu Ram University of Science & Technology, Murthal (Sonepat), Haryana, India

Haren B. Gosai   Department of Biosciences, School of Sciences, Indrashi lUniversity, Rajpur-Kadi, Mehasana, Gujarat, India

Haresh Z. Panseriya   Gujarat Ecology Society, 3rd Floor, Synergy House, Subhanpura, Vadodara, Gujarat, India

Harsha Nirvan   Department of Biotechnology, Deenbandhu Chhotu Ram University of Science & Technology, Murthal (Sonepat), Haryana, India

Hiral B. Trivedi   Department of Biosciences, School of Sciences, Indrashi lUniversity, Rajpur-Kadi, Mehasana, Gujarat, India

Karunakaran Rohini	Faculty of Medicine, AIMST University, Semeling-08100, Kedah, Malaysia
Lham Dorjee	Division of Plant Pathology, Indian Agricultural Research Institute, Pusa, New Delhi-110012, India
Lorena Patricia Licón Trillo	Facultad de Ciencias Agrícolas y Forestales, Universidad Autónoma de Chihuahua, Km 2.5, carretera Delicias-Rosales, campus Delicias, CD. Delicias, Chihuahua, CP 33000, Mexico
Madan L. Verma	Department of Biotechnology, Indian Institute of Information Technology Una, Himachal Pradesh-177220, India
Manjit K. Selwal	DST Women Scientist Scheme (WoS)-B, Deenbandhu Chhotu Ram University of Science & Technology, Murthal (Sonepat), Haryana, India
María Esther González Vega	Instituto Nacional de Ciencias Agrícolas (INCA), Carretera a Tapaste, Km 3,5. San José de las Lajas, Mayabeque, CP 32700, Mexico
Meenu Thakur	Department of Biotechnology, Shoolini Institute of Life Sciences and Business Management, Solan-173212, Himachal Pradesh, India
Neelam R. Yadav	Department of Molecular Biology, Biotechnology and Bioinformatics, CCS Haryana Agricultural University, Hisar-125004, India
Prem Lal Kashyap	ICAR-Indian Institute of Wheat and Barley Research Karnal, Haryana-132001, India
R.C. Yadav	Department of Molecular Biology, Biotechnology and Bioinformatics, CCS Haryana Agricultural University, Hisar-125004, India
Rajinder S. Beniwal	Department of Plant, Pathology, CCS Haryana Agricultural University, Hisar-125004, India
Rekha Kushwaha	Division of Biological Sciences, University of Missouri, Columbia Missouri-(65211), USA
Renuka Agrawal	Department of Botany, Miranda House, University of Delhi, Delhi, india
Rizwana Rehsawla	Department of Molecular Biology, Biotechnology and Bioinformatics, CCS Haryana Agricultural University, Hisar-125004, India
S. Rajeshkumar	Department of Pharmacology, Saveetha Dental College and Hospitals, Saveetha Institute Medical and Technical Sciences, Chennai-600077, India
Saloni Bahri	Department of Botany, Miranda House, University of Delhi, Delhi, India
Sandra Pérez Álvarez	Facultad de Ciencias Agrícolas y Forestales, Universidad Autónoma de Chihuahua, Km 2.5, carretera Delicias-Rosales, campus Delicias, CD. Delicias, Chihuahua, CP 33000, Mexico
Santosh Kumar	Donald Danforth Plant Science Center, 975 North Warson Raod, Saint Louis, Missouri-63132, USA
Shweta Meshram	Division of Plant Pathology, Indian Agricultural Research Institute, Pusa, New Delhi-110012, India
Sofia Agriopoulou	Department of Food Science and Technology, University of the Peloponnese, Antikalamos, 24100 Kalamata, Greece
Somdutta Sinha Roy	Department of Botany, Miranda House, University of Delhi, Delhi, India
Sonali Singhal	ICAR-National Research Centre for Integrated Pest Management, New Delhi-110012, India

Sunaina Bisht   Department of Plant Pathology, College of Agriculture, Rani Lakshmi Bai Central Agricultural University, Jhansi, Uttar Pradesh-284003, India

Tanzeel Ahmed   School of Biotechnology, IFTM University, Lodhipur-Rajput, NH-24, Delhi Road, Moradabad, Uttar Pradesh-244001, India

Theodoros Varzakas   Department of Food Science and Technology, University of the Peloponnese, Antikalamos, 24100 Kalamata, Greece

Vasiliki Skiada   Department of Food Science and Technology, University of the Peloponnese, Antikalamos, 24100 Kalamata, Greece

Veerasamy Ravichandran   Faculty of Pharmacy, AIMST University, Semeling-08100, Kedah, Malaysia

CHAPTER 1

Fungi and Nanotechnology: History and Scope

Haresh Z. Panseriya[1], Haren B. Gosai[2], Hiral B. Trivedi[2], Anjana K. Vala[3,*] and Bharti P. Dave[2]

[1] *Gujarat Ecology Society, 3rd Floor, Synergy House, Subhanpura, Vadodara, Gujarat, India*

[2] *Department of Biosciences, School of Sciences, Indrashi University, Rajpur-Kadi, Mehasana, Gujarat, India*

[3] *Department of Life Sciences, Maharaja Krishnakumarsinhji Bhavnagar University, Bhavnagar, Gujarat, India*

Abstract: Nanotechnology is one of the most fascinating areas of research, it is the cutting-edge technology that has a great impact on various application fields. Nanoparticles have been under consideration due to their applicability in almost every field. There are many methods used for the synthesis of nanoparticles but biological methods have proved to be superior. Among various biological sources, microorganisms have gained attention recently. Bacterial nanoparticle syntheses from terrestrial as well as from marine habitats have been frequently studied as compared to fungal counterparts. Recently, Fungal Nanotechnology has received much attention as it has a big role to play in future as well. During the last decades, marine fungi have been observed to exhibit novel nanotechnological application potentialities. This chapter deals with the history and emergence of myconanotechnology, focusing on terrestrial as well as marine fungal resources. Fungal nanoproducts have noteworthy scope in diverse fields. This chapter also discusses the scope of myconanotechnology in future.

Keywords: Fungi, History, Marine, Nanotechnology, Scope, Terrestrial.

INTRODUCTION

Nanotechnology involves designing, synthesizing and characterizing application-oriented materials and devices' smallest functional organization of which at least in one dimension is on nanometer scale. Materials at this scale exhibit unique properties than their bulk counterparts [1].

Nanotechnology is increasingly being used in agriculture, medicine, environment, textiles and opto-electronics for over the last two decades due to the tunable properties of nanomaterials [2]. Specifically, metal nanoparticles have applicati-

---

* **Corresponding author Anjana K. Vala**: Department of Life Sciences, Maharaja Krishnakumarsinhji Bhavnagar University, Bhavnagar, Gujarat, India; Tel: +912782519824; Email: akv@mkbhavuni.edu.in, anjana_vala@yahoo.co.in

Savita, Anju Srivastava, Reena Jain & Pratap Kumar Pati (Eds.)

ons in resistance of arthropods [3], high prevalence of antimicrobial agents in different microorganisms [4], antitumor therapy [5]. Apart from that, metal nanoparticles have also increased interests of researchers as an alternative class of agents with antiviral, larvicidal, antiprotozoal, acaricidal, *etc* [6 - 8].

Numerous methods of nanoparticles synthesis, such as chemical, biological and physical, have been explored. Mostly, chemical methods are used in the fabrication of nanoparticles because it facilitates large quantities production in relatively short time with a good control on the shape, size and distribution [9]. In chemical methods, the variety of sizes and shape could be adjusted by controlling the reaction conditions and concentration of reacting chemicals. But, most of these methods are employ toxic chemicals, produce hazardous waste and energy intensive in nature. Similarly, physical methods such as microwave-assisted synthesis, laser ablation, sputter deposition, *etc.*, are available for synthesis of nanoparticles. Due to involvement of high temperature, radiations, pressures in physical methods biogenic synthesis of nanoparticles is gaining interest [10].

Investigation and development of optimized protocol for the synthesis of nanoparticles of tailored size and shape is required advancement of nanotechnology [11]. Hence, scientists around the globe are foreseeing biological systems that can be used as an efficient system for the synthesis of metal nanoparticles. These biological systems include microorganisms *i.e.* bacteria, cyanobacteria, fungi, algae, *etc.* and plants for intracellular and extracellular synthesis of nanoparticles [12 - 14]. Among these microorganisms, fungi are most commonly used due to their wide distribution in nature and hence, play an important role in synthesis of nanoparticles. In 2009, Mahendra Rai and his co-workers have proposed the term "Myconanotechnology" for integrated research on mycology and nanotechnology [15]. Since then in last one decade myconanotechnology is gaining interest of scientists, policymakers and entrepreneurs due to many advantages and revenue generations in pharmaceuticals, chemical and healthcare industries.

FUNGI AS AN EFFICIENT SYSTEM FOR SYNTHESIS OF NANOPARTICLES

Fungi have various enzymatic and protein machinery which can be used as reducing agents, subsequently used in synthesizing nanoparticles from their salts. Fungi secrete a large amount of protein at a very high rate and therefore, the conversion of salts to nanoparticles is at a very high rate compared to bacteria. Additionally, fungal biomass also grows faster compared to bacteria under the same conditions. Mycelia of fungi offer many folds larger surface area for interaction and facilitate synthesis at a higher rate compared to bacteria [16 - 19].

Mycelia of fungi can tolerate high flow pressure and agitation in bioreactors compared to other microbes and plants [20]. The cell wall of fungi gives mechanical strength to tolerate osmotic pressure and environmental stress [21]. Another advantage is economic viability of fungi as large-scale synthesis is possible with small amount of biomass [22, 23]. These characteristics make fungi an efficient system for the synthesis of nanoparticles.

HISTORY, OCCURRENCE OF FUNGI AND EMERGENCE OF MYCONANOTECHNLOGY

The fungi have ancient origin and evidence that indicate the fungi possibly initially appeared billion years ago, therefore, record from fossil is very little. This was confirmed by the fungal hyphae evidence within the from oldest plant fossils. At the present knowledge on fungi called *Prototaxites* from terrestrial plantlike fossil source were common in all parts of the world throughout in Devonian period (419.2 million to 358.9 million years ago). Filamentous fungi from fossil of *Tortotubus protuberans* till the date oldest fungal species on the globe from Silurian period (440 million years ago) from fossil of terrestrial source [24]. During 20th century, classification of fungi was done and in the middle of the 20th century fungi were solely recognized as a distinct kingdom. This is based on their mode of nutrition (using excretion of digestive enzymes and absorption of externally digested nutrients).

Fungi are ubiquitous and occur abundantly in **(a)** terrestrial as well as in **(b)** marine environment.

Terrestrial Systems and Various Sources of Fungi

A terrestrial ecosystem involves the interaction of biotic and abiotic components and land-based community of organism in specific area, *e.g* of the terrestrial ecosystem are tundra, taigas, temperate deciduous forests, tropical rainforests, grasslands, and deserts. The types of terrestrial ecosystem were dependent on major four parameters *i.e.* temperature range in the area, average annual precipitation received, type of the soil and the light intensity received in the area.

Fungi are one of the largest terrestrial and aquatic organisms which play various roles in ecological and environmental balance. Till date, approximately 1.5 million fungal species were recognized by scientific communities [25, 26]. In worldwide, a minimum of 712285 extant fungal species are found, out of which 600,000 fungal species are found associated with terrestrial plants [27]. The use of current cultivation techniques has led to10% of the total fungi have been cultured [28].

Literature regarding fungal collection indicates still need formal description on isolation and cultivation, methods, physiology of fungal species, diversity in fungi [29]. About 90% of the higher fungi species are recognized worldwide [30]. Especially in China, near about 5000 fungal species have been recognized out of 1200 genera [31].These fungal species are either lichenized or macrofungi, whereas other half represent microfungi *i.e.* from aquatic fungi, soil-inhabiting fungi, terrestrial plant-associated fungi, and arthropod-associated fungi [27]. Therefore, a source of higher fungus represents megadiverse bioresources. Due to this reason, fungi have special metabolism and produce various functional metabolites with different chemical structure can able to stop cell proliferation and differentiation to achieve the self-defense mechanism which has potentially used drug delivery. Moreover, more fungal diversity also helps in the development of new drugs [32].

Fungi are unique eukaryotic organisms. As per the current knowledge, about 70000 fungal species have been explored and identified from 1.5 million fungal species in the globe. Recently identification processes were done using the help of high throughput sequencing methods, and 5.1 million fungal species were identified [33].

The Terrestrial ecosystem, fungi play a crucial role and maintain various processes as decomposers of plant detritus and mutualistic partners of most of the higher terrestrial multicellular organisms. This decomposer fungi diversity is very high in forest and other ecosystems particularly where high grazing, and less fire and human harvesting activities were observed in terrestrial ecosystem. Annually plant forest ecosystem can produce and add organic matter around 5-33 t/ha, and 73 petagrams of estimated global carbon pool is bound in dead wood [34]. These lignocelluloses, recalcitrant organic matter were efficiently decomposed by only fungi [35]. The entire process was crucial for energy and release of nutrients, so using the basis of soil food chains and is grazed on directly or indirectly by a wide range of invertebrate and vertebrate taxa by fungi [36].

As mentioned earlier, the list of the mutualistic relations of fungi such as lichens and mycorrhizae is very high in the ecosystems. Primary production and nitrogen fixation in terrestrial environment with adverse conditions have been observed to be associated with highly stress tolerant fungi and green algae or cyanobacteria [37]. In case of the other habitat or climate zone *i.e.* forest ecosystem, fungi form microhabitats and use of tree trunk, rock surface and living leaves of the forest trees [38]. Almost all plants rely on mycelial network of fungi for the nutrients and uptake of crucial minerals such as N, P and other minerals from soil and the fungi can receive enough amount of the sugar from their partner and produce around 15–30% of the net primary production [39]. In forest ecosystem, mycorrhizal fungi control most of the natural environmental cycles, *e.g.* nutrient cycling, mineral weathering and carbon storage [40, 41]. The fungi have a diverse group of species and each fungus has different enzymatic activities according to the native environmental changes, either natural or anthropogenic pressure, which can shift ecosystem processes and be closely involved in plant growth and competition [42].

More specifically, all vascular plants' internal tissues host diverse communities of fungal endophytes. Some of these fungal endophytes prevent attacks of herbivores and pathogens, whereas others have decomposer strategies [43]. Globally these fungal endophytes represent hyperdiverse group in case of known species and unknown bioactive compounds [44]. Exceptionally due to environmental changes or evolutionary processes beneficial fungi could be converted into pathogenic also.

Among that, the alternative approach has been emerged as biological synthesis of nanoparticles by using physical and chemical methods. This new field recognized as 'green nanotechnology' or 'nanobiotechnology'. The uses of biological principles with physical and chemical procedures with all three together develop eco-friendly nanoparticles for specified functions. Now days, plants, fungi, algae, bacteria, yeast and viruses used for the synthesis of nanoparticles. Researchers now focused on the synthesis of nanoparticles, especially by fungi, as a potential source that can lead to various bioactive properties for application in the biomedicine field.

Nowadays an extensive research on fungi and their enzymes, their mode of action was carried out by the researchers. Selection of fungi is majorly due to the ease to handle different scales, therefore it is widely used in the synthesis of nanoparticles by using thin solid substrate fermentation technique. At industrial level to achieve large amount of target enzymes, production was feasible due to secretion by the fungi [45]. The economic feasibility and facility of implying biomass for the fungus species is another advantage for the utilization of the green approach to synthesize metallic nanoparticles.

In addition, fungi are easy to culture and maintain due to their fast growth including high wall-binding and metal uptake properties in laboratory [45]. Along with that, fungi also produced metal nanoparticles through the enzymatic activity of intracellularly or extracellularly, *i.e.* silver nanoparticles. The developmental processes for silver nanoparticles were almost similar with slight modification in the processes depending on the type of fungi species used for the synthesis of nanoparticles [46 - 48].

Previous studies indicated various research groups working on fungi during the last decade. Chemical and physical approaches for the development of nanoparticles have some limitations, such as more expensive, though the scientific community is running towards clean, ecofriendly and economically beneficial approach [49 - 51].

Nowadays, the production or synthesis of silver nanoparticles is on the top list due to their various applications in the biomedical field. Various countries, including India, are also working and developing cost-effective silver nanoparticles from various sources.

Mukherjee focused on the synthesis of silver and gold nanoparticles as a natural extracellular product by filamentous fungal species *Verticillium*. This was the first evidence of harnessing a filamentous fungus for nanotechnological studies [52, 53]. Again filamentous fungi were used to check biocatalytic properties of produced nanoparticles by Mukherjee [53] and Sawle [54]. Additionally, Mohanpuria *et al.*, 2008 investigated heavy metal tolerance, the productivity of proteins and amount of nanoparticle synthesis with minimum time interval [55]. Various literature indicates synthesis pathways and methods for constructing nanoparticles by using mycosynthesis [20, 46, 56, 57]. Still throughout the globe, there is continuous effort to uncover synthesis of nanoparticles from different geological terrestrial ecosystem.

The synthesis of nanoparticles can be of both types: intra- and extra-cellularly [58]. Intracellular synthesis is subject to treatment of metal salt solution and incubation for 24 h in the dark condition as synthesis will take place inside the cell wall. Whereas, in extracellular synthesis, the fungal filtrate is treated with a metal salt solution and incubated till formation of nanoparticles. Extracellular nanoparticles synthesis is much faster compared to intracellular nanoparticles synthesis [59]. Additionally, nanoparticles synthesized from extracellular methods are much larger compared to the intracellularly synthesized nanoparticles [60]. These differences in size may be due to the nucleation of particles inside the fungus. Moreover, in intracellular methods of synthesis, downstream processing becomes very tedious and difficult as synthesis takes place inside the cell [61]. In contrast to that, in extracellular methods of synthesis, extensive downstream process is not necessary, which offers easier and cost effective synthesis of nanoparticles.

There are some reports on intra- and extra-cellular synthesis of nanoparticles that have been briefly discussed below.

Sastry *et al.* have demonstrated synthesis of silver nanoparticles (size range 2-25nm) from *Verticillium sp.* using intracellular method [62]. They have observed clear deposition of the metal on the surface of cytoplasmic membrane. In another similar study, gold nanoparticles have been synthesized using the same fungus [52]. Some other studies were also carried out to synthesized gold nanoparticles using fungi such as *Trichothecium spp., Penicillium chrysogenum,* and *Verticillium luteoalbum* [63 - 65]. Similarly, silver nanoparticles were intra-cellularly synthesized using *Aspergillus flavus* [23].

Extracellular synthesis of gold nanoparticles using *Alternaria sp.* has been reported by Dhanasekar [66]. They have used different concentrations of chloroaurate solution to optimize the size of the nanoparticles. They have optimized 1mN chloroaurate solution concentration for the synthesis of spherical, rod, pentagonal shape nanoparticles and the same was confirmed using TEM analysis. Apart from these shapes, quasi-spherical and heart like morphologies of gold nanoparticles were also achieved at lower concentration (0.3-0.5 mM). Devi and Joshi have isolated three endophytic fungi *Pencillium ochrochloron PFR8, Aspergillus tamarii PFL2,* and *Aspergillus niger PFR6,* from medicinal plant *Potentillafulgens L.,* for the synthesis of silver nanoparticles. They have reported *A. tamarii PFL2* synthesized nanoparticles have the smallest average particle size (3.5 nm) compared to the other two fungi [67].

Literature suggested extracellular metabolites from biomass of natural product exposed with metallic ion solution [52]. This was supported by Ahmed, utilized fungus with CdS metabolites with other natural products such as PbS, ZnS and MoS_2. In aqueous solution, production of natural products and the presence of proteins were confirmed by sulfate-diminishing enzyme-based procedure. Hence, fungi can transform their morphology and change into 5-50 nm size nanoparticles [12].

Tarafdar [68] isolated *Aspergillius tubingensis* TFR-5 from agricultural farm of Central Arid Zone Research Institute (CAZRI) located in Jodhpur, India and confirmed synthesis of phosphorous nanoparticles from tri-calcium phosphate by *Aspergillius tubingensis* TFR-5.

In the terrestrial system, Arbuscular mycorrhizal fungi (AMF) and 90% of land plant root have a mutualistic symbiosis [69]. Feng investigated and revealed the responses of mycorrhizal clover (*Trifoliumrepens*) to silver nanoparticles (AgNPs) and iron oxide nanoparticles (FeONPs) using each concentration gradient of each silver nanoparticles (AgNPs) and iron oxide nanoparticles (FeONPs). The study suggested mycorrhizal clover biomass was 34% significantly reduced by FeONPs at 3.2 [69].

Proceedings of ISBN documented potential fungi synthesis nanoparticles using nanotechnology [70]. Ahmed showed *Fusarium oxysporum* able to synthesized extracellular biosynthesis of silver nanoparticles from ionic silver [12]. Moreover, nanoparticles can catalyze certain enzymes. Similarly, two fungal species *i.e.* *Aspergillus* and *Neurospora* synthesized gold microwires [71]. The reports indicate that fungi were treated with gold for a week and that were surface functionalized with glutamate, aspartate, and polyethylene glycol. The consumption of this organic compound by the growing fungi resulted in the assembly of the gold microwires. Finally, endocytosis process can enter nanoparticles into the fungal cells [72].

Neethu investigated and explained the synthesis of silver nanoparticles using cell filtrate of *Penicillium polonicum* with an optimum concentration of silver nanoparticles, fungal mycelium, pH, optimum time and effect of light. Characterization of nanoparticles and establishment of antibacterial activity was checked against *Acenetobacter baumanii*. Besides that, using TEM, interaction of *A. baumanii* with silver nanoparticles was studied [73].

Ahmed also detected and reported amide I and amide II responsible for the stability of AgNPs in the aqueous solution. This has also medicinal application [74].

Šebesta isolated filamentous fungus *Aspergillus niger* (Tiegh.), strain CBS 140837 from the mercury-contaminated soil. The results confirmed that the influence of soil fungus *A. niger* transformation was carried out of ZnO nanoparticles to biogenic mineral phases and the process took place in soil environment under the right circumstances [75].

Fungi *Thielaviopsis basicola* and *Phytophthora nicotianae* were isolated from black root- and black stem-infected tobacco plants in continuously 10 years growing field from Chongqing. A study was conducted to check the effect of nanoparticles MgO and soil borne pathogens *P. nicotianae* and *T. bacicola* in two different systems *i.e. in vitro* and in a green house. The study illustrated nanoparticles have a significant inhibition effect on spore germination, sporangium formation, and hyphal development [76]. Edible mushrooms also have been found to biosynthesize gold and silver nanoparticles [77].

Not only filamentous fungi but yeasts also have been found to be a very good source for biosynthesizing various nanoparticles. To the best of authors' knowledge, yeasts had been harnessed even before their filamentous fungal counterparts for biosynthesizing nanoparticles. Dameron discovered biosynthesis of quantum CdS crystallites in yeasts *Candida glabrata* and *Schizosaccharomyces pombe* and the role of short chelating peptides in controlling nucleation and growth of CdS crystallites [78]. *Saccharomyces cerevisae* had been observed to carry out extracellular synthesis of gold and silver nanoparticles. Various mechanisms involved in yeast-based biosynthesis of nanoparticles include sorption, chelation, enzymatic reduction and controlled cell membrane transport of heavy metals [79, 80].

Fungi in Marine Environment

Oceans are reservoirs of rich biodiversity. Fungi from marine habitats were first described in the middle of 19th century [81]. According to their biogeochemical distribution, they can be categorized as temperate, tropical, subtropical and cosmopolitan species. Based on their ability to grow and sporulate, these fungi can be classified as obligate marine fungi and facultative marine fungi, where the former grow and sporulate exclusively in marine or estuarine habitat while the latter have terrestrial or freshwater origin having the ability to grow and possibly sporulate in marine habitats [82].

Marine-derived fungi have been isolated from nearly all habitats, including seawater, sediments, marine plants, sponges and other microorganisms. They play an important role in biodegradation of pollutants [83 - 85] and produce a number of bioactive compounds with novel traits and application potentialities [86].

Kathiresan carried out pioneering work on the synthesis of nanoparticles using marine-derived fungi *Penicillium fellutanum* (isolated from mangrove sediment of south India). They observed the formation of silver nanoparticles in the culture filtrates. Kathiresan reported extracellular synthesis of silver nanoparticles (5-35nm) by *Aspergillus niger* obtained from coastal mangrove sediment and observed the presence of 70KDa protein in culture filtrate by carrying out SDS-PAGE analysis [87].

Pioneering work on mycosynthesis of silver and gold nanoparticles has been undertaken at four institutions in India. National chemical Laboratory (NCL), Pune carried out studies on biosynthesis using fungi from other sources while Annamalai University initiated work on silver nanoparticle synthesis by marine-derived filamentous fungi, Maharaja Krishnakumarsinghji Bhavnagar University initiated work on gold and silver nanoparticles by marine-derived filamentous fungi and University of Pune worked on gold and silver nanoparticles using marine-derived yeasts.

Marine-derived filamentous fungi from Bhavnagar Coast, Gulf of Khambhat, West Coast of India were examined for their silver and gold nanoparticle potentials [11, 18, 88 - 90] and revealed mycobiota of Bhavnagar coast as efficient myco-nanofactories for silver and gold nanoparticles with application potentialities. Plate 1 displays silver nanoparticle synthesis by a marine-derived

Aspergillus sp. Moreover, Zomorodian also synthesized silver nanoparticles from *Aspergillus niger* [91].

Plate (1). Biosynthesis of silver nanoparticles by a marine-derived *Aspergillus* sp. (left: before, right after biosynthesis).

Recently, Hulikere and Joshi reported silver nanoparticle synthesis by a marine endophytic fungus *Cladosporium cladosporoides*. The authors suggested NADPH-dependent reductase as the responsible enzyme for the formation of AgNPs [92].

Vala examined three marine-derived fungal isolates viz. *Aspergillus candidus, Aspergillus flavus* and *Aspergillus niger* for their gold nanoparticle biosynthesis potential. It was observed that all the test isolates generally synthesized gold nanoparticles extracellularly, *A. candidus* showed concentration dependent change in mode of biosynthesis [90]. Concentration dependent change in behaviour was also observed in a marine-derived *Rhizopusoryzae* by Vala [90] and in marine-derived *Aspergillus sydowii* [18].

Dave reported the formation of gold nanoparticles when a marine-derived *Aspergillus niger* was exposed to Au(III) at different pH (7-10) [19]. The authors observed a decrease in particle size with increasing pH. Gold nanoparticle biosynthesis had been observed in tropical marine yeast *Yarrowialipolytica* NCIM 3589 [93].

Marine yeast *Pichia capsulata* were observed to be effective in synthesizing silver nanoparticles [94]. Manivannan reported silver nanoparticles by a marine yeast *Pichia capsulate* [95]. *Rhodosporidium diobovatum* was reported to synthesize lead nanoparticles (2-5nm) by Seshadri [50].

SCOPE OF MYCO-DERIVED NANOPARTICLES

Nanotechnologies have been recently used in various fields of science such as agriculture, pharmaceutical, material science, chemistry, physics and medicine. The promising results in above fields opened lots of scope of nanotechnology recently. According to the European Union, farming management concept in agriculture and their output from available resources has widely accepted new technologies, including nanotechnology. Therefore, nanotechnology is widely used in terrestrial system *i.e.* modern farming. In modern research, nanotechnology has emerged as new area by using nanoparticles with size smaller than 100 nm for the synthesis and application. Moreover, using nanoparticles, nanostructured materials and devices are also created.

Metal nanoparticles synthesized by fungi have numerous potential applications in the areas of agriculture, healthcare and pest control. Synthesis of fungal nanoparticles is advantageous in terms of large production of metabolites. Additionally, fungi have the capacity to produce antibiotics that could be contained in the capping and act in synergy with the nanoparticle core. Many reports have suggested controlling the pathogenic fungi and bacteria, providing larvicidal and insecticidal activities and combating cancer (Fig. **1**).

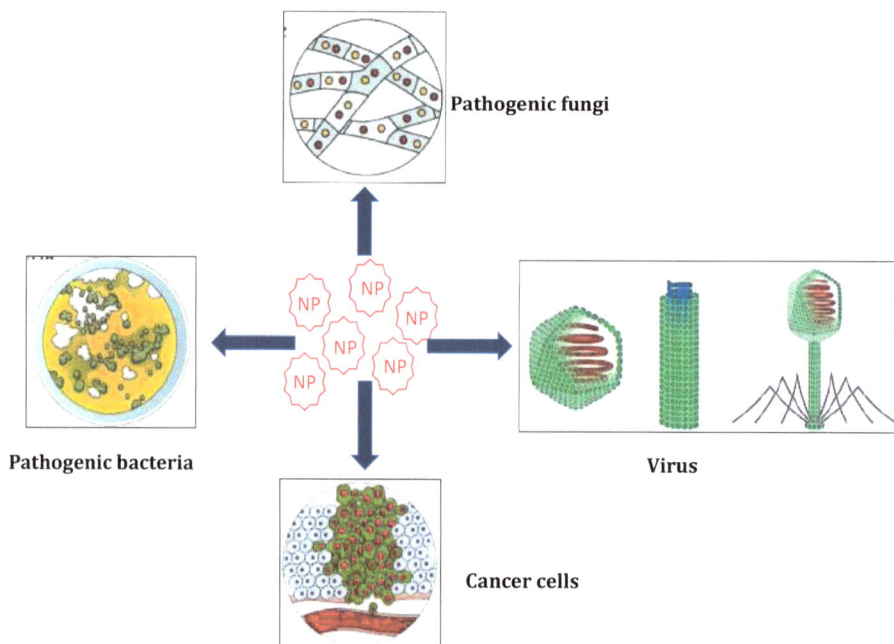

Fig. (1). Applications of myco-derived nanoparticles.

SCOPE IN AGRICULTURE

There are many studies to investigate the potential of metal nanoparticles synthesized by fungi for the control of phytopathogenic fungi in agriculture. There have been few studies to evaluate the potential of metal nanoparticles synthesized using biogenic methods for the control of phytopathogenic fungi in agriculture and pests. Table **1** shows some studies in which silver nanoparticles synthesized from different fungal species were employed in agriculture and pests control.

Table 1. Myco-derived fungal nanoparticles and its applications in agriculture.

Fungal Source	Activity	Nanoparticles	Target Organism	Effective Concentration	References
Trichoderma harzianum	Antifungal	Silver	*Sclerotinia sclerotiorum*	0.15 × 1012 and 0.31 × 1012 NPs/mL	[102]
Arthrodermafulvum	Antifungal	Silver	*Fusarium sp.*	0.125–4.00µg/mL	[103]
Alternaria alternata	Antifungalcombinedwith Fluconazol	Silver	*Phomaglomerata, Phomaherbarum, and Fusarium semitectum*	20 µL/disk	[98]
Trichoderma harzianum	Insecticide	Silver	*Aedes aegypti*	0.2–1.0%	[104]
Beauveria bassiana	Insecticide	Silver	*Aedes aegypti*	0.06–1.0 ppm	[100]
A. niger	Antibacterial	Silver	-	-	[105]
F. oxysporum	Live cell imaging and diagnostics	Cadmium	-	-	[106]
Rhizopus oryzae	Water hygiene management	Gold	-	-	[107]
Phytophthora infestans	Antimicrobial activity	Silver	-	-	[108]

Silver nanoparticles synthesized from *Aspergillus versicolor* used as an antifungal against fungi such as *Sclerotinia sclerotiorum* and *Botrytis cinereal* in strawberry plants. Elgorban has observed the maximum effect of these nanoparticles against *Botrytis cinerea* [96]. Similarly, Qian has also synthesized silver nanoparticles using fungus *Epicoccum nigrum* and reported antifungal activity against pathogenic fungi *Candida albicans, Sporothrix schenckii, Aspergillus flavus* and *Aspergillus fumigates* [97]. Apart from antifungal activity, many studies have been reported in the combination of biogenic particles and conventional biocides. Silver nanoparticles synthesized from *Alternaria alternata* are used in combination with antifungal compound fluconazole, against the phytopathogenic fungi *Fusarium semitectum, Phomaherbarum* and *Phoma glomerate* [98]. These nanoparticles were also effective against the biological control agent *Trochoderma sp.* and the human pathogenic fungus *Candida albicans.* One more study reported that silver nanoparticles in combination with antifungal compound triclabendazole for controlling the parasite *Fasciola sp.,* which affects sheep and cattle. In this study, it was observed that nanoparticles combined with triclabendazole had inhibited egg hatching by 90.6%, while only triclabendazole has inhibited 70.6%. These studies suggested that the use of the nanoparticles together with the drug could be a way to overcome the resistance that the target organisms has developed toward the drug.

Banu and Balasubramaniam have also synthesized silver nanoparticles using the fungus *Isaria fumosorosea* and they have applied these nanoparticles against the mosquito species *Culex quinquefasciatus* and *Adedes aegypti* for mortality test. The results suggested the greatest effectiveness against *Aedes aegypti*, for which the mortality of 1st instar larvae reached 100% within 24h. Apart from that, the 4th instar larvae of both mosquito species was lower susceptible to the nanoparticles. Thus, these nanoparticles could be considered as potential larvicides for mosquito control [99]. The same authors also synthesized silver nanoparticles using the mycelial extract of the *Beauveria bassiana.* They have also obtained 100% mortality of the 1st and 2nd instar larvae of *Aedes aegypti* within 21h of exposure to the nanoparticles. These nanoparticles could be considered for the environmentally safe strategy for vector control, followed by field applications [100].

Apart from its major application as antifungal and antibacterial agents for the management of plant pathogens, nanoparticles can serve as nanofertilizers, nano-pesticides and nano-insecticides also. Additionally, they are also used in the development of nanobiosensors which can be useful for the preparation of devices, useful in precision farming. Excessive use of chemical fertilizers and water resources leads to the decrement in fertility of the soil and eventually in crop production. Nanofertilizers could be an attractive alternative for the regaining and protection of fertility of the soil with minimum damage to the soil. These nanofertilizers can be used for targeted delivery or slow and controlled release or conditional release according to environmental condition and biological demands. These controlled and targeted delivery of nanofertilizers into the damaged soil lead to increment in nutrients, reduces soil toxicity and minimizes the potential negative effects associated to the chemical fertilizers. This approach could be an attractive tool for the development of sustainable agriculture at the large scale level in the developing countries of Asia and Africa. Chinnamuthu and

Boopathi have suggested the use of naturally occurring minerals, such as nano clays and zeolites, as nanofertilizers [101].

SCOPE OF FUNGAL NANOPARTICLES IN BIOMEDICAL AND HEALTHCARE

There are many studies reported describing the use of nanoparticles synthesized from fungi for applications in the area of health, involving the control of bacteria, fungi and tumor cells. Growth of bacteria is directly inhibited by the nanoparticles, which contact the cell wall and give metabolic responses, with the production of reactive oxygen species [109]. The smaller nanoparticles have greater effects as compared to bigger nanoparticles because they can penetrate the bacterial cell membrane and damage the respiratory chain, cause DNA and RNA damage, affect cell division, eventually lead to the death of cell [15]. Especially, interaction of nanoparticles with thiol groups of essential enzymes leads to the release of silver ion, eventually formation of complexes with nucleotides and damaging the DNA.

Nanoparticles synthesized from fungi have also potential to become an alternative approach for the treatment of fungal infections [110]. The large surface area of metal nanoparticles, especially silver nanoparticles can contribute to high antimicrobial activity. The toxic ions bind to sulfur containing proteins and affect cell permeability, leading to a change in DNA replication process. The interaction of nanoparticles with thiol groups of some enzymes influences electron transport and protein oxidation [111, 112]. There are some studies that reported fungal nanoparticles in biomedical and healthcare applications as per given Table **2**.

Table 2. Myco-derived nanoparticles and their applications in biomedical and healthcare.

Synthesis Source	Nanoparticles	Bioactivity	Target Organism	Effective Concentration	References
Aspergillus tubingiensis and Bionectriaochroleuca	Silver	Antibacterial	*Candida sp.* and *Pseudomonas aeruginosa*	0.11–1.75µg/mL (*Candida sp.*) and 0.28µg/mL (*P. aeruginosa*)	[117]
Fusarium verticillioides	Silver	Antibacterial	*S. aureus* and *E. coli*	5 and 10mM	[118]
Inonotus obliquus	Silver	Antibacterial and antiproliferative in cancer cells	*E. coli KCTC 2441 A549* (lung cancer) and MCF-7 (breast cancer)	30 µL/disk	[119]
Pleurotusdjamor var. roseus	Silver	Antiproliferative in cancer cells	PC3 cells (prostate cancer)	10–40µg/mL	[120]

(Table 2) cont.....

Synthesis Source	Nanoparticles	Bioactivity	Target Organism	Effective Concentration	References
Penicillium chrysogenum and Aspergillus oryzae	Silver	Antifungal	*Trichophyton rubrum*	*P. chrysogenum* 0.5µg/mL and *A. oryzae* >7.5µg/mL	[121]
Penicillium italicum	Silver	Antibacterial and antifungal	Multidrug-resistant *S.aureus, Shewanella putrefaciens,* and *C. albicans*	25 µL/disk	[122]
Fusarium oxysporum	Silver	Antifungal	*Candida spp.* and *Cryptococcus spp.*	*Candida spp.* 0.84-1.68µg/mL and *Cryptococcus spp.* 0.42–0.84µg/mL	[123]
T. viride	Gold	Antibacterial	*S. aureus, E.coli, VRSA*	-	[124]
A. niger	Silver	Wound healing activity	-	-	[125]

Silver nanoparticles synthesized from fungus *Guignardia mangiferae* were observed for antibacterial activity against gram-negative bacteria. These nanoparticles had given effects such as increased permeability, the release of nucleic acids and changes in membrane transport [113]. In contrast to that, these silver nanoparticles were not effective against gram-positive bacteria due to the layer of peptidoglycans that act as a barrier and prevent penetration of the nanoparticles [114]. However, some studies have shown exceptional results with gram-positive bacteria and shown inhibitory effects against gram-positive bacteria.

Ahluwalia has synthesized silver nanoparticles using *Trichoderma harzianum* to control growth of *Klebsiella pneumonia* and *Staphylococcus aureus* in *in vitro*. The rates of inhibition were observed dependent on concentration with gram-negative bacterium (*K. pneumoniae*) and showed higher sensitivity [115]. Whereas, Gade has reported inhibitory effect against gram positive bacteria *S. aureus* using silver nanoparticles synthesized from *A. niger*. These inhibitory effects were equivalent to those of the antibiotic gentamicin against *S. aureus* [105].

Some of the fungal synthesized metal nanoparticles have also been used in combination with antifungals and antibiotics, for possible solution to the problem of resistance towards these drugs used in health area. Silver nanoparticles synthesized from *Candida albicans* used alone and in combination with the ciprofloxacin against *S. aureus, E. coli, B. cereus and V. cholerae*. The results showed that antibiotic activity was higher when nanoparticles were used in combination with antibiotics [116]. One more study has reported antimicrobial and antifungal activity of silver nanoparticles synthesized from filtrate of *Aspergillus flavus*.

These nanoparticles were used to control the growth of *Bacillus subtilis, Escherichia coli, Staphylococcus aureus, Enterobacter aerogenes, Bacillus cereus*. Results have shown higher sensitivity of silver nanoparticles against *B. subtilis and E. coli*. As per the earlier studies, in this study also, activity was concentration-dependent. Better results were obtained using nanoparticles in combination with tetracycline rather than on their own.

Bao reported the production of cadmium telluride quantum dots using yeast *Saccharomyces cerevisae*. The synthesized cadmium telluride quantum dots offered good biocompatibility and potential for application in bioimaging and biolabeling [126].

In addition to the antimicrobial activity, fungal synthesized nanoparticles used to exert effects on tumor cells. Silver nanoparticles synthesized from *Fusarium oxysporum* have been used for antibacterial and antitumor activity. The nanoparticles were effective against *E. coli* and *S. aureus* as well as tumor cell line. A low IC50 value (121.23μg cm^3) for human breast adenocarcinoma MCF-7 cells was obtained by exposure of the nanoparticles. This indicates high cytotoxicity and potential tumor control. These results were obtained by disrupting mitochondrial respiratory chain, leading to the production of reactive oxygen species, subsequently delaying the synthesis of ATP and damaging the nucleic acids [127]. El-sonabaty also carried out a similar study, in which they have reported antitumor potential using *Agaricus bisporus* synthesized silver nanoparticles [128]. They have studied this activity *in vitro* against MCF-7 tumor cells and *in vivo* against Ehrlich carcinoma in mice. Although, the application of fungi synthesized nanoparticles for the antitumor and cytotoxicity is of considerable interest and has shown promising results, this technique still requires further investigation and the use of clinical trials.

Scope of Fungi Synthesized Nanoparticles in Environmental Protection

Environmental protection is an important area for the existence of life on the earth. Because of industrial and anthropogenic activities of human air, water and soil have been polluted by many pollutants. Carbon monoxides, heavy metals such as arsenic, mercury and nickel, various hydrocarbons are responsible for pollution in air, water and soil [84, 85, 129 - 132]. Many techniques are being implemented to remove contaminants and restore environment. But because of time consuming, expensive and lacking in efficacy, new techniques are required. Nanotechnology has great prospects in this area as many researchers around the globe have used various kinds of nanomaterials for the environmental protection.

Zhang has reported iron nanoparticles are being used to transform and detoxify environmental pollutants, such as pesticides, polychlorinated biophenyls and chlorinated organic solvents [133]. Rajan has suggested that iron nanoparticles can be used for remediation of groundwater so as to make it potable [134].

Heavy metals, such as mercury, nickel, chromium and arsenic, can interact with various biomacromolecules, leading to a change in their structure and functions. Farrukh and Telling have reported that magnetic nanoparticles can remove mercury ions and chromium ions from contaminated water [135, 136]. Chong also found promising results using TiO_2 nanoparticles. He observed that these nanoparticles are a potential candidate for photocatalytic degradation of several pollutants [137]. Mycobiota having metal nanoparticle biosynthesis ability can be useful in the recovery of precious metals as well.

CONCLUSION

Green synthesis of various nanoparticles by fungi offers an environment friendly approach and is superior to other sources, hence, it has gained attention worldwide. A number of applications in various fields are increasing day by day. Filamentous fungi and yeasts from diverse habitats have shown impressive records, however, exploring newer habitats may prove even more beneficial. A number of issues have been resolved, however, there is still a long way to go. Process centric and datacentric approaches can be harnessed for optimization of biosynthesis of nanoparticles and scaling up of the process would facilitate large scale production. Molecular characterization of capping agents would help understand the process better and would help increasing the stability of particles where required. Fungi with nanoparticle biosynthetic ability can be harnessed for the recovery of precious metals.

CONSENT FOR PUBLICATION

Not applicable.

CONFLICT OF INTEREST

The authors declare no conflict of interest, financial or otherwise.

ACKNOWLEDGEMENT

Declared none.

REFERENCES

[1]   Khandel P, Shahi SK. Mycogenic nanoparticles and their bio-prospective applications: current status and future challenges. J Nanostructure Chem 2018; 8(4): 369-91.
        [http://dx.doi.org/10.1007/s40097-018-0285-2]

[2]   Nanda A, Majeed S. "Enhanced antibacterial efficacy of biosynthesized AgNPs from *Penicillium glabrum* (MTCC1985) pooled with different drugs". Int J Pharm Tech Res 2014; 6: 217-23.

[3]   Jayaseelan C, Rahuman AA. "Acaricidal efficacy of synthesized silver nanoparticles using aqueous leaf extract of *Ocimum canum* against *Hyalomma anatolicum* and *Hyalomma marginatum* isaaci (Acari: Ixodidae)". Parasitol Res 2012; 111(3): 1369-78.
        [http://dx.doi.org/10.1007/s00436-011-2559-1] [PMID: 21789583]

[4]   Shelar GB, Chavan AM. Fungus-mediated biosynthesis of silver nanoparticles and its antibacterial activity. Arch Appl Sci Res 2014; 6: 111-4.

[5]   Daenen LGM, Houthuijzen JM, Cirkel GA, Roodhart JML, Shaked Y, Voest EE. Treatment-induced host-mediated mechanisms reducing the efficacy of antitumor therapies. Oncogene 2014; 33(11): 1341-7.
        [http://dx.doi.org/10.1038/onc.2013.94] [PMID: 23524584]

[6]   Kim HJ, Kim U, Kim HM, *et al.* High mobility in a stable transparent perovskite oxide. Appl Phys Express 2012; 5(6): 061102.
        [http://dx.doi.org/10.1143/APEX.5.061102]

[7]   Ben Said M, Galai Y, Mhadhbi M, Jedidi M, de la Fuente J, Darghouth MA. "Molecular characterization of Bm86 gene orthologs from *Hyalomma excavatum, Hyalomma dromedarii* and *Hyalomma marginatum* and comparison with a vaccine candidate from *Hyalomma scupense*". Vet Parasitol 2012; 190(1-2): 230-40.
        [http://dx.doi.org/10.1016/j.vetpar.2012.05.017] [PMID: 22683299]

[8]   Naseem T, Farrukh MA. "Antibacterial activity of green synthesis of iron nanoparticles using *Lawsoniainermis* and *Gardenia jasminoides* leaves extract". J Chem 2015.

[9]   He Q, Yang H, Wu L, Hu C. Effect of light intensity on physiological changes, carbon allocation and neutral lipid accumulation in oleaginous microalgae. Bioresour Technol 2015; 191: 219-28.
        [http://dx.doi.org/10.1016/j.biortech.2015.05.021] [PMID: 25997011]

[10]   Dzido G, Markowski P, Małachowska-Jutsz A, Prusik K, Jarzębski AB. Rapid continuous microwave-assisted synthesis of silver nanoparticles to achieve very high productivity and full yield: from mechanistic study to optimal fabrication strategy. J Nanopart Res 2015; 17(1): 27.
        [http://dx.doi.org/10.1007/s11051-014-2843-y] [PMID: 25620882]

[11]   Vala AK, Shah S, Patel R. "Biogenesis of silver nanoparticles by marine-derived fungus *Aspergillus flavus* from Bhavnagar Coast, Gulf of Khambhat, India". J Mar Biol Oceanogr 2014; 3(1): 1-3.

[12]   Ahmed S. Saifullah, Ahmad M, Swami BL, Ikram S. "Green synthesis of silver nanoparticles using *Azadirachtaindica* aqueous leaf extract". J Radiat Res Appl Sci 2015; 9(1): 1-7.

[13]   Chen F, Ehlerding EB, Cai W. Theranostic Nanoparticles. J Nucl Med 2014; 55(12): 1919-22.
        [http://dx.doi.org/10.2967/jnumed.114.146019] [PMID: 25413134]

[14]   Adil SF, Assal ME, Khan M, Al-Warthan A, Siddiqui MRH, Liz-Marzán LM. Biogenic synthesis of metallic nanoparticles and prospects toward green chemistry. Dalton Trans 2015; 44(21): 9709-17.
[http://dx.doi.org/10.1039/C4DT03222E] [PMID: 25633046]

[15]   Mahendra R, Alka Y, Bridge P, Aniket G. Myconanotechnology: a new and emerging science. Applied mycology 2009; 258-67.
[http://dx.doi.org/10.1079/9781845935344.0258]

[16]   Taherzadeh M, Fox M, Hjorth H, Edebo L. Production of mycelium biomass and ethanol from paper pulp sulfite liquor by *Rhizopus oryzae*". Bioresour Technol 2003; 88(3): 167-77.
[http://dx.doi.org/10.1016/S0960-8524(03)00010-5] [PMID: 12618037]

[17]   Pantidos N, Horsfall LE. Biological synthesis of metallic nanoparticles by bacteria, fungi and plants. J Nanomed Nanotechnol 2014; 5(5): 1.
[http://dx.doi.org/10.4172/2157-7439.1000233]

[18]   Vala AK. Exploration on green synthesis of gold nanoparticles by a marine-derived fungus *Aspergillus sydowii*. Environ Prog Sustain Energy 2015; 34(1): 194-7.
[http://dx.doi.org/10.1002/ep.11949]

[19]   Dave V, Vala AK, Patel R. Observation of weak localization of light in gold nanofluids synthesized using the marine derived fungus *Aspergillus niger. RSC Adv.* 2015, 5: 16780-16784 Soni N, Prakash S. Efficacy of fungus mediated silver and gold nanoparticles against Aedes aegypti larvae. Parasitol Res 2012; 110(1): 175-84.
[PMID: 21647674]

[20]   Bowman SM, Free SJ. The structure and synthesis of the fungal cell wall. BioEssays 2006; 28(8): 799-808.
[http://dx.doi.org/10.1002/bies.20441] [PMID: 16927300]

[21]   Ingle A, Rai M, Gade A, Bawaskar M. "*Fusarium solani:* a novel biological agent for the extracellular synthesis of silver nanoparticles". J Nanopart Res 2009; 11(8): 2079-85.
[http://dx.doi.org/10.1007/s11051-008-9573-y]

[22]   Vala AK, Trivedi H, Gosai H, Panseriya H. Dave BP Biosynthesized Silver Nanoparticles and Their Therapeutic Applications. In: Biosynthesized Nanomaterials Volume 100 (Eds Verma SK, Das AK). Elsevier 2020.

[23]   Smith MR. Cord-forming Palaeozoic fungi in terrestrial assemblages. J Linn Soc Bot 2016; 180(4): 452-60.

[24]   Hawksworth DL. The magnitude of fungal diversity: the 1.5 million species estimate revisited. Mycol Res 2001; 105(12): 1422-32.
[http://dx.doi.org/10.1017/S0953756201004725]

[25]   Hawksworth DL. Fungal diversity and its implications for genetic resource collections. Stud Mycol 2004; 50(1)

[26]   Schmit JP, Mueller GM. An estimate of the lower limit of global fungal diversity. Biodivers Conserv 2007; 16(1): 99-111.
[http://dx.doi.org/10.1007/s10531-006-9129-3]

[27]   Manoharachary C, Sridhar K, Singh R, *et al.* Fungal biodiversity: distribution, conservation and prospecting of fungi from India. Curr Sci 2005; 58-71.

[28]   Hawksworth DL, Rossman AY. Where are all the undescribed fungi? Phytopathology 1997; 87(9): 888-91.
[http://dx.doi.org/10.1094/PHYTO.1997.87.9.888] [PMID: 18945058]

[29]   Kirk PM, Cannon PF, David JC, Stalpers JA. Ainsworth and Bisby's Dictionary of the Fungi. CABI publishing 2001.

[30]   Zhuliang Y, Mu Z. Tropical affinities of higher fungi in southern China. Yunnan Zhi Wu Yan Jiu

2003; 25(2): 129-44.

[31]   Zhong JJ, Xiao JH. Secondary metabolites from higher fungi: discovery, bioactivity, and bioproduction InBiotechnology in China I. Berlin, Heidelberg: Springer 2009; pp. 79-150.

[32]   Blackwell M. The Fungi: 1, 2, 3 … 5.1 million species? Am J Bot 2011; 98(3): 426-38.
[http://dx.doi.org/10.3732/ajb.1000298] [PMID: 21613136]

[33]   Boddy L, Heilmann-Clausen J. Basidiomycete community development in temperate angiosperm wood.British Mycological Society Symposia Series. Academic Press 2008; Vol. 28: pp. 211-37.

[34]   Pan Y, Birdsey RA, Fang J, *et al.* A large and persistent carbon sink in the world's forests. Science 2011; 333(6045): 988-93.
[http://dx.doi.org/10.1126/science.1201609] [PMID: 21764754]

[35]   Stokland JN, Siitonen J, Jonsson BG. Biodiversity in dead wood. Cambridge university press 2012.
[http://dx.doi.org/10.1017/CBO9781139025843]

[36]   Shaver GR, Chapin FS III. Production: biomass relationships and element cycling in contrasting arctic vegetation types. Ecol Monogr 1991; 61(1): 1-31.
[http://dx.doi.org/10.2307/1942997]

[37]   Scheidegger C, Werth S. Conservation strategies for lichens: insights from population biology. Fungal Biol Rev 2009; 23(3): 55-66.
[http://dx.doi.org/10.1016/j.fbr.2009.10.003]

[38]   Smith SE, Read DJ. Mycorrhizal symbiosis.,(Academic Press: Amsterdam).

[39]   Courty PE, Buée M, Diedhiou AG, *et al.* The role of ectomycorrhizal communities in forest ecosystem processes: New perspectives and emerging concepts. Soil Biol Biochem 2010; 42(5): 679-98.
[http://dx.doi.org/10.1016/j.soilbio.2009.12.006]

[40]   Clemmensen KE, Bahr A, Ovaskainen O, *et al.* Roots and associated fungi drive long-term carbon sequestration in boreal forest. Science 2013; 339(6127): 1615-8.
[http://dx.doi.org/10.1126/science.1231923] [PMID: 23539604]

[41]   Averill C, Turner BL, Finzi AC. Mycorrhiza-mediated competition between plants and decomposers drives soil carbon storage. Nature 2014; 505(7484): 543-5.
[http://dx.doi.org/10.1038/nature12901] [PMID: 24402225]

[42]   Rodriguez RJ, White JF Jr, Arnold AE, Redman RS. Fungal endophytes: diversity and functional roles. New Phytol 2009; 182(2): 314-30.
[http://dx.doi.org/10.1111/j.1469-8137.2009.02773.x] [PMID: 19236579]

[43]   Arnold AE, Lutzoni F. Diversity and host range of foliar fungal endophytes: are tropical leaves biodiversity hotspots? Ecology 2007; 88(3): 541-9.
[http://dx.doi.org/10.1890/05-1459] [PMID: 17503580]

[44]   Castro-Longoria E, Moreno-Velázquez SD, Vilchis-Nestor AR, Arenas-Berumen E, Avalos-Borja M. "Production of platinum nanoparticles and nanoaggregates using *Neurospora crassa*". J Microbiol Biotechnol 2012; 22(7): 1000-4.
[http://dx.doi.org/10.4014/jmb.1110.10085] [PMID: 22580320]

[45]   Ahmad A, Mukherjee P, Senapati S, *et al.* "Extracellular biosynthesis of silver nanoparticles using the fungus *Fusarium oxysporum*". Colloids Surf B Biointerfaces 2003; 28(4): 313-8.
[http://dx.doi.org/10.1016/S0927-7765(02)00174-1]

[46]   Ahmad A, Senapati S, Khan MI, *et al.* Intracellular synthesis of gold nanoparticles by a novel alkalotolerant actinomycete, *Rhodococcus* species. Nanotechnology 2003; 14(7): 824-8.
[http://dx.doi.org/10.1088/0957-4484/14/7/323]

[47]   Durán N, Marcato PD, Alves OL, De Souza GIH, Esposito E. Mechanistic aspects of biosynthesis of silver nanoparticles by several *Fusarium oxysporum* strains. J Nanobiotechnology 2005; 3(1): 8.
[http://dx.doi.org/10.1186/1477-3155-3-8] [PMID: 16014167]

[48]   Rai M, Yadav A, Gade A. Silver nanoparticles as a new generation of antimicrobials. Biotechnol Adv 2009; 27(1): 76-83.
[http://dx.doi.org/10.1016/j.biotechadv.2008.09.002] [PMID: 18854209]

[49]   Seshadri S, Saranya K, Kowshik M. Green synthesis of lead sulfide nanoparticles by the lead resistant marine yeast, *Rhodosporidium diobovatum*. Biotechnol Prog 2011; 27(5): 1464-9.
[http://dx.doi.org/10.1002/btpr.651] [PMID: 21710608]

[50]   Wei Y, Han B, Hu X, Lin Y, Wang X, Deng X. Synthesis of Fe3O4 nanoparticles and their magnetic properties. Procedia Eng 2012; 27: 632-7.
[http://dx.doi.org/10.1016/j.proeng.2011.12.498]

[51]   Mukherjee P, Ahmad A, Mandal D, *et al.* Fungus-mediated synthesis of silver nanoparticles and their immobilization in the mycelial matrix: a novel biological approach to nanoparticle synthesis. Nano Lett 2001; 1(10): 515-9.
[http://dx.doi.org/10.1021/nl0155274]

[52]   Mukherjee P, Patra CR, Ghosh A, Kumar R, Sastry M. Characterization and catalytic activity of gold nanoparticles synthesized by autoreduction of aqueous chloroaurate ions with fumed silica. Chem Mater 2002; 14(4): 1678-84.
[http://dx.doi.org/10.1021/cm010372m]

[53]   Sawle BD, Salimath B, Deshpande R, Bedre MD, Prabhakar BK, Venkataraman A. Biosynthesis and stabilization of Au and Au–Ag alloy nanoparticles by fungus, *Fusarium semitectum*. Science and technology of advanced materials STAM 2008; 9(3): 035012.

[54]   Mohanpuria P, Rana NK, Yadav SK. Biosynthesis of nanoparticles: technological concepts and future applications. J Nanopart Res 2008; 10(3): 507-17.
[http://dx.doi.org/10.1007/s11051-007-9275-x]

[55]   Mishra A, Tripathy SK, Yun SI. Bio-synthesis of gold and silver nanoparticles from *Candida guilliermondii* and their antimicrobial effect against pathogenic bacteria. J Nanosci Nanotechnol 2011; 11(1): 243-8.
[http://dx.doi.org/10.1166/jnn.2011.3265] [PMID: 21446434]

[56]   Gopinath PM, Narchonai G, Dhanasekaran D, Ranjani A, Thajuddin N. Mycosynthesis, characterization and antibacterial properties of AgNPs against multidrug resistant (MDR) bacterial pathogens of female infertility cases. Asian Journal of Pharmaceutical Sciences 2015; 10(2): 138-45.
[http://dx.doi.org/10.1016/j.ajps.2014.08.007]

[57]   Li G, He D, Qian Y, *et al.* Fungus-mediated green synthesis of silver nanoparticles using *Aspergillus terreus*. Int J Mol Sci 2011; 13(1): 466-76.
[http://dx.doi.org/10.3390/ijms13010466] [PMID: 22312264]

[58]   Mukherjee B, Santra K, Pattnaik G, Ghosh S. Preparation, characterization and *in-vitro* evaluation of sustained release protein-loaded nanoparticles based on biodegradable polymers. Int J Nanomedicine 2008; 3(4): 487-96.
[http://dx.doi.org/10.2147/IJN.S3938] [PMID: 19337417]

[59]   Thakkar KN, Mhatre SS, Parikh RY. Biological synthesis of metallic nanoparticles. Nanomedicine 2010; 6(2): 257-62.
[http://dx.doi.org/10.1016/j.nano.2009.07.002] [PMID: 19616126]

[60]   Zhang X, Lin S, Chen Z, Megharaj M, Naidu R. Kaolinite-supported nanoscale zero-valent iron for removal of Pb2+ from aqueous solution: Reactivity, characterization and mechanism. Water Res 2011; 45(11): 3481-8.
[http://dx.doi.org/10.1016/j.watres.2011.04.010] [PMID: 21529878]

[61]   Sastry M, Ahmad A, Khan MI, Kumar R. Biosynthesis of metal nanoparticles using fungi and actinomycete. Curr Sci 2003; 85(2): 162-70.

[62]   Ahmad A, Senapati S, Khan MI, Kumar R, Sastry M. Extra-/intracellular biosynthesis of gold

nanoparticles by an alkalotolerant fungus, *Trichothecium sp.* J Biomed Nanotechnol 2005; 1(1): 47-53.
[http://dx.doi.org/10.1166/jbn.2005.012]

[63] Sheikhloo Z, Salouti M. Intracellular biosynthesis of gold nanoparticles by the fungus *Penicillium chrysogenum*. Int J Nanosci Nanotech 2011; 7(2): 102-5.

[64] Gericke M, Pinches A. Biological synthesis of metal nanoparticles. Hydrometallurgy 2006; 83(1-4): 132-40.
[http://dx.doi.org/10.1016/j.hydromet.2006.03.019]

[65] Dhanasekar NN, Rahul GR, Narayanan KB, Raman G, Sakthivel N. Green chemistry approach for the synthesis of gold nanoparticles using the fungus *Alternaria sp.* J Microbiol Biotechnol 2015; 25(7): 1129-35.
[http://dx.doi.org/10.4014/jmb.1410.10036] [PMID: 25737119]

[66] Joshi SR, Devi LS. Ultrastructures of silver nanoparticles biosynthesized using endophytic fungi. J Microsc Ultrastruct 2015; 3(1): 29-37.
[http://dx.doi.org/10.1016/j.jmau.2014.10.004] [PMID: 30023179]

[67] Tarafdar JC, Raliya R, Rathore I. Microbial synthesis of phosphorous nanoparticle from tri-calcium phosphate using *Aspergillus tubingensis* TFR-5. Journal of Bionanoscience 2012; 6(2): 84-9.
[http://dx.doi.org/10.1166/jbns.2012.1077]

[68] Feng Y, Cui X, He S, *et al.* The role of metal nanoparticles in influencing arbuscular mycorrhizal fungi effects on plant growth. Environ Sci Technol 2013; 47(16): 9496-504.
[http://dx.doi.org/10.1021/es402109n] [PMID: 23869579]

[69] Mandal D, Bolander ME, Mukhopadhyay D, Sarkar G, Mukherjee P. The use of microorganisms for the formation of metal nanoparticles and their application. Appl Microbiol Biotechnol 2006; 69(5): 485-92.
[http://dx.doi.org/10.1007/s00253-005-0179-3] [PMID: 16317546]

[70] Sugunan A, Melin P, Schnürer J, Hilborn JG, Dutta J. Nutrition-Driven Assembly of Colloidal Nanoparticles: Growing Fungi Assemble Gold Nanoparticles as Microwires. Adv Mater 2007; 19(1): 77-81.
[http://dx.doi.org/10.1002/adma.200600911]

[71] Fischer-Parton S, Parton RM, Hickey PC, Dijksterhuis J, Atkinson HA, Read ND. Confocal microscopy of FM4-64 as a tool for analysing endocytosis and vesicle trafficking in living fungal hyphae. J Microsc 2000; 198(3): 246-59.
[http://dx.doi.org/10.1046/j.1365-2818.2000.00708.x] [PMID: 10849201]

[72] Neethu S, Midhun SJ, Radhakrishnan EK, Jyothis M. Green synthesized silver nanoparticles by marine endophytic fungus *Penicillium polonicum* and its antibacterial efficacy against biofilm forming, multidrug-resistant Acinetobacter baumanii. Microb Pathog 2018; 116: 263-72.
[http://dx.doi.org/10.1016/j.micpath.2018.01.033] [PMID: 29366864]

[73] Ahmed AA, Hamzah H, Maaroof M. Analyzing formation of silver nanoparticles from the filamentous fungus *Fusarium oxysporum* and their antimicrobial activity. Turk J Biol 2018; 42(1): 54-62.
[http://dx.doi.org/10.3906/biy-1710-2] [PMID: 30814870]

[74] Šebesta M, Urík M, Bujdoš M, *et al.* Fungus *Aspergillus niger* Processes Exogenous Zinc Nanoparticles into a Biogenic Oxalate Mineral. J Fungi (Basel) 2020; 6(4): 210.
[http://dx.doi.org/10.3390/jof6040210] [PMID: 33049947]

[75] Chen J, Wu L, Lu M, Lu S, Li Z, Ding W. Comparative Study on the Fungicidal Activity of Metallic MgO Nanoparticles and Macroscale MgO Against Soilborne Fungal Phytopathogens. Front Microbiol 2020; 11: 365.
[http://dx.doi.org/10.3389/fmicb.2020.00365] [PMID: 32226420]

[76] Philip D. Biosynthesis of Au, Ag and Au–Ag nanoparticles using edible mushroom extract. Spectrochim Acta A Mol Biomol Spectrosc 2009; 73(2): 374-81.

[http://dx.doi.org/10.1016/j.saa.2009.02.037] [PMID: 19324587]

[77]   Dameron CT, Reese RN, Mehra RK, *et al.* Biosynthesis of cadmium sulphide quantum semiconductor crystallites. Nature 1989; 338(6216): 596-7.
[http://dx.doi.org/10.1038/338596a0]

[78]   Breierová E, Vajcziková I, Sasinková V, *et al.* Biosorption of cadmium ions by different yeast species. Z Naturforsch C J Biosci 2002; 57(7-8): 634-9.
[http://dx.doi.org/10.1515/znc-2002-7-815] [PMID: 12240989]

[79]   Prasad R, Pandey R, Barman I. Engineering tailored nanoparticles with microbes: *quo vadis?* Wiley Interdiscip Rev Nanomed Nanobiotechnol 2016; 8(2): 316-30.
[http://dx.doi.org/10.1002/wnan.1363] [PMID: 26271947]

[80]   Raghukumar C. Marine fungal biotechnology: an ecological perspective. 2008.

[81]   Kohlmeyer J, Kohlmeyer E. Marine mycology, the higher fungi. New York: Academic Press 1979.

[82]   Vala AK. Tolerance and removal of arsenic by a facultative marine fungus *Aspergillus candidus.* Bioresour Technol 2010; 101(7): 2565-7.
[http://dx.doi.org/10.1016/j.biortech.2009.11.084] [PMID: 20022490]

[83]   Gosai HB, Sachaniya BK, Dudhagara DR, Rajpara RK, Dave BP. Concentrations, input prediction and probabilistic biological risk assessment of polycyclic aromatic hydrocarbons (PAHs) along Gujarat coastline. Environ Geochem Health 2018; 40(2): 653-65.
[http://dx.doi.org/10.1007/s10653-017-0011-x] [PMID: 28801833]

[84]   Gosai HB, Sachaniya BK, Panseriya HZ, Dave BP. Functional and phylogenetic diversity assessment of microbial communities at Gulf of Kachchh, India: An ecological footprint. Ecol Indic 2018; 93: 65-75.
[http://dx.doi.org/10.1016/j.ecolind.2018.04.072]

[85]   Bonugli-Santos RC, dos Santos Vasconcelos MR, Passarini MRZ, *et al.* Marine-derived fungi: diversity of enzymes and biotechnological applications. Front Microbiol 2015; 6: 269.
[http://dx.doi.org/10.3389/fmicb.2015.00269] [PMID: 25914680]

[86]   Kathiresan K, Alikunhi NM, Pathmanaban S, Nabikhan A, Kandasamy S. Analysis of antimicrobial silver nanoparticles synthesized by coastal strains of *Escherichia coli* and *Aspergillus niger.* Can J Microbiol 2010; 56(12): 1050-9.
[http://dx.doi.org/10.1139/W10-094] [PMID: 21164575]

[87]   Vala AK, Shah S. Rapid Synthesis of Silver Nanoparticles by a Marine-derived Fungus *Aspergillusniger* and their Antimicrobial Potentials Int. J Nanosci Nanotechnol 2012; 8(4): 197-206.

[88]   Vala AK, Chudasama B, Patel RJ. Green synthesis of silver nanoparticles using marine-derived fungus *Aspergillus niger.* Micro & Nano Lett 2012; 7(8): 859-62.
[http://dx.doi.org/10.1049/mnl.2012.0403]

[89]   Vala AK. Investigations on gold nanoparticles biosynthesis potential of marine –derived fungi. Biotechnology (Rajkot) 2014; 9(5): 206-9. a

[90]   Vala AK. Intra-and extracellular biosynthesis of gold nanoparticles by a marine-derived fungus *Rhizopus oryzae.* Synth React Inorg Met-Org Nano-Met Chem 2014; 44(9): 1243-6. b
[http://dx.doi.org/10.1080/15533174.2013.799492]

[91]   Zomorodian K, Pourshahid S, Sadatsharifi A, *et al.* Biosynthesis and characterization of silver nanoparticles by *Aspergillus species.* BioMed Res Int 2016; 2016: 1-6.
[http://dx.doi.org/10.1155/2016/5435397] [PMID: 27652264]

[92]   Manjunath Hulikere M, Joshi CG. Characterization, antioxidant and antimicrobial activity of silver nanoparticles synthesized using marine endophytic fungus- *Cladosporium cladosporioides.* Process Biochem 2019; 82: 199-204.
[http://dx.doi.org/10.1016/j.procbio.2019.04.011]

[93] Pimprikar PS, Joshi SS, Kumar AR, Zinjarde SS, Kulkarni SK. Influence of biomass and gold salt concentration on nanoparticle synthesis by the tropical marine yeast *Yarrowia lipolytica* NCIM 3589. Colloids Surf B Biointerfaces 2009; 74(1): 309-16.
[http://dx.doi.org/10.1016/j.colsurfb.2009.07.040] [PMID: 19700266]

[94] Subramanian M, Alikunhi NM, Kandasamy K. *In vitro* synthesis of silver nanoparticles by marine yeasts from coastal mangrove sediment. Adv Sci Lett 2010; 3(4): 428-33.
[http://dx.doi.org/10.1166/asl.2010.1168]

[95] Subramanian M, Alikunhi NM, Kandasamy K. *In vitro* synthesis of silver nanoparticle by marine yeasts from coastal mangrove sediment. Adv Sci Lett 2010; 3(4): 428-33.
[http://dx.doi.org/10.1166/asl.2010.1168]

[96] Elgorban AM, El-Samawaty AERM, Yassin MA, *et al.* Antifungal silver nanoparticles: synthesis, characterization and biological evaluation. Biotechnol Biotechnol Equip 2016; 30(1): 56-62.
[http://dx.doi.org/10.1080/13102818.2015.1106339]

[97] Qian Y, Yu H, He D, *et al.* Biosynthesis of silver nanoparticles by the endophytic fungus *Epicoccum nigrum* and their activity against pathogenic fungi. Bioprocess Biosyst Eng 2013; 36(11): 1613-9.
[http://dx.doi.org/10.1007/s00449-013-0937-z] [PMID: 23463299]

[98] Gajbhiye M, Kesharwani J, Ingle A, Gade A, Rai M. Fungus-mediated synthesis of silver nanoparticles and their activity against pathogenic fungi in combination with fluconazole. Nanomedicine 2009; 5(4): 382-6.
[http://dx.doi.org/10.1016/j.nano.2009.06.005] [PMID: 19616127]

[99] Banu AN, Balasubramanian C. Myco-synthesis of silver nanoparticles using Beauveria bassiana against dengue vector, Aedes aegypti (Diptera: Culicidae). Parasitol Res 2014; 113(8): 2869-77. a
[http://dx.doi.org/10.1007/s00436-014-3948-z] [PMID: 24861012]

[100] Banu AN, Balasubramanian C. Optimization and synthesis of silver nanoparticles using Isaria fumosorosea against human vector mosquitoes. Parasitol Res 2014; 113(10): 3843-51. b
[http://dx.doi.org/10.1007/s00436-014-4052-0] [PMID: 25085201]

[101] Chinnamuthu CR, Boopathi PM. Nanotechnology and agroecosystem. Madras Agric J 2009; 96(1/6): 17-31.

[102] Guilger M, Pasquoto-Stigliani T, Bilesky-Jose N, *et al.* Biogenic silver nanoparticles based on trichoderma harzianum: synthesis, characterization, toxicity evaluation and biological activity. Sci Rep 2017; 7(1): 44421.
[http://dx.doi.org/10.1038/srep44421] [PMID: 28300141]

[103] Xue B, He D, Gao S, Wang D, Yokoyama K, Wang L. Biosynthesis of silver nanoparticles by the fungus Arthroderma fulvum and its antifungal activity against genera of *Candida, Aspergillus* and *Fusarium.* Int J Nanomedicine 2016; 11: 1899-906.
[PMID: 27217752]

[104] Sundaravadivelan C, Padmanabhan MN. Effect of mycosynthesized silver nanoparticles from filtrate of *Trichoderma harzianum* against larvae and pupa of dengue vector Aedes aegypti L. Environ Sci Pollut Res Int 2014; 21(6): 4624-33.
[http://dx.doi.org/10.1007/s11356-013-2358-6] [PMID: 24352539]

[105] Gade AK, Bonde P, Ingle AP, Marcato PD, Durán N, Rai MK. Exploitation of Aspergillus niger for synthesis of silver nanoparticles. J Biobased Mater Bioenergy 2008; 2(3): 243-7.
[http://dx.doi.org/10.1166/jbmb.2008.401]

[106] Kumar SA, Ansary AA, Ahmad A, Khan MI. Extracellular biosynthesis of CdSe quantum dots by the fungus, *Fusarium oxysporum.* J Biomed Nanotechnol 2007; 3(2): 190-4.
[http://dx.doi.org/10.1166/jbn.2007.027]

[107] Das SK, Dickinson C, Lafir F, Brougham DF, Marsili E. Synthesis, characterization and catalytic activity of gold nanoparticles biosynthesized with *Rhizopus oryzae* protein extract. Green Chem 2012;

14(5): 1322-34.
[http://dx.doi.org/10.1039/c2gc16676c]

[108] Thirumurugan G, Shaheedha SM, Dhanaraju MD. *In vitro* evaluation of antibacterial activity of silver nanoparticles synthesised by using Phytophthora infestans. Int J Chemtech Res 2009; 1(3): 714-6.

[109] Gudikandula K, Vadapally P, Singara Charya MA. Biogenic synthesis of silver nanoparticles from white rot fungi: Their characterization and antibacterial studies. OpenNano 2017; 2: 64-78.
[http://dx.doi.org/10.1016/j.onano.2017.07.002]

[110] Ashajyothi C, Harish KH, Dubey N, Chandrakanth RK. Antibiofilm activity of biogenic copper and zinc oxide nanoparticles-antimicrobials collegiate against multiple drug resistant bacteria: a nanoscale approach. J Nanostructure Chem 2016; 6(4): 329-41.
[http://dx.doi.org/10.1007/s40097-016-0205-2]

[111] Fatima M, Zaidi NSS, Amraiz D, Afzal F. *In vitro* antiviral activity of Cinnamomum cassia and its nanoparticles against H7N3 influenza a virus. J Microbiol Biotechnol 2016; 26(1): 151-9.
[http://dx.doi.org/10.4014/jmb.1508.08024] [PMID: 26403820]

[112] Reidy B, Haase A, Luch A, Dawson K, Lynch I. Mechanisms of silver nanoparticle release, transformation and toxicity: a critical review of current knowledge and recommendations for future studies and applications. Materials (Basel) 2013; 6(6): 2295-350.
[http://dx.doi.org/10.3390/ma6062295] [PMID: 28809275]

[113] Balakumaran MD, Ramachandran R, Kalaichelvan PT. Exploitation of endophytic fungus, Guignardia mangiferae for extracellular synthesis of silver nanoparticles and their *in vitro* biological activities. Microbiol Res 2015; 178: 9-17.
[http://dx.doi.org/10.1016/j.micres.2015.05.009] [PMID: 26302842]

[114] Shrivastava S, Bera T, Roy A, Singh G, Ramachandrarao P, Dash D. Characterization of enhanced antibacterial effects of novel silver nanoparticles. Nanotechnology 2007; 18(22): 225103.
[http://dx.doi.org/10.1088/0957-4484/18/22/225103]

[115] Ahluwalia V, Kumar J, Sisodia R, Shakil NA, Walia S. Green synthesis of silver nanoparticles by Trichoderma harzianum and their bio-efficacy evaluation against Staphylococcus aureus and Klebsiella pneumonia. Ind Crops Prod 2014; 55: 202-6.
[http://dx.doi.org/10.1016/j.indcrop.2014.01.026]

[116] Bhat MA, Nayak BK, Nanda A. Evaluation of bactericidal activity of biologically synthesised silver nanoparticles from Candida albicans in combination with ciprofloxacin. Mater Today Proc 2015; 2(9): 4395-401.
[http://dx.doi.org/10.1016/j.matpr.2015.10.036]

[117] Rodrigues AG, Ping LY, Marcato PD, *et al.* Biogenic antimicrobial silver nanoparticles produced by fungi. Appl Microbiol Biotechnol 2013; 97(2): 775-82.
[http://dx.doi.org/10.1007/s00253-012-4209-7] [PMID: 22707055]

[118] Mekkawy A, El-Mokhtar M, Nafady N, *et al. In vitro* and *in vivo* evaluation of biologically synthesized silver nanoparticles for topical applications: effect of surface coating and loading into hydrogels. Int J Nanomedicine 2017; 12: 759-77.
[http://dx.doi.org/10.2147/IJN.S124294] [PMID: 28176951]

[119] Nagajyothi PC, Sreekanth TVM, Lee J, Lee KD. Mycosynthesis: Antibacterial, antioxidant and antiproliferative activities of silver nanoparticles synthesized from Inonotus obliquus (Chaga mushroom) extract. J Photochem Photobiol B 2014; 130: 299-304.
[http://dx.doi.org/10.1016/j.jphotobiol.2013.11.022] [PMID: 24380885]

[120] Raman J, Reddy GR, Lakshmanan H, *et al.* Mycosynthesis and characterization of silver nanoparticles from Pleurotus djamor var. roseus and their *in vitro* cytotoxicity effect on PC3 cells. Process Biochem 2015; 50(1): 140-7.
[http://dx.doi.org/10.1016/j.procbio.2014.11.003]

[121] Pereira L, Dias N, Carvalho J, Fernandes S, Santos C, Lima N. Synthesis, characterization and antifungal activity of chemically and fungal-produced silver nanoparticles against *Trichophyton rubrum*. J Appl Microbiol 2014; 117(6): 1601-13.
[http://dx.doi.org/10.1111/jam.12652] [PMID: 25234047]

[122] Nayak BK, Nanda A, Prabhakar V. Biogenic synthesis of silver nanoparticle from wasp nest soil fungus, Penicillium italicum and its analysis against multi drug resistance pathogens. Biocatal Agric Biotechnol 2018; 16: 412-8.
[http://dx.doi.org/10.1016/j.bcab.2018.09.014]

[123] Ishida K, Cipriano TF, Rocha GM, *et al.* Silver nanoparticle production by the fungus Fusarium oxysporum: nanoparticle characterisation and analysis of antifungal activity against pathogenic yeasts. Mem Inst Oswaldo Cruz 2013; 109(2): 220-8.
[http://dx.doi.org/10.1590/0074-0276130269] [PMID: 24714966]

[124] Mohammed Fayaz A, Girilal M, Mahdy SA, Somsundar SS, Venkatesan R, Kalaichelvan PT. Vancomycin bound biogenic gold nanoparticles: A different perspective for development of anti VRSA agents. Process Biochem 2011; 46(3): 636-41.
[http://dx.doi.org/10.1016/j.procbio.2010.11.001]

[125] Aguilar-Méndez MA, San Martín-Martínez E, Ortega-Arroyo L, Cobián-Portillo G, Sánchez-Espíndola E. Synthesis and characterization of silver nanoparticles: effect on phytopathogen Colletotrichum gloesporioides. J Nanopart Res 2011; 13(6): 2525-32.
[http://dx.doi.org/10.1007/s11051-010-0145-6]

[126] Bao H, Hao N, Yang Y, Zhao D. Biosynthesis of biocompatible cadmium telluride quantum dots using yeast cells. Nano Res 2010; 3(7): 481-9.
[http://dx.doi.org/10.1007/s12274-010-0008-6]

[127] Husseiny SM, Salah TA, Anter HA. Biosynthesis of size controlled silver nanoparticles by Fusarium oxysporum, their antibacterial and antitumor activities. Beni Suef Univ J Basic Appl Sci 2015; 4(3): 225-31.
[http://dx.doi.org/10.1016/j.bjbas.2015.07.004]

[128] El-Sonbaty SM. Fungus-mediated synthesis of silver nanoparticles and evaluation of antitumor activity. Cancer Nanotechnol 2013; 4(4-5): 73-9.
[http://dx.doi.org/10.1007/s12645-013-0038-3] [PMID: 26069502]

[129] Panseriya HZ, Gosai HB, Sachaniya BK, Vala AK, Dave BP. Marine microbial mettle for heavy metal bioremediation: a perception. Marine pollution: current status, impacts and remedies 2019; 1: 409-34.
[http://dx.doi.org/10.2174/9789811412691119010021]

[130] Panseriya HZ, Gosai HB, Sankhwal AO, Sachaniya BK, Gavali DJ, Dave BP. Distribution, speciation and risk assessment of heavy metals: geochemical exploration of Gulf of Kachchh, Gujarat, India. Environ Earth Sci 2020; 79(10): 213.
[http://dx.doi.org/10.1007/s12665-020-08972-x]

[131] Sachaniya BK, Gosai HB, Panseriya HZ, Vala AK, Dave BP. Polycyclic aromatic hydrocarbons (PAHs): occurrence and bioremediation in the marine environment. Marine pollution: current status, impacts and remedies 2019; 1: 435-66.

[132] Sachaniya BK, Gosai HB, Panseriya HZ, Dave BP. Bioengineering for multiple PAHs degradation for contaminated sediments: Response surface methodology (RSM) and artificial neural network (ANN). Chemom Intell Lab Syst 2020; 202: 104033.
[http://dx.doi.org/10.1016/j.chemolab.2020.104033]

[133] Zhang W. Nanoscale iron particles for environmentalremediation: an overview. J Nanopart Res 2003; 5(3/4): 323-32.
[http://dx.doi.org/10.1023/A:1025520116015]

[134] Rajan CS. Nanotechnology in groundwater remediation. Int J Environ Sci Technol 2011; 2(3): 182.

[135]	Farrukh A, Akram A, Ghaffar A, *et al.* Design of polymer-brush-grafted magnetic nanoparticles for highly efficient water remediation. ACS Appl Mater Interfaces 2013; 5(9): 3784-93.
[http://dx.doi.org/10.1021/am400427n] [PMID: 23570443]

[136]	Telling ND, Coker VS, Cutting RS, *et al.* Remediation of Cr(VI) by biogenic magnetic nanoparticles: An x-ray magnetic circular dichroism study. Appl Phys Lett 2009; 95(16): 163701.
[http://dx.doi.org/10.1063/1.3249578]

[137]	Chong MN, Jin B, Chow CWK, Saint C. Recent developments in photocatalytic water treatment technology: A review. Water Res 2010; 44(10): 2997-3027.
[http://dx.doi.org/10.1016/j.watres.2010.02.039] [PMID: 20378145]

CHAPTER 2

Fungal Nanobionics - Principle, Advances and Applications

Sonali Singhal[1,*] and **Babita Singh**[1]

[1] *ICAR-National Research Centre for Integrated Pest Management, New Delhi-110012, India*

Abstract: Nanotechnology is the science of research and development at the nanoscale (*i.e.* 10^{-9} m) at least in one dimension. The capability of nanotechnology is often known to revolve around nanoparticles. The core versatility of the nanoparticles is the fact that they exhibit more significant properties than that of bulk counterparts. Nanobionics is the structural and functional study of biological systems which serve as a model for the design and engineering of materials and machines at the cistron level. Fungi have emerged as important systems for the synthesis of nanoparticles due to the production of extracellular enzymes which can utilize heavy metal ions and produce nanoparticles, easy to isolate and subculture on synthetic media due to low nutritional requirements, high wall binding capacity, simpler biomass handling and extracellular synthesis of nanoparticle help in easy downstream processing. Fungi can produce nanoparticles both extracellularly as well as intracellularly. For example, the synthesis of silver nanoparticles has been reported utilizing many ubiquitous fungal species including *Trichoderma, Fusarium, Penicillium, Rhizoctonia, Pleurotus* and *Aspergillus*.

Extracellular synthesis has been shown by *Trichoderma viridae* while intracellular synthesis was shown to occur in a *Verticillium* species, and in *Neurospora crassa* whereas the synthesis of gold nanoparticles has been reported utilizing *Fusarium, Neurospora, Verticillium, yeasts,* and *Aspergillus*. In this chapter, we further focus on the applications of fungal nanobionics in various industries.

Keywords: Extracellular synthesis, Fungal nanotechnology, Nanobionics, Nanoparticles.

INTRODUCTION

Nanotechnology finds its applications in many fields like medicine, cosmetics, device development, *etc*. Nanotechnology relies on the development of nano-particles from various sources. Based on the source of the nanoparticles, they are classified mainly as organic and inorganic.

* **Corresponding author Sonali Singhal:** ICAR-National Research Centre for Integrated Pest Management, New Delhi-110012 India; Tel: +91 8586939864; Email: singhalr58@gmail.com

Savita, Anju Srivastava, Reena Jain & Pratap Kumar Pati (Eds.)

Organic/Carbon-based nanoparticles: Carbon-based nanoparticles are made up of two materials carbon nanotubes (CNTs), made up of graphene and fullerenes. CNTs are further classified into single-walled carbon nanotubes and (SWCNTs) and multi-walled carbon nanotubes (MWCNTs). CNTs thermally conduct along the length and non-conductive across the tube and have higher strength. CNTs are being utilized as adsorbents for toxic chemicals from the manufacturing industry or pharmaceutical wastewater and microbial fuel cells [1].

Inorganic nanoparticles: Nanoparticles with magnetic, noble metal (*e.g.* palladium, gold and silver) and semiconductor nanoparticles (*e.g.* titanium dioxide and zinc oxide). Metal nanoparticles are produced from metal precursors with the help of various chemical, electrochemical, or photochemical methods. Metal nanoparticles can absorb small molecules and have high surface energies. These nanoparticles can be synthesized and modified with various chemical functional groups, which allow them to be linked with antibodies, ligands, and drugs of interest, resulting in opening a wide range of potential applications in biotechnology, magnetic separation, preconcentration of target analytes, targeted drug delivery and vehicles for gene and drug delivery [2]. Metal and metal oxide-based nano-adsorbents play a significant role in the removal of pollutants from wastewater, coating of magnetic nanoparticles with other supports resulting in an increased adsorption efficiency. For example, 6.16% higher by coating silver. Similarly, a Co_3O_4-SiO_2 magnetic nanoparticle coated with nylon 6 enables the adsorption of 666.67 mg/g Pb (II) from wastewater at 298K. The iron oxide nanoparticles implanted on hyperbranched polyglycerol were able to remove nickel, copper, and aluminium from wastewater in 35s successfully.

These nanoparticles can be prepared by different methods like physical and chemical methods. Physical methods used for the synthesis of nanoparticles are energy-intensive and expensive, while chemical methods are based on hazardous chemicals. Hence, there is a need to develop eco-friendly and biocompatible [3] techniques for nanoparticles synthesis. Biological synthesis of nanoparticles can be an alternative to physical/chemical methods of nanoparticle synthesis *i.e.* nanobionics. These methods provide rapid synthesis, better control over size and shape, less toxicity, cost benefits and are eco-friendly [4]. Plant, bacteria, yeast and fungi, can be utilized as effective producers of nanoparticles using principles of nanobionics. Thus, biomimetic nanoparticles formations are known to be much more 'cleaner' and 'greener', owing great relevance to the field of nano-biotechnology. The synthesized nanoparticles with a larger proportion of surface area, higher surface energy, spatial confinements and perfect dimensions are known to have potential benefits in the field based on catalytic, magnetic, electronic and optical properties. The realm of nanotechnology limits its boundaries with bio-based nanoparticles, which gives a promising benefit on protecting the environment through pollution prevention, treatment, and cleanup of contaminates, among others. Also, the broader spectrum of synthesized nanoparticles morphologies and considerably faster biosynthesis rate in cell-free filtrate (due to the higher number of proteins secreted in fungi) make this a specifically exciting route.

Fungi have evolved as an important system for the synthesis of nanoparticles due to their ability to produce extracellular enzymes which can utilize heavy metal ions and can effectively produce nanoparticles. It was easy to isolate and subculture on synthetic media due to low nutritional requirements, high wall binding capacity, simpler biomass handling and extracellular synthesis of nanoparticles which help in easy downstream processing [4], tolerance to high agitation and low pressure as compared to bacteria and plants [5]. There are two strategies for the synthesis of nanoparticles by fungi (1) top-down and (2) bottom-up. In the former approach, the process starts from bulk material and is scaled down to produce nanoparticles, while in the latter, from the small particles (atoms) nanoparticles are assembled.

Green Synthesis of Nanoparticles

Mechanisms of Mycosynthesis of Nanoparticles

Studies on the understanding of mechanisms of synthesis of nanoparticles are underway and a lot is to be deciphered. Briefly, the following mechanisms are proposed for mycosynthesis of nanoparticles. Fungi can produce nanoparticles both extracellularly as well as intracellularly. During intracellular synthesis, fungal biomass is treated with the metal ion solution and incubated in the dark for 24 hrs [3]. Heavy metal attaches to the fungal cell wall with the help of proteins or enzymes present on it *via* electrostatic interactions. Enzymes present in the cell wall reduce the metal ions, leading to aggregation of metal ions and formation of nanoparticles [5, 6]. Extracellular synthesis assumes the interaction of metal ions and releases enzyme mainly reductase with subsequent formation of nanoparticles in solution. Extracellular synthesis of nanoparticles has advantages as it does not require lysis of fungal cells, downstream processing for recovery and purification of nanoparticles [7] whereas, in the case of intracellular synthesis recovery and purification of nanoparticles from fungal biomass is a laborious task that requires analytical equipment and long processing techniques.

Various types of nanoparticles such as silver, gold, *etc.* have been successfully produced by various groups of fungal genera. For example, the synthesis of silver nanoparticles has been reported utilizing many ubiquitous fungal species including *Trichoderma, Fusarium, Penicillium, Rhizoctonia, Pleurotus and Aspergillus.* Extracellular synthesis has been produced by *Trichodermaviride* while intracellular synthesis was shown to occur in a *Verticillium* species, and in *Neurosporacrassa* whereas the synthesis of gold nanoparticles has been reported utilizing *Fusarium, Neurospora, Verticillium, yeasts, and Aspergillus.* Extra-cellular gold nanoparticle synthesis was observed with the help of *Fusarium-oxysporum, Aspergillusniger,* and cytosolic extracts from *Candida albicans.* Intracellular gold nanoparticle synthesis has been explained by a *Verticillium species, V. luteoalbum.* In the search for nanoparticles utilization for applications in biological systems, iron oxide nanoparticles have proven to be a good candidate due to their chemical stability and low toxicity. Thus, materials such as nanocata-lysts, nanocomposites, and bioactive nanoparticles have been continuously utilized in the production of processes and methods for remediation and monitoring of systems contaminated with chemical wastes. Importantly, biogenic fungal nanoparticles target the significant synergistic characteristics when combined with antibiotics and fungicides to offer substantially higher resistance to microbial growth and applications in nanomedicine ranging from topical ointments and bandages for wound healing to coated stents. The technique of nanobionics is employed in various chemical, pharmaceutical, cosmetic industries, such as in nanomedicine.

APPLICATIONS OF NANOBIONICS

Environmental Role

Sustainable development requires the development and promotion of environmental management and a constant search for green technologies to treat a wide range of aquatic and terrestrial habitats contaminated by increasing anthropogenic activities. Bioremediation is an alternative to conventional methods for treating waste compounds and media with the possibility to degrade contaminants using natural microbial activities mediated by different consortia of microbial strains. With the growing environmental problems, decontamination will be achieved by bioremediation which is required for a safer environment. Many studies about bioremediation have been recorded and the scientific literature has shown the progressive emergence of various bioremediation techniques. Bioremediation by utilizing microorganisms such as fungi or the enzymes which are developed by the fungi at the nanoscale level to degrade toxic substances into non-toxic substances can be a viable alternative for fungi mediated bioremediation. The linkage of nanotechnology with the different types of particles and organisms, like fungi can provide sustainable environment-friendly alternatives for bioremediation [8].

Nanoparticles act as adsorbents to remove harmful contaminants from industrial wastewater. Nano-adsorbents can readily remove organic and inorganic pollutants. Also, nanofiltration (NF) membranes (Nanofiltration (NF) is a membrane liquid-separation technology sharing many characteristics with reverse osmosis (RO). Unlike RO, which has the high rejection of virtually all dissolved solutes, NF provides high rejection of multivalent ions, such as calcium, and low rejection of monovalent ions, such as chloride). It plays a vital role in the recovery of nutrients from industrial effluents. It is beneficial to prepare such a membrane due to its increased flux speed, immense water stabilities, and high rejection abilities. For example, (1) NF90 was shown to give the highest rejection (70%) of phosphorous from pulp and paper industry effluent. But the problem in the fouling of membrane is observed due to the high phosphorous content, (2) Ceramic supported graphene oxide (GO)/Attapulgite (ATP) composite membrane can completely remove metal ions of copper, nickel, lead, and cadmium [1], (3)The nanoparticles synthesized from microbes were able to remove more than 90% of heavy metals [Pb (II), Ni (II), Cu (II), and Zn (II)] from wastewater with a regeneration capability of up to five cycles.Mahanty *et al* . (2020) biofabricated iron oxide nanoparticles from *Aspergillus tubingensis* (STSP 25) developed from the rhizospheres of *Avicennia officinalis* in Sundarbans, India [9]. The metal ions were chemically adsorbed on the surface of the nanoparticles in endothermic reactions. In another study, exopolysaccharides (EPS) obtained from *Chlorella vulgaris* were used to co-precipitate with iron oxide nanoparticles. These nanocomposites were able to remove 91% of PO_4^{3-} and 85% of NH_4^+.

Nanobionics in Cosmetics

Nanotechnology can be employed in cosmetics and dermatological products, such as soaps, anti-wrinkle creams, perfumes, kinds of toothpaste, lipsticks, moisturizers, sunscreens, hair care products, skin cleansers, and nail care products. According to Lohani *et al*., NPs are generally classified into eight product classes in terms of their size and functionality; these are liposomes, nanocapsules, solid lipid nanoparticles, nanocrystals, dendrimers, cubosomes, niosomes, and nanogold and nanosilver [10]. Recently, considerable attention has focused on eco-friendly new technologies for the production of metal nanoparticles such as gold, silver, and platinum. The technology is called eco-friendly because the agents used, such as bacteria, fungi, yeasts, and plants, are the biofactories for the NPs.

Nanobionics in Drug Delivery

New drug delivery systems based on nanotechnology have been applied in the treatment of human diseases such as cancer, diabetes, microbial infections and gene therapy. The benefits of these treatments are that the drug is targeted to diseased cells, and its safety profile is enhanced by the reduced toxic side effects to normal cells. In general, NPs can be conjugated with different types of drugs to deliver bioactive compounds to the target site by various methods such as the use of nanotubes, liposomes, quantum dots, nanopores and dendrimers. For example, because of their safety in terms of toxicities and immune compatibilities, Au-NPs are suitable for the preparation of drug delivery scaffolds. Various drugs traverse through these pathways which are mediated by nanomaterials.

These are certain examples of a few drugs which are mediated through nanoparticles. Their mechanism has been altered through various nanomaterials such as gold and silver nanomaterials due to which drugs are easily designed to effectively deliver the required drugs to the system.

Other Applications

Metal nanoparticles synthesized by fungi have great potential to be used as sensors for optical and electronic devices. It was reported by Fayaz *et al* that *Trichoderma viride* synthesized Ag-NPs that were successfully used in biosensor and bio-imaging applications [11]. These Ag-NPs were used for blue orange light emission at wavelengths of 320–520 nm and full characterizations were carried out by EDX (Energy DispersiveX-ray) and XRD analyses. A study by Zheng *et al* described the synthesis of Au-Ag alloy nanoparticles by yeast; the application of these Ag-NPs as a novel vanillin sensor showed that they were five times more sensitive than other methods [12]. This study revealed the high potential of Ag-NPs as sensors in the quantitative determination of vanillin production from the vanilla bean and vanilla tea. A study by Thibault *et al.* showed that Au-NPs enhanced the enzyme activity of glucoseoxidase (GOx) as an indicator for the determination of glucose content in commercial glucose injections. The action of this Au-NP-GOx-based biosensor is based on the highly sensitive detection exhibited by Au-NPs [13].

CONCLUSION

From this chapter, we came to know about the existing importance of Fungal Nanobionics which emerged as a potential system in almost every field of health area. It can be concluded that with the help of various fungus extracted from different sources, nanomaterials can be obtained which further targets the cell in the drug delivery system or in various cosmetic products and can modify in various forms according to its need.

CONSENT FOR PUBLICATION

Not applicable.

CONFLICT OF INTEREST

The authors declare no conflict of interest, financial or otherwise.

ACKNOWLEDGEMENT

Declared none.

REFERENCES

[1] Mandeep and Shukla Pratyoosh. Microbial Nanotechnology for Bioremediation of Industrial Wastewater mandeep and pratyooshshuklaFront. Microbiol 2020.
[http://dx.doi.org/10.3389/fmicb.2020.590631]

[2] Mody V, Siwale R, Singh A, Mody H. Introduction to metallic nanoparticles. J Pharm Bioallied Sci 2010; 2(4): 282-9.
[http://dx.doi.org/10.4103/0975-7406.72127] [PMID: 21180459]

[3] Yadav A, Kon K, Kratosova G, Duran N, Ingle AP, Rai M. Fungi as an efficient mycosystem for the synthesis of metal nanoparticles: progress and key aspects of research. Biotechnol Lett 2015; 37(11): 2099-120.
[http://dx.doi.org/10.1007/s10529-015-1901-6] [PMID: 26164702]

[4] Saxena J, Sharma MM, Gupta S, Singh A. Emerging role of fungi in nanoparticle synthesis and their applications. World J Pharm Pharm Sci 2014; 3(9): 1586-613.

[5] Saha S, Sarkar J, Chattopadhyay D, Patra S, Chakraborty A, Acharya K. Production ofsilver nanoparticles by a phytopathogenic fungus Bipolarisnodulosa and its antimicrobialactivity. Dig J Nanomater Biostruct 2010; 5(4): 887-95.

[6] Kashyap PL, Kumar S, Srivastava AK, Sharma AK. Myconanotechnology in agriculture: a perspective. World J Microbiol Biotechnol 2013; 29(2): 191-207.
[http://dx.doi.org/10.1007/s11274-012-1171-6] [PMID: 23001741]

[7] Gade AK, Bonde P, Ingle AP, Marcato PD, Durán N, Rai MK. Exploitation of *Aspergillusniger* for synthesis of silver nanoparticles. J Biobased Mater Bioenergy 2008; 2(3): 243-7.
[http://dx.doi.org/10.1166/jbmb.2008.401]

[8] JuwarkarAsha A, Singh Sanjeev K, Mudhoo Ackmez. A comprehensive overview of elements in bioremediationReviews in Environmental Science and Bio/Technology 2010; 9: 215-88.

[9] Mahanty S, Chatterjee S, Ghosh S, Tudu P, Gaine T. Synergistic approach towards the sustainable management of heavy metals in wastewater using mycosynthesized iron oxide nanoparticles, Biofabrication, adsorptive dynamics and chemometric modeling study. Journal of Water Process Engineering 2020; 37: 101426.
[http://dx.doi.org/10.1016/j.jwpe.2020.101426]

[10] Lohani A, Verma A, Joshi H, Yadav N, Karki N. Nanotechnology-Based Cosmeceuticals. ISRN Dermatol 2014; 2014: 1-14.
[http://dx.doi.org/10.1155/2014/843687] [PMID: 24963412]

[11] Fayaz M, Tiwary CS, Kalaichelvan PT, Venkatesan R. Blue orange light emission from biogenic synthesized silver nanoparticles using Trichoderma viride. Colloids Surf B Biointerfaces 2010; 75(1): 175-8.
[http://dx.doi.org/10.1016/j.colsurfb.2009.08.028] [PMID: 19783414]

[12]   Zheng D, Hu C, Gan T, Dang X, Hu S. Preparation and application of a novel vanillin sensor based on biosynthesis of Au–Ag alloy nanoparticles. Sens Actuators B Chem 2010; 148(1): 247-52.
[http://dx.doi.org/10.1016/j.snb.2010.04.031]

[13]   Thibault S, Aubriet H, Arnoult C, Ruch D. Gold nanoparticles and a glucose oxidase based biosensor: an attempt to follow-up aging by XPS. Mikrochim Acta 2008; 163(3-4): 211-7.
[http://dx.doi.org/10.1007/s00604-008-0028-z]

CHAPTER 3

Green Synthesis of Fungal Mediated Silver Nanoparticles with Potential Biocontrol Application

Madan L. Verma[1,*], Meenu Thakur[2], Anamika Das[3], Santosh Kumar[4] and **Rekha Kushwaha[5]**

[1] *Department of Biotechnology, Indian Institute of Information Technology Una, Himachal Pradesh-177220, India*

[2] *Department of Biotechnology, Shoolini Institute of Life Sciences and Business Management, Solan-173212, Himachal Pradesh, India*

[3] *Department of Paramedical Sciences, Guru Kashi University, Talwandi Sabo, Bathinda, Punjab-151302, India*

[4] *Donald Danforth Plant Science Center, 975 North Warson Raod, Saint Louis, Missouri 63132, USA*

[5] *Division of Biological Sciences, University of Missouri, Columbia, Missouri 65211, USA*

Abstract: Era of nanotechnology has played a significant role in various aspects of our life. These are minute particles having vast roles. Numerous techniques have been employed for its synthesis. Previously chemical and physical approaches were exploited for their synthesis but nowadays researchers are leaning on biological entities for their creation. And this terminology is green chemistry which does not harm the environment. Mycogenesis plays a terrific role in producing nanoparticles as they contain various enzymes and proteins playing a role in reducing the metal nanoparticles. Metal nanoparticles such as silver nanoparticles act as an efficient biocontrol agent. They were explored to control different types of pests, pathogens, microbes, *etc*. Different mechanisms were used for controlling the pathogens. They are effective due to broad-spectrum efficiency and ruin the cells by binding with phosphorus and sulphur present in the structure of protein and DNA. They are toxic to a wide range of microorganisms. This chapter focuses on the synthesis of silver nanoparticles using different fungal agents and the processes involved. Further, how the prepared particles have prospective application in the control of living organisms, description of all the pathogens against which silver nanoparticles were effective has been provided. This review will comprehensively provide some knowledge regarding the biocontrol application of myconanotechnology.

[*] **Corresponding author Madan L. Verma**: Department of Biotechnology, Indian Institute of Information Technology Una, Himachal Pradesh-177209, India; Tel: 01975-257917; Email: madanverma@iiitu.ac.in, madanverma@gmail.com

Savita, Anju Srivastava, Reena Jain & Pratap Kumar Pati (Eds.)

Keywords: Biocontrol agents, Fungi, Nanoparticles, Silver.

INTRODUCTION

Nanoparticles made up of metals have appealed to researchers for two decades due to the unveiling of distinctive properties. Numerous chemical and physical procedures were employed to synthesize the desired kind of nanoparticles. Though these techniques were providing fruitful results but they involved hazardous chemicals. Reduction of extremity necessitated developing a technique benign towards our environment and minor toxicity. Here comes the green chemistry or the procedure involving the synthesis of nanoparticles using biological bodies (BB) which is also termed biogenic synthesis. Antimicrobial activity is of utmost importance while synthesizing nanoparticles using green chemistry and among metal nanoparticles, those made up of silver hugely act as antimicrobials. They possess both antibacterial and antifungal activity [1]. Silver nanoparticles have gained attraction towards its implications in various fields. Various bacteria [2, 3], algae [4, 5], fungi [6, 7], and plants [8, 9] were studied or used for silver nanoparticle synthesis. The involvement of BB in a procedure shows efficacious results and compassion towards the environment. Exploration has led to the expansion of nanotechnology, generating 'myconanotechnology'. In this area, the fungus is explored to obtain nanoparticles. Numerous researchers have reported the synthesis of various metal nanoparticles by a diverse range of fungi [7, 10 - 12]. Several pragmatic and beneficial aspects can be achieved using fungus as a source of nanoparticles than other microbial agents. One such advantage is the simplicity of the scale-up process making the whole process a lucrative one [13]. Fig. (**1**) shows the schematic diagram summarizing the topics of this chapter.

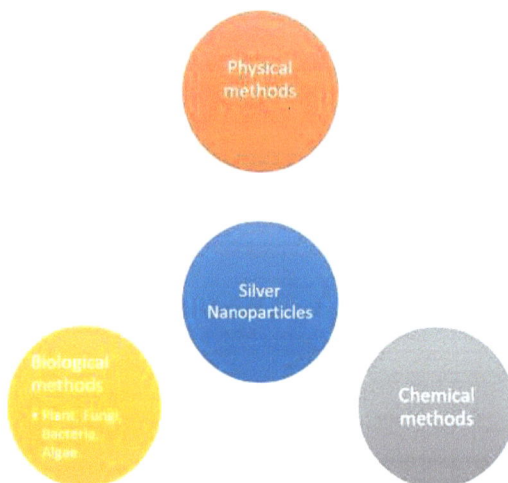

(Fig. 1) contd.....

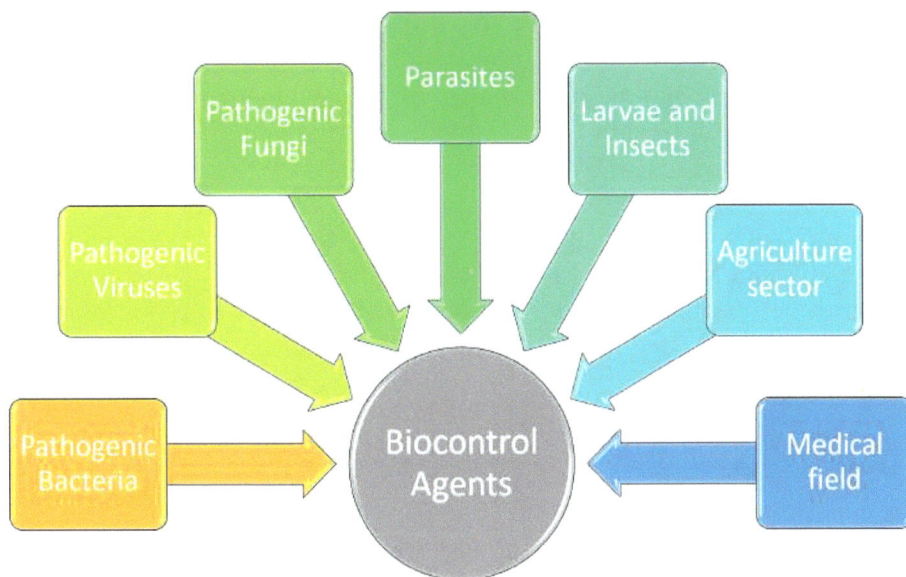

Fig. (1a and b). Schematic diagram of the silver nanoparticles synthesis and their application as biocontrol agents.

Nowadays, it has been reported and observed that the chemical agents used against the harmful pests in the agriculture sector have reported severe ill effects on living beings and our environment. The agricultural sector has used nanotechnology for various aspects. It has been reported that nanoparticles can play a vital role as pesticides, plant growth regulators, and in the process of fertilization when produced using green chemistry [14]. This technology helps in the augmentation of crop productivity by providing novel agrochemicals and efficient ways to deliver. The promising role alleviates the hope of reduced usage of chemical agents. There are groups of researchers inclined towards the synthesis of eco-friendly pesticides and they are herbicides and nanoparticles-based pesticides. They are exploring them for their antimicrobial activity to safeguard the crops against infections. For the exclusion of chemical-based pesticides in farming, a widespread study on nanoparticle-based systems is obligatory [15]. There are reports that nanosilver is effective against phytopathogen *Colletotrichum gloeosporioides* [16]. *Trichoderma longibrachiatum* has been explored to prepare silver nanoparticles and further, the prepared nanoparticles were used against nine phytopathogenic fungi showing effective results [17].

Not only agriculture, even the medical field saw the expansion of nanotechnology applications. Various infections caused by microbial agents have become resistant to antimicrobial treatment available so biogenic silver nanoparticles gained pervasive usage due to their broad-spectrum antimicrobial properties, safety, and inexpensive way of production [18 - 20]. So, to name a few involve the synthesis of silver nanoparticles using *Candida albicans* [13] and *Arthroderma fulvum* [21]. Significant antimicrobial activity was observed against *Escherichia coli*, *Staphylococcus aureus* (Rahimi *et al* ., 2016), genera of *Candida, Aspergillus,* and *Fusarium* [21]. They cause the inactivation of enzymes by binding to thiol groups causing oxidative stress which affects protein oxidation and electron transport [22]. And they alter the replication process when they bind to protein molecules containing sulphur, which also affects cell permeability [23]. There are various reports on the effective role of silver nanoparticles against numerous infectious agents which has been described elaborately later in the section.

The chapter discusses the methodologies explored to synthesize the silver nanoparticles using fungal bodies and which genera or species are more suitable. Further, the application of synthesized silver nanoparticles is elaborated against various pests, insects, *etc*. Further, the era of nanotechnology allows us to explore environment-friendly ways of producing silver nanoparticles having desirable potential.

Fungal Mediated Synthesis of Silver Nanoparticles

Biological synthesis of the nanoparticle by fungi is a rapid, eco-friendly, and easily scaled-up technology. Metal bioaccumulation and tolerance capability of fungi have attracted more attention for the biogenic synthesis of nanomaterials. Fungi are widely being used as nano factories for the large-scale production of nanoparticles with controlled size and morphology [24 - 26]. Digestion of extracellular food, discharge of particular enzymes for hydrolysis of large molecules to smaller molecules that can be utilized as an energy resource is the important abilities of fungi and are very helpful in the production of nanoparticles [27]. The synthesis of a large number of proteins and enzymes by fungi is another advantage of fungi to be used for the fast and sustainable synthesis of nanoparticles [28 - 30]. Several kinds of fungi have been used for the synthesis of nanoparticles such as *Aspergillus flavus, A. fumigatus, Fusarium oxysporum, Fusarium acuminatum, F. culmorum, F. solani, Metarhiziumanisopliae, Phomaglomerata, Phytophthora infestans, Trichoderma viride, Verticillium sp.*

Mechanism of Synthesis of Nanoparticle through Fungi

The synthesis of nanoparticles occurs when the salt (metal) solution is allowed to react with mycelium. The exposure stimulates them to produce metabolites and enzymes for the existence of fungus itself [31]. For biosynthesis of the silver specific fungi were allowed to grow on the solid media followed by the transfer to a liquid medium. These cultural techniques and media depend upon the requirements of the fungal isolate involved in the process. For an increase in the magnitude of nanoparticles production, the enzymatic activity was stimulated providing a modified medium [32]. Besides media, other factors such as agitation, temperature, light, and culture and synthesis times also affect the biosynthesis process and need to be standardized according to the need of the experiment so that desired nanoparticle characteristics can be achieved [33 - 36]. After transferring to the liquid media culture was allowed to grow under specific conditions of temperature, time, and rotation (Table 1). The biomass produced in liquid media is further transferred to water so that all the compounds required for the synthesis get released. This process is followed by filtration. After filtration silver nitrate is added to the filtrate [11, 23, 31, 37, 38]. The ion solution is then monitored for 2 to 3 days or even more for the formation of nanoparticles. Silver nitrate is the most widely used source of silver ions. The Ag^+ ions were reduced by nitrate reductase. Enzymes, especially the reductases, play a vital role in the reduction and stabilization of nanoparticles [12]. Alpha NADPH-dependent nitrate reductase is the main enzyme involved in silver nanoparticle synthesis [39, 40]. Recently, Hietzschold and his colleagues [41] reported that the presence/assistance of nitrate reductase for nanoparticle synthesis is not obligatory. Researchers have also linked proteinaceous substances to silver nanoparticle synthesis and their stabilization [42].

Intracellular Biosynthesis

The biogenic synthesis of nanoparticles using fungi may be intracellular or extracellular. During intracellular synthesis, the metal precursor is added to the mycelial culture. These metal precursors bind to the fungal cell wall by proteins or enzymes present on it *via* electrostatic interactions. The enzymes present on the cell wall further reduce the metal ions and further lead to aggregation of metal ions and formation of nanoparticles [26, 43]. Extraction of the nanoparticles after the synthesis, requires chemical treatment, centrifugation, and filtration to disrupt the biomass and release the nanoparticles [34, 44]. For *e.g.* Mukherjee *et al.* synthesized 25 ±12 nm-sized silver nanoparticles using the fungus *Verticillium* by the intracellular reduction of the metal ions [29].

Extracellular Biosynthesis

During extracellular biosynthesis, a metal precursor is added to the aqueous filtrate containing only the fungal biomolecules. Thus, the interaction of metal ions and the release of an enzyme (reductase) with the subsequent formation of nanoparticles happen in solution. This extracellular synthesis of nanoparticles does not require lysis of fungal cells for further recovery and purification process of nanoparticles. Simple filtration, membrane filtration, gel filtration, dialysis, and ultracentrifugation can be used to eliminate fungal residues and impurities [45 -

47]. Due to the time, cost, and efficiency, this method is preferred by the researchers.

Table 1. Biosynthesis of Silver Nanoparticles by different fungi.

S.NO.	Fungal Organism	Localization	Size Nanometer (nm)/Shape	Liquid Fungal Culture Media	Fungal Culture Temp/ Time	Nanoparticle Synthesis Fungi (g) / Ag NO₃ (M)/ Temp/time/ rpm/Light	Refs
1.	*Verticillium* sp.	Intracellular	25/ Spherical	100 mL MGYP media (malt extract (0.3%), glucose (1.0%), yeast extract (0.3%), and peptone (0.5%)	25-28°C/96 h	10 grams/ 10-4 M / 28 °C / 200 rpm/ 72 h	[29]
2.	*Aspergillus flavus*	Cell wall surface	8.92 /Spherical	Yeast malt broth (dextrose, 10 g/L; peptic digest of animal tissue, 5 g/L; yeast extract, 3 g/L; and malt extract, 3 g/L.	37 C/ 120h	5g/1mM / 37°C/ 200rpm/-/ dark	[48]
3.	*Aspergillus fumigatus*	Extracellular	5–25/ Mostly Spherical, Some Triangular	Liquid media containing (g/l): KH_2PO4, 7.0: K_2HPO_4, 2.0; $MgSO_4 \cdot 7H_2O$, 0.1; $(NH_4)_2SO_4$, 1.0; yeast extract, 0.6; and glucose, 10.0.	25 °C /72h	20 g/1mM/ 25 °C / Dark	[49]
4.	*Aspergillus fumigatus*	-	15 -45 /Spherical	MGYP medium	28°C/72h	10g/ 1 mM/ 28°C/ 150 rpm /24 h /dark	[50]
5.	*Aspergillus niger*	Extracellular	3–30/ Spherical	Liquid medium with (g/L) $NaNO_3$-2.0; KH_2PO_4–0.1; $MgSO_4 \cdot 7H_2O$–0.5; KCl–0.5; sucrose–3.0; streptomycin–0.03	28°C/72h	10 g/1mM/ 28 ± 4 °C/100 rpm/-/-	[40]
6.	*Aspergillus terreus*	Extracellular	1–20 / Spherical	Potato dextrose broth liquid medium (PDB).	28°C/96h	25g/10 mmol/28 °C /-/ 24h/ dark	[51]
7.	Aspergillus *terreus* CZR-1	Extracellular	2.5/Spherical	=	-	-	[72]
8.	*Aspergillus clavitus*	Extracellular	550–650/-	Liquid media containing (g/l) KH_2PO_4, 7.0; K_2HPO_4, 2.0; $MgSO_4 \cdot 7H_2O$, 0.1; $(NH_4)_2SO_4$, 1.0; yeast extract, 0.6; and glucose, 10.0	25°C/72h	20 g /1mM/25 °C - /120h/dark	[53]

(Table 1) cont.....

S.NO.	Fungal Organism	Localization	Size Nanometer (nm)/Shape	Liquid Fungal Culture Media	Fungal Culture Temp/ Time	Nanoparticle Synthesis Fungi (g) / Ag NO$_3$ (M)/ Temp/time/ rpm/Light	Refs
9.	*Aspergillus flavus NJP08*	-	17/ Spherical	MGYP medium (0.3% malt extract, 1.0% glucose, 0.3% yeast extract, 0.5% peptone; pH 7.0)	28°C/96h	10 g /1.0 mM/ 72h 28 C/ 150rpm/ dark	[52]
10.	*Amylomycesrouxii KSU-09*	-	5–27/ Spherical	Potato dextrose broth liquid medium (PDB)	28°C/72h	10 g/1 mM/28°C/200rpm/72 h/-.	[54]
11.	*Cladosporium cladosporioides*	Extracellular	10-100/ Mostly Spherical	liquid media containing (g/L) K$_2$HPO$_4$, 2.5; KNO$_3$, 5.0; MgSO$_4$·7H$_2$O, 1.00; MnSO$_4$·H$_2$O, 0.001; CuSO4·5H$_2$O, 0.003; ZnSO4·7H$_2$O, 0.01; Na2MoO$_3$·2H$_2$O, 0.0015; FeCl$_3$, 0.02; Na$_2$SO$_4$, 1.00; glucose, 40.00	27°C/168h	-	[55]
12.	*Coriolis versicolor*	Extra- and intracellular/	20–100, 100-300/Spherical and Ellipsoidal	-	37°/168h	8g/ 1mM/RT/	[56]
13.	*Humicolasp*	Extracellular	5-25/Spherical	MGYP medium	50 °C/ 96 h	20g/1mM/50°C/200rpm/ 96h/dark	[59]
14.		Extracellular	5-15/ Highly Variable	-	-	10g/ 10^{-3} M//Dark	[58]
15.	*Fusarium oxysporum*	Extracellular	20-50/ Spherical	Media with malt extract 2% and yeast extract 0.5%	25-28°C/144 h	10g/ 10^{-3} M/Dark	[57]
16.		Extracellular	10-25 /Aggregates	MGYP medium	26 ± 1°C /96 h.	-	[39]
17.	*Macrophominaphaseolina*	Cell free	5-40/Spherical	potato dextrose broth (PDB)	28°C/ 120h	1mM/28°C/120rpm/ 24,48,72 h/dark	[71]
18.	*Penicillium fellutanum*	Extracellular	5–25/Mostly Spherical	Liquid media containing (g/l) KH$_2$PO$_4$–7.0; K$_2$HPO$_4$–2.0; MgSO$_4$–0.1; (NH$_4$)$_2$SO$_4$–1.0; yeast extract–0.6; and glucose–10.0.	25 °C/72 h	20g/ 1mM/25°C/Dark	[60]

(Table 1) cont.....

S.NO.	Fungal Organism	Localization	Size Nanometer (nm)/Shape	Liquid Fungal Culture Media	Fungal Culture Temp/ Time	Nanoparticle Synthesis Fungi (g) / Ag NO$_3$ (M)/ Temp/time/ rpm/Light	Refs
19.	*Penicillium strain J3*	Cell Free	10–100/ Mostly Spherical	liquid medium containing (g/l): KH$_2$PO$_4$ -7; K$_2$HPO$_4$ -2.0; MgSO4 × 7H$_2$O 0.1; NH$_4$NO$_3$ 1.0; yeast extract 0.6; glucose 20.0.	25 °/72 -96h	10g/1mM/25°C-/dark	[61]
20.	*Phomaglomerata*	Extracellular	60–80/ Spherical	Potato Dextrose agar	25°C/48h	-	[62]
21.	*Pleurotussajorcaju*	Extracellular	30.5/Spherical	-	37°C/=	-	[63]
22.	*Rhizopus stolonifer*	-	25–30/ Quasi-spherical	liquid media containing (g/l) KH2PO4, 7.0: K2HPO4, 2.0; MgSO4·7H2O, 0.1; (NH$_4$)2SO$_4$, 1.0; yeast extract, 0.6; and glucose	25°C/72h	-/1mM/ 25°C/ -/ dark	[64]
23.	*Trichoderma viride*	-	5–40/ Spherical, rod-like		27°C/72h	20 g/ 10^{-3} M/ 27 °C /- /dark	[65]
24.		-	2–4, 10–40, 80–100/ Spherical			25g/10^{-3} M/ (10, 27 and 40 °C)/dark	[66]
25.		-	2–4/ Mostly Spherical			20 g/ 10^{-3} M/ 27 °C /dark	[68]
26.		Extracellular	5–40/ Spherical, Rod-Like			20 g/ 10^{-3} M/ 27 °C /dark	[67]
27.		-	2–5/ Spherical; 40–65/ Rectangular; 50–100 Penta/Hexagonal	Potato dextrose broth	28°C/96h	20, 30 and 40°C/ 24, 48 and 72 h/-/dark	[73]
28.	*Trichoderma asperellum*	Cell free	13–18/ Nanocrystalline	Liquid media containing (g/l) KH$_2$PO$_4$, 7.0: K$_2$HPO$_4$, 2.0; MgSO$_4$·7H$_2$O, 0.1; (NH$_4$)$_2$SO4, 1.0; yeast extract, 0.6; and glucose, 10.0.	-	5g/1 mM/25 ˙C)/180 rpm/-dark	[69]
29.	*Volvariellavolvaceae*	Extracellular	15/ Spherical	-	-	-	[70]

Different types of fungal species have been used by researchers for the synthesis of extracellular biosynthesis of silver nanoparticles (Table **1**).

Aspergillus sp.

Vignashwara *et al.* [48] used the fungus *Aspergillus flavus* for biological synthesis of the nanoparticle of size 8.92 nm with spherical shape and those were found to be stable in water for more than 3 months. *Aspergillus flavus* is also used by Bhainsa and D'souza [49] for the synthesis of spherical and triangular silver nanoparticles of 5–25 nm size. Another interesting study by Alani *et al.* [50] compared the biosynthesis of nanoparticles of *Streptomyces sp.* with *Aspergillus fumigatus* and identified the localization of the primary factors (enzyme and/or other) produced by the two strains for the nanoparticle synthesis. According to this study, the primary factor responsible for the reduction of $AgNO_3$ to Ag metal nanoparticles is intracellular. The role of extracellular enzyme-nitrate reductase during the synthesis of 3–30 nm silver nanoparticles by *Aspergillus niger* has been studied by Jaidev *et al.* [40]. Polydisperse spherical silver nanoparticles ranging in size from 1 to 20 nm have been created by Guangquan *et al.* [51]. These authors used a reduction of aqueous Ag+ with the culture supernatants of *Aspergillus terreus* [51]. The synthesized silver nanoparticles were stabilized in the solution. *Aspergillus flavus* NJP08 has been used by Jain *et al.* [52] for the synthesis of 17 nm silver nanoparticles. Saravanan and Nanda [53] reported the soil fungus *Aspergillus clavatus* for the extracellular synthesis of silver nanoparticles with a size range of 550–650 nm (Table **1**).

Amylomycesrouxii KSU-09

Amylomycesrouxii KSU-09 is a fungal strain isolated from the roots of the date palm. Musarrat *et al* [54]. used this fungus for the biosynthesis of the spherical silver nanoparticles (AgNPs) of the size from 5–27 nm.

Cladosporium Cladosporioides

Cladosporium cladosporioides, a darkly pigmented mould, was used by Balaji *et al.* [55] for the extracellular biosynthesis of functionalized silver nanoparticles. The biosynthesized nanoparticles vary from 10-100 nm in size and are spherical.

Coriolus Versicolor

A controlled and up-scalable method for the biosynthesis of silver nanoparticles by using fungal proteins of *Coriolus versicolor* showed that alkaline conditions for biosynthesis can accelerate the reaction even at room temperature [56].

Fusarium Oxysporum

Several strains of the fungus *Fusarium oxysporum* have been used for the extracellular biosynthesis of silver nanoparticles [57]. Silver hydrosol was formed after exposure of aqueous silver ions to strains of *Fusarium oxysporum*. The silver nanoparticles were in the range of 20–50 nm in dimensions. According to Durán [57], this reduction of the metal ions occurs by a nitrate-dependent reductase and a shuttle quinone extracellular process. The technique of fungal mediated biosynthesis of nanoparticles has several technical applications, including their high potential as antimicrobial materials. In another study by Ahmad *et al.* [58], biosynthesis of the silver nanoparticle of the range of 5/15 nm in dimensions was observed by using *Fusarium oxysporum* and the proteins secreted by the fungus helped in the stabilization of nanoparticles in the solution. *Fusarium oxysporum* synthesized the silver nanoparticles using an NADPH-dependent nitrate reductase and phytochelatin *in vitro*. This led to the formation of a stable silver hydrosol 10–25 nm diameter. Stabilization is achieved by a capping peptide using the enzyme nitrate reductase phytochelatin, and 4-hydroxyquinoline in the presence of a co-factor (NADPH). The absence of any of these particles leads to the non-formation of nanoparticles [39].

Humicola sp.

Sayed *et al*. [59] reported the extracellular biosynthesis of silver nanoparticles by the thermophilic fungus *Humicola sp.*

Penicillium sp.

Kathiresan *et al*. [60] studied the biosynthesis of the silver nanoparticles by using *Penicillium fellutanum,* a marine fungus. Maliszewska *et al*. [61] used the cell-free extract of *Penicillium* strain isolated from the soil for the synthesis of silver nanoparticles.

Phomaglomerata

Extracellular synthesis of silver nanoparticles from *Phomaglomerata* was carried out and its efficacy against *Escherichia coli, Staphylococcus aureus,* and *Pseudomonas aeruginosa* Birla [62] was compared.

Rhizopus Stolonifer

Nadanathangam *et al*. [63] synthesized the silver-protein (Core-Shell) nanoparticle by using a spent mushroom substrate; the average size of nanoparticles was reported to be 30.5 nm. These silver nanoparticles were found to be stable in solution for more than 6 months. Cell filtrate from inactive biomass of *Rhizopus stolonifer* was also used for the production of 25–30 nm near Ag nanoparticles at room temperature [64].

Trichoderma sp.

Trichoderma viride has been used by many researchers for the synthesis of Ag-nanoparticles and is further used in many applications. Fayaz *et al*. [65] synthesized the silver nanoparticles using *Trichoderma viride* and then incorporated them into sodium alginate for vegetable and fruit preservation. The authors also studied the effect of temperature on controlling the size of silver nanoparticles. An increase in reaction temperature leads to a decrease in the size of silver nanoparticles and an increase in monodispersity [66]. In another study, authors investigated the use of the fungus *Trichoderma viride* for the extracellular biosynthesis of 5–40 nm silver nanoparticles from silver nitrate solution [67]. The silver nanoparticles of size range 2-4 nm which can emit Blue orange light were synthesized using *Trichoderma viride* [68]. *Trichoderma viride* was also used for the biosynthesis of different shapes of silver nanoparticles such as spherical, rectangular, penta, and hexagonal (42). These changes in shape were related to change in the pH, temperature, and reaction time. *Trichoderma asperellum* and *Trichoderma asperellum* non-pathogenic and commercially viable biocontrol agents have been utilized in a controlled and up-scalable biosynthetic route for the biosynthesis of nanocrystalline silver particles of size range 13–18 nm [69].

Volvariellavolvacea

Philips [70] reported an extracellular synthesis method for the preparation of silver nanoparticles 15 nm in size in water, using the extract *of Volvariella volvacea*.

Macrophomina Phaseolina

Phytopathogenic soil-borne fungi *Macrophomina phaseolina* used for biosynthesis of silver nanoparticles of size range 5 to 40 nm [71].

APPLICATIONS OF FUNGAL NANOPARTICLES AS BIOCONTROL AGENTS

Rapid urbanization, industrialization, and competition to produce more have led to the large-scale use of chemical pesticides. Pesticides are chemicals that can control the spread of various crop diseases. Piles of these chemicals are not only burdening our ecosystem but also polluting our environment with their residues leading to serious health impacts to humans. Moreover, excessive use of these chemicals for more products is not an option for sustainable agriculture [74]. Pesticide is a broader term that can be classified as fungicides, herbicides, rodenticides, and insecticides [75]. Some of the pesticides have been used in the health sector to prevent the spread of some vector-borne diseases such as in shampoos of pets [76, 77].

Most pesticides are known to have serious effects on the environment and human health [78, 79]. Some serious health hazards are respiratory, reproductive, gastrointestinal, neurological, carcinogenic, and endocrine effects [80, 81]. Various types of chemical pesticides such as carbamates, triazines, organochlorines, organophosphates, pyrethroids, and glyphosate cause ill effects to the environment [82]. Most of these pesticides have chronic effects on human health, such as endocrine disruption [83, 84], affect reproductive behaviour [85] neurotoxic [86, 87] cytotoxicity/carcinogenic [88]. Pesticide residues can be detected in food, beverage, milk samples. There is an utmost need for searching for alternative solutions for sustainable agriculture. Innovative ideas have been adopted to develop sustainable strategies for Agriculture with fewer health hazards. Biogenic nanomaterials are one such novel idea where biogenic particles can be associated with nanoparticles to increase the dimensions of applications. One such strategy is to use fungal cells as biocontrol agents (Table **2**). However, using fungi for controlling plant diseases is an age-old practice, but using it along with nanomaterials has increased its applications by reducing the negative effects of chemical pesticides on human health and the environment. Silver metal possesses various antimicrobial properties with a wide spectrum of other useful characteristics [89, 90]. Their antimicrobial activity involves the attachment of the particle with the cell membrane to reach inside the cells. The mechanism involves the production of reactive oxygen species upon attaching to the cells of pathogenic/target [91]. Biogenic synthesis of nanoparticles can be produced using metabolic activities of microorganisms that act as stabilizing agents exhibit better antimicrobial activities [42].

Table 2. Examples of biogenic metallic nanoparticles with biocontrol activity.

Sr. No.	Name of the Fungal Organism	Target Organism	Application and Crops	References
1.	*Trichoderma viride*	*Rhizoctonia solani , Fusariu m moniliforme*	Prevent crop losses in Rice plant.	[93]
2.	*Calothrixelenkinii*	*Alternaria alternata*	Inhibit fungal infection in tomato plants	[118]
3.	*Aspergillus niger*	Fungal isolates from vegetables and fruits like strawberry, orange, cucumber and lemon	Prevents yield losses in vegetables and fruit crops.	[119]
4.	*Trichoderma harizanum*	*Sclerotia sclerotiorum*	Soybean, tomato, lettuce and beans	[36, 101, 102]
5.	*Penicillium* endophyte of *Curcuma longa*	*Streptococcus aureus*	Effective in infections caused by resistant strains.	[103]
6.	*Trichoderma longibrachiatum* isolated from cucumber	*Trichoderma viride, Trichoderma harzianum, Alternaria alternata, Rhizoctoniastolonifer*	Prevents crop losses in vegetable crops.	[17]
7.	*Beauveria bassiana* isolated from *Brassica compestris*	*Lipophiserysimi*	Prevents crop and oil losses in the mustard plant from aphid attack	[104]

(Table 2) cont.....

Sr. No.	Name of the Fungal Organism	Target Organism	Application and Crops	References
8.	*Aspergillus versicolor*	*Sclerotoniasclerotiorum, Botrytis cinerea*	Prevents crop losses in Strawberry plants	[108]
9.	*Fusarium solani*	Pathogenic against 13 isolates from grains	Crop losses in wheat, barley, and maize	[109]
10.	*Guignardiamangiferae*	*Colletotrichum, Rhizoctonia solani , Curvala rialunata*	Biocontrol agents	[110]
11.	*Epicoccum nigrum*	*Candida albicans, Fusarium solani*	Biocontrol agents	[111]
12.	*Trichoderma atroviride*	*Phonopsis*Canker disease	Economic losses in tea plants	[113]
13.	*Trichoderma atroviride*	Birds eye spot disease	Yield losses in tea plants	[114]
14.	*Pleurotusostreatus*	*E. coli, Klebsiella pneumonia, S. aureus, Vibrio cholerae*	Antibacterial	[115]
15.	*Trichoderma sp.* isolated from Brazil nut	Antibacterial activity in Gram –vebacteria	Biocontrol agents	[116]

Applications of these fungal-mediated silver nanoparticles in biocontrol/pest control have been evaluated and reviewed by a few previous researchers [23]. Some of the major strategies are using fungal species with antifungal activities, use of these nanoparticles with conventional insecticides, and another way is to control the vectors for insects. In this chapter, an attempt has been made to compile all the applications of fungal-mediated silver nanoparticles as biocontrol agents.

Mycoparasitism is one such activity in which one fungal organism possesses antifungal activity against other harmful and pathogenic fungi [92]. The potential of silver nanoparticles mediated by *Trichoderma viride* isolated from the soil as biocontrol agents against *Rhizoctonia solani* and *Fusarium moniliforme* has been checked and evaluated [93]. The mechanism behind mycoparasitism (Fig. **2**) is the expression of catalase and copper-induced superoxide dismutase (SOD) genes that induces the production of lytic enzymes such as chitinase, glucanases, proteases, cellulases, xylanases, and mannanases [94]. Thus, *Trichoderma viride* not only acts as biocontrol agents but enhances overall productivity. Metallic nanoparticles containing silver along with *T. viride* have been synthesized and characterized by Fourier transform infrared spectroscopy and scanning electron microscopy (SEM) analysis. Mycoparasitism potential was checked using discs and seeded with *Rhizoctonia solani* and *Fusarium moniliforme* affecting rice. The development of a halo zone around the cavity is an indication of biocontrol action when 50 µg/mL silver nitrate suspension has been used and 9.2mm and 3.8 mm zone of inhibition was observed in *Fusarium moniliforme* and *Rhizoctonia solani* respectively.

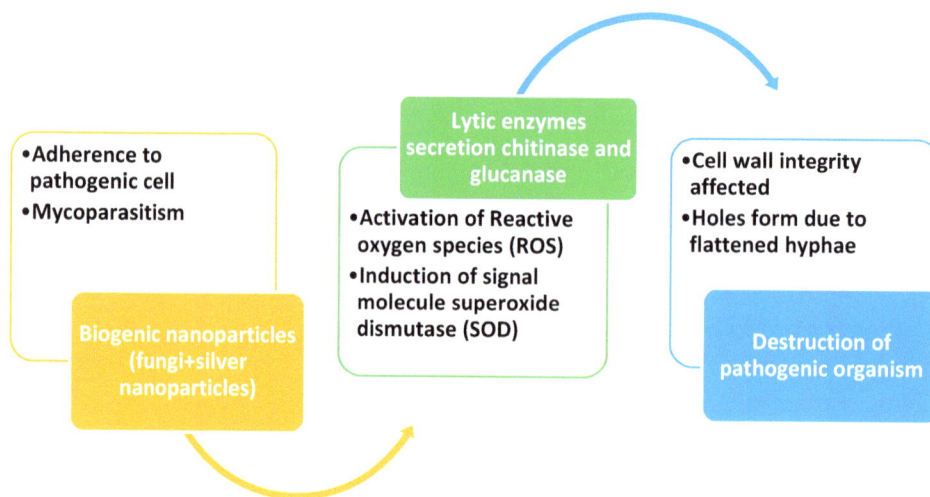

Fig. (2). Mycoparasitic mechanism of silver nanoparticles against fungal pathogens.

One of the most important vegetable crops is tomato (*Lycopersicon esculentum* Mill.), world's second most grown vegetable [95]. Early blight caused by *Alternaria alternata* is an airborne pathogen responsible for major losses. Silver nanoparticles combined with fungi provide an excellent alternative against early blight pathogens over chemical fungicides [96]. Excessive use of fungicides in agriculture has led to the emergence of resistant strains of pathogens that poses a major upcoming challenge for sustainable agriculture [97]. *Calothrix elenkinii* associated with silver particles has shown a synergistic effect of inhibition against *Alternaria alternata* [98].

In another similar work, antifungal activity of *Aspergillus niger* associated with silver nanoparticles has been tested against various fungal isolates affecting fresh market vegetables and fruits such as orange (*Citrus sinensis*), lemon (*Citrus limon*), tomatoes (*Lycopersicon esculentum*), grapes (*Vitis vinifera*), strawberries (*Fragaria ananassa*), cucurbita (*Cucurbita pepo*), cucumbers (*Cucumis sativus*), eggplants (*Solanum melongena*), bell pepper (*Capsicum annuum*) and soft dates (*Phoenix dactylifera* Lin.) of Saudi Arabia [99]. The biogenic synthesis of silver nanoparticles has raised attention, offering a low-cost alternative with less energy consumption and high yields [36]. *Trichoderma harzianum* is a mycoparasitic fungus that can be used for the production of fungicides against various fungi. The main mechanism behind phytopathogenesis is the synthesis of hydrolytic enzymes such as chitinase [100]. Mycoparasitic activity has been evaluated against *S. sclerotiorum* effecting crop losses of soybean, tomato, lettuce, beans, and sunflower [36, 101, 102]. Moreover, cytotoxic and genotoxic evaluation of nanoparticles has been checked by MTT assay against V79, 3T3, and HaCat cell lines [36]. *Curcuma longa* (turmeric) is a medicinal plant and one of its endophytic fungi *Penicillium sp.* has been characterized by FTIR technique against multiple drug-resistant strains such as *E. coli*. and *Streptococcus aureus* [103].

Phytopathogenic fungi are responsible for major crop losses due to damage to the crops. In one of the important studies, *Trichoderma longibrachiatum* isolated from cucumber along with silver nanoparticles has been evaluated for its mycotoxic activities against phytopathogens such as *T. viride, T. harzianum, F. verticillioides, Alternaria alternata* and *Rhizopus stolonifer* isolated from cucumber, tomato, pepper, tomato, and pepper, respectively [102]. Various physical parameters such as temperature, incubation time, agitation conditions and biomass concentration has been optimized for finding best condition. The stability with silver nanoparticles was observed to be 2 months. Significant activities (90%) were noted against 9 fungal isolates such as *Fusarium verticilloides, Fusarium moniliforme, Penicillium brevicompactum, Helonominthosperium oryzae Penicillium glabrum, Aspergillus flavus, Penicllium griseae* and *Pyricularia grisea* by *in vitro* studies [100].

Mustard (*Brassica campestris*) is one of the major oil crops used for the production of edible oil. Significant yield losses have been observed by the action of aphid *Lipophis erysimi* Kalt with 9-96% loss to yield and 5-6% oil yield. *Beauveria bassiana* entomo pathogenic fungi were isolated from *Brassica campestris* and its impact on aphids has been studied [104]. The efficiency of 25 different isolates of *B. Bassiana* has been evaluated against aphids. Maximum percent mortality has been reported with B4 (90%) and B13 (64%) isolates. The nanoparticles have been characterized by FTIR and X-ray diffraction studies. This study has provided new insights into the efficacy of fungal activity against aphids and can prevent the major losses to the oil crop *Brassica campestris* [104].

Green nanotechnology has a significant effect due to the effective synthesis of valuable products with minimum harm to the environment [105]. *Justicia spicigera* has antidiabetic properties owing to its antioxidant potential and can be used for various biotechnological applications [106]. In one of the previous studies, silver nanoparticles derived from *J. spicigera* have been evaluated for controlling the growth of phytopathogenic fungi and food-borne bacteria [107]. Scanning electron microscopic studies confirmed the size of nanoparticles to be in the range of 86-100nm. Antibacterial activity was checked against *Bacillus cereus, Klebsiella pneumoniae,* and *Enterobacter aerogenes* by disc diffusion method and found to be effective at 100mg/mL concentration. However, antifungal activity has been observed to be 79.77 and 60.1% radial growth inhibition against *M. phaseolina*. *Alternaria alternata* comparatively a weaker response has been reported against *Colletotrichum sp.*(40%) and *Fusarium solani* (35%), respectively.

Strawberry plants are prone to fungal diseases and the efficacy of silver nanoparticles combined with *Aspergillus versicolor* was checked against *Sclerotonia sclerostiorum* and *Bortytis cineaea* [108]. In another study on fungal isolates from wheat, *Fusarium solani* is effective for the treatment of barley, maize, and wheat seeds [109]. In one similar study, endophytic fungi isolated from leaves of medicinal plants have been tested for mycoparasitic potential. Total 13 fungal isolates have been used for the synthesis of silver nanoparticles. Out of them, 4 fungal particles have resulted in stable nanoparticles. The physicochemical conditions have been optimized for the biosynthesis of nanoparticles, followed by characterization with TEM, SEM, X-ray diffraction (XRD), and Electron dispersion spectroscopy (EDX). Among these *Guignardia mangiferae* synthesized most stable silver nanoparticles and exhibited antifungal activity against *Colletotrichum sp.*, *Rhizoctonia solani*, *Curvalaria lunata* [110]. Effects of inhibitory concentrations of *Epicoccum nigrum* against *Candida albicans*, *Fusarium solani*, *Sporothrix schenckii*, *Cryptococcus neoformans*, *Aspergillus flavus,* and *Aspergillus fumigatus* have been checked and found to be concentration-dependent [111].

Biosynthesized nanoparticles have raised much interest due to the beneficial effects of phytopathogenic metabolites. So, in one of the recent studies, eco-friendly, cost-effective silver nanoparticles with aqueous leaf extracts of *Cucumis prophetarum* have been fabricated. Its antibacterial activity against *Salmonella typhi* and *Staphylococcus aureus* and cytotoxic antiproliferative activity was checked against HepG2, MCF-7, respectivelyS [112]. *Trichoderma* is a fungus with diverse mycotoxic effects. Tea is one of the important plants and phytopathogenic activities of metallic nanoparticles (silver and gold) with *Trichoderma atroviride* against biological control of Phomopsis canker disease in tea plants [113]. In one similar studies, silver nanoformulations using *Trichoderma atroviride* have been developed as a foliar spray against bird's eye spot disease in tea plantations [114].

Antibiotic resistance in bacteria is a very common and serious issue these days. Because a very limited number of antibiotics are available for existing resistant strains, fungal nanoparticles containing silver have been evaluated for antibacterial properties against resistant strains. In one similar study, silver nanoparticles (Ag-NPs) combined with *Pleurotus ostreatus* and their antibacterial potential have been checked against *E. coli*, *Klebsiella pneumonia*, *Pseudomonas aeruginosa*, and *Vibrio cholera* [115]. In another similar study, silver nanoparticles have been fabricated using *Trichoderma spp.* isolated from Brazillian nut (*Bertholletia excelsa*) from forest areas of Brazilian amazon and were further evaluated for antibacterial activities against Gram –ve bacteria. Moreover, these particles were characterized using various physical techniques [116].

As silver nanoparticles have a significant effect on controlling phytopathogenic fungi and can be used as biocontrol agents but their impact on beneficial fungi present in the rhizosphere needs to be explored further. However, in very few studies, negative impact has been reported [117]. Thus, nanotechnology-based applications are becoming an integral part of our environmental protection [120 - 124].

CONCLUSION

Green chemistry involving fungal agents is the solution to the approaches causing environmental hazards. For a better polydispersity of nanoparticles and to obtain the required size, an understanding of molecular mechanisms involved during the green formation of nanoparticles is needed. Silver nanoparticles illustrate antagonistic activity against various pests and pathogens. SN is responsible for inhibiting their growth. Various constraints are still prevailing for the commercialization of nanoformulations in field trials which have to be looked after through policymaking. Nanofungicides and nano pesticides synthesized using green chemistry have a bright future in agriculture. Agriculture is the backbone of any country which needs the implication of newer techniques for various aspects. While against the microbial agents in the medical field requires suitable formulations and application of biosynthesized silver nanoparticles exploration by the pharmaceutical industry.

CONSENT FOR PUBLICATION

Not applicable.

CONFLICT OF INTEREST

The authors declare no conflict of interest, financial or otherwise.

ACKNOWLEDGEMENT

One of the authors (Dr. Madan L. Verma) would like to thank the Director of Indian Institute of Information Technology Una, Himachal Pradesh, India for providing the necessary facility to pursue the recent study.

REFERENCES

[1]   Nakamura S, Sato M, Sato Y, *et al.* Synthesis and application of silver nanoparticles (Ag NPs) for the prevention of infection in healthcare workers. Int J Mol Sci 2019; 20(15): 3620.
[http://dx.doi.org/10.3390/ijms20153620] [PMID: 31344881]

[2]   Singh P, Singh H, Kim YJ, Mathiyalagan R, Wang C, Yang DC. Extracellular synthesis of silver and gold nanoparticles by Sporosarcina koreensis DC4 and their biological applications. Enzyme Microb Technol 2016; 86: 75-83.
[http://dx.doi.org/10.1016/j.enzmictec.2016.02.005] [PMID: 26992796]

[3]   Khan I, Saeed K, Khan I. Nanoparticles: Properties, applications and toxicities. Arab J Chem 2019; 12(7): 908-31.
[http://dx.doi.org/10.1016/j.arabjc.2017.05.011]

[4]   Hamouda RA, Hussein MH, Abo-elmagd RA, Bawazir SS. Synthesis and biological characterization of silver nanoparticles derived from the cyanobacterium Oscillatoria limnetica. Sci Rep 2019; 9(1): 13071.
[http://dx.doi.org/10.1038/s41598-019-49444-y] [PMID: 31506473]

[5]   Bhuyar P, Rahim MHA, Sundararaju S, Ramaraj R, Maniam GP, Govindan N. Synthesis of silver nanoparticles using marine macroalgae Padina sp. and its antibacterial activity towards pathogenic bacteria. Beni Suef Univ J Basic Appl Sci 2020; 9(1): 3.
[http://dx.doi.org/10.1186/s43088-019-0031-y]

[6]   Gudikandula K, Vadapally P, Singara Charya MA. Biogenic synthesis of silver nanoparticles from white rot fungi: Their characterization and antibacterial studies. OpenNano 2017; 2: 64-78.
[http://dx.doi.org/10.1016/j.onano.2017.07.002]

[7]   Tyagi S, Tyagi PK, Gola D, Chauhan N, Bharti RK. Extracellular synthesis of silver nanoparticles using entomopathogenic fungus: characterization and antibacterial potential. SN Applied Sciences 2019; 1(12): 1545.
[http://dx.doi.org/10.1007/s42452-019-1593-y]

[8]   Masum MMI, Siddiqa MM, Ali KA, *et al.* Biogenic synthesis of silver nanoparticles using *phyllanthus emblica* fruit extract and its inhibitory action against the pathogen *acidovorax oryzae* strain RS-2 of rice bacterial brown stripe. Front Microbiol 2019; 10: 820.
[http://dx.doi.org/10.3389/fmicb.2019.00820] [PMID: 31110495]

[9]   Pirtarighat S, Ghannadnia M, Baghshahi S. Green synthesis of silver nanoparticles using the plant extract of Salvia spinosa grown *in vitro* and their antibacterial activity assessment. J Nanostructure Chem 2019; 9(1): 1-9.
[http://dx.doi.org/10.1007/s40097-018-0291-4]

[10]   Saravanan M, Arokiyaraj S, Lakshmi T, Pugazhendhi A. Synthesis of silver nanoparticles from Phenerochaete chrysosporium (MTCC-787) and their antibacterial activity against human pathogenic bacteria. Microb Pathog 2018; 117: 68-72.
[http://dx.doi.org/10.1016/j.micpath.2018.02.008] [PMID: 29427709]

[11]   Guilger M, Pasquoto-Stigliani T, Bilesky-Jose N, *et al.* Biogenic silver nanoparticles based on trichoderma harzianum: synthesis, characterization, toxicity evaluation and biological activity. Sci Rep 2017; 7(1): 44421.
[http://dx.doi.org/10.1038/srep44421] [PMID: 28300141]

[12]   Zomorodian K, Pourshahid S, Sadatsharifi A, *et al.* Biosynthesis and characterization of silver nanoparticles by *aspergillus* species. BioMed Res Int 2016; 2016: 1-6.
[http://dx.doi.org/10.1155/2016/5435397] [PMID: 27652264]

[13]   Rahimi G, Alizadeh F, Khodavandi A. Mycosynthesis of silver nanoparticles from candida albicans and its antibacterial activity against *escherichia coli* and *staphylococcus aureus*. Trop J Pharm Res 2016; 15(2): 371.
[http://dx.doi.org/10.4314/tjpr.v15i2.21]

[14]   Ghidan A, Awwad A, Ghidan O, Araj S, Ateyyat M. Comparison of different green synthesized nanomaterials on green peach aphid as aphicidal potential. Fresenius Environ Bull 2018; 27: 7009-16.

[15]   Baruah S, Dutta J. Nanotechnology applications in pollution sensing and degradation in agriculture: a review. Environ Chem Lett 2009; 7(3): 191-204.
[http://dx.doi.org/10.1007/s10311-009-0228-8]

[16]   Madan HR, Sharma SC, Udayabhanu , *et al.* Facile green fabrication of nanostructure ZnO plates, bullets, flower, prismatic tip, closed pine cone: Their antibacterial, antioxidant, photoluminescent and photocatalytic properties. Spectrochim Acta A Mol Biomol Spectrosc 2016; 152: 404-16.
[http://dx.doi.org/10.1016/j.saa.2015.07.067] [PMID: 26241826]

[17]   Elamawi RM, Al-Harbi RE, Hendi AA. Biosynthesis and characterization of silver nanoparticles using Trichoderma longibrachiatum and their effect on phytopathogenic fungi. Egypt J Biol Pest Control 2018; 28(1): 28.
[http://dx.doi.org/10.1186/s41938-018-0028-1]

[18]   C A, C P, Handral HK, Kelmani R C. Investigation of antifungal and anti-mycelium activities using biogenic nanoparticles: An eco-friendly approach. Environ Nanotechnol Monit Manag 2016; 5: 81-7.
[http://dx.doi.org/10.1016/j.enmm.2016.04.002]

[19]   Jiravova J, Tomankova KB, Harvanova M, *et al.* The effect of silver nanoparticles and silver ions on mammalian and plant cells *in vitro.* Food Chem Toxicol 2016; 96: 50-61.
[http://dx.doi.org/10.1016/j.fct.2016.07.015] [PMID: 27456126]

[20]   Gopinath PM, Narchonai G, Dhanasekaran D, Ranjani A, Thajuddin N. Mycosynthesis, characterization and antibacterial properties of AgNPs against multidrug resistant (MDR) bacterial pathogens of female infertility cases. Asian Journal of Pharmaceutical Sciences 2015; 10(2): 138-45.
[http://dx.doi.org/10.1016/j.ajps.2014.08.007]

[21]   Wang L, He D, Gao S, Wang D, Xue B, Yokoyama K. Biosynthesis of silver nanoparticles by the fungus Arthroderma fulvum and its antifungal activity against genera of Candida, Aspergillus and Fusarium. Int J Nanomedicine 2016; 1899: 1899.
[http://dx.doi.org/10.2147/IJN.S98339]

[22]   Fatima F, Verma SR, Pathak N, Bajpai P. Extracellular mycosynthesis of silver nanoparticles and their microbicidal activity. J Glob Antimicrob Resist 2016; 7: 88-92.
[http://dx.doi.org/10.1016/j.jgar.2016.07.013] [PMID: 27689341]

[23]   Guilger-Casagrande M, Lima R. Synthesis of Silver Nanoparticles Mediated by Fungi: A Review. Front Bioeng Biotechnol 2019; 7: 287.
[http://dx.doi.org/10.3389/fbioe.2019.00287] [PMID: 31696113]

[24]   Sastry M, Ahmad A, Islam Khan M, Kumar R. Biosynthesis of metal nanoparticles using fungi and actinomycete. Curr Sci 2003; 85: 162-70.

[25]   Raliya R, Rathore I, Tarafdar JC. Development of microbial nanofactory for zinc, magnesium, and titanium nanoparticles production using soil fungi. Journal of Bionanoscience 2013; 7(5): 590-6.
[http://dx.doi.org/10.1166/jbns.2013.1182]

[26]   Saxena J, Mohan Sharma M, Gupta S, Singh A. Emerging role of fungi in nanoparticle synthesis and their applications. World J Pharm Pharm Sci 2014; 3: 1586-613.

[27]   Blackwell M. The Fungi: 1, 2, 3 … 5.1 million species? Am J Bot 2011; 98(3): 426-38.
[http://dx.doi.org/10.3732/ajb.1000298] [PMID: 21613136]

[28]   Rai M, Ingle AP, Birla S, Yadav A, Santos CAD. Strategic role of selected noble metal nanoparticles in medicine. Crit Rev Microbiol 2015; 42(5): 1-24.
[http://dx.doi.org/10.3109/1040841X.2015.1018131] [PMID: 26089024]

[29]   Mukherjee P, Ahmad A, Mandal D, *et al.* Fungus-mediated synthesis of silver nanoparticles and their immobilization in the mycelial matrix: a novel biological approach to nanoparticle synthesis. Nano

Lett 2001; 1(10): 515-9.
[http://dx.doi.org/10.1021/nl0155274]

[30]   Boroumand Moghaddam A, Namvar F, Moniri M, Md Tahir P, Azizi S, Mohamad R. Nanoparticles biosynthesized by fungi and yeast: a review of their preparation, properties, and medical applications. Molecules 2015; 20(9): 16540-65.
[http://dx.doi.org/10.3390/molecules200916540] [PMID: 26378513]

[31]   Costa Silva LP, Pinto Oliveira J, Keijok WJ, *et al.* Extracellular biosynthesis of silver nanoparticles using the cell-free filtrate of nematophagous fungus *Duddingtonia flagrans*. Int J Nanomedicine 2017; 12: 6373-81.
[http://dx.doi.org/10.2147/IJN.S137703] [PMID: 28919741]

[32]   Hamedi S, Ghaseminezhad M, Shokrollahzadeh S, Shojaosadati SA. Controlled biosynthesis of silver nanoparticles using nitrate reductase enzyme induction of filamentous fungus and their antibacterial evaluation. Artif Cells Nanomed Biotechnol 2017; 45(8): 1588-96.
[http://dx.doi.org/10.1080/21691401.2016.1267011] [PMID: 27966375]

[33]   Birla SS, Gaikwad SC, Gade AK, Rai MK. Rapid synthesis of silver nanoparticles from Fusarium oxysporum by optimizing physicocultural conditions. ScientificWorldJournal 2013; 2013: 1-12.
[http://dx.doi.org/10.1155/2013/796018] [PMID: 24222751]

[34]   Rajput S, Werezuk R, Lange RM, McDermott MT. Fungal isolate optimized for biogenesis of silver nanoparticles with enhanced colloidal stability. Langmuir 2016; 32(34): 8688-97.
[http://dx.doi.org/10.1021/acs.langmuir.6b01813] [PMID: 27466012]

[35]   Saxena J, Sharma PK, Sharma MM, Singh A. Process optimization for green synthesis of silver nanoparticles by Sclerotinia sclerotiorum MTCC 8785 and evaluation of its antibacterial properties. Springerplus 2016; 5(1): 861.
[http://dx.doi.org/10.1186/s40064-016-2558-x] [PMID: 27386310]

[36]   Ma L, Su W, Liu JX, *et al.* Optimization for extracellular biosynthesis of silver nanoparticles by Penicillium aculeatum Su1 and their antimicrobial activity and cytotoxic effect compared with silver ions. Mater Sci Eng C 2017; 77: 963-71.
[http://dx.doi.org/10.1016/j.msec.2017.03.294] [PMID: 28532117]

[37]   Mekkawy A, El-Mokhtar M, Nafady N, *et al. In vitro* and *in vivo* evaluation of biologically synthesized silver nanoparticles for topical applications: effect of surface coating and loading into hydrogels. Int J Nanomedicine 2017; 12: 759-77.
[http://dx.doi.org/10.2147/IJN.S124294] [PMID: 28176951]

[38]   Ottoni CA, Simões MF, Fernandes S, *et al.* Screening of filamentous fungi for antimicrobial silver nanoparticles synthesis. AMB Express 2017; 7(1): 31.
[http://dx.doi.org/10.1186/s13568-017-0332-2] [PMID: 28144889]

[39]   Anil Kumar S, Abyaneh MK, Gosavi SW, *et al.* Nitrate reductase-mediated synthesis of silver nanoparticles from AgNO3. Biotechnol Lett 2007; 29(3): 439-45.
[http://dx.doi.org/10.1007/s10529-006-9256-7] [PMID: 17237973]

[40]   Jaidev LR, Narasimha G. Fungal mediated biosynthesis of silver nanoparticles, characterization and antimicrobial activity. Colloids Surf B Biointerfaces 2010; 81(2): 430-3.
[http://dx.doi.org/10.1016/j.colsurfb.2010.07.033] [PMID: 20708910]

[41]   Hietzschold S, Walter A, Davis C, Taylor AA, Sepunaru L. Does nitrate reductase play a role in silver nanoparticle synthesis? evidence for NADPH as the sole reducing agent. ACS Sustain Chem& Eng 2019; 7(9): 8070-6.
[http://dx.doi.org/10.1021/acssuschemeng.9b00506]

[42]   Ballottin D, Fulaz S, Souza ML, *et al.* Elucidating protein involvement in the stabilization of the biogenic silver nanoparticles. Nanoscale Res Lett 2016; 11(1): 313.
[http://dx.doi.org/10.1186/s11671-016-1538-y] [PMID: 27356560]

[43] Kashyap PL, Kumar S, Srivastava AK, Sharma AK. Myconanotechnology in agriculture: a perspective. World J Microbiol Biotechnol 2013; 29(2): 191-207.
[http://dx.doi.org/10.1007/s11274-012-1171-6] [PMID: 23001741]

[44] Castro-Longoria E, Vilchis-Nestor AR, Avalos-Borja M. Biosynthesis of silver, gold and bimetallic nanoparticles using the filamentous fungus Neurospora crassa. Colloids Surf B Biointerfaces 2011; 83(1): 42-8.
[http://dx.doi.org/10.1016/j.colsurfb.2010.10.035] [PMID: 21087843]

[45] Ashrafi SJ, Rastegar MF, Ashrafi M, Yazdian F, Pourrahim R, Suresh AK. Influence of external factors on the production and morphology of biogenic silver nanocrystallites. J Nanosci Nanotechnol 2013; 13(3): 2295-301.
[http://dx.doi.org/10.1166/jnn.2013.6791] [PMID: 23755682]

[46] Qidwai A, Pandey A, Kumar R, Shukla SK, Dikshit A. Advances in biogenic nanoparticles and the mechanisms of antimicrobial effects. Indian J Pharm Sci 2018; 80(4)
[http://dx.doi.org/10.4172/pharmaceutical-sciences.1000398]

[47] Yahyaei B, Pourali P. One step conjugation of some chemotherapeutic drugs to the biologically produced gold nanoparticles and assessment of their anticancer effects. Sci Rep 2019; 9(1): 10242.
[http://dx.doi.org/10.1038/s41598-019-46602-0] [PMID: 31308430]

[48] Vigneshwaran N, Ashtaputre NM, Varadarajan PV, Nachane RP, Paralikar KM, Balasubramanya RH. Biological synthesis of silver nanoparticles using the fungus Aspergillus flavus. Mater Lett 2007; 61(6): 1413-8.
[http://dx.doi.org/10.1016/j.matlet.2006.07.042]

[49] Bhainsa KC, D'Souza SF. Extracellular biosynthesis of silver nanoparticles using the fungus Aspergillus fumigatus. Colloids Surf B Biointerfaces 2006; 47(2): 160-4.
[http://dx.doi.org/10.1016/j.colsurfb.2005.11.026] [PMID: 16420977]

[50] Alani F, Moo-Young M, Anderson W. Biosynthesis of silver nanoparticles by a new strain of Streptomyces sp. compared with Aspergillus fumigatus. World J Microbiol Biotechnol 2012; 28(3): 1081-6.
[http://dx.doi.org/10.1007/s11274-011-0906-0] [PMID: 22805829]

[51] Li G, He D, Qian Y, *et al.* Fungus-mediated green synthesis of silver nanoparticles using Aspergillus terreus. Int J Mol Sci 2011; 13(1): 466-76.
[http://dx.doi.org/10.3390/ijms13010466] [PMID: 22312264]

[52] Jain N, Bhargava A, Majumdar S, Tarafdar JC, Panwar J. Extracellular biosynthesis and characterization of silver nanoparticles using Aspergillus flavusNJP08: A mechanism perspective. Nanoscale 2011; 3(2): 635-41.
[http://dx.doi.org/10.1039/C0NR00656D] [PMID: 21088776]

[53] Saravanan M, Nanda A. Extracellular synthesis of silver bionanoparticles from Aspergillus clavatus and its antimicrobial activity against MRSA and MRSE. Colloids Surf B Biointerfaces 2010; 77(2): 214-8.
[http://dx.doi.org/10.1016/j.colsurfb.2010.01.026] [PMID: 20189360]

[54] Musarrat J, Dwivedi S, Singh BR, Al-Khedhairy AA, Azam A, Naqvi A. Production of antimicrobial silver nanoparticles in water extracts of the fungus Amylomyces rouxii strain KSU-09. Bioresour Technol 2010; 101(22): 8772-6.
[http://dx.doi.org/10.1016/j.biortech.2010.06.065] [PMID: 20619641]

[55] Balaji DS, Basavaraja S, Deshpande R, Mahesh DB, Prabhakar BK, Venkataraman A. Extracellular biosynthesis of functionalized silver nanoparticles by strains of Cladosporium cladosporioides fungus. Colloids Surf B Biointerfaces 2009; 68(1): 88-92.
[http://dx.doi.org/10.1016/j.colsurfb.2008.09.022] [PMID: 18995994]

[56] Sanghi R, Verma P. Biomimetic synthesis and characterisation of protein capped silver nanoparticles.

Bioresour Technol 2009; 100(1): 501-4.
[http://dx.doi.org/10.1016/j.biortech.2008.05.048] [PMID: 18625550]

[57]   Durán N, Marcato PD, Alves OL, De Souza GIH, Esposito E. Mechanistic aspects of biosynthesis of silver nanoparticles by several Fusarium oxysporum strains. J Nanobiotechnology 2005; 3(1): 8.
[http://dx.doi.org/10.1186/1477-3155-3-8] [PMID: 16014167]

[58]   Ahmad A, Mukherjee P, Senapati S, *et al.* Extracellular biosynthesis of silver nanoparticles using the fungus Fusarium oxysporum. Colloids Surf B Biointerfaces 2003; 28(4): 313-8.
[http://dx.doi.org/10.1016/S0927-7765(02)00174-1]

[59]   Syed A, Saraswati S, Kundu GC, Ahmad A. Biological synthesis of silver nanoparticles using the fungus Humicola sp. and evaluation of their cytoxicity using normal and cancer cell lines. Spectrochim Acta A Mol Biomol Spectrosc 2013; 114: 144-7.
[http://dx.doi.org/10.1016/j.saa.2013.05.030] [PMID: 23770500]

[60]   Kathiresan K, Manivannan S, Nabeel MA, Dhivya B. Studies on silver nanoparticles synthesized by a marine fungus, Penicillium fellutanum isolated from coastal mangrove sediment. Colloids Surf B Biointerfaces 2009; 71(1): 133-7.
[http://dx.doi.org/10.1016/j.colsurfb.2009.01.016] [PMID: 19269142]

[61]   Maliszewska I, Szewczyk K, Waszak K. Biological synthesis of silver nanoparticles. J Phys Conf Ser 2009; 146: 012025.
[http://dx.doi.org/10.1088/1742-6596/146/1/012025]

[62]   Birla SS, Tiwari VV, Gade AK, Ingle AP, Yadav AP, Rai MK. Fabrication of silver nanoparticles by *Phoma glomerata* and its combined effect against *Escherichia coli*, *Pseudomonas aeruginosa* and *Staphylococcus aureus*. Lett Appl Microbiol 2009; 48(2): 173-9.
[http://dx.doi.org/10.1111/j.1472-765X.2008.02510.x] [PMID: 19141039]

[63]   Vigneshwaran N, Kathe AA, Varadarajan PV, Nachane RP, Balasubramanya RH. Silver-protein (core-shell) nanoparticle production using spent mushroom substrate. Langmuir 2007; 23(13): 7113-7.
[http://dx.doi.org/10.1021/la063627p] [PMID: 17518485]

[64]   Binupriya AR, Sathishkumar M, Yun SI. Biocrystallization of silver and gold ions by inactive cell filtrate of Rhizopus stolonifer. Colloids Surf B Biointerfaces 2010; 79(2): 531-4.
[http://dx.doi.org/10.1016/j.colsurfb.2010.05.021] [PMID: 20627484]

[65]   Mohammed Fayaz A, Balaji K, Girilal M, Kalaichelvan PT, Venkatesan R. Mycobased synthesis of silver nanoparticles and their incorporation into sodium alginate films for vegetable and fruit preservation. J Agric Food Chem 2009; 57(14): 6246-52.
[http://dx.doi.org/10.1021/jf900337h] [PMID: 19552418]

[66]   Mohammed Fayaz A, Balaji K, Kalaichelvan PT, Venkatesan R. Fungal based synthesis of silver nanoparticles—An effect of temperature on the size of particles. Colloids Surf B Biointerfaces 2009; 74(1): 123-6.
[http://dx.doi.org/10.1016/j.colsurfb.2009.07.002] [PMID: 19674875]

[67]   Fayaz AM, Balaji K, Girilal M, Yadav R, Kalaichelvan PT, Venketesan R. Biogenic synthesis of silver nanoparticles and their synergistic effect with antibiotics: a study against gram-positive and gram-negative bacteria. Nanomedicine 2010; 6(1): 103-9.
[http://dx.doi.org/10.1016/j.nano.2009.04.006] [PMID: 19447203]

[68]   Fayaz M, Tiwary CS, Kalaichelvan PT, Venkatesan R. Blue orange light emission from biogenic synthesized silver nanoparticles using Trichoderma viride. Colloids Surf B Biointerfaces 2010; 75(1): 175-8.
[http://dx.doi.org/10.1016/j.colsurfb.2009.08.028] [PMID: 19783414]

[69]   Mukherjee P, Roy M, Mandal BP, *et al.* Green synthesis of highly stabilized nanocrystalline silver particles by a non-pathogenic and agriculturally important fungus *T. asperellum*. Nanotechnology 2008; 19(7): 075103.
[http://dx.doi.org/10.1088/0957-4484/19/7/075103] [PMID: 21817628]

[70]   Philip D. Biosynthesis of Au, Ag and Au–Ag nanoparticles using edible mushroom extract. Spectrochim Acta A Mol Biomol Spectrosc 2009; 73(2): 374-81.
[http://dx.doi.org/10.1016/j.saa.2009.02.037] [PMID: 19324587]

[71]   Chowdhury S, Basu A, Kundu S. Green synthesis of protein capped silver nanoparticles from phytopathogenic fungus Macrophomina phaseolina (Tassi) Goid with antimicrobial properties against multidrug-resistant bacteria. Nanoscale Res Lett 2014; 9(1): 365.
[http://dx.doi.org/10.1186/1556-276X-9-365] [PMID: 25114655]

[72]   Raliya R, Tarafdar JC. Novel approach for silver nanoparticle synthesis using <I>Aspergillus terreus</I> CZR-1: Mechanism perspective. Journal of Bionanoscience 2012; 6(1): 12-6.
[http://dx.doi.org/10.1166/jbns.2012.1073]

[73]   Kumari M, Pandey S, Giri VP, *et al.* Tailoring shape and size of biogenic silver nanoparticles to enhance antimicrobial efficacy against MDR bacteria. Microb Pathog 2017; 105: 346-55.
[http://dx.doi.org/10.1016/j.micpath.2016.11.012] [PMID: 27889528]

[74]   Nicolopoulou-Stamati P, Maipas S, Kotampasi C, Stamatis P, Hens L. Chemical pesticides and human health: the urgent need for a new concept in agriculture. Front Public Health 2016; 4: 148.
[http://dx.doi.org/10.3389/fpubh.2016.00148] [PMID: 27486573]

[75]   Townson H. Public health impact of pesticides used in agriculture. Trans R Soc Trop Med Hyg 1992; 86(3): 350.
[http://dx.doi.org/10.1016/0035-9203(92)90345-D]

[76]   CanadianCancerSociety. Cosmetic pesticides. Inf Br 2013.

[77]   Hoffman RS, Capel PD, Larson SJ. Comparison of pesticides in eight U.S. urban streams. Environ Toxicol Chem 2000; 19(9): 2249-58.
[http://dx.doi.org/10.1002/etc.5620190915]

[78]   Pimentel D, Burgess M. Environmental and economic costs of the application of pesticides primarily in the United States. In: Peshin R, Pimentel D, Eds. Integr Pest Manag. New York, Heidelberg, Dordrecht, London: Springer Netherlands 2014; pp. 47-71.
[http://dx.doi.org/10.1007/978-94-007-7796-5_2]

[79]   Zheng S, Chen B, Qiu X, Chen M, Ma Z, Yu X. Distribution and risk assessment of 82 pesticides in Jiulong River and estuary in South China. Chemosphere 2016; 144: 1177-92.
[http://dx.doi.org/10.1016/j.chemosphere.2015.09.050] [PMID: 26461443]

[80]   Mostafalou S, Abdollahi M. Pesticides and human chronic diseases: Evidences, mechanisms, and perspectives. Toxicol Appl Pharmacol 2013; 268(2): 157-77.
[http://dx.doi.org/10.1016/j.taap.2013.01.025] [PMID: 23402800]

[81]   Khot R, Joshi PP, Pandharipande M, Nagpure K, Thakur DS. Glyphosate poisoning with acute pulmonary edema. Toxicol Int 2014; 21(3): 328-30.
[http://dx.doi.org/10.4103/0971-6580.155389] [PMID: 25948977]

[82]   Baylis AD. Why glyphosate is a global herbicide: strengths, weaknesses and prospects. Pest Manag Sci 2000; 56(4): 299-308.
[http://dx.doi.org/10.1002/(SICI)1526-4998(200004)56:4<299::AID-PS144>3.0.CO;2-K]

[83]   Pandey SP, Mohanty B. The neonicotinoid pesticide imidacloprid and the dithiocarbamate fungicide mancozeb disrupt the pituitary–thyroid axis of a wildlife bird. Chemosphere 2015; 122: 227-34.
[http://dx.doi.org/10.1016/j.chemosphere.2014.11.061] [PMID: 25496744]

[84]   Breckenridge CB, Sawhney Coder P, Tisdel MO, *et al.* Effect of age, duration of exposure, and dose of atrazine on sexual maturation and the luteinizing hormone surge in the female sprague–dawley rat. Birth Defects Res B Dev Reprod Toxicol 2015; 104(5): 204-17.
[http://dx.doi.org/10.1002/bdrb.21154] [PMID: 26439775]

[85]   Jaensson A, Scott AP, Moore A, Kylin H, Olsén KH. Effects of a pyrethroid pesticide on endocrine

responses to female odours and reproductive behaviour in male parr of brown trout (Salmo trutta L.). Aquat Toxicol 2007; 81(1): 1-9.
[http://dx.doi.org/10.1016/j.aquatox.2006.10.011] [PMID: 17174415]

[86]   Shafer TJ, Meyer DA, Crofton KM. Developmental neurotoxicity of pyrethroid insecticides: critical review and future research needs. Environ Health Perspect 2005; 113(2): 123-36.
[http://dx.doi.org/10.1289/ehp.7254] [PMID: 15687048]

[87]   Moore A, Waring CP. The effects of a synthetic pyrethroid pesticide on some aspects of reproduction in Atlantic salmon (Salmo salar L.). Aquat Toxicol 2001; 52(1): 1-12.
[http://dx.doi.org/10.1016/S0166-445X(00)00133-8] [PMID: 11163426]

[88]   Caron-Beaudoin É, Denison MS, Sanderson JT. Effects of neonicotinoids on promoter-specific expression and activity of aromatase (CYP19) in human adrenocortical carcinoma (H295R) and primary umbilical vein endothelial (HUVEC) cells. Toxicol Sci 2016; 149(1): 134-44.
[http://dx.doi.org/10.1093/toxsci/kfv220] [PMID: 26464060]

[89]   Gupta RK, Kumar V, Gundampati RK, Malviya M, Hasan SH, Jagannadham MV. Biosynthesis of silver nanoparticles from the novel strain of Streptomyces Sp. BHUMBU-80 with highly efficient electroanalytical detection of hydrogen peroxide and antibacterial activity. J Environ Chem Eng 2017; 5(6): 5624-35.
[http://dx.doi.org/10.1016/j.jece.2017.09.029]

[90]   Loo YY, Rukayadi Y, Nor-Khaizura MAR, *et al.* *In Vitro* antimicrobial activity of green synthesized silver nanoparticles against selected gram-negative foodborne pathogens. Front Microbiol 2018; 9: 1555.
[http://dx.doi.org/10.3389/fmicb.2018.01555] [PMID: 30061871]

[91]   Dakal TC, Kumar A, Majumdar RS, Yadav V. Mechanistic basis of antimicrobial actions of silver nanoparticles. Front Microbiol 2016; 7: 1831.
[http://dx.doi.org/10.3389/fmicb.2016.01831] [PMID: 27899918]

[92]   van Lenteren JC, Bolckmans K, Köhl J, Ravensberg WJ, Urbaneja A. Biological control using invertebrates and microorganisms: plenty of new opportunities. BioControl 2018; 63(1): 39-59.
[http://dx.doi.org/10.1007/s10526-017-9801-4]

[93]   Manikandaselvi S, Sathya V, Vadivel V, Sampath N, Brindha P. Evaluation of bio control potential of AgNPs synthesized from Trichoderma viride. Advances in Natural Sciences: Nanoscience and Nanotechnology 2020; 11(3): 035004.
[http://dx.doi.org/10.1088/2043-6254/ab9d16]

[94]   Linger JG, Taylor LE II, Baker JO, *et al.* A constitutive expression system for glycosyl hydrolase family 7 cellobiohydrolases in Hypocrea jecorina. Biotechnol Biofuels 2015; 8(1): 45.
[http://dx.doi.org/10.1186/s13068-015-0230-2] [PMID: 25904982]

[95]   Adhikari P, Oh Y, Panthee D. Current status of early blight resistance in tomato: an update. Int J Mol Sci 2017; 18(10): 2019.
[http://dx.doi.org/10.3390/ijms18102019] [PMID: 28934121]

[96]   Pane C, Zaccardelli M. Evaluation of Bacillus strains isolated from solanaceous phylloplane for biocontrol of Alternaria early blight of tomato. Biol Control 2015; 84: 11-8.
[http://dx.doi.org/10.1016/j.biocontrol.2015.01.005]

[97]   Khan N, Mishra A, Nautiyal CS. Paenibacillus lentimorbus B-30488r controls early blight disease in tomato by inducing host resistance associated gene expression and inhibiting Alternaria solani. Biol Control 2012; 62(2): 65-74.
[http://dx.doi.org/10.1016/j.biocontrol.2012.03.010]

[98]   Al-Zubaidi S, Al-Ayafi A, Abdelkader H. Biosynthesis, characterization and antifungal activity of silver nanoparticles by aspergillus niger isolate. Journal of Nanotechnology Research 2019; 1(1)
[http://dx.doi.org/10.26502/jnr.2688-8521002]

[99]   Qualhato TF, Lopes FAC, Steindorff AS, Brandão RS, Jesuino RSA, Ulhoa CJ. Mycoparasitism studies of Trichoderma species against three phytopathogenic fungi: evaluation of antagonism and hydrolytic enzyme production. Biotechnol Lett 2013; 35(9): 1461-8.
[http://dx.doi.org/10.1007/s10529-013-1225-3] [PMID: 23690037]

[100]   Elamawi R, El-Shafey R. Inhibition effects of silver nanoparticles against rice blast disease caused by Magnapor thegrisea. Egypt J Agric Res 2013; 4: 1271-81.

[101]   Fall ML, Willbur JF, Smith DL, Byrne AM, Chilvers MI. Spatiotemporal distribution pattern of *Sclerotinia sclerotiorum* apothecia is modulated by canopy closure and soil temperature in an irrigated soybean field. Plant Dis 2018; 102(9): 1794-802.
[http://dx.doi.org/10.1094/PDIS-11-17-1821-RE] [PMID: 30125202]

[102]   Willbur J, McCaghey M, Kabbage M, Smith DL. An overview of the Sclerotinia sclerotiorum pathosystem in soybean: impact, fungal biology, and current management strategies. Trop Plant Pathol 2019; 44(1): 3-11.
[http://dx.doi.org/10.1007/s40858-018-0250-0]

[103]   Singh D, Rathod V, Ninganagouda S, Hiremath J, Singh AK, Mathew J. Optimization and characterization of silver nanoparticle by endophytic fungi *Penicillium sp.* isolated from *Curcuma longa* (Turmeric) and application studies against MDR *E. coli* and *S. aureus*. Bioinorg Chem Appl 2014; 2014: 1-8.
[http://dx.doi.org/10.1155/2014/408021] [PMID: 24639625]

[104]   Kamil D, Thokala P, Ganesh S, Narayanasamy P. KUmar R, Thomas S. Green synthesis of silver nanoparticles by entomopathogenic fungus Beauveria bassiana and their bioefficacy against mustard aphid (Lipaphis erysimi Kalt.). Indian J Exp Biol 2017; 55: 555-61.

[105]   Percival SL, Bowler PG, Russell D. Bacterial resistance to silver in wound care. J Hosp Infect 2005; 60(1): 1-7.
[http://dx.doi.org/10.1016/j.jhin.2004.11.014] [PMID: 15823649]

[106]   Baqueiro-Peña I, Guerrero-Beltrán JÁ. Physicochemical and antioxidant characterization of Justicia spicigera. Food Chem 2017; 218: 305-12.
[http://dx.doi.org/10.1016/j.foodchem.2016.09.078] [PMID: 27719914]

[107]   Bernardo-Mazariegos E, Valdez-Salas B, González-Mendoza D, *et al.* Silver nanoparticles from Justicia spicigera and their antimicrobial potentialities in the biocontrol of foodborne bacteria and phytopathogenic fungi. Rev Argent Microbiol 2019; 51(2): 103-9.
[http://dx.doi.org/10.1016/j.ram.2018.05.002] [PMID: 30029815]

[108]   Elgorban AM. Extracellular synthesis of silver nanoparticles using Aspergillus versicolor and evaluation of their activity on plant pathogenic fungi. Mycosphere 2016; 7(6): 844-52.
[http://dx.doi.org/10.5943/mycosphere/7/6/15]

[109]   El-aziz A, Mahmoud MA, Metwaly HA. Biosynthesis of silver nanoparticles using fusarium solani. Dig J Nanomater Biostruct 2015; 10: 655-62.

[110]   Balakumaran MD, Ramachandran R, Kalaichelvan PT. Exploitation of endophytic fungus, Guignardia mangiferae for extracellular synthesis of silver nanoparticles and their *in vitro* biological activities. Microbiol Res 2015; 178: 9-17.
[http://dx.doi.org/10.1016/j.micres.2015.05.009] [PMID: 26302842]

[111]   Qian Y, Yu H, He D, *et al.* Biosynthesis of silver nanoparticles by the endophytic fungus Epicoccum nigrum and their activity against pathogenic fungi. Bioprocess Biosyst Eng 2013; 36(11): 1613-9.
[http://dx.doi.org/10.1007/s00449-013-0937-z] [PMID: 23463299]

[112]   Hemlata , Meena PR, Singh AP, Tejavath KK. Biosynthesis of silver nanoparticles using *Cucumis prophetarum* aqueous leaf extract and their antibacterial and antiproliferative activity against cancer cell lines. ACS Omega 2020; 5(10): 5520-8.
[http://dx.doi.org/10.1021/acsomega.0c00155] [PMID: 32201844]

[113] Ponmurugan P. Biosynthesis of silver and gold nanoparticles using *Trichoderma atroviride* for the biological control of *Phomopsis* canker disease in tea plants. IET Nanobiotechnol 2017; 11(3): 261-7.
[http://dx.doi.org/10.1049/iet-nbt.2016.0029] [PMID: 28476983]

[114] Mythili Gnanamangai B, Ponmurugan P, Jeeva SE, *et al.* Biosynthesised silver and copper nanoformulation as foliar spray to control bird's eye spot disease in tea plantations. IET Nanobiotechnol 2017; 11(8): 917-28.
[http://dx.doi.org/10.1049/iet-nbt.2017.0023] [PMID: 29155390]

[115] Devika R, Elumalai S, Manikandan E, Eswaramoorthy D. Biosynthesis of silver nanoparticles using the fungus Pleurotus ostreatus and their antibacterial activity. Open Access Sci Reports 2012; 1: 1-5.
[http://dx.doi.org/10.4172/scientificreports.5]

[116] Ramos MM, Dos S Morais E, da S Sena I, *et al.* Silver nanoparticle from whole cells of the fungi Trichoderma spp. isolated from Brazilian Amazon. Biotechnol Lett 2020; 42(5): 833-43.
[http://dx.doi.org/10.1007/s10529-020-02819-y] [PMID: 32026287]

[117] Oktarina H, Singleton I. Can nano-silver products endanger beneficial soil fungi? IOP Conf Ser Earth Environ Sci 2020; 425(1): 012070.
[http://dx.doi.org/10.1088/1755-1315/425/1/012070]

[118] Mahawar H, Prasanna R, Gogoi R, Singh SB, Chawla G, Kumar A. Synergistic effects of silver nanoparticles augmented Calothrix elenkinii for enhanced biocontrol efficacy against Alternaria blight challenged tomato plants. 3 Biotech 2020.
[http://dx.doi.org/10.1007/s13205-020-2074-0]

[119] Guilger-Casagrande M, Germano-Costa T, Pasquoto-Stigliani T, Fraceto LF, Lima R. Biosynthesis of silver nanoparticles employing Trichoderma harzianum with enzymatic stimulation for the control of Sclerotinia sclerotiorum. Sci Rep 2019; 9(1): 14351.
[http://dx.doi.org/10.1038/s41598-019-50871-0] [PMID: 31586116]

[120] Verma ML, Dhanya BS, Sukriti , *et al.* Carbohydrate and protein based biopolymeric nanoparticles: Current status and biotechnological applications. Int J Biol Macromol 2020; 154: 390-412.
[http://dx.doi.org/10.1016/j.ijbiomac.2020.03.105] [PMID: 32194126]

[121] Verma ML, Naebe M, Barrow CJ, Puri M. Enzyme immobilisation on amino-functionalised multi-walled carbon nanotubes: structural and biocatalytic characterisation. PLoS One 2013; 8(9): e73642.
[http://dx.doi.org/10.1371/journal.pone.0073642] [PMID: 24069216]

[122] Chamundeeswari M, Jeslin J, Verma ML. Nanocarriers for drug delivery applications. Environ Chem Lett 2019; 17(2): 849-65.
[http://dx.doi.org/10.1007/s10311-018-00841-1]

[123] Verma ML. Nanobiotechnology advances in enzymatic biosensors for the agri-food industry. Environ Chem Lett 2017; 15(4): 555-60.
[http://dx.doi.org/10.1007/s10311-017-0640-4]

[124] Verma ML, Rajkhowa R, Wang X, Barrow CJ, Puri M. Exploring novel ultrafine Eri silk bioscaffold for enzyme stabilisation in cellobiose hydrolysis. Bioresour Technol 2013; 145: 302-6.
[http://dx.doi.org/10.1016/j.biortech.2013.01.065] [PMID: 23462595]

CHAPTER 4

Microbial Mediated Synthesis, Characterisation and Application of Selenium Nanoparticles

Veerasamy Ravichandran[1,*], **Karunakaran Rohini**[2], **Anitha Roy**[3] and **S. Rajeshkumar**[3,*]

[1] *Faculty of Pharmacy, AIMST University, Semeling-08100, Kedah, Malaysia*

[2] *Faculty of Medicine, AIMST University, Semeling-08100, Kedah, Malaysia*

[3] *Department of Pharmacology, Saveetha Dental College and Hospitals, Saveetha Institute Medical and Technical Sciences, Chennai-600077, India*

Abstract: The development in nanotechnology, specifically the nanoparticulate system, has a great impact on medicine, engineering and other scientific areas. Inorganic nanoparticles such as silver, gold, zinc oxide, selenium, iron, lead, platinum and copper, *etc.* were found to exhibit antimicrobial, antioxidant and other biological activities, used as biosensors and also used in different fields of engineering. In the 21st century, microorganisms and plant parts are playing a major role in the synthesis of inorganic nanoparticles. Green synthesis of inorganic nanoparticles becomes preferable to other approaches because of its eco-friendly and non-toxic approach. Additionally, the active molecules of plants (Tannins, flavonoids, terpenoids, saponins, proteins and glycosides) which act as capping and reducing agents in the synthesis of metal nanoparticles could make them most suitable for biomedical applications. This green approach fascinated researchers across the globe to explore the potential of different microorganisms and plants in the synthesis of inorganic nanoparticles. Selenium nanoparticles are one of the inorganic nanoparticles which are widely used in the area of medicine and engineering. In this chapter, we discussed the green synthesis using microorganism and Agri based products, characterisation and various applications of selenium nanoparticles.

Keywords: Applications, Characterisation, Green synthesis, Nanoparticles, Selenium.

---

[*] **Corresponding authors Veerasamy Ravichandran & S. Rajeshkumar:** Faculty of Pharmacy, AIMST University, Semeling-08100, Kedah, Malaysia and Department of Pharmacology, Saveetha Dental College and Hospitals, Saveetha Institute Medical and Technical Sciences, Chennai-600077, India; Tel: 006-0164581626; Emails: sameshyaravi@gmail.com & ssrajeshkumar@hotmail.com

INTRODUCTION

Nanoparticles

In the 21st century, nanotechnology is known as an emerging interdisciplinary field, involving biology, physics, medicine, chemistry and material science. Nano is derived from the Greek word 'Nanos' that refers to things of one-billionth (10^{-9}) in size. Richard Feynman is the one who has presented the primary concept of nanotechnology in his lecture topic entitled "There's plenty of room at the bottom" at the American Institute of Technology in 1959. Nanoparticles have typical physical, chemical, magnetic, electronic, mechanical, thermal, optical and biological properties than bulk materials. This is due to the large number of surface atoms, large surface energy, spatial confinement and reduced imperfections in NPs because of decreasing the dimension of particles to the nanoscale.

Nanoparticles are usually 0.1 to 100 nm in each spatial dimension. Hence, nanoparticles are considered building blocks for the next generation of optoelectronic devices and chemical and biochemical sensors [1, 2]. Monodispersed nanoparticles with particular shapes have wide applications in the areas of optics, computation, medical diagnostics, cancer therapy and drug delivery. Therefore, interest in developing monodispersed nanoparticles with different sizes and shapes has increased in the past decade.

Conventionally, nanoparticles were synthesised *via* physical, chemical, and biological methods. There are two approaches, which play very significant roles in nanoscience and nanotechnology, for the synthesis of nanomaterials, namely the "top-down" and the "bottom-up" methods (Fig. **1**). The top-down approach is mainly used by physicists, and the bottom-up approach is used by chemists. There are advantages and disadvantages to both approaches. In the bottom-up approach, the nanostructured building blocks are synthesised and then assembled into the final structures, *i.e.* atom by atom, molecule by molecule or cluster by cluster, often by self-assembly. The colloidal dispersion is an example of a bottom-up approach in the synthesis of nanoparticles. The bottom-up method is suited for generating uniform particles of distinct size, shape, and structure.The major challenge in the bottom-up approach is the problem of stability because in most cases, the particles are formed in a solvent medium and thus have a better chance to collidewith each other and undergo aggregation [3, 4].

The top-down approach begins with suitable bulk material being broken down into particles of nano dimension by various physical processes like casting, moulding, rolling, forging, extruding, machining, etching, *etc*. During these operations, there is no control over the positions of individual atoms. This results in defects and impurities and is not suited for the synthesizing monodispersed particles. The top-down techniques are most apparent in the computer industry in the fabrication of microprocessors by lithography [5].

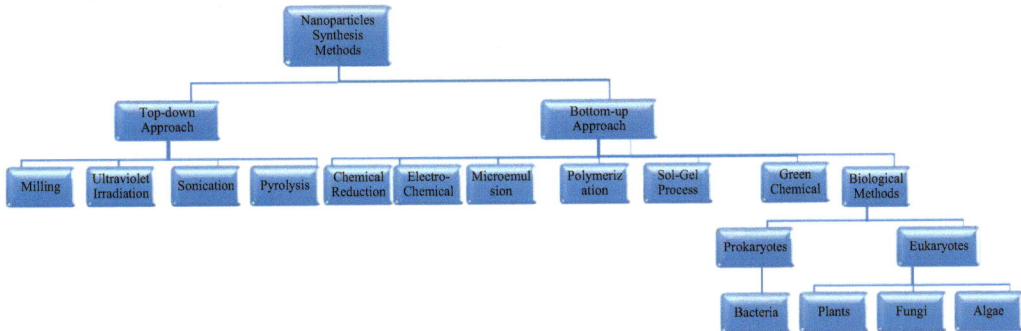

Fig. (1). Top-down and Bottom-up approaches in nanoparticle synthesis.

Various studies indicated that the size, shape, stability, as well as physical and chemical properties of the metal nanoparticles, are largely influenced by the experimental conditions such as the kinetics of interaction of metal ions with reducing agents, and interaction of the stabilising agent with metal nanoparticles [6 - 8]. Hence, the design of a synthesis method in which the size, morphology, stability and properties can be controlled has become a significant field of interest.

Selenium

Selenium (Se) is a non-metal and it has strong antagonistic biological interactions with several toxic metals, like cadmium (Cd), mercury (Hg) and silver (Ag) and showing less effectiveness against the toxic effects of lead (Pb). Less effect over Pb because PbSe is more easily oxidised compared to other metal selenium compounds. Hence, Se could be used as protection against the toxic effects of these metals [9, 10]. Underwood studied the effect of Se against chalcophile toxic metals, such as Hg and Cd and reported that the high intake levels of Se have a strong protective effect against the toxicity [11]. Clinical trials from the United States stated that Se supplementation had been shown to protect against prostate, colorectal, lung and liver cancer [12]. Nevertheless, a high intake of Se is needed for optimal protection against HIV disease, some forms of cancer and cardiovascular diseases than the intake needed for prevention of manifest Se deficiency diseases [13, 14]. Human daily intake of Se should not be more than 400 µg. Se has the transitional properties of both metal and non-metal, and is also available in the organic and inorganic state. The organic forms of Se are selenocysteine, selenomethionine, and the inorganic forms are selenite, selenide and selenite. Se is amorphous, and the toxicity of Se is lower when it is in the elemental (Se^0) form [15, 16].

Selenium (Se) has unique properties and great potential in the field of physics, chemistry and biology [17 - 19]. Selenium has been successfully used in the production of photovoltaic cells, photographic exposure meters and xerography [20, 21]. In the past decades, many stable organic selenium compounds have been synthesised, which are used as antioxidants, enzyme inhibitors, anti-tumour and anti-infective agents, cytokine inducers and immuno-modulators [22, 23]. The selenium at nano size acts as a potential chemo-preventive agent with reduced toxicity [24]. The research on Se nanoparticles is gaining greater importance because of its applications in technology and life science.

Comparison of Elemental Selenium and Nano Selenium

Se nanoparticles, when compared with several extensively studied Se compounds, *i.e.*, sodium selenite, selenomethionine, and methyl selenocysteine, were reported to exhibit markedly lower acute toxicity and significantly lower short-term and subchronic toxicities. Additionally, it was equivalently efficacious in its ability to increase selenoenzyme. Furthermore, Se nanoparticles were reported to be more efficient than sodium selenite and selenomethionine in increasing glutathioneS-transferase activity. Wang *et al.* reported a comparative study of determining the toxicity of elemental Se at nanosize (SeNP) and selenomethionine (SeMet) in mice. The LD_{50} of SeMet and nano-Se were reported to be 25.6 mg Se/kg body weight and 92.1 mg Se/kg body weight, respectively. These values suggest that it takes almost four times the concentration of Se NP to reach LD50 values compared to SeMet. SeMet caused 90% mortality at a dose of 32 mg Se/kg body weight; however, Se NP caused only 10% mortality at a dose of 36 mg Se/kg body weight and 70% mortality at a dose as high as 150 mg Se/kg body weight [25].

Furthermore, it was reported that the biomarkers of liver injury *via* hepatic metabolising enzymes alanine aminotransferase(ALT), aspartate aminotransferase (AST), and lactate dehydrogenase (LDH) were severely elevated in the SeMet-treated group as compared to SeNP-treated mice; additionally, the SeMet-treated group exhibited severe liver injury, which was not observed in SeNP-treated mice. The levels of ALT, AST, and LDH were sharply elevated at 12 hin SeMet-treated mice, but not in SeNP-treated mice. The increase in these liver enzymes in serum over time clearly showed that an acute and severe liver injury occurred in SeMet-treated mice, but not in SeNP-treated mice [25].

GENERAL METHODS FOR THE SYNTHESIS OF INORGANIC NANOPARTICLES

Numerous approaches, using physical, chemicals or electrochemical, are used in the preparation of nanoparticles. Generally, many of these strategies are having difficulty in the purification stage since the used chemicals or the by-products formed are hazardous and high energy required for the preparation. Apart from the toxicity problem, controlling the size/shape and attaining the monodispersity are the other major challenges that are frequently encountered by researchers in the preparation of nanoparticles. To address all these problems, plant extracts have been evolved as one of the most promising options in the synthesis of nanoparticles. The plants' constituents have both protective and reduction properties, which are essential for the reduction of metal ions to their corresponding nanoparticles. The major advantages of this method for the synthesis of nanoparticles could be as related to the absence of any intricate processes like complex purification steps and maintenance of the microbial cell cultures [26].

The Physical Method of Nanoparticle Synthesis

UV irradiation, sonochemistry, laser ablation and radiolysis, *etc.* are the different physical methods used in the synthesis of nanoparticles. The first step during the process is the evaporation of metal atoms and followed by condensation of various supports. During this process, the metallic atoms are rearranged and aggregated as small clusters of metallic nanoparticles [27]. Physical approaches produce nanoparticles with high purity and definite shape. Usually, the physical method of nanoparticles synthesis requires more sophisticated instruments, chemicals and high-power consumption, even though we get high purity with definite shape nanoparticles by this method.

Chemical Method of Nanoparticle Synthesis

Another method is using chemicals to reduce the metal ions for the synthesis of the nanoparticle. Metal undergo either the process of nucleation or aggregation to form small clusters based on the condition of the reaction mixture. Hydrazine, sodium borohydride, and hydrogen are some generally used chemicals for the reduction of metals [28]. Some stabilising agents like synthetic or natural polymers such as cellulose, natural rubber, chitosan, and copolymers are also used. Organic solvents such as ethane, dimethyl, formaldehyde, toluene, and chloroform are needed since these chemicals are hydrophobic.

Biological Method of Nanoparticle Synthesis

In the last two decades, the biogenic synthesis of nanoparticles has attracted considerable attention. Microorganisms and plants are used in the biogenic synthesis process of nanoparticles [29]. Better-defined size and morphology are the advantages of nanoparticles obtained from biogenic methods. The microbial-based synthesis process is readily scalable and eco-friendly, but the purification process is a tedious one. It is found that microbe-mediated synthesis is not applicable for large scale production, since it requires high aseptic conditions and a tedious scale-up process [30].

Plant-based synthesis approaches are eco-friendly, cheaper, and easily scale-up processes than chemical and physical methods. There is no need for high temperature, pressure, and toxic chemicals in plant-based synthesis [31]. In plant-mediated synthesis secondary metabolites are involved in the formation of nanoparticles as a single-step, whereas in microorganisms-based nanoparticle synthesis, the microorganism may lose their ability to synthesise nanoparticles due to mutation; thus, plant research is expanding rapidly [32].

Synthesis of nanoparticles using the plant is compatible with green chemistry approaches in which the biomolecules secreted by the plant can act as both reducing and stabilising agents during the reaction. In the 21st century, the plant-based synthesis of nanoparticles has got much attention due to its rapidity and its simplicity [33].

GENERAL CHARACTERISATION OF NANOPARTICLES

Some of the spectroscopy and microscopy techniques routinely used for the characterisation of nanoparticles include UV-visible spectroscopy (UV-vis), dynamic light scattering (DLS), atomic force microscopy (AFM), transmission electron microscopy (TEM), scanning electron microscopy (SEM), energy dispersive spectroscopy (EDS), powder X-ray diffraction (XRD), Fourier transform infrared spectroscopy (FT-IR), and Raman spectroscopy.

Microscopy based techniques such as AFM, SEM and TEM are considered direct methods of obtaining data from images taken of the nanoparticles. In particular, both SEM and TEM have been extensively used to determine the size and morphological features of nanoparticles [34]. The major difference between TEM and SEM is that SEM generates the image with the help of secondary electrons that gives the impression of three dimensions while TEM projects electron through a thin slice of the specimen and produces a two-dimension image. Selected area (electron) diffraction (abbreviated as SAD or SAED) is a crystallographic experimental technique that can be performed inside TEM. In

TEM, a thin crystalline sample is subjected to a parallel beam of high-energy electrons. SAED pattern to see (i) sample is amorphous (diffuse rings), crystalline (bright spots), polynanocrystalline (small spots are making up a ring, each spot arising from Bragg's reflection from an individual crystallite). AFM is used to study the crystallinity nature and size of the nanoparticles.

Indirect methods used to determine the data related to composition, structure, crystal phase, and properties of nanoparticles are UV-Vis, DLS, XRD, EDS, FT-IR, and Raman spectroscopy. UV-Vis absorption measurements for silver nanoparticles are usually between 400 and 450 nm; gold nanoparticles are between 500 and 550 nm, ZnO nanoparticles are between 350 and 380 nm [35], UV spectra centred between 200 and 300 nm is due to the formation and surface plasmon vibration of Se nanoparticles [36].

DLS spectroscopy can be used to determine size distribution and quantify the surface charge of nanoparticles [37]. The average size, particle size distribution and polydispersity index of the molecules in the solution were obtained by measuring the time scale of light intensity fluctuations at a fixed scattering angle (typically 90°). The average size revealed by DLS is the hydrodynamic surface of the particles, which means the overall size of the particle with a coating or capping agent. Polydispersity index value should be near to zero for better monodispersity, if the value is near 1.0, then particles are polydispersed. Z-potential value gives information about the compounds that stabilise nanoparticles. Still, the absolute value refers to the degree of stabilisation under the studied conditions due to Van der Waals interactions. It is well known that absolute values of Z-potential higher than +30 mV or less than -30 mV result in highly stable nanoparticles in this context.

EDS is used to determine the elemental composition of nanoparticles, whereas the diffraction pattern of nanoparticles is examined by XRD [38]. From XRD diffraction, we can get abundant information including phase formation (crystalline or amorphous), the type of crystal structure, phase composition, lattice parameters, interplanar distance, preferred orientation (growth orientation), crystallite size, *etc.* Bragg's reflection of Se nanoparticles should be comparable with standard XRD spectra of trigonal-phase use from JCPDs file no 06-0362.

FT-IR spectroscopy can be used to investigate the surface functional groups like carbonyls, amines and hydroxyls attached to nanoparticles. FT-IR data was used to confirm the surface functional groups and chemicals involved in the reduction and to confirm the capping agents. Although, XPS (X-ray photoelectron spectroscopy) technique has also been used to determine the oxidation states and elemental composition of surfaces of nanoparticles. The characteristic Raman resonance absorption band for a-selenium and monoclinic selenium are centred at 264 and 256 cm^{-1}, respectively [39].

Rajalakshmi and Arora have reported that the modes of selenium appear at A1 that is only one resonance mode at around 235 cm^{-1}and this is attributed to the stretching vibration of selenium which exists for the hexagonal phase. In addition to the aforementioned characterisation techniques, the gas adsorption/desorption technique is often used to collect information about the specific surface areas of nanoparticles and to determine their pore sizes and volumes. Such information is particularly useful for characterising porous nanoparticles, and for correlating their structure with their aspect of catalytic activity [40].

SELENIUM NANOPARTICLES

The basic reaction involved in the synthesis of Se nanoparticles is:

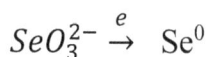

$$SeO_3^{2-} \xrightarrow{e} Se^0$$

The selenite ion accepts electrons from the electrons donor and gets reduced into Se atoms. Lots of research are focussing on Se nanoparticles due to their life science and technological applications. Se nanoparticles are synthesised by physical, chemical and biogenesis methods.

In a study, Se nanoparticles were produced by the sonication method with optimum temperature and pressure throughout the process [41]. In another method, the microwave was used to synthesize Se nanoparticles and a high temperature of $\leq 150°C$ was maintained to manipulate the shape [42, 43]. In addition, the ionic liquid-induced method was used to prepare Se nanoparticles [44]. However, Bartunck *et al.* used stabilisers like polysorbate 80, sodium dodecyl sulphate and tryptone in the synthesis of Se nanoparticles [45]. Further, an electrochemical method was reported for the production of Se nanoparticles by using sucrose, polyvinylpyrrolidone, and sodium dodecyl sulphate as modifiers, and cetane trimethyl ammonium bromide as a shape-changer [46]. The drawback of the above-mentioned physical and chemical methods of Se nanoparticle synthesis is expensive, maintains high temperature, consumption of toxic chemicals and has to use a separate stabilising and capping agent.

GREEN SYNTHESIS OF SELENIUM NANOPARTICLES

The plants or microbes mediated Se nanoparticles are called green synthesis or biological synthesis (Fig. **2**). Sometimes green chemicals such as carbohydrates or sugar are also used in the green synthesis of nanoparticles. All the parts of the plant, such as leaf, flower, bark, stem, fruit, seed, and root extracts, are used in the preparation of nanoparticles. On the other hand, bacteria, yeast, and fungi are the microbes involved in the synthesis of nanoparticles. Both in plants and microbes, the nanoparticles are synthesised intra and extracellularly.

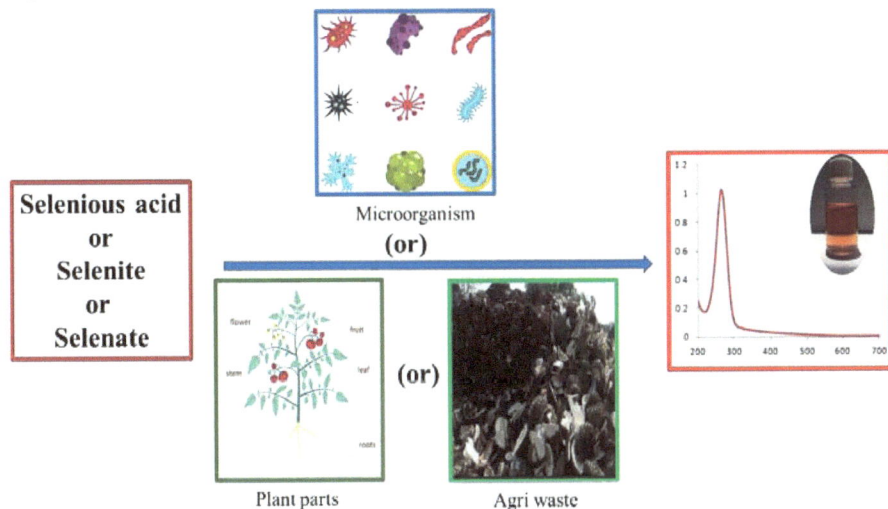

Fig. (2). Green synthesis of selenium nanoparticles.

The main drawback of plant-mediated nanoparticle synthesis is low yield, and microbes mediated synthesis is a time-consuming process and needs a high purification process. Both plant and microbes mediated synthesis, the depicted parameters in Fig. (3) to be optimised, and the characterisation parameters depicted in Fig. (4) to be carried out. Stability studies should be carried out at room temperature for both methods.

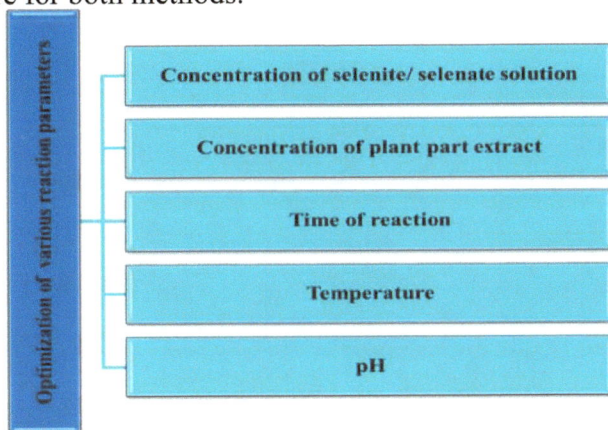

Fig. (3). Various reaction parameters to be optimized for green synthesis of selenium nanoparticles.

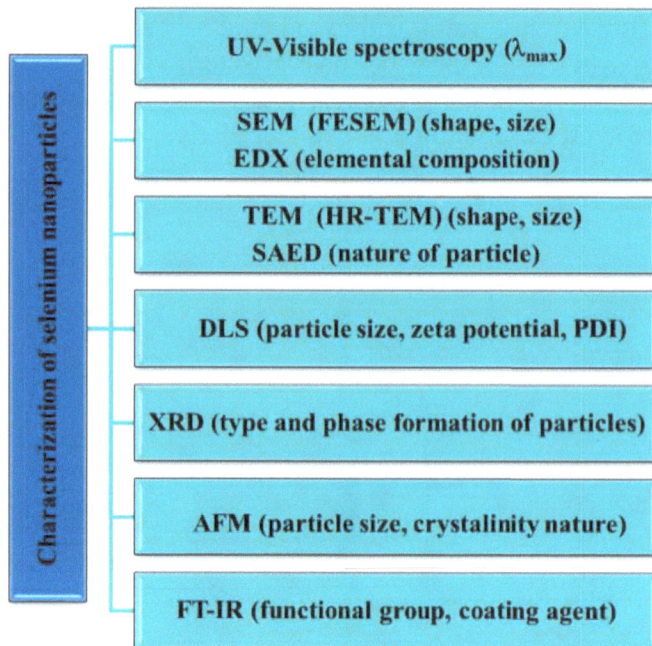

Fig. (4). Various characterization methods for green synthesis of selenium nanoparticles.

Green Synthesis of Se Nanoparticles using Green Chemical Agents

Recently, green chemicals, which are easily biodegradable such as reducing sugars, glucose, gallic acid, a natural polyphenol present in plants or chitosan, have been widely used in the Se nanoparticle synthesis (Fig. **5**). Coating of these compounds over Se nanoparticles could help to improve the efficacy of nanoparticles in their application, such as anticancer therapy, improving the immune system, regulating diabetes and cardiovascular diseases treatment [47]. Examples for the green synthesis of Se nanoparticles using green chemical agents are given in Table **1**.

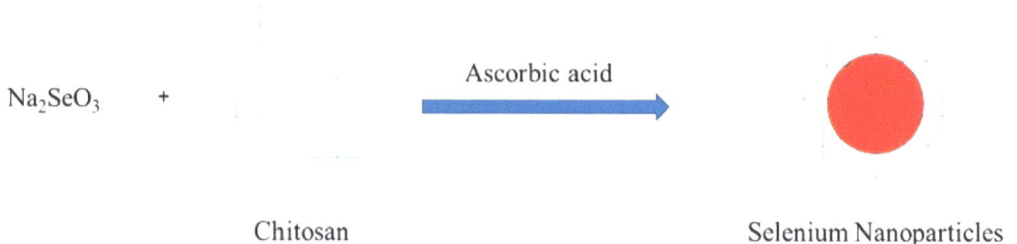

Fig. (5). The reaction involved in the green synthesis of selenium nanoparticles using a green chemical agent.

Table 1. Chemical agents mediated synthesis of Se nanoparticles.

Precursor	Reducing Agents	Stabilizers	Size (nm)/ Shape (SEM or TEM)	References
Sod. Selenite pentahydrate	Acetic acid	Chitosan	20–70	[48]
Sod. Selenite	Ascorbic acid	Anisomycin	56–185	[49]
	Glutathione	BSA	25–70	[50, 51]
	Glucose	Glucose	280–295	[52]
	Ascorbic acid	5-fluorouracil	70	[53]
	Guar gum	Triton X-100	69–173	[54]
Sod. Selenosulfate	Different organic carboxylic acids	Poly vinyl alcohol	40–100	[55]
	Ionic liquid	Polyvinyl alcohol	76–150	[44]
Selenium dioxide	Ascorbic acid (Vc)	Glucose, sucrose, chitosan	50–65	[56, 57]
	Acetamide	Vitamin A	5–200	[58]
Selenious acid	Glutathione	BSA	12–23	[59 - 61]
	Ascorbic acid	Sodium alginate	20–70	[62]
	Sod. borohydride	Triton X-100	25	[63]
	Ascorbic acid	Silk fibroin	100–500	[64]
Gray Se powder	Polyethylene glycol (PEG 200)	PEG 200	28	[65 - 67]

Green Synthesis of Se Nanoparticles Using Bacteria

The bacteria are grown in media containing both nutrients and selenite to produce Se nanoparticles (Fig. **6**). Various optimised conditions such as pH, temperature, concentration, *etc.* to be maintained to maximise the Se nanoparticle production. The changes in colour from colourless to characteristic colour indicate the formation of Se nanoparticles, and the Se nanoparticles are collected after centrifugation at high rpm. Siderophore mediated reduction, sulphide mediated reduction, thioredoxin reductase system, Painter-type reaction, and dissimilarity reduction are the possible mechanism involved in the reduction of selenite to elemental selenium [68]. The details of bacteria used in the synthesis of Se nanoparticles, characterisation parameters and applications are given in Table **2**.

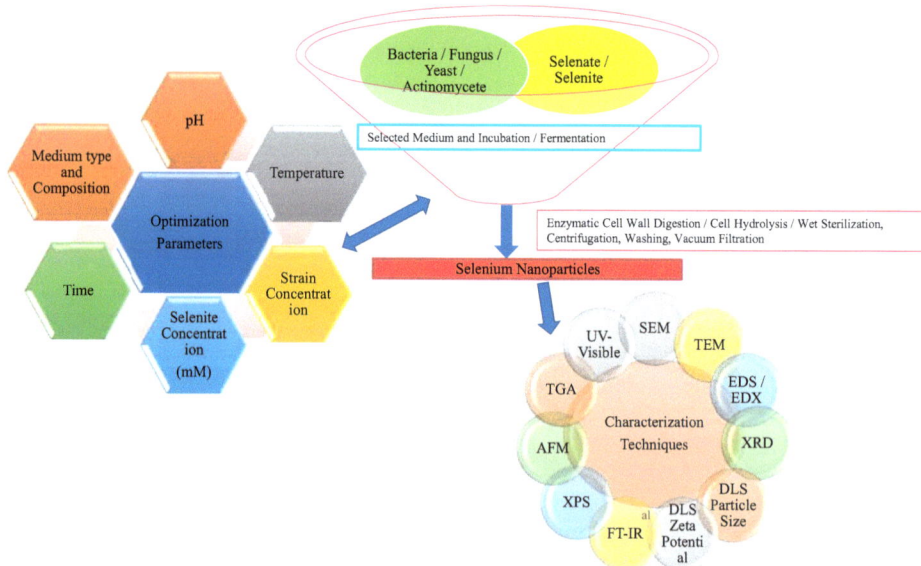

Fig. (6). Process of green synthesis of selenium nanoparticles using bacteria/fungus/yeast/actinomycete.

Table 2. Bacteria, fungus and yeast mediated synthesis of Se nanoparticles.

Bacteria							
Source	Reaction Condition	Colour / UV-Vis wavelength (nm)	Size (nm)/ Shape (Sem or TEM)	DLS size / PDI	Functional Group	Activity	References
Lactobacillus sporogenesin	Aq. Spores of *Lactobacillus sporogenesin* + Na$_2$SeO$_3$ Shaking incubator, 36-48 h, at 37°C	----	11-23 (TEM)	---	---	Biofilm scavenging	[71]
Bacillus species	*Bacillus species* + Na$_2$SeO$_3$ + ascorbic acid Incubation for 5 h	Red / 266.5	209-748	--	-OH, -CH	Antimicrobial Antifungal Antioxidant	[72]
Bacillus species Msh-1	SeO$_2$ + *Bacillus species* Msh-1 shaker incubator (30°C, 150 rpm) for 14 h	--	80-120 (TEM) / spherical	--	--	Antifungal	[73]
Azoarcus sp. CIB	*Azoarcus* sp. CIB + Na$_2$SeO$_3$ incubated anaerobically without cells for 7 days at 30 °C	--	88 ± 40 (TEM) / spherical	---	----	----	[74]

(Table 2) cont.....

Bacteria							
Source	**Reaction Condition**	**Colour / UV-Vis wavelength (nm)**	**Size (nm)/ Shape (Sem or TEM)**	**DLS size / PDI**	**Functional Group**	**Activity**	**References**
Bacillus laterosporus	*Bacillus laterosporus* + HNaO$_3$Se 48 h, 37°C Rotary shaker, 150 rpm	--- / 360	40-70 (TEM) / spherical	74.8 / ---	-NH, -CH, -C=O, -CN	Biofilm inhibition Antimicrobial	[75]
Bacillus subtilis	*Bacillus subtilis* + Na$_2$SeO$_3$ Incubation at 32°C for 24 h in a shaker	Red	spherical	334	-CO, -NH, -C-C	--	[76]
Bacillus tropicus Ism 2	*Bacillus tropicus* Ism 2 + Na$_2$SeO$_3$ incubated for 7 h at 180 rpm and 35°C.	Red / 270	60-125 (TEM)	60-125 / 0.544	---	Antibacterial Cytotoxicity	[77]
Idiomarina sp. PR58-8	*Idiomarina* sp. PR58-8 + Na$_2$SeO$_3$ 37°C, 110 rpm for 48 h The Se nanoparticles were extracted from the cells by wet heat sterilization process in a lab-oratory autoclave at 121°C, 15 psi for 20 min.	Brick red / 320	150-350 (TEM) / spherical	--	---	Antineoplastic	[78]
Acinetobacter sp. SW30	Cell suspension of *Acinetobacter sp.* SW30+ Na$_2$SeO$_3$ incubated at 30°C, 180 rpm, 24 h. Intracellularly	Orange red / 300 and 540	< 100 (TEM) / rod	---	---	Anticancer (Breast cancer)	[79]
Klebsiella pneumoniae	*Klebsiella pneumoniae* + selenium chloride incubated at 37°C for 24 h *K. pneumoniae* cells containing the red selenium particles disrupted using a wet heat sterilization process in a laboratory autoclave at 121°C, 17 psi for 20 min.	Red / 218 and 248	100-550 / spherical	--	--	--	[80]

Bacteria							
Source	**Reaction Condition**	**Colour / UV-Vis wavelength (nm)**	**Size (nm)/ Shape (Sem or TEM)**	**DLS size / PDI**	**Functional Group**	**Activity**	**References**
Pseudomonas aeruginosa	*Pseudomonas aeruginosa +* Na_2SeO_3 incubated at 37°C for 24 h	Red / 420 and 550	47-165/ rod	251.8 ± 26.3	-OH, -NH, -CH, -C-O	--	[81]
Escherichia coli	*Escherichia coli +* Na_2SeO_3 37°C for 48 h	Red / 550	100-183 / spherical	375.6 ± 38.8	-OH, -NH, -CH, -C-O	--	[82]
Sulfurospirillum barnesii, Bacillus selenitireducens, and Selenihalanaerobacter shriftii	Bacteria + Na_2SeO_3 + sodiumlactate 25°C shaking, 140 rpm, an incubator for 3 days	Red / > 600 (Amorphous, Nanospheres)	200-400	--	--	--	[83]
Pseudomonas stutzeri strain NT-I	*Pseudomonas stutzeri strain NT-I* + selenate or selenite Under N_2 gas, 24 h, anaerobic condition	Red	< 200	--	--	--	[84]
Bacillus cereus	*Bacillus cereus +* Na_2SeO_3 37°C, 200 rpm, 24 h, innoculation	Red / 590	150-200	--	--	--	[85]
Klebsiella pneumonia and *Clostridium perfringens*	*Klebsiella pneumonia* and *Clostridium perfringens* + selenium oxide incubated at 37°C for 24 h, wet sterilization	Red / 210	43-531(Kp) 28-300(Cp)	--	--	--	[86]
Lactobacillus acidophilus	*Lactobacillus acidophilus +* Na_2SeO_3 shaking incubator for 48 h at 37°C.	Red / 280-353	15-50 (TEM) / spherical	---	--	--	[87]
Bacillus mycoides SeITE01	*Bacillus mycoides* SeITE01 + Na_2SeO_3 48 h	--	50-400	--	--	---	[70]
Enterobacter cloacae Z0206	*Enterobacter cloacae* Z0206 + Na_2SeO_3 incubation at 32°C, 250 rpm for 96 h.	Red	--	--	--	--	[68]

(Table 2) cont.....

Bacteria							
Source	**Reaction Condition**	**Colour / UV-Vis wavelength (nm)**	**Size (nm)/ Shape (Sem or TEM)**	**DLS size / PDI**	**Functional Group**	**Activity**	**References**
Rhodopseudomonas palustris	*Rhodopseudomonas palustris* strain N + Na$_2$SeO$_3$ incubated at 30°C in the presence of incandescent light (1500 Lux) for 8 days	Red	80-200 / spherical	--	--	--	[88]
Burkholderia fungorum DBT1 and Burkholderia fungorum 95	*Burkholderia fungorum DBT1 and Burkholderia fungorum 95* + Na$_2$SeO$_3$ incubated aerobically at 27°C for 24	--	170 200 / spherical	--	---	---	[89]
Fungus							
Source	**Reaction Condition**	**Colour / UV-Vis wavelength (nm)**	**Size (nm)/ Shape (Sem or TEM)**	**DLS size / PDI**	**Functional Group**	**Activity**	**References**
Alternaria alternata	Na$_2$SeO$_3$ + culture filtrate of *Alternaria alternata*, RT, agitate for 24 h	Dark red / --	90 ± 10 (TEM) / ---	30-150	-OH, -CONH, -CN, -COO-	--	[90]
Aspergillus terreus	SeO$_2$ + *Aspergillus terreus* Incubate RT for 60 min	--- / 245	100 / spherical	18-105 / ---	--	----	[91]
Trichoderma atroviride	*Trichoderma atroviride* (cell lysate, cell wall debris, culture filtrate) + Na$_2$SeO$_3$ 23°C for 24 h and with constant rotation at 200 rpm	Brick orange / 260	60.48-123.16 (TEM) / spherical	93-98 / ---	-OH, -NH,	Antifungal	[92]
Fusarium semitectum	*Fusarium semitectum* + Na$_2$SeO$_3$ incubated for 24 h at 28 ± 2°C.	Red / 262	32.80-103.82 (TEM) / spherical	92.33 ± 48.5 / 0.3	-OH, -NH, -CONH, -CH	Antibacterial Antioxidant	[93]

(Table 2) cont.....

Bacteria							
Source	**Reaction Condition**	**Colour / UV-Vis wavelength (nm)**	**Size (nm)/ Shape (Sem or TEM)**	**DLS size / PDI**	**Functional Group**	**Activity**	**References**
Trichoderma	Culture filtrate/cell lysate/crude cell wall of *T. asperellum/ T. harzianum/T. atroviride/T. virens/T. longibrachiatum/T. brevicompactum* + Na$_2$SeO$_3$ Incubation 24 h, RT	Orange-red / 200-400	49-312 / Spherical and irregular	--	-OH, -NH, -CONH, -CH, -CN	Anti-mildew activity	[94]
Pleurotus ostreatus, Lentinus edodes, Ganoderma lucidum, and Grifola frondosa	Phosphate buffer solution of *Pleurotus ostreatus, Lentinus edodes, Ganoderma lucidum, and Grifola frondosa* + Na$_2$SeO$_3$ + Incubated	Red / 350	50-150 (TEM) 20-50 (TEM)	--	---	Cytotoxicity	[95]
Yeast /Actinomycetes							
Source	**Reaction Condition**	**Colour / UV-Vis wavelength (nm)**	**Size (nm)/ Shape (Sem or TEM)**	**DLS size / PDI**	**Functional Group**	**Activity**	**References**
Saccharomyces cerevisiae	Na$_2$SeO$_3$ + *Saccharomyces cerevisiae* 32°C, 4 days, Intracellular	--- / 350	50 (TEM) / spherical	75 / 0.189 (5 μg Se) 709 / 0.189 (25 μg Se)	---	Antioxidant	[96]
Halococcus salifodinae BK18	*Halococcus salifodinae BK18* + Na$_2$SeO$_3$ incubated at 37°C, 110 rpm for 7 days	Brick red / 270	129 / rod	--	--	Antiproliferative	[97]

The Intracellular Mechanism for Se Nanoparticle Synthesis

The bacterial cell membrane reductase is assumed to be involved in the reduction of selenium oxyanions (selenate/selenite) with the help of low molecular weight thiols such as bacilithiol (BSH) or the Trx/TrxRed system to precipitate Se nanoparticles in the cytoplasm or periplasm of the bacterium. The synthesised Se nanoparticles are released after the cell lysis [69].

The Extracellular Mechanism for Se Nanoparticle Synthesis

The bacterial cell membrane reductase is also playing an essential role in the reduction of selenite. The protein or peptide or components containing a thiol group released from bacterial cells react with selenite, and it gets reduced. Meanwhile, NADH, along with other protein components released from a bacterium, plays an important role in the formation of elemental Se. The formed Se nanoparticles are not stable since they are small and have a larger surface area. But, the Se nanoparticles are aggregated to become more extensive in size, to attain a lower energy state, by Ostwald ripening [70] (Fig. **7**).

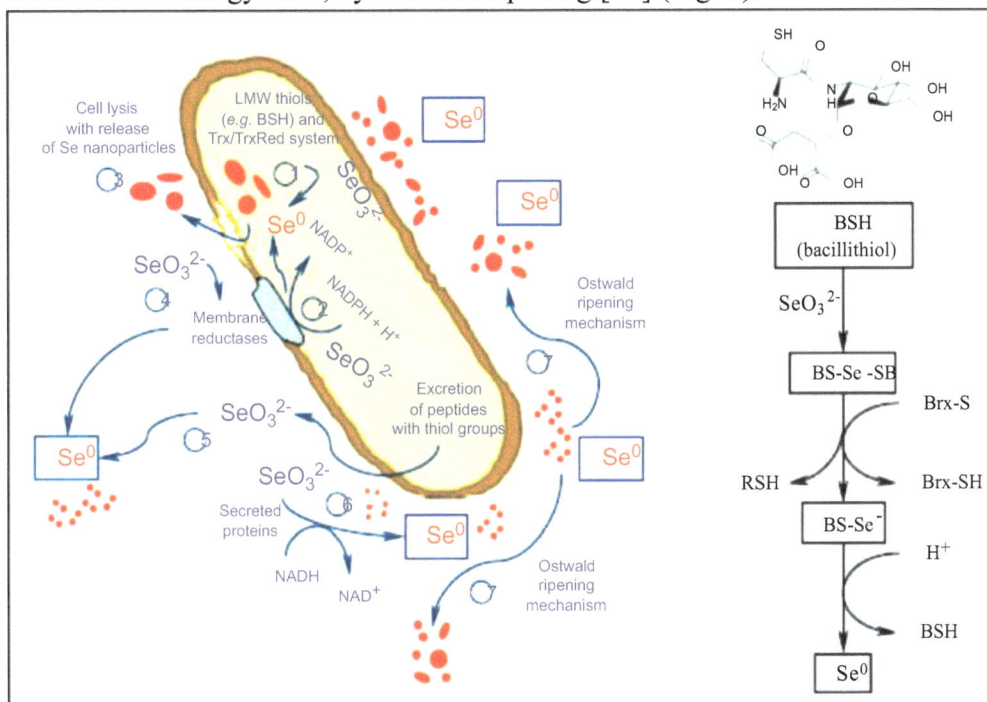

Fig. (7). The hypothesis of Se nanoparticles formation in *Bacillus mycoides* SeITE01.

A - Synoptic schematization of proposed biogenesis mechanisms of zero-valent selenium nanoparticles in Bacillus mycoides SeITE01. B - Suggested mechanism of selenite detoxification in Bacillus sp. involving Brx-like proteins. (Adapted from Lampis *et al.* 2014) [70]. © 2014 Lampis *et al.*; licensee BioMed Central Ltd.

Green Synthesis of Se Nanoparticles using Fungus, Yeast and Actinomycetes

Like bacteria, some of the fungal species, yeasts and actinomycetes are also being used in the synthesis of Se nanoparticles. The method of preparation of Se nanoparticles using fungus, yeast and actinomycetes is the same as the method used in bacteria mediated synthesis of Se nanoparticles, only the difference is the composition of medium and pH. The details of fungus, yeast and actinomycetes used in the synthesis of Se nanoparticles, characterisation parameters and applications are given in Table **2**.

Green Synthesis of Se Nanoparticles using Plants

Synthesis of Se nanoparticles using different parts of plants is a rapid process; however, the process utilises non-hazardous material. The active constituents of plant parts act as a reducing agent as well as stabilising or capping agent which could control the aggregation of nanoparticles when it is stored for a long duration. The plant components are also playing a major role in heavy metal detoxification and accumulation [98, 99]. Plant active constituents like alkaloids, flavonoids, phenolics, tannins, and terpenoids act as good reducing agents for the preparation of Se nanoparticles. It has been established that the Se nanoparticles prepared from biological material are less toxic than the bulk Se nanoparticles prepared from chemicals. The flow of plant-mediated synthesis of Se nanoparticles is shown in Fig. (**8**), and the details of plant and plant parts used in the synthesis of Se nanoparticles, characterisation parameters and applications are given in Table **3**.

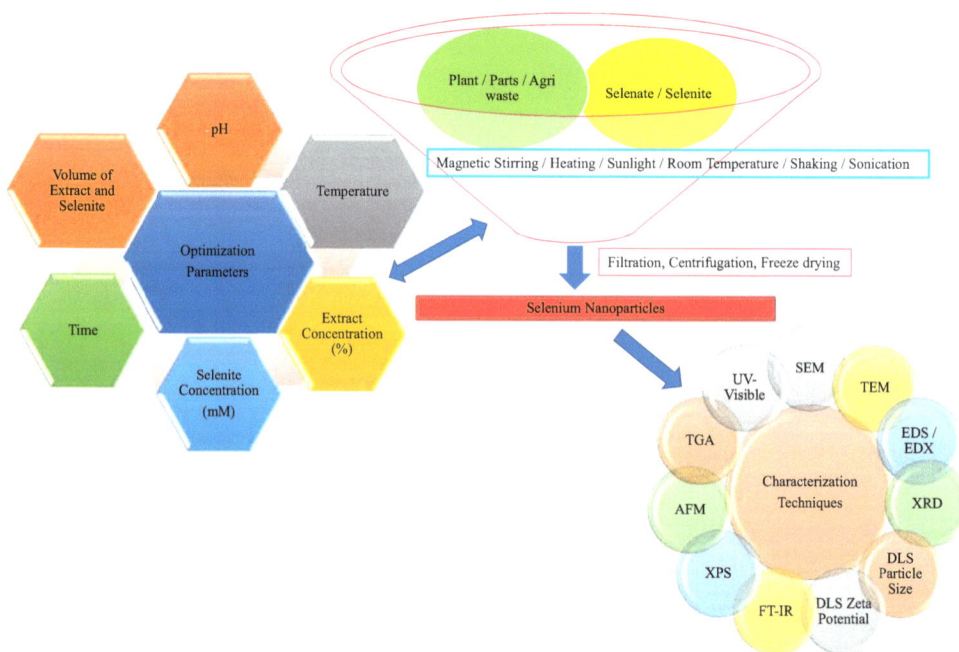

Fig. (8). Process of green synthesis of selenium nanoparticles using plant/Agri waste.

Table 3. Plant parts mediated synthesis of Se nanoparticles.

Source	Reaction Condition	Colour / UV-Vis wavelength (nm)	Size (nm)/ Shape (SEM or TEM)	DLS size / PDI	Functional Group	Activity	References
Vitis vinifera	H_2SeO_3 + Aq. Fruit extract Refluxed for 15 min	Brick red / --	3–18 / spherical	8.12 / 0.212	-OH, -CH, Phenolic OH, -OCH$_3$	--	[100]
Aloevera	Na_2SeO_3 + Aq. Leaf extract Autoclave at 121°C, 1.5 bar for 15 min	Red / 323	50 / spherical	136 / 0.321	-OH, -COOH, -CONH	Antibacterial, Antifungal	[101]
Pelargonium zonale	Na_2SeO_3 + Aq. Leaf extract Microwave for 4 min, 800W	Red / 319	50 / spherical	136 / 0.321	-OH, -CONH	Antimicrobial	[102]
Cassia auriculata	Na_2SeO_3 + Aq. Leaf extract RT, shaker for 48 h	--- / 252	10-20 / --	--	-CH, -CN, -NH	Anticancer	[103]
Leucas lavandulifolia Sm	Na_2SeO_3 + Aq. Leaf and stem extract + Ascorbic acid	Ruby red / 293	56-75 / spherical	--	-OH alcohol and phenol, -NH, -C=--, -CH, ---O-C-	Antibacterial	[104]
Allium sativum	Aq. *Allium sativum* buds + Na_2SeO_3 Magnetic stirring, 72 h	--- / 400	8-52 (TEM) / spherical	--	-NH, -CH, -C=O, -C=N	Antimicrobial Antioxidant	[105, 106]
Aloevera	Aq. *Aloevera* leaf + Na_2SeO_3 shaker, 72 h	Dark pink / 350	7-48 (TEM) / --	---	-NH, -CH, -C=O, -C=N	-- Antioxidant	[107, 108]
Withania somnifera	Aq. *Withania somnifera* leave + H_2SeO_3 RT, stir for 24 h	Ruby red / 310	45-90 / spherical		-OH alcohol and phenol, -CH, -C-C, -CN, -C-N-C-	Antioxidant Antiproliferative Antibacterial Photocatalytic	[109]

Source	Reaction Condition	Colour / UV-Vis wavelength (nm)	Size (nm)/ Shape (SEM or TEM)	DLS size / PDI	Functional Group	Activity	References
Green tea	*Lycium barbarum* polysaccharides + green tea + Na_2SeO_3 RT, 12 h	Brown / --	83-160 (TEM) / spherical	258.7 / 0.241	-C-H, -OH,	Antioxidant	[110]
Bougainvillea spectabilis Wild	Na_2SeO_3 + Aq. *Bougainvillea spectabilis* Wild flower incubator cum shaker with 250 rpm at 36°C for 120 h.	Dark red / 326	47, 18-35 (TEM) / spherical	--	-CH, -NH, -OH, -C=O	---	[111]
Catharanthus roseus	Aq. *Catharanthus roseus* flower + Na_2SeO_3 incubator cum shaker with 250 rpm at 36°C for 7 days.	Pinkish red / 335	17-34 (TEM) / hollow spherical	--	-COO, -CONH, -NH	--	[112]
Peltophorum pterocarpum	Aq. *Peltophorum pterocarpum* flower + Na_2SeO_3 incubator cum shaker with 250 rpm at 36°C for 7 days.	Light brown / 325	21-42 (TEM) / hollow spherical	--	-C=O, -NH, -OH	--	[112]
Potatoes	Potatoes + Na_2SeO_3 Autoclave 121°C for 30 min	Deep red / 200-300	100 / hollow spherical	115	-OH, - CH, -C=O	Antioxidant Antibacterial	[113]
Clausena dentata	Aq. *Clausena dentata* plant + selenium powder	Brown / 420	46.32 to 78.88 / sphericsl	--	-C=C, -OH, -NH	Larvicidal activity	[114]

(Table 3) cont.....

Source	Reaction Condition	Colour / UV-Vis wavelength (nm)	Size (nm)/ Shape (SEM or TEM)	DLS size / PDI	Functional Group	Activity	References
Fenugreek	Aq. Fenugreek seed + H$_2$SeO$_3$ + ascorbic acid Incubate 24 h	Ruby red / 200-400	50-150 / oval	--	-C=C, -NH$_2$, -COOH, -CH, -C=O	Anticancer (Breast cancer)	[115]
Emblica officinalis	Aq. *Emblica officinalis* fruit + Na$_2$SeO$_3$ reduction in dark condition at 27°C and 120 rpm on orbital shaker for 24 h	Brick red / 270	15-40 (TEM) / spherical	20-60 / 0.2	-C=C, -NH$_2$, -COOH, -CH, -C=O, -C-O-C-	Antioxidant Antimicrobial Antifungal	[116]
Petroselinum crispum	Aq. *Petroselinum crispum* leaf + Na$_2$SeO$_3$ Overnight, RT	Red / 270	5-100 (AFM) / spherical	400	-CH, -NH, -OH, -C=O, Se-O, -OCH$_3$	--	[117]
Moringa oleifera	Aq. *Moringa oleifera* leaf + Na$_2$SeO$_3$ incubation for 24 h at RT	Brick red / 299 and 400	23-35 (TEM) / spherical	--	-CH, -NH, -OH, -C=O	Anticancer Photocatalutic	[118]
Hordeum vulgare L.	Aq. *Hordeum vulgare L.* leave + Na$_2$SeO$_3$ Incubation 24 h	Red / 275	50-200	--	--	Promote the growth of barley seedlings under salt stress	[119]
Crataegus hupehensis Sarg	Aq. *Crataegus hupehensis* fruit + Na$_2$SeO$_3$ Magnetic stirring for 12 h	Red / 280	spherical	113	--	Anticancer (liver)	[120]
Asteriscus graveolens	Aq. green aerial part of *Asteriscus graveolens* + H$_2$SeO$_3$, incubated for 24 h	Red / 350	21.6 / spherical	20.06 / 1.0	-OH, -C=O, Se-O, -NH	Antiproliferation	[121]

(Table 3) cont.....

Source	Reaction Condition	Colour / UV-Vis wavelength (nm)	Size (nm)/ Shape (SEM or TEM)	DLS size / PDI	Functional Group	Activity	References
Diospyros montana	Aq. *Diospyros montana* leaf + H_2SeO_3, Ascorbic acid, 5 min	Brick red / 261	4-16 / spherical	4-16	-CH, -NH, -OH, -C=O	Antioxidant, Antimicrobial, Cytotoxicity	[122]
Ficus benghalensis	Aq. *Ficus benghalensis* leaf + Selenium metal + sodium sulphide, stir for 10 min.	Reddish orange / 268	20-140 / spherical	40-95 / 0.096	--	Photocatalytic	[123]
Orthosiphon stamineus	Aq. *Orthosiphon stamineus* leave + H_2SeO_3, Ascorbic acid, incubated, RT, 24 h	Darkish brownish orange / 209	88-141 / balls like	--	-OH alcohol and phenol, -CH, -C-C, -CN	Cytotoxicity	[124]
Spermacoce hispida	Aq. *Spermacoce hispida* leaf + H_2SeO_3, 40°C for 1 h	Red / 350	--	--	--	Hepatoprotective, Nephroprotective against acetaminophen toxicity	[125]
Spermacoce hispida	Aq. *Spermacoce hispida* leaf + H_2SeO_3, pH 9, incubated at 40 °C for 10 min	Red / 350	120 ± 15	--	-OH, -CH, C=O,-C=C, -N–O, - C–N	Antioxidant, Antibacterial, Anticancer, Anti-inflammatory	[126]

Green Synthesis of Se Nanoparticles using Agriwaste

Agri waste is a renewable and sustainable source. As per the literature, the annual production of agricultural waste globally is 998 million tonnes [127]. The disposal of agricultural waste is a significant burden for farmers and industries. The improper agricultural waste management system is a major problem in agricultural activities. Thus, research is needed to focus on Sustainable Agricultural Waste Management System, mainly to protect public health (SDG 3 - Ensure healthy lives and promote well-being for all at all ages & SDG 11-Make cities and human settlements inclusive, safe, resilient and sustainable) and environment (SDG 6-7-Sustainable management of water & modern energy; SDG 11-Sustainable cities & SDG 15-Protect ecosystem) [128].

In the aspect of biodiversity for sustainability, the use of Agri waste in the synthesis of nanoparticles is the best choice, even though there is a lot of chemical, physical and edible plant parts mediated methods are available to synthesize nanoparticles (Fig. **8**). The atmosphere gets polluted when we use physical or chemical methods for nanoparticle synthesis. We have to follow the tedious purification process if we use microorganisms in the green synthesis. The use of edible plant parts is not a suitable approach since we are already facing a food crisis as globally. However, the economic crisis and health issues caused by the current pandemic COVID-19 are going to create a lot of impact on the human lifestyle. Agri waste mediated bio/green synthesis of nanoparticles is better to approach when considering all these factors, such as sustainability, waste management and biodiversity management. Agri waste material contains various organic compounds such as polyphenols, flavonoids, carotenoids and vitamins, which can act as reducing or capping agents during the synthesis of nanoparticles [129, 130].

Selenium nanoparticles can be synthesised by using Agri wastes like a shell, fruit peel, straw, husk, *etc.* as simple, non-toxic, eco-friendly, greener materials. In this section, we have not included any plant leaves, flowers and fruit since they are essential parts of the plant for photosynthesis and crops; moreover, this part is already covered under the plant-mediated synthesis of Se nanoparticles. The crushed, air-dried peel powder extracts are used to reduce metal solutions. A variety of nanoparticles are formed when the reaction conditions are altered concerning pH, concentration and incubation temperature. The details of Agri waste used in the synthesis of Se nanoparticles, characterisation parameters and applications are given in Table **4**.

SELENIUM NANOPARTICLES APPLICATIONS

Nanoparticles assist to diminishing toxicity, improving bioactivity, improving targeting, and providing versatile means to control the release profile of the encapsulated moiety. Among different nanoparticles, inorganic nanoparticles of metals like Ag, Au, Ce, Fe, Se, Ti and Zn possess a significant place owing to their unique bioactivities in nanoforms. Selenium (Se) is an essential trace element. It is incorporated into selenoproteins such as thioredoxin reductase, glutathione peroxidase, glutathione hydroxy peroxidase, and phospholipids. Selenocysteine represents the most important part of the active centre of their enzymatic activities. Many selenoproteins regulate the redox balance in physiology due to their oxidoreductase activity. The toxicity margin of selenium is very delicate since it has a narrow therapeutic window, whereas the Se nanoparticles possess remarkably reduced toxicity.

Table 4. Agri waste mediated synthesis of Se nanoparticles.

Source	Reaction Condition	Colour / UV-Vis wavelength (nm)	Size (nm)/ Shape (Sem or TEM)	DLS size / PDI	Functional Group	Activity	References
Banana leaf powder and groundnut oil cake	Banana leaf powder and groundnut oil cake fermented with *Saccharomyces cerevisiae* + Na$_2$SeO$_3$ Incubated, RT for 24 h	Red / 540	170-240 / flower like	---	----	Antibacterial Anticancer Antioxidant Photocatalytic	[131]
Theobroma cacao L.	Aq. *Theobroma cacao* L. + Na$_2$SeO$_3$ Mircowave 400 rpm	-- / 276	1-3 (TEM) / --	---	-OH, -C=O, Se-O	Antioxidant	[132]
Citrus reticulata	Aq. *Citrus reticulata* peel + Na$_2$SeO$_3$ + 40°C for 1 h	Red / 265	70	--	--	Anti-algal	[133]

Se nanoparticles have been explored in various oxidative stress and inflammation-mediated disorders like arthritis, cancer, diabetes and nephropathy with potential therapeutic benefits. Se nanoparticles also act as a transport carrier for various drugs to the site of action. Herein we have discussed the role of Se nanoparticles in pharmacological and biological action (Fig. 9). However, it is principally unknown how Se nanoparticles may affect the pharmacokinetics and pharmacodynamics of selenoproteins. We have described the application of only a few biosynthesised Se nanoparticles in the following headings, and the remaining are included in Tables **2-4**.

Antioxidant Activity

Free radicals are regulatory and signalling molecules in their optimum concentration in the body, but they can cause damage to the cellular components if they are excess in their quantity. It may be causing agent for oxidative stress diseases, neurodegenerative diseases and cardiovascular diseases. Antioxidants play an essential role in protection against these diseases [134].

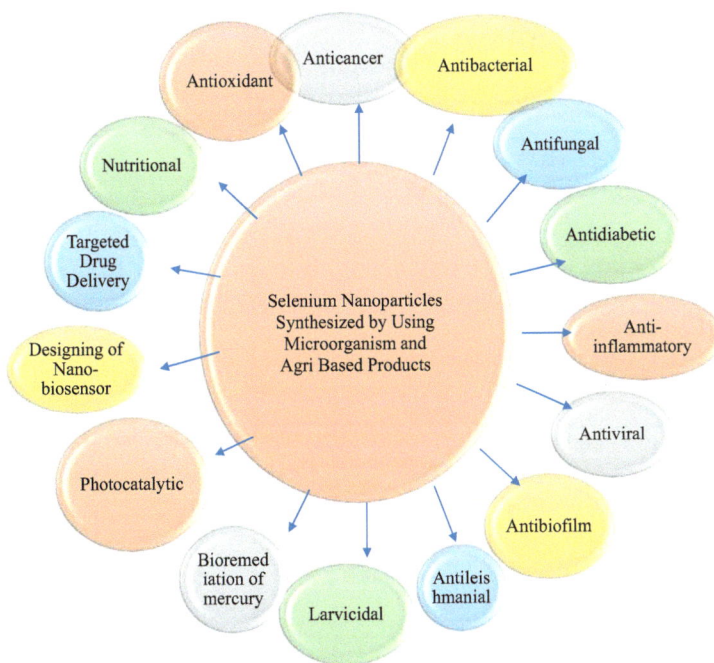

Fig. (9). Application of green synthesized selenium nanoparticles.

Horkey *et al.* studied the effect of Se nanoparticle solution (0.06 mg of Se per kg of body weight/day) and selenium nanoparticles bound with glucose (SeN-GLU) (0.06 mg of Se and 0.3 mg of glucose per kg of body weight/day) in rats. Glutathione peroxidase and superoxide dismutase in rat blood was considered as an antioxidant marker, and it was found that SeN-GLU showed a maximum antioxidant potential [135]. In another study, the antioxidant activity of Agri-waste (Banana leaf powder and groundnut oil cake fermented with *Saccharomyces cerevisiae*) mediated Se nanoparticles was reported. It was found that the maximum free radical scavenging activity of Se nanoparticles is 79% at 500 µg/ml, and IC_{50} value 236 µg/ml, by DPPH assay. Ferric reducing antioxidant power (FRAP) was used to find out the reducing power, and it was found Se nanoparticles showed better reducing power than ascorbic acid [131]. To improve nutritional value, broccoli sprouts were supplemented with Se nanoparticles, and the antioxidant activity of broccoli and the mixture was determined by DPPH assay. It was found that sprouts supplemented with 100 ppm of Se nanoparticles showed the highest antioxidant activity [136].

Recently, the antioxidant activity of microwave-assisted Se nanoparticles stabilised and capped with *Theobroma cacao L.* bean shell extract was reported by [132]. The antioxidant performance was tested by the 2,2'-azino-bis-3-ethylbenzothiazoline-6-sulphonic acid) (ABTS) and FRAP methods. It was found that Se nanoparticles with high antioxidant potential than the extract itself.

Moreover, the antioxidant capacity of Se nanoparticles remained stable for 55 days with no significant differences ($p > 0.05$) between samples at different times, indicating the high stability of synthesized Se nanoparticles. In another study, Qi *et al.* analysed the antioxidant activity of quercetin mediated Se nanoparticles. This method involves combining Que and Na_2SeO_3 to obtain selenium nanoparticles, and then these nanoparticles are modified using a combination of acacia and polysorbate 80. Also concluded that the nanocomposite showed excellent antioxidant activity in the DPPH method [137].

Further, the protective and antioxidant role of Se nanoparticles and vitamin C was studied against acrylamide induced hepatotoxicity in male mice. The study concluded that Se and vitamin C administration declined acrylamide elicited increment in aminotransferase level and oxidative stress. Additionally, it was stated that the combination of vitamin C and Se nanoparticles had synergetic antioxidant activity, and they prohibited liver damage in mice [138].

Antimicrobial Activity

The antimicrobial activity of Se nanoparticles and different Se compounds was attributed to the formation of free radicals. They react with intracellular thiols and cause oxidative stress. Se nanoparticles have a higher potential to generate these free radicals irrespective of their morphology, but toxicity varied with the microbial populations [139].

In a study, the antibacterial activity of agro-waste (Banana leaf powder and groundnut oil cake fermented with *Saccharomyces cerevisiae*) mediated Se nanoparticles was tested against different gram-positive and gram-negative bacteria by standard broth microdilution method. Se nanoparticles showed complete inhibition of *S. aureus* at a concentration of 250µg/ml, whereas no significant inhibition of growth was observed for the other test organisms [131]. The probiotic bacteria *Lactobacillus sp.* mediated Se nanoparticles showed inhibition of *Candida albicans* [140]. Antimicrobial activity of *Aloe vera* leaf extract mediated Se nanoparticles were assessed by Fardsadegh and Jafarizadeh-Malmiri, against Gram-positive (*Staphylococcus aureus*) and Gram-negative (*Escherichia coli*) bacteria., fungal strain *Colletotrichum coccodes* and *Penicillium digitatum*. It was found that Se nanoparticles showed better anti-bacterial activity against *S. aureus*, antifungal activity against *P. Digitatum* [101].

Further, Fardsadegh *et al.* reported the antimicrobial activities of *Pelargonium zonale* leaf extract mediated Se nanoparticles. In this study, they stated that the antibacterial activities of the synthesised Se NPs against *Escherichia coli* and *Staphylococcus aureus* indicated that the created NPs had higher antibacterial activities toward the Gram-positive bacteria. Besides, the synthesised Se NPs indicated higher antifungal activities against *Colletotrichum coccodes* and *Penicillium digitatum* [102].

Anticancer Activity

Chemotherapeutic agents have undesirable side effects, and also resistance developed by the cancer cell against these chemotherapeutic agents leads the researcher to combine nanotechnology with cancer treatments, which makes drugs more specific, highly effective and biocompatible. The surface-modified or functionalised nanoparticles with linkers are conjugated with anticancer drugs to increase the affinity towards the specific target and prolong the drug release in a controlled manner. The surface coating is done by using either plant or microbes or biodegradable polymers [141]. Antioxidant defense, carcinogen detoxification, anti-angiogenesis, and cell invasion can be the probable function responsible for the anticancer activity of Se nanoparticles [81]. In a study, the anticancer activity of Agri-waste (Banana leaf powder and groundnut oil cake fermented with *Saccharomyces cerevisiae*) mediated Se nanoparticles was tested using 3-(4,--dimethylthiazol-2-yl)-2,5-diphenyltetrazolium bromide (MTT) assay against SKOV3, MCF7, A549 and CHO cancer cells. In comparison, 100µg/ml of Se nanoparticles exhibited intense cytotoxicity against SKOV3, MCF7, A549 cells and significantly less cytotoxicity against normal cells [131].

Ahmad *et al.* studied and reported that novel strain Ess_amA-1 of *Streptomyces bikiniensis* mediated Se nanorods induces cell death of Hep-G2 and MCF-7 human cancer cells. The lethal dose ($LD_{50\%}$) of Se nanorods on Hep-G2 and MCF-7 cells was recorded at 75.96 µg/ml and 61.86 µg/ml, respectively. It was concluded that *S. bikiniensis* strain Ess_amA-1 could be used as renewable bioresources of biosynthesis of anticancer Se nanorods [142].

Green synthesis and characterisation of selenium nanoparticles using *Acinetobacter* sp. SW30 and its anticancer activity in breast cancer cells were reported by Wadhwani *et al.* In their research, they exposed Breast cancer cells (4T1, MCF-7) and non-cancer cells (NIH/3T3, HEK293) to different concentrations of Se nanoparticles and reported that the chemically synthesised Se nanoparticles showed better anticancer activity than *Acinetobacter* sp. mediated Se nanoparticles, but the earlier one is toxic to non-cancer cell also [79]. Sonkusre and Cameotra demonstrated that *Bacillus licheniformis* JS2 derived biogenic selenium nanoparticles induce non-apoptotic cell death in the prostate adenocarcinoma cell line, PC-3, at a minimal concentration of 2 μg Se/ml. The Se nanoparticles did not cause toxicity to the primary cells. However, the mechanism behind its anticancer activity was intangible [143]. In another research, Patel *et al.* synthesised Se nanoparticles by an eco-friendly method using whole cells of *Saccharomyces boulardii* [144].

Moreover, the cytoprotective role of selenium nanoparticles has been shown by cell viability in MTT assay. Leaf extracts of drumstick mediated Se nanoparticles were analysed by Hassanein *et al.*, and they summarised that Se nanoparticles were effective against three types of human cancers (Caco-2 cells, HepG2 cells, and Mcf-7 cells) [118]. Cui *et al.* used dried hawthorn fruits to synthesis Se nanoparticles and reported that Se nanoparticles showed obvious antitumor activities towards HepG2 cells with an IC_{50} of 19.22 ± 5.3 μg/ml. Meanwhile, they said that the Se nanoparticles up-regulated the intracellular ROS levels and reduced the mitochondrial membrane potential. In addition, it's also induced the up-regulation of caspase-9 and down-regulation of Bcl-2, which are identified by Western blot [120]. Very recently, Anu *et al.* reported the anti-leukaemia activity of leaves of *Cassia auriculata* mediated Se nanoparticles. Additionally, *in vitro* anticancer results against Human leukaemia (HL-60) cell lines were also reported [103].

Other Medicinal Properties

A study by Krishnan *et al* described the hepatoprotective and nephroprotective activity of green synthesised selenium nanoparticles using *Spermacoce hispida* against acetaminophen toxicity. In this study, it was found that the pre-treatment with Se nanoparticles attenuated the acetaminophen toxicity-induced elevation of kidney and liver injury markers in the blood circulation. Histological observation showed that nanoparticles pre-treatment protected the morphology of liver and kidney tissue [125].

Recently, Keyhani *et al* reported the prophylactic activity of biogenic selenium nanoparticles against chronic *Toxoplasma gondii* infection in mice models. In this study male, BALB/c mice were orally treated with Se nanoparticles at the doses of 2.5, 5, 10 mg/kg once a day for 14 days. On the 15th day, the mice were infected with the intraperitoneal inoculation of 20-25 tissue cysts from the Tehran strain of *Toxoplasma gondii*. The study indicates that the prophylactic use of biogenic selenium nanoparticles in concentrations of 2.5-10 mg/kg for two weeks was able to prevent severe symptoms of toxoplasmosis in the mice model [145].

Photocatalytic Activity

The bandgap (Eg) of a semiconductor and the generation of the photoinduced e−/h+ pairs are vital parameters for the photocatalytic behaviour of semiconductors. In general, the photocatalyst absorbs some energy when light (hυ ≥ Bandgap) strikes it, leading to the excitation of electrons from the valence band (vb) to the conduction band (cb) of photocatalyst, generating a positively charged hole in the valence band ($h_{vb}+$) and negative charge in the conduction band ($e_{cb}−$). The so formed conduction band electrons and the valence band holes then migrate to the surface of photocatalyst where they react with the chemisorbed O_2 and/or H_2O molecules to form reactive oxygen species such as $O_2 \cdot −$, OH• radicals, *etc.* These hyperactive radicals (*e.g.*, •OH, $O_2 \cdot −$) and/or photogenerated holes which are considered capable of directly oxidising many organic pollutants, attack dye molecules successively to break down into smaller fragments that finally decompose into simple inorganic minerals [146].

In a study, the photocatalytic activity of agro-waste (Banana leaf powder and groundnut oil cake fermented with *Saccharomyces cerevisiae*) mediated Se nanoparticles were checked against methylene blue (MB), and the degradation of MB was visualised by a decrease in peak intensity within three hours of irradiation. However, there was no considerable shift in the peak position of the control sample without exposure to Senanoparticles [131]. Photocatalytic degradation of sunset yellow azo dye by leaf extracts of drumstick mediated Se nanoparticles were analysed by Hassanein *et al.* and reported that Se nanoparticles have the efficiency in degrading sunset yellow dye and proposed the mechanism of degradation [118]. Tripathi *et al* synthesised Se nanoparticles using *Ficus benghalensis* leaves extract and reported for the photocatalytic activity. They demonstrated that the photocatalytic degradation of methylene blue by Se nanoparticle was up to 57.63% in 40 min. as a photocatalyst [123].

Other Applications

Sowndarya *et al* reported the green synthesis and insecticidal potential against mosquito vectors of selenium nanoparticles conjugated *Clausena dentata* plant leaf extract. In their research, they found that the synthesised Se nanoparticles showed LC_{50} values of 240.714 mg/l, 104.13 mg/l, and 99.602 mg/l for *A. stephensi*, *A. Aegypti*, and *C. quinquefasciatus*, respectively. Results of the study suggest that the *C. dentata* leaf extract-mediated Se nanoparticles have the potential to control mosquito vectors at early stages [114]. Sasidharan *et al.* described peel extract of citrus reticulata mediated biosynthesis of Se nanoparticles using the effect of pH and temperature. They showed Se nanoparticles could possess anti-algal activity [133].

Sawant *et al.* synthesised rod-shaped Se nanoparticles using lemon fruit extract and developed a naked eye cost-effective spectrometric sensing method for the detection of H_2O_2 [147]. Adsorption removal of multiple dyes using biogenic selenium nanoparticles from an *Escherichia coli* strain overexpressed selenite reductase CsrF was studied by Xia *et al.* They reported the maximum adsorption capacity for anionic dye (Congo red) at acidic pH and cationic dyes (safranine T and methylene blue) at alkaline pH. Adsorption kinetics, adsorption isotherms and adsorption thermodynamics studies showed that the adsorption capacities of Congo red, safranine T and methylene blue were 1577.7, 1911.0 and 1792.2 mg/g, respectively. In addition, Se nanoparticles can be effectively reused by 200 mmol/l NaCl. The adsorption capacities of Se nanoparticles for Congo red, safranine T and methylene blue were reported as 6.8%, 25.2% and 49.0% higher than that for traditional bio-based materials, respectively [148]. Prasad and Selvarat studied the effect of *Terminalia arjuna* leaf extract mediated Se nanoparticles on As(III)-induced toxicity on human lymphocytes. Studies on cell viability using MTT assay and DNA damage using comet assay revealed that synthesised selenium nanoparticles showed a protective effect against As(III)-induced cell death and DNA damage [149].

TOXICITY STUDIES ON BIOGENIC SELENIUM NANOPARTICLES

Acute and subacute toxicity of novel biogenic selenium nanoparticles, mediated by *Bacillus* sp. MSh-1, in mice, was studied and reported by Shakibaie *et al.* The LD_{50} of Se nanoparticles and SeO_2 were determined and the subacute toxicity was evaluated by oral administration of 0, 2.5, 5, 10 and 20 mg/kg of Se nanoparticles to male mice for 14 consecutive days. AST, ALT, ALP, creatinine, BUN, cholesterol, bilirubin, triglyceride and CPK were also experimentally measured. It was found that the LD_{50} of SeO_2 and Se nanoparticles were 4.39–10.38 mg/kg and 152.4–242.29 mg/kg (range with 95% confidence limits), respectively [150].

CONCLUSION

Synthesis of Se nanoparticles has been the subject of many studies due to its importance in pharmaceuticals, therapeutics, diagnostics, electronics, sustainable and renewable energy as well as other commercial products. The basic advantage of the green approach using plants and Agri based products in the synthesis of selenium nanoparticles can overcome the time-consuming system of culture of microbes and tedious purification techniques. On the other hand, there is a need to find a productive solution for the synthesis of nanoparticles from plant materials for a commercially viable approach. Meanwhile, when the same species of plants from different parts of the world are used for the synthesis of selenium nanoparticles leading to different results, which is because of variation in the chemical composition of plant compounds. This is the major downside of the synthesis of nanoparticles using plant extracts as reducing and stabilising agents, and thus there is a serious need to resolve this issue.

Moreover, the effect of biological source and physicochemical parameters such as salt concentration, pH, the temperature of the reaction, reaction time, *etc.*, should be studied and optimized to synthesize Se nanoparticles with smaller than 100 nm size and desired properties. There are very few reports available (Tables **2-3**) with all the above said parameters; hence it is still a great challenge for researchers and scientists.

However, the laboratory scale production of nanoparticles using plants have a wide perspective, but their production methods need to be altered specifically for large scale production. Also, the functional mechanism against the pathogenic microorganisms needs to be exposed so that the green chemistry method would attain a much broader approach. The green chemistry approach is helpful in environmental protection and bioremediation. However, plant uptake and utilisation of selenium nanoparticles require more detailed research on many issues such as uptake potential of various species, process and translocation, as well as the activities of the green synthesised nanoparticles at the cellular and molecular levels.

CONSENT FOR PUBLICATION

Not applicable.

CONFLICT OF INTEREST

The authors declare no conflict of interest, financial or otherwise.

ACKNOWLEDGEMENT

Declared none.

REFERENCES

[1]   Wong TS, Schwaneberg U. Protein engineering in bioelectrocatalysis. Curr Opin Biotechnol 2003;
        14(6): 590-6.
        [http://dx.doi.org/10.1016/j.copbio.2003.09.008] [PMID: 14662387]

[2]   Ramanavicius A, Kausaite A, Ramanaviciene A. Polypyrrole coated glucose oxidase nanoparticles for
        biosensor design. Sens Actuators B Chem 2005; 111: 532-9.
        [http://dx.doi.org/10.1016/j.snb.2005.03.038]

[3]   Fendler JH. Nano particles and nanostructured films: Preparation, characterization and applications.
        New York: John Wiley & Son 1998.
        [http://dx.doi.org/10.1002/9783527612079]

[4]   Wang Y, Xia Y. Bottom-up and top-down approaches to the synthesis of monodispersed spherical colloids of low melting-point metals. Nano Lett 2004; 4(10): 2047-50.
[http://dx.doi.org/10.1021/nl048689j]

[5]   Datta M. Electrochemical processing technologies in chip fabrication: challenges and opportunities. Electrochim Acta 2003; 48(20-22): 2975-85.
[http://dx.doi.org/10.1016/S0013-4686(03)00363-3]

[6]   Ghorbani HR, Safekordi AA, Attar H, Rezayat Sorkhabadi SM. Biological and non-biological methods for silver nanoparticles synthesis. Chem Biochem Eng Q 2011; 25(3): 317-26.
https://hrcak.srce.hr/71947

[7]   Zhang D, Qi L, Yang J, Ma J, Cheng H, Huang L. Wet chemical synthesis of silver nanowire thin films at ambient temperature. Chem Mater 2004; 16(5): 872-6.
[http://dx.doi.org/10.1021/cm0350737]

[8]   Zhang W, Qiao X, Chen J, Wang H. Preparation of silver nanoparticles in water-in-oil AOT reverse micelles. J Colloid Interface Sci 2006; 302(1): 370-3.
[http://dx.doi.org/10.1016/j.jcis.2006.06.035] [PMID: 16860816]

[9]   Frost DV. Selenium and vitamin E as antidotes against heavy metal toxicities. In: Spallholz JE, Martin JL, Ganther HE, Eds. Selenium in biology and medicine. Westport, Connecticut: AVI Publishing Company, Inc. 1981; pp. 490-8.

[10]   Whanger PD. Selenium and heavy metal toxicity. In: Spallholz JE, Martin JL, Ganther HE, Eds. Selenium in biology and medicine. Westport, Connecticut: AVI Publishing Company, Inc. 1981; pp. 230-55.

[11]   Underwood EJ. Trace elements in human and animal nutrition. New York: Academic Press 1977; pp. 302-46.
[http://dx.doi.org/10.1016/B978-0-12-709065-8.50016-X]

[12]   Clark LC, Combs GF Jr, Turnbull BW, *et al.* Effects of selenium supplementation for cancer prevention in patients with carcinoma of the skin. A randomized controlled trial. JAMA 1996; 276(24): 1957-63.
[http://dx.doi.org/10.1001/jama.1996.03540240035027] [PMID: 8971064]

[13]   Vinceti M, Maraldi T, Bergomi M, Malagoli C. Risk of chronic low-dose selenium overexposure in humans: insights from epidemiology and biochemistry. Rev Environ Health 2009; 24(3): 231-48.
[http://dx.doi.org/10.1515/REVEH.2009.24.3.231] [PMID: 19891121]

[14]   Yu SY, Zhu YJ, Li WG, *et al.* A preliminary report on the intervention trials of primary liver cancer in high-risk populations with nutritional supplementation of selenium in China. Biol Trace Elem Res 1991; 29(3): 289-94.
[http://dx.doi.org/10.1007/BF03032685] [PMID: 1726411]

[15]   Skalickova S, Milosavljevic V, Cihalova K, Horky P, Richtera L, Adam V. Selenium nanoparticles as a nutritional supplement. Nutrition 2017; 33: 83-90.
[http://dx.doi.org/10.1016/j.nut.2016.05.001] [PMID: 27356860]

[16]   El-Sonbaty SM, El-Arab WE, Barakat M. Ameliorative effect of selenium nanoparticles and ferulic acid on acrylamide-induced neurotoxicity in rats. Ann Med Biomed Sci 2017; 3(2): 35-45.

[17]   Drake EN. Cancer chemoprevention: selenium as a prooxidant, not an antioxidant. Med Hypotheses 2006; 67(2): 318-22.
[http://dx.doi.org/10.1016/j.mehy.2006.01.058] [PMID: 16574336]

[18]   Manna L, Scher EC, Alivisatos AP. Synthesis of soluble and processable rod-, arrow-, teardrop-, and tetrapod-shaped CdSe nanocrystals. J Am Chem Soc 2000; 122(51): 12700-6.
[http://dx.doi.org/10.1021/ja003055+]

[19]   Poborchii VV, Kolobov AV, Tanaka K. Photomelting of selenium at low temperature. Appl Phys Lett

1999; 74(2): 215-7.
[http://dx.doi.org/10.1063/1.123297]

[20] Wang G, Yang X, Qian F, Zhang JZ, Li Y. Double-sided CdS and CdSe quantum dot co-sensitized ZnO nanowire arrays for photoelectrochemical hydrogen generation. Nano Lett 2010; 10(3): 1088-92.
[http://dx.doi.org/10.1021/nl100250z] [PMID: 20148567]

[21] Leonard KA, Hall JP, Nelen MI, *et al.* A selenopyrylium photosensitizer for photodynamic therapy related in structure to the antitumor agent AA1 with potent *in vivo* activity and no long-term skin photosensitization. J Med Chem 2000; 43(23): 4488-98.
[http://dx.doi.org/10.1021/jm000154p] [PMID: 11087573]

[22] Parnham MJ, Graf E. Pharmacology of synthetic organic selenium compounds. Prog Drug Res 1991; 36: 9-47.
[http://dx.doi.org/10.1007/978-3-0348-7136-5_1] [PMID: 1876711]

[23] Sies H, Masumoto H. Ebselen as a glutathione peroxidase mimic and as a scavenger of peroxynitrite. Adv Pharmacol 1997; 38: 229-46.

[24] Zhang JS, Gao XY, Zhang LD, Bao YP. Biological effects of a nano red elemental selenium. Biofactors 2001; 15(1): 27-38.
[http://dx.doi.org/10.1002/biof.5520150103] [PMID: 11673642]

[25] Wang H, Zhang J, Yu H. Elemental selenium at nano size possesses lower toxicity without compromising the fundamental effect on selenoenzymes: comparison with selenomethionine in mice. Free Radic Biol Med 2007; 42(10): 1524-33.
[http://dx.doi.org/10.1016/j.freeradbiomed.2007.02.013] [PMID: 17448899]

[26] Narayanan KB, Sakthivel N. Green synthesis of biogenic metal nanoparticles by terrestrial and aquatic phototrophic and heterotrophic eukaryotes and biocompatible agents. Adv Colloid Interface Sci 2011; 169(2): 59-79.
[http://dx.doi.org/10.1016/j.cis.2011.08.004] [PMID: 21981929]

[27] Hurst SJ, Lytton-Jean AKR, Mirkin CA. Maximizing DNA loading on a range of gold nanoparticle sizes. Anal Chem 2006; 78(24): 8313-8.
[http://dx.doi.org/10.1021/ac0613582] [PMID: 17165821]

[28] Egorova EM, Revina AA. Synthesis of metallic nanoparticles in reverse micelles in the presence of quercetin. Colloids Surf A Physicochem Eng Asp 2000; 168(1): 87-96.
[http://dx.doi.org/10.1016/S0927-7757(99)00513-0]

[29] Mukunthan KS, Balaji S. Cashew apple juice (*Anacardium occidentale* L.) speeds up the synthesis of silver nanoparticles. Int J Grn Nanotech 2012; 4(2): 71-9.
[http://dx.doi.org/10.1080/19430892.2012.676900]

[30] Dhuper S, Panda D, Nayak PL. Green synthesis and characterization of zero valent iron nanoparticles from the leaf extract of *Mangifera indica*. Nano Trends J Nanotech App 2012; 13: 16-22.

[31] Shankar SS, Rai A, Ahmad A, Sastry M. Rapid synthesis of Au, Ag, and bimetallic Au core-Ag shell nanoparticles using Neem (*Azadirachta indica*) leaf broth. J Colloid Interface Sci 2004; 275(2): 496-502.
[http://dx.doi.org/10.1016/j.jcis.2004.03.003] [PMID: 15178278]

[32] Narayanan KB, Sakthivel N. Coriander leaf mediated biosynthesis of gold nanoparticles. Mater Lett 2008; 62(30): 4588-90.
[http://dx.doi.org/10.1016/j.matlet.2008.08.044]

[33] Kalishwaralal K, Deepak V, Ram Kumar Pandian S, *et al.* Biosynthesis of silver and gold nanoparticles using *Brevibacterium casei*. Colloids Surf B Biointerfaces 2010; 77(2): 257-62.
[http://dx.doi.org/10.1016/j.colsurfb.2010.02.007] [PMID: 20197229]

[34] Eppler AS, Rupprechter G, anderson EA, Somorjai GA. Thermal and chemical stability and adhesion strength of Pt nanoparticle arrays supported on silica studied by transmission electron microscopy and

atomic force microscopy. J Phys Chem B 2000; 104(31): 7286-92.
[http://dx.doi.org/10.1021/jp0006429]

[35]   Kelly KL, Coronado E, Zhao LL, Schatz GC. The optical properties of metal nanoparticles: Influence of size, shape and dielectric environment. J Phys Chem B 2003; 107(3): 668-77.
[http://dx.doi.org/10.1021/jp026731y]

[36]   Yang LB, Shen YH, Xie AJ, Liang JJ, Zhang BC. Synthesis of Se nanoparticles by using TSA ion and its photocatalytic application for decolorization of congo red under UV irradiation. Mater Res Bull. 2008; 43: pp. 572-82.

[37]   Jiang J, Oberdörster G, Biswas P. Characterization of size, surface charge, and agglomeration state of nanoparticle dispersions for toxicological studies. J Nanopart Res 2009; 11(1): 77-89.
[http://dx.doi.org/10.1007/s11051-008-9446-4]

[38]   Klug HP, Alexander LE. X-ray diffraction procedures for poly-crystallite and amorphous materials. New York, NY, USA: Wiley 1974.

[39]   Lin ZH, Wang CRC. Evidence on the size-dependent absorption spectral evolution of selenium nanoparticles. Mater Chem Phys 2005; 92(2-3): 591-4.
[http://dx.doi.org/10.1016/j.matchemphys.2005.02.023]

[40]   Rajalakshmi M, Arora AK. Optical properties of selenium nanoparticles dispersed in polymer. Solid State Commun 1999; 110(2): 75-80.
[http://dx.doi.org/10.1016/S0038-1098(99)00055-1]

[41]   Mayers BT, Liu K, Sunderland D, Xia Y. Sonochemical synthesis of trigonal selenium nanowires. Chem Mater 2003; 15(20): 3852-8.
[http://dx.doi.org/10.1021/cm034193b]

[42]   Hostetler EB, Kim KJ, Oleksak RP, *et al.* Synthesis of colloidal PbSe nanoparticles using a microwave-assisted segmented flow reactor. Mater Lett 2014; 128: 54-9.
[http://dx.doi.org/10.1016/j.matlet.2014.04.089]

[43]   Harpeness R, Gedanken A, Weiss AM, Slifkin MA. Microwave-assisted synthesis of nanosized MoSe$_2$. J Mater Chem 2003; 13(10): 2603-6.
[http://dx.doi.org/10.1039/b303740a]

[44]   Langi B, Shah C, Singh K, Chaskar A, Kumar M, Bajaj PN. Ionic liquid-induced synthesis of selenium nanoparticles. Mater Res Bull 2010; 45(6): 668-71.
[http://dx.doi.org/10.1016/j.materresbull.2010.03.005]

[45]   Bartůněk V, Junková J, Babuněk M, Ulbrich P, Kuchař M, Sofer Z. Synthesis of spherical amorphous selenium nano and microparticles with tunable sizes. Micro & Nano Lett 2016; 11(2): 91-3.
[http://dx.doi.org/10.1049/mnl.2015.0353]

[46]   Cao GS, Juan Zhang X, Su L, Yang Ruan Y. Hydrothermal synthesis of selenium and tellurium nanorods. J Exp Nanosci 2011; 6(2): 121-6.
[http://dx.doi.org/10.1080/17458081003774677]

[47]   Mary TA, Shanthi K, Vimala K, Soundarapandian K. PEG functionalized selenium nanoparticles as a carrier of crocin to achieve anticancer synergism. RSC Advances 2016; 6(27): 22936-49.
[http://dx.doi.org/10.1039/C5RA25109E]

[48]   Dagmar H, Kristyna C, Pavel K, Vojtech A, Rene K. Selenium nanoparticles and evaluation of their antimicrobial activity on bacterial isolates obtained from clinical specimens. Nanocon. 2015 Oct; 14th–16th.

[49]   Xia Y, You P, Xu F, Liu J, Xing F. Novel functionalized selenium nanoparticles for enhanced anti-hepatocarcinoma activity in vitro. Nanoscale Res Lett 2015; 10(1): 1051.
[http://dx.doi.org/10.1186/s11671-015-1051-8] [PMID: 26334544]

[50]   Guo L, Huang K, Liu H. Biocompatibility selenium nanoparticles with an intrinsic oxidase-like

activity. J Nanopart Res 2016; 18(3): 74.
[http://dx.doi.org/10.1007/s11051-016-3357-6]

[51]   Li H, Zhang J, Wang T, Luo W, Zhou Q, Jiang G. Elemental selenium particles at nano-size (Nano-Se) are more toxic to Medaka (*Oryzias latipes*) as a consequence of hyper-accumulation of selenium: a comparison with sodium selenite. Aquat Toxicol 2008; 89(4): 251-6.
[http://dx.doi.org/10.1016/j.aquatox.2008.07.008] [PMID: 18768225]

[52]   Nie T, Wu H, Wong KH, Chen T. Facile synthesis of highly uniform selenium nanoparticles using glucose as the reductant and surface decorator to induce cancer cell apoptosis. J Mater Chem B Mater Biol Med 2016; 4(13): 2351-8.
[http://dx.doi.org/10.1039/C5TB02710A] [PMID: 32263230]

[53]   Liu W, Li X, Wong YS, *et al.* Selenium nanoparticles as a carrier of 5-fluorouracil to achieve anticancer synergism. ACS Nano 2012; 6(8): 6578-91.
[http://dx.doi.org/10.1021/nn202452c] [PMID: 22823110]

[54]   Soumya RS, Ghosh SK, Abraham ET. Preparation and characterization of guar gum nanoparticles. Int J Biol Macromol 2010; 46: 267-9.
[http://dx.doi.org/10.1016/j.ijbiomac.2009.11.003]

[55]   Dwivedi C, Shah CP, Singh K, Kumar M, Bajaj PN. An organic acid-induced synthesis and characterization of selenium nanoparticles. J Nanotechnol 2011.
[http://dx.doi.org/10.1155/2011/651971]

[56]   Bai Y, Wang Y, Zhou Y, Li W, Zheng W. Modification and modulation of saccharides on elemental selenium nanoparticles in liquid phase. Mater Lett 2008; 62(15): 2311-4.
[http://dx.doi.org/10.1016/j.matlet.2007.11.098]

[57]   Sheng-Yi Z, Zhang J, Wang H, Chen H. Synthesis of selenium nanoparticles in the presence of polysaccharides. Mater Lett 2004; 58: 2590-4.

[58]   Refat MS, El-Sabawy KM. Infrared spectra, Raman laser, XRD, DSC/TGA and SEM investigations on the preparations of selenium metal, (Sb$_2$O$_3$, Ga$_2$O$_3$, SnO and HgO) oxides and lead carbonate with pure grade using acetamide precursors. Bull Mater Sci 2011; 34(4): 873-81.
[http://dx.doi.org/10.1007/s12034-011-0208-z]

[59]   El-Batal AI, Sidkey NM, Ismail AA, Arafa RA, Fathy RM. Impact of silver and selenium nanoparticles synthesized by gamma irradiation and their physiological response on early blight disease of potato. J Chem Pharm Res 2016; 8(4): 934-51.

[60]   Thabet NM, Osman A, Abdel Ghaffar AB, Azab KS. Amelioration of oxidative damage induced in gamma irradiated rats by nano selenium and lovastatin mixture. World Appl Sci J 2012; 19(7): 962-71.
[http://dx.doi.org/10.5829/idosi.wasj.2012.19.07.2778]

[61]   Huang B, Zhang J, Hou J, Chen C. Free radical scavenging efficiency of nano-Se *in vitro*. Free Radical Biol Med 2003; 35(7): 805-13.
[http://dx.doi.org/10.1016/S0891-5849(03)00428-3]

[62]   Tan L, Jia X, Jiang X, *et al.* In vitro study on the individual and synergistic cytotoxicity of adriamycin and selenium nanoparticles against Bel7402 cells with a quartz crystal microbalance. Biosens Bioelectron 2009; 24(7): 2268-72.
[http://dx.doi.org/10.1016/j.bios.2008.10.030] [PMID: 19101136]

[63]   Nath S, Ghosh SK, Panigahi S, Thundat T, Pal T. Synthesis of selenium nanoparticle and its photocatalytic application for decolorization of methylene blue under UV irradiation. Langmuir 2004; 20(18): 7880-3.
[http://dx.doi.org/10.1021/la0493181] [PMID: 15323543]

[64]   Xia Y. Synthesis of selenium nanoparticles in the presence of silk fibroin. Mater Lett 2007; 61(21): 4321-4.
[http://dx.doi.org/10.1016/j.matlet.2007.01.095]

[65]   Deng Z, Cao L, Tang F, Zou B. A new route to zinc-blende CdSe nanocrystals: mechanism and synthesis. J Phys Chem B 2005; 109(35): 16671-5.
[http://dx.doi.org/10.1021/jp052484x] [PMID: 16853121]

[66]   Yang F, Tang Q, Zhong X, *et al.* Surface decoration by Spirulina polysaccharide enhances the cellular uptake and anticancer efficacy of selenium nanoparticles. Int J Nanomedicine 2012; 7: 835-44.
[http://dx.doi.org/10.2147/IJN.S28278] [PMID: 22359460]

[67]   Zheng S, Li X, Zhang Y, *et al.* PEG-nanolized ultrasmall selenium nanoparticles overcome drug resistance in hepatocellular carcinoma HepG2 cells through induction of mitochondria dysfunction. Int J Nanomedicine 2012; 7: 3939-49.
[http://dx.doi.org/10.2147/IJN.S30940] [PMID: 22915845]

[68]   Song D, Li X, Cheng Y, *et al.* Aerobic biogenesis of selenium nanoparticles by *Enterobacter cloacae* Z0206 as a consequence of fumarate reductase mediated selenite reduction. Sci Rep 2017; 7(1): 3239.
[http://dx.doi.org/10.1038/s41598-017-03558-3] [PMID: 28607388]

[69]   Lampis S, Zonaro E, Bertolini C, *et al.* Selenite biotransformation and detoxification by *Stenotrophomonas maltophilia* SeITE02: Novel clues on the route to bacterial biogenesis of selenium nanoparticles. J Hazard Mater 2017; 324(Pt A): 3-14.
[http://dx.doi.org/10.1016/j.jhazmat.2016.02.035] [PMID: 26952084]

[70]   Lampis S, Zonaro E, Bertolini C, Bernardi P, Butler CS, Vallini G. Delayed formation of zero-valent selenium nanoparticles by *Bacillus mycoides* SeITE01 as a consequence of selenite reduction under aerobic conditions. Microb Cell Fact 2014; 13(1): 35.
[http://dx.doi.org/10.1186/1475-2859-13-35] [PMID: 24606965]

[71]   Kaur R, Kaudal T, Sharma A. Probiotic mediated synthesis of selenium nanoparticles: characterization and biofilm scavenging analysis. Res J Life Sci Bioinform Pharm Chem Sci 2018; 4(3): 291-304.
[http://dx.doi.org/10.26479/2018.0403.26]

[72]   Greeshma BC, Mahesh M. Biosynthesis of selenium nanoparticles from Bacillus species and its applications. J Appl Nat Sci 2019; 11(4): 810-5.
[http://dx.doi.org/10.31018/jans.v11i4.2188]

[73]   Shakibaie M, Salari Mohazab N, Ayatollahi Mousavi SA. Antifungal activity of selenium nanoparticles synthesized by *Bacillus* species Msh-1 against *Aspergillus fumigatus* and *Candida albicans.* Jundishapur J Microbiol 2015; 8(9): e26381.
[http://dx.doi.org/10.5812/jjm.26381] [PMID: 26495111]

[74]   Fernández-Llamosas H, Castro L, Blázquez ML, Díaz E, Carmona M. Biosynthesis of selenium nanoparticles by *Azoarcus sp.* CIB. Microb Cell Fact 2016; 15(1): 109.
[http://dx.doi.org/10.1186/s12934-016-0510-y] [PMID: 27301452]

[75]   El-Batal AI, Essam TM, El-Zahaby DA, Amin MA. Synthesis of selenium nanoparticles by *Bacillus laterosporus* using gamma radiation. Br J Pharm Res 2014; 4(11): 1364-86.
[http://dx.doi.org/10.9734/BJPR/2014/10412]

[76]   Chandramohan S, Sundar K, Muthukumaran A. Monodispersed spherical shaped selenium nanoparticles (SeNPs) synthesized by Bacillus subtilis and its toxicity evaluation in zebrafish embryos. Mater Res Exp 2018; 5(2): 025020.
[http://dx.doi.org/10.1088/2053-1591/aaabeb]

[77]   Abdallah OM. EL-Baghdady KZ, Khalil MM, El Borhamy MI, Meligi G. Biosynthesis, characterization and cytotoxicity of selenium nanoparticles using Bacillus tropicus Ism 2. Egypt Acad J Biol Sci G Microbiol 2019; 11(1): 47-57.

[78]   Srivastava P, Kowshik M. Anti-neoplastic selenium nanoparticles from Idiomarina sp. PR58-8. Enzyme Microb Technol 95: 192-200.2016;

[79]   Wadhwani SA, Gorain M, Banerjee P, *et al.* Green synthesis of selenium nanoparticles using *Acinetobacter* sp. SW30: optimization, characterization and its anticancer activity in breast cancer

cells. Int J Nanomedicine 2017; 12: 6841-55.
[http://dx.doi.org/10.2147/IJN.S139212] [PMID: 28979122]

[80]   Fesharaki PJ, Nazari P, Shakibaie M, *et al.* Biosynthesis of selenium nanoparticles using *Klebsiella pneumoniae* and their recovery by a simple sterilization process. Braz J Microbiol 2010; 41(2): 461-6.
[http://dx.doi.org/10.1590/S1517-83822010000200028] [PMID: 24031517]

[81]   Kora AJ, Rastogi L. Biomimetic synthesis of selenium nanoparticles by *Pseudomonas aeruginosa* ATCC 27853: An approach for conversion of selenite. J Environ Manage 2016; 181: 231-6. a
[http://dx.doi.org/10.1016/j.jenvman.2016.06.029] [PMID: 27353373]

[82]   Kora AJ, Rastogi L. Bacteriogenic synthesis of selenium nanoparticles by *Escherichia coli* ATCC 35218 and its structural characterisation. IET Nanobiotechnol 2017; 11(2): 179-84. b
[http://dx.doi.org/10.1049/iet-nbt.2016.0011] [PMID: 28477001]

[83]   Oremland RS, Herbel MJ, Blum JS, *et al.* Structural and spectral features of selenium nanospheres produced by Se-respiring bacteria. Appl Environ Microbiol 2004; 70(1): 52-60.
[http://dx.doi.org/10.1128/AEM.70.1.52-60.2004] [PMID: 14711625]

[84]   Kuroda M, Notaguchi E, Sato A, *et al.* Characterization of *Pseudomonas stutzeri* NT-I capable of removing soluble selenium from the aqueous phase under aerobic conditions. J Biosci Bioeng 2011; 112(3): 259-64.
[http://dx.doi.org/10.1016/j.jbiosc.2011.05.012] [PMID: 21676651]

[85]   Dhanjal S, Cameotra SS. Aerobic biogenesis of selenium nanospheres by Bacillus cereus isolated from coalmine soil. Microb Cell Fact 2010; 9(1): 52.
[http://dx.doi.org/10.1186/1475-2859-9-52] [PMID: 20602763]

[86]   Saljoghi H, Mohamadi M. Biogenic producing selenium nanoparticles by *Clostridium perfringens.* J Appl Environ Biol Sci 2014; 4: 245-8.

[87]   Visha P, Nanjappan K, Selvaraj P, Jayachandran S, Elango A, Kumaresan G. Biosynthesis and structural characteristics of selenium nanoparticles using *Lactobacillus acidophilus* bacteria by wet sterilization process. Int J Adv Vet Sci Technol 2015; 4(1): 178-83.
[http://dx.doi.org/10.23953/cloud.ijavst.183]

[88]   Li B, Liu N, Li Y, *et al.* Reduction of selenite to red elemental selenium by *Rhodopseudomonas palustris* strain N. PLoS One 2014; 9(4): e95955.
[http://dx.doi.org/10.1371/journal.pone.0095955] [PMID: 24759917]

[89]   Khoei NS, Lampis S, Zonaro E, Yrjälä K, Bernardi P, Vallini G. Insights into selenite reduction and biogenesis of elemental selenium nanoparticles by two environmental isolates of *Burkholderia fungorum.* New Biotechnol 2017; 34: 1-11.

[90]   Sarkar J, Dey P, Saha S, Acharya K. Mycosynthesis of selenium nanoparticles. Micro & Nano Lett 2011; 6(8): 599-602.
[http://dx.doi.org/10.1049/mnl.2011.0227]

[91]   Zare B, Babaie S, Setayesh N, Shahverdi AR. Isolation and characterization of a fungus for extracellular synthesis of small selenium nanoparticles. Nanomed J 2013; 1(1): 13-9.
[http://dx.doi.org/10.7508/NMJ.2013.01.002]

[92]   Joshi SM, De Britto S, Jogaiah S, Ito SI. Mycogenic selenium nanoparticles as potential new generation broad spectrum antifungal molecules. Biomolecules 2019; 9(9): 419.
[http://dx.doi.org/10.3390/biom9090419] [PMID: 31466286]

[93]   Abbas H, Abou Baker D. Biological Evaluation of Selenium Nanoparticles biosynthesized by *Fusarium semitectum* as antimicrobial and anticancer agents. Egypt J Chem 2020; 63(4): 18-9.

[94]   Nandini B, Hariprasad P, Prakash HS, Shetty HS, Geetha N. Trichogenic-selenium nanoparticles enhance disease suppressive ability of Trichoderma against downy mildew disease caused by *Sclerospora graminicola* in pearl millet. Sci Rep 2017; 7(1): 2612.
[http://dx.doi.org/10.1038/s41598-017-02737-6] [PMID: 28572579]

[95]   Vetchinkina E, Loshchinina E, Kupryashina M, Burov A, Pylaev T, Nikitina V. Green synthesis of nanoparticles with extracellular and intracellular extracts of basidiomycetes. PeerJ 2018; 6: e5237.
[http://dx.doi.org/10.7717/peerj.5237] [PMID: 30042892]

[96]   Faramarzi S, Anzabi Y, Jafarizadeh-Malmiri H. Nanobiotechnology approach in intracellular selenium nanoparticle synthesis using *Saccharomyces cerevisiae*-fabrication and characterization. Arch Microbiol 2020; 202(5): 1203-9.
[http://dx.doi.org/10.1007/s00203-020-01831-0] [PMID: 32077990]

[97]   Srivastava P, Braganca JM, Kowshik M. *In vivo* synthesis of selenium nanoparticles by *Halococcus salifodinae* BK18 and their anti-proliferative properties against HeLa cell line. Biotechnol Prog 2014; 30(6): 1480-7.
[http://dx.doi.org/10.1002/btpr.1992] [PMID: 25219897]

[98]   Makarov VV, Love AJ, Sinitsyna OV, *et al.* "Green" nanotechnologies: synthesis of metal nanoparticles using plants. Acta Nat (Engl Ed) 2014; 6(1): 35-44.
[http://dx.doi.org/10.32607/20758251-2014-6-1-35-44] [PMID: 24772325]

[99]   Mittal J, Batra A, Singh A, Sharma MM. Phytofabrication of nanoparticles through plant as nanofactories. Adv Nat Sci: Nanosci Nanotech 2014; 5(4): 043002.
[http://dx.doi.org/10.1088/2043-6262/5/4/043002]

[100]   Sharma G, Sharma AR, Bhavesh R, *et al.* Biomolecule-mediated synthesis of selenium nanoparticles using dried *Vitis vinifera* (raisin) extract. Molecules 2014; 19(3): 2761-70.
[http://dx.doi.org/10.3390/molecules19032761] [PMID: 24583881]

[101]   Fardsadegh B, Jafarizadeh-Malmiri H. Aloe vera leaf extract mediated green synthesis of selenium nanoparticles and assessment of their in vitro antimicrobial activity against spoilage fungi and pathogenic bacteria strains. Green Process Syn 2019; 8(1): 399-407.
[http://dx.doi.org/10.1515/gps-2019-0007]

[102]   Fardsadegh B, Vaghari H, Mohammad-Jafari R, Najian Y, Jafarizadeh-Malmiri H. Biosynthesis, characterization and antimicrobial activities assessment of fabricated selenium nanoparticles using *Pelargonium zonale* leaf extract. Green Process Syn 2019; 8(1): 191-8.
[http://dx.doi.org/10.1515/gps-2018-0060]

[103]   Anu K, Devanesan S, Prasanth R, AlSalhi MS, Ajithkumar S, Singaravelu G. Biogenesis of selenium nanoparticles and their anti-leukemia activity. J King Saud Univ Sci 2020; 32(4): 2520-6.
[http://dx.doi.org/10.1016/j.jksus.2020.04.018]

[104]   Kirupagaran R, Saritha A, Bhuvaneswari S. Green synthesis of selenium nanoparticles from leaf and stem extract of *Leucas lavandulifolia* sm. and their application. J Nanosci Technol 2016; pp. 224-6.

[105]   Vyas J, Rana S. Synthesis of selenium nanoparticles using *Allium sativum* extract and analysis of their antimicrobial property against gram positive bacteria. Pharma Innov Int J 2018; 7(9): 262-6. a

[106]   Vyas J, Rana S. Antioxidant activity and green synthesis of selenium nanoparticles using *Allium sativum* extract. Int J Phytomed 2017; 9(4): 634. a
[http://dx.doi.org/10.5138/09750185.2185]

[107]   Vyas J, Rana S. Biosynthesis of selenium nanoparticles using *Aloe vera* leaf extract. Int J Adv Res (Indore) 2018; 6(1): 104-10. b
[http://dx.doi.org/10.21474/IJAR01/6191]

[108]   Vyas J, Rana S. Antioxidant activity and biogenic synthesis of selenium nanoparticles using the leaf extract of *Aloe vera.* Int J Curr Pharm Res 2017; 9(4): 147-52. b
[http://dx.doi.org/10.22159/ijcpr.2017v9i4.20981]

[109]   Alagesan V, Venugopal S. Green synthesis of selenium nanoparticle using leaves extract of *withania somnifera* and its biological applications and photocatalytic activities. Bionanoscience 2019; 9(1): 105-16.
[http://dx.doi.org/10.1007/s12668-018-0566-8]

[110] Zhang W, Zhang J, Ding D, *et al.* Synthesis and antioxidant properties of *Lycium barbarum* polysaccharides capped selenium nanoparticles using tea extract. Artif Cells Nanomed Biotechnol 2018; 46(7): 1463-70.
[http://dx.doi.org/10.1080/21691401.2017.1373657] [PMID: 28880681]

[111] Deepa B, Ganesan V. Biogenic synthesis and characterization of selenium nanoparticles using the flower of *Bougainvillea spectabilis* Willd. Int J Sci Res 2015; 4: 690-5. a

[112] Deepa B, Ganesan V. Bioinspiredsynthesis of selenium nanoparticles using flowers of *Catharanthus roseus* (L.) G. Don. and *Peltophorum pterocarpum* (DC.) Backer ex Heyne–a comparison. Int J Chem Technol Res 2015; 7: 725-33. b

[113] Chandramohan S, Sundar K, Muthukumaran A. Hollow selenium nanoparticles from potato extract and investigation of its biological properties and developmental toxicity in zebrafish embryos. IET Nanobiotechnol 2018; 13(3): 275-81.

[114] Sowndarya P, Ramkumar G, Shivakumar MS. Green synthesis of selenium nanoparticles conjugated *Clausena dentata* plant leaf extract and their insecticidal potential against mosquito vectors. Artif Cells Nanomed Biotechnol 2017; 45(8): 1490-5.
[http://dx.doi.org/10.1080/21691401.2016.1252383] [PMID: 27832715]

[115] Ramamurthy Ch, Sampath KS, Arunkumar P, *et al.* Green synthesis and characterization of selenium nanoparticles and its augmented cytotoxicity with doxorubicin on cancer cells. Bioprocess Biosyst Eng 2013; 36(8): 1131-9.
[http://dx.doi.org/10.1007/s00449-012-0867-1] [PMID: 23446776]

[116] Gunti L, Dass RS, Kalagatur NK. Phytofabrication of selenium nanoparticles from *Emblica officinalis* fruit extract and exploring its biopotential applications: antioxidant, antimicrobial, and biocompatibility. Front Microbiol 2019; 10: 931.
[http://dx.doi.org/10.3389/fmicb.2019.00931] [PMID: 31114564]

[117] Fritea L, Laslo V, Cavalu S, Costea T, Vicas SI. Green biosynthesis of selenium nanoparticles using parsley (*Petroselinum crispum*) leaves extract. Studia Universitatis "Vasile Goldis" Arad. Seria Stiintele Vietii 2017; 27(3): 203-8. [Life Sciences Series].

[118] Hassanien R, Abed-Elmageed AA, Husein DZ. Eco-friendly approach to synthesize selenium nanoparticles: Photocatalytic degradation of Sunset yellow azo dye and anticancer activity. ChemistrySelect 2019; 4(31): 9018-26.
[http://dx.doi.org/10.1002/slct.201901267]

[119] Habibi G, Aleyasin Y. Green synthesis of Se nanoparticles and its effect on salt tolerance of barley plants. Int J Nanodimens 2020; 11(2): 145-57.

[120] Cui D, Liang T, Sun L, *et al.* Green synthesis of selenium nanoparticles with extract of hawthorn fruit induced HepG2 cells apoptosis. Pharm Biol 2018; 56(1): 528-34.
[http://dx.doi.org/10.1080/13880209.2018.1510974] [PMID: 30387372]

[121] Zeebaree SYS, Zeebaree AYS, Zebari OIH. Diagnosis of the multiple effect of selenium nanoparticles decorated by Asteriscus graveolens components in inhibiting HepG2 cell proliferation. Sustainable Chem Pharm 2020; 15: 100210.

[122] Kokila K, Elavarasan N, Sujatha V. *Diospyros montana* leaf extract-mediated synthesis of selenium nanoparticles and their biological applications. New J Chem 2017; 41(15): 7481-90.
[http://dx.doi.org/10.1039/C7NJ01124E]

[123] Tripathi RM, Hameed P, Rao RP, Shrivastava N, Mittal J, Mohapatra S. Biosynthesis of highly stable fluorescent selenium nanoparticles and the evaluation of their photocatalytic degradation of dye. Bionanoscience 2020; 10(2): 389-96.
[http://dx.doi.org/10.1007/s12668-020-00718-0]

[124] Sivakumar C, Jeganathan K. In-vitro cytotoxicity of Java tea mediated selenium nanoballs against L6 cell lines. J Drug Deliv Ther 2018; 8(6): 195-200.

[http://dx.doi.org/10.22270/jddt.v8i6.2046]

[125] Krishnan V, Loganathan C, Thayumanavan P. Green synthesized selenium nanoparticles using *Spermacoce hispida* as carrier of s-allyl glutathione: to accomplish hepatoprotective and nephroprotective activity against acetaminophen toxicity. Artif Cells Nanomed Biotechnol 2019; 47(1): 56-63.
[http://dx.doi.org/10.1080/21691401.2018.1543192] [PMID: 30669860]

[126] Vennila K, Chitra L, Balagurunathan R, Palvannan T. Comparison of biological activities of selenium and silver nanoparticles attached with bioactive phytoconstituents: green synthesized using *Spermacoce hispida* extract. Adv Nat Sci: Nanosci Nanotechnol 2018; 9(1): 015005.
[http://dx.doi.org/10.1088/2043-6254/aa9f4d]

[127] Agamuthu P. http://www.env.go.jp/recycle/3r/en/forum_asia/results/pdf/20091111/08.pdf [Accessed on 20th Jan 2020]

[128] Neh A. Agricultural Waste Management System [AWMS] in Malaysian. Open Access J Waste Management & Xenobiotics.
[http://dx.doi.org/https://doi.org/10.23880/oajwx-16000140]

[129] Kim H, Kim H, Mosaddik A, Gyawali R, Ahn KS, Cho SK. Induction of apoptosis by ethanolic extract of mango peel and comparative analysis of the chemical constitutes of mango peel and flesh. Food Chem 2012; 133(2): 416-22.
[http://dx.doi.org/10.1016/j.foodchem.2012.01.053]

[130] Heim KE, Tagliaferro AR, Bobilya DJ. Flavonoid antioxidants: chemistry, metabolism and structure-activity relationships. J Nutr Biochem 2002; 13(10): 572-84.
[http://dx.doi.org/10.1016/S0955-2863(02)00208-5] [PMID: 12550068]

[131] Goud KG, Veldurthi NK, Vithal M, Reddy G. Characterization and evaluation of biological and photocatalytic activities of selenium nanoparticles synthesized using yeast fermented broth. Appl Nanomed 2016; 1(1): 12-9.

[132] Mellinas C, Jiménez A, Garrigós MDC. Microwave-assisted green synthesis and antioxidant activity of selenium nanoparticles using *Theobroma cacao* L. Bean shell extract. Molecules 2019; 24(22): 4048.
[http://dx.doi.org/10.3390/molecules24224048] [PMID: 31717413]

[133] Sasidharan S, Sowmiya R, Balakrishnaraja R. Biosynthesis of selenium nanoparticles using *Citrus reticulata* peel extract. World J Pharm Res 2014; 4: 1322-30.

[134] Thilagavathi T, Kathiravan G, Srinivasan K. Antioxidant activity and synthesis of silver nanoparticles using the leaf extract of *Limonia acidissima*. Int J Pharma Bio Sci 2016; 7(4): 201-5.
[http://dx.doi.org/10.22376/ijpbs.2016.7.4.b201-205]

[135] Horky P, Ruttkay-Nedecky B, Nejdl L, *et al.* Electrochemical methods for study of influence of selenium nanoparticles on antioxidant status of rats. Int J Electrochem Sci 2016; 11(4): 2799-824.
[http://dx.doi.org/10.20964/110402799]

[136] Vicas SI, Cavalu S, Laslo V, Costea TO, Moldovan L. Effect of nano-Se particles supply on assimilator pigments, phenols content and antioxidant capacity of *Broccoli sprouts*. Wulfenia 2017; 24: 37-46.

[137] Qi Y, Yi P, He T, *et al.* Quercetin-loaded selenium nanoparticles inhibit amyloid-β aggregation and exhibit antioxidant activity. Colloids and Surfaces A: Physicochem Eng Aspects 2020; 125058.
[http://dx.doi.org/10.1016/j.colsurfa.2020.125058]

[138] Hamza RZ, Al-Motaani SE, Malik N. Protective and antioxidant role of selenium nanoparticles and vitamin C against acrylamide induced hepatotoxicity in male mice. Int J Pharmacol 2019; 15(6): 664-74.
[http://dx.doi.org/10.3923/ijp.2019.664.674]

[139] Zonaro E, Lampis S, Turner RJ, Qazi SJ, Vallini G. Biogenic selenium and tellurium nanoparticles

synthesized by environmental microbial isolates efficaciously inhibit bacterial planktonic cultures and biofilms. Front Microbiol 2015; 6: 584.
[http://dx.doi.org/10.3389/fmicb.2015.00584] [PMID: 26136728]

[140]   Kheradmand E, Rafii F, Yazdi MH, Sepahi AA, Shahverdi AR, Oveisi MR. The antimicrobial effects of selenium nanoparticle-enriched probiotics and their fermented broth against *Candida albicans.* Daru 2014; 22(1): 48.
[http://dx.doi.org/10.1186/2008-2231-22-48] [PMID: 24906455]

[141]   Rameshthangam P, Chitra JP. Synergistic anticancer effect of green synthesized nickel nanoparticles and quercetin extracted from *Ocimum sanctum* leaf extract. J Mater Sci Technol 2018; 34(3): 508-22.
[http://dx.doi.org/10.1016/j.jmst.2017.01.004]

[142]   Ahmad MS, Yasser MM, Sholkamy EN, Ali AM, Mehanni MM. Anticancer activity of biostabilized selenium nanorods synthesized by *Streptomyces bikiniensis* strain Ess_amA-1. Int J Nanomedicine 2015; 10: 3389-401.
[PMID: 26005349]

[143]   Sonkusre P, Cameotra SS. Biogenic selenium nanoparticles induce ROS-mediated necroptosis in PC-3 cancer cells through TNF activation. J Nanobiotechnology 2017; 15(1): 43.
[http://dx.doi.org/10.1186/s12951-017-0276-3] [PMID: 28592284]

[144]   Patel N, Kaler A, Jain S, Chand Banerjee U. Biosynthesis of selenium nanoparticle by whole cells of *Saccharomyces boulardii* and its evaluation as anticancer agent. Curr Nanosci 2013; 9(4): 463-8.
[http://dx.doi.org/10.2174/15734137113099990058]

[145]   Keyhani A, Shakibaie M, Mahmoudvand H, *et al.* Prophylactic activity of biogenic selenium nanoparticles against chronic *Toxoplasma gondii* infection. Recent Patents Anti-Infect Drug Disc 2020; 15(1): 75-84.
[http://dx.doi.org/10.2174/1574891X15666200604115001] [PMID: 32496989]

[146]   Veldurthi NK, Palla S, Velchuri R, Guduru P, Muga V. Degradation of mixed dyes in aqueous wastewater using a novel visible light driven LiMgO. 5MnO. 5O$_2$ photocatalyst. Mater Express 2015; 5(5): 445-50.
[http://dx.doi.org/10.1166/mex.2015.1255]

[147]   Sawant VJ, Sawant VJ. Biogenic capped selenium nano rods as naked eye and selective hydrogen peroxide spectrometric sensor. Sens Biosensing Res 2020; 27: 100314.
[http://dx.doi.org/10.1016/j.sbsr.2019.100314]

[148]   Xia X, Zhou Z, Wu S, Wang D, Zheng S, Wang G. Adsorption removal of multiple dyes using biogenic selenium nanoparticles from an *Escherichia coli* strain overexpressed selenite reductase CsrF. Nanomaterials (Basel) 2018; 8(4): 234.
[http://dx.doi.org/10.3390/nano8040234] [PMID: 29649129]

[149]   Prasad KS, Selvaraj K. Biogenic synthesis of selenium nanoparticles and their effect on As(III)-induced toxicity on human lymphocytes. Biol Trace Elem Res 2014; 157(3): 275-83.
[http://dx.doi.org/10.1007/s12011-014-9891-0] [PMID: 24469678]

[150]   Shakibaie M, Shahverdi AR, Faramarzi MA, Hassanzadeh GR, Rahimi HR, Sabzevari O. Acute and subacute toxicity of novel biogenic selenium nanoparticles in mice. Pharm Biol 2013; 51(1): 58-63.
[http://dx.doi.org/10.3109/13880209.2012.710241] [PMID: 23035822]

CHAPTER 5

Green Synthesis of Nanoparticles using Fungal Extracts

Krishan K. Selwal[1,*], Garima Deswal[1], Harsha Nirvan[1] and **Manjit K. Selwal[2]**

[1] *Department of Biotechnology, Deenbandhu Chhotu Ram University of Science & Technology, Murthal (Sonepat), Haryana, India*

[2] *DST Women Scientist Scheme (WoS)-B, Deenbandhu Chhotu Ram University of Science & Technology, Murthal (Sonepat), Haryana, India*

Abstract: Nanotechnology involves the synthesis of nanoparticles (NPs) and paved the way for the possibility of applications in different fields such as pharmaceutical science, industry, environment and biosensor technology. The metal nanoparticles synthesis using fungal extract is gaining momentum due to their novel chemical, optical, electrical, and magnetic properties. The mycelial biomass is found to be more resistant against pH, temperature, agitation and pressure compared to bacterial and plant extract and thus more appropriate for industrial production. The nano-sized particles synthesized by green chemistry are of better quality than the ones made by chemical reduction methods such as laser ablation, metallic wire explosion, photochemical or radiation reduction and sonochemical method. The chemical methods can pose a risk to environmental and animal health due to release of the hazardous toxic component. Therefore, nanoparticles synthesis using fungal extract could be an ecofriendly alternative to chemical-based methods as green synthesis has the lesser possibility of such component release. The fungal extract comprises a plethora of secreted extracellular proteins, enzymes, vitamins and ions which are responsible for the reduction and stability of nano-size metallic particles. The biogenic nanoparticles thus produced have attained much interest due to their composition, shape and size, photochemical, optical and chemical properties. The nanomaterials have applications in various fields such as biosensor technology, DNA based techniques, metabolomics, antimicrobial agents, cancer cell treatment, protein engineering, purification of water and degradation of pesticides, synthetic biology, downstream processing and delivery of therapeutic compounds.

Keywords: Biogenic synthesis, FTIR, Fungal Extract, Fungi, Metal nanoparticles, TEM, XRD.

---

[*] **Corresponding Author Krishan K. Selwal**: Department of Biotechnology, Deenbandhu Chhotu Ram University of Science & Technology, Murthal (Sonepat), Haryana, India; Tel: +91-9466069167; Fax: 0130-2484005; Email: krishan.kselwal@gmail.com

Savita, Anju Srivastava, Reena Jain & Pratap Kumar Pati (Eds.)

INTRODUCTION

The concept of nanotechnology, though considered to be a modern science, has its history dating as back as the 9th century. Nanotechnology involves the production of metal nanoparticles (NPs), nowadays is a vastly studied area because of their photoelectrochemical, chemical, optical and electronic properties, particularly in noble metals such as silver, gold and copper [1]. Nanoparticles of gold and silver were used by the artisans of Mesopotamia to generate a glittering effect on pots. The foremost scientific description of nanoparticles properties was reported in 1857 by Michael Faraday in his famous article "Experimental relations of gold (and other metals) to light" [2]. The revolution of nano work conceptually started in the early 1980s, with the first paper on nanotechnology being published in 1981 by K. Eric Drexler of Space Systems Laboratory, Massachusetts Institute of Technology [3]. Nanoparticles are particulate dispersions or solid particles with a size in the range of 10-1000nm. Nanotechnology is being utilized in offering many new developments in the fields of biosensors, biomedicine, bionanotechnology diagnosis, therapeutic drug delivery and the development of treatments for many diseases and disorders [4].

Recently there are various biological approaches of using fungal, plant or bacterial extract to reduce precursor metallic ions to elemental state. Fungal species specifically are known to be widely used as a promoter for the synthesis of nanoparticles because of high biomass, the large amount of enzyme production and easy handling. Their capability of transforming or degrading complex hazardous environmental compounds renders them, excellent bioremediation agents. The NPs synthesized by biological materials possess various unique properties and thus have many advantages over other methods that produce different shapes, sizes and voids. This chapter is mainly devoted to the preparation of various nanoparticles using fungi as biological materials and their applications.

APPROACHES FOR SYNTHESIS OF METAL NANOPARTICLES

The development of reliable/suitable processes for the synthesis of nanoparticles is an utmost aspect of current nanotechnology research. Both physical and chemical methods have been commonly used for various types of nanoparticles synthesis with particular sizes and shapes (Fig. 1). Despite the high efficiency of chemical methods in producing NPs, they generally release toxic and hazardous pollutants in the environment during NPs synthesis process. Besides these chemical methods, the bottom-up or top-down techniques are generally used for NPs synthesis. The former approach entails mechanical breakdown or grinding of the larger piece of metallic material to mini scale particles, whereas the latter approach comprises the biological or chemical mediated synthesis of the nanostructure. The above-mentioned processes cause controlled condensation of solute molecules that formed NPs of desired shape and size. The following are the examples of commonly used methods for nanomaterial synthesis such as thermal synthesis, sputtering, Inert gas condensation, lithography, laser ablation, aerosol, radiolysis, reduction, sol gel technique, *etc*.

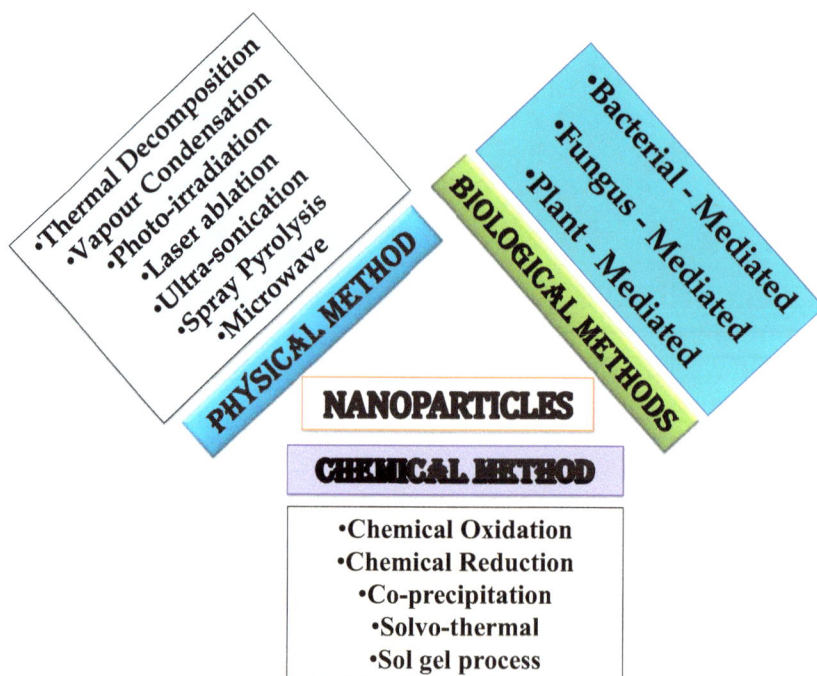

Fig. (1). Different methods for NPs synthesis.

Evaporation-Condensation

Various kinds of metal nanomaterials such as Ag, Au, PbS, WO_3, SnO_2, ZnO and fullerene, have previously been produced using the evaporation/condensation method [5, 6]. The technique involves the organic or inorganic material evaporated at a very high temperature usually in the furnace and the gaseous phase would be converted into homogenous nucleation by coalescence of atoms. The vapour pressure and pressure of the gas will determine the size and stability of NPs [7]. This method of synthesis of various kinds of NPs has several shortcomings, such as the requirement of large space for adjustment of the furnace, requiring more consumption of electricity (~10 kilowatts) for preheating for a long time to attain a stable operating temperature. Apart from this, the agglomeration of NPs is the major problem faced by the method [8].

Sputtering

Another similar technique that involves evaporation and condensation approach is known as sputtering, which comprises vaporizing materials from a solid surface by bombarding a target with high-velocity ions of an inert gas, which causes the ejection of atoms and clusters. Previously Gunther and Kumpmann [9] utilized an electron beam of an inert gas with pressure up to 5m bar to generate Al_2O_3 and SiO_2 particles of 5-10 nm sizes. The main advantage of this technique is the non-deviation of target material and the composition of sputtered material from the target material.

Inert Gas Method

This technique is very common for nanoparticles synthesis and was developed in Japan in 1960. The method synthesizes NPs by evaporation of substrate and then subsequently condensation to produce required size particles normally in the range of 10-100nm. An inert gas such as Helium or Argon is introduced in the chamber with a pressure of 1-40 Torr, leading to the formation of NPs.

Laser Ablation

The process involves removing material from a solid surface by irradiating with a laser beam. The material is heated by absorbed laser energy at low laser flux which evaporates or sublimates the material. The same is converted to plasma at higher laser flux. Carbon nanotubes can be produced by this method. Other investigators also used this technique for the production of silver nanoparticles and carbon nanotubes [10, 11]. The optical properties of metallic materials and laser wavelength affect the amount of material utilized and the depth over which laser energy is observed [12]. The complete absence of chemical constituents in solution is the main advantage of laser ablation when compared to other conventional methods.

Chemical Method

Chemical reduction is the most frequently applied method for the preparation of metallic NPs as stable, colloidal dispersions in water or organic solvents. The process requires a reaction between a substrate, reducing agent and stabilizing agents resulting in the formation of uniform shape, size and stable nanoparticles. Commonly used reducing agents are ascorbate, elemental hydrogen, N, N-dimethylformamide (DMF), sodium borohydride ($NaBH_4$), polyol process, Tollens reagent, sodium citrate, and Polyvinylpyrrolidone (PVP) with ethylene glycol andoleyl amine-liquid paraffin system. These chemical species will reduce metallic ions such as silver ions (Ag+) in an aqueous solution into colloidal silver with particle diameters of several nanometers. The colloidal silver was then eventually converted into colloidal silver metals ranging from monodisperse to spherical. Protective agents such as polyols may also be added into the solution to protect and stabilize the NPs. Various chemical methods employed for NP synthesis include the Tollen method, photo-induced method, irradiation method, microwave-assisted method, electrochemical synthetic method, and micro-emulsion technique and discussion on these are beyond the scope of the topic.

DRAWBACKS OF PHYSICAL AND CHEMICAL METHODS

The physical and chemical processes for nanoparticles synthesis have several advantages such as large-scale production with no purification is required. However, all these methods are high energy-consuming, costly and laborious. The use of these synthesis methods requires both strong and weak reducing agents and protective agents like sodium borohydride, sodium citrate and alcohols. These mostly used agents are toxic, flammable, cannot be easily degraded which pose a serious environmental issue. For instance, wave assisted methods used surfactants in coordination with ultrasonic waves for NPs production. Furthermore, these methods produced nanoparticles of undesired size, required special equipment and needed stabilizing agents to prevent the NPs from agglomeration. So, there is a necessity to find out a green technology comprising microorganisms or plant extracts or fungal species for the synthesis of NPs.

GREEN METHODS FOR NPS PRODUCTION

Though synthesis of nanoparticles by physical and chemical methods offers well distinct and pure nanoparticles, these NPs are quite expensive and may not be environment friendly [13]. The use of microorganisms and plant materials could be an alternative to the physical and chemical synthesis of nanoparticles [14]. Nature's "Bio-laboratory" that includes bacteria, algae, fungi, plants, *etc.* provide precursors for the synthesis of nanoparticles. The naturally occurring biomolecules of these biological matters have been used in the formation of NPs of different shapes, sizes and compositions with unique physiochemical properties. This route of nanoparticles synthesis is relatively simple, environment friendly with cost-effective and minimum energy requirement [15 - 17].

FUNGAL ASSISTED NPS SYNTHESIS

Fungi are generally categorized into decomposer groups present on earth and recent techniques of molecular biology such as high-throughput sequencing methods estimated that approximately 5.1 million fungal species are found. Fungi are well known for a large amount of enzymes production and so widely used in various degradation studies as their ability to transform or mineralize a wide range of environmentally hazardous compounds. Fungi can be simply cultivated on a minimal medium for a large scale and able to produce nanosize elements with controlled size and morphology [18]. Fungi are preferred over other microorganisms for NPs synthesis as they produce large quantities of proteins and enzymes used for the fast and sustainable synthesis of NPs. Furthermore, the fungal species have been studied as promoters for the synthesis of nanoparticles due to their high biomass production and easy handling. Apart from the above, fungal species require a simple growth medium, culture procedure, tolerance to the high concentration of NPs and improved stability of high concentration of nanoparticles [19]. The cell wall of fungus provides mechanical strength to overcome osmotic pressure and environmental fluctuation and it is easily scaled up without cell disruption. The mycelia provide a large surface area for interaction and the enzymes and proteins secreted by fungi act as reducing agents for the synthesis of nanoparticles. There are several fungi reported for nanoparticles synthesis of different shapes and sizes, as shown in Table **1**. Generally, the fungi-mediated NPs synthesis process categorizes into two categories: *in-vivo* and *in-vitro* methods. In an *"in-vivo"* method, the nanoparticles synthesis occur intracellularly and the synthesized NPs required chemical treatment for breakage of the cell wall, centrifugation and filtration to collect released particles [20].While in the case of the *"in-vitro"* method, the mycelia were first grown, washed and the fungal cell-free extract is prepared for NPs formation. The extract containing biomolecules is normally mixed with precursor metal ions solution resulting in the formation of particles of nano size. The process of nanoparticle synthesis employing fungal species comprises the following steps as depicted below:

Table 1. Metal Nanoparticles synthesis by different fungi.

Fungi	Nanoparticle	Shape	Production site	Size min and max (nm)	References
Schizosaccharomyces pombe	Cadminumsulphide	hexagonal	Intracellular	1-1.5	[42]
Fusarium oxysporum	Cadminumsulphide	spherical	Extracellular	5-20	[43]
Colletotrichum sp	Gold	spherical	Extracellular	8-40	[44]
Fusarium oxysporum	Gold	nanotriangular	Extracellular	50-100	[44]
Trichotheciumsp	Gold	Spherical, rods, triangular	Extracellular and Intracellular	5-200	[45]
Aspergillus flavus	Silver	spherical	Extracellular	5-25	[46]
Phanerochaetechrysosporium	Silver	Face centred cubic	Extracellular	10-34	[47]

(Table 1) cont.....

Fungi	Nanoparticle	Shape	Production site	Size min and max (nm)	References
Verticillumluteoalbum	Gold	All shapes	Intracellular	Upto100	[48]
Schizosaccharomyces pombe	Cadminumsulphide	Quantum dots	Intracellular	1-2	[49]
Aspergillus niger	Gold	Nanowalls, spiral	Extracellular	12.8-20	[50]
Fusarium semitectum	Gold	spherical	Extracellular	18-80	[51]
Corious versicolor	Gold	Spherical, elliptical	Extracellular and intracellular	20-300	[52]
Cochlibolus lunatus	Silver	spherical	Extracellular	3-21	[53]
Neurospora crassa	Silver	spherical	Extracellular and intracellular	11	[54]
Aspergillus clavatus	Gold	Triangular, spherical, hexagonal	Extracellular	24.4	[55]
Epicoccum nigrum	Gold	Spherical, rods	Extracellular	5-50	[56]
Aspergillus terreus	Silver	spherical	Extracellular	1-20	[57]
Rhizopousoryzae	Gold	Hexagonal, pentagonal	Extracellular	-	[58]
Aspergillus flavus	Titanium dioxide	Spherical, oval	Extracellular	62-74	[59]
Saccharomyces boulardii	silver	spherical	Extracellular	3-10	[60]
Cylindrocladiumfloridanum	Silver	spherical	Extracellular	25	[61]
Epicoccum nigrum	Silver	spherical	Extracellular	1-22	[62]
Aspergillus sp	Gold	Triangular, spherical, hexagonal	Extracellular and intracellular	17-26	[63]
Aspergillus oryzae	Iron	spherical	Extracellular	10-24.6	[64]
Aspergillus fumigatus	ZnO	Spherical and hexagonal	Extracellular	1.2-6.8	[65]
Macrophominaphaseolina	Silver	spherical	Extracellular	5-40	[66]
Penicillium chrysogenum	Silver	Spherical, polyhedric	Extracellular	100.6	[67]
Trichoderma harzianum	Silver	oval	Extracellular	6.78	[68]
Botrytis cinerea	Gold	Triangular, Hexagonal, Spherical, Pyramidal	Extracellular	1-100	[69]

(Table 1) cont.....

Fungi	Nanoparticle	Shape	Production site	Size min and max (nm)	References
Alternaria alternata	ZnO		Extracellular	45-150	[70]
Aspergillus tubingensis, Aspergillus flavus	MgO	spherical	Extracellular	1-5.8	[71]
Phanerochaetechrysosporium	Cadmium sulphide	-	Extracellular		[72]
Guignardiamangiferae	Silver	spherical	Extracellular	5-30	[73]
Monascuspurpureus	Silver	Spherical, cuboidal	Extracellular	2-8	[74]
Penicillium expansum	Silver	-	Extracellular	14-25	[75]
Aspergillus terreus	Gold	various	Extracellular	10-50	[76]
Rhizopousoryzae	Gold	spherical	Extracellular	16-25	[77]
Rhizopus Stolonifer	Silver	spherical	Extracellular	9.47	[39]
Duddingtonia flagrans	Silver	spherical	Extracellular	38	[78]
Penicillium aculeatum	Gold	spherical	Extracellular	35-85	[79]
Aspergillus niger	Zinc	spherical	Extracellular	53-69	[41]
Aspergillus nidulans	Cobalt oxide	-	Extracellular	20.29	[80]
Trichoderma longibrachiatum	Silver	-	Extracellular	-	[81]
Botryosphaeria rhodina	Silver	Spherical, rectangular, triangular	-	2-50	[82]
Aspergillus sp.	Zinc	Hexagonal	Extracellular	30-100	[83]
Fusarium solani	Gold	spindle	-	40-45	[84]
Periconium sp.	ZnO	Quasi-spherical	-	16-78	[85]
Alternaria tenuissima	Zinc	spherical	-	15.45	[86]
Tritirachium oryzae	Silver	Spherical, oval	Extracellular	7-75	[87]
Fusarium oxysporum	Gold	Triangular and spherical	Intracellular	20	[88]
Rhizoctonia solani	Gold	spherical	Intracellular	30	[88]
Verticillium dahliae	Gold	polygonal	Intracellular	40	
Aspergillus flavus	Gold	spherical	Intracellular	30	

FUNGAL EXTRACT PREPARATION

The first important step in nanomaterial synthesis is the fungal extract that contains a plethora of enzymes and metabolites present that reduce substrate ions into nanoscale metal particles. The fungal biomass production is achieved by incubating the mycelium culture on an appropriate medium such as Potato Dextrose agar or Czapek Dox agar. The medium should keep under constant shaking to maximize biomass production. The biomass generated is then harvested by filtration and washed with distilled water to remove any medium component. The specific quantity of biomass is mixed with double deionized water and agitated and incubated at 25°C for 24-72 hrs for release of compounds. After incubation, the fungal filtrate was obtained by passing through suitableWhatman filter paper. In our previous study, the extract of *Penicillium atramentosum* KM was prepared in the same fashion by mixing already washed fungal biomass in distilled water followed by incubation at 30°C for 72 h under constant agitation at a speed of 120 rpm. The fungal biomass was then separated using Whatman No. 1 filter paper and the fungal extract thus obtained is ready to use. There is a fair advantage of cell-free extract preparation when compared with bacterial extract preparation as the former provides good biomass production and do not require additional steps to extract the filtrate as in later case [21]. Similarly, the extract preparation is more cumbersome in plant cases as it also requires additional steps. Fungal mycelial biomass, on the other hand, is resistant to pressure and agitation and more apt for nanomaterial synthesis [22].

BIOSYNTHESIS MECHANISM

In the process of biosynthesis of NPs, reduction of precursor (mainly metal salts ion) by reducing agents (an enzyme or biological agent) normally results in accumulation ofreduced ions and formation of NPs. The synthesis of NPs in fungi may be intracellular or extracellular and the underlying mechanism of nanoparticles biosynthesis is yet to be fully elucidated. However, it is well known that the enzymes and biomolecules present in the fungal system can reduce the different precursor ions such as Ag, Cu, Fe, Co into elemental at a nanometric scale (Fig. **2**). Various investigators demonstrated that fungal extract contains diverse biomolecules like amino acids, proteins, peptides, cofactors, naphtho-quinones, anthraquinones, mono and polysaccharides, melanin and others that are collectively able to reduce different substrates into NPs [23 - 25]. Anil Kumar *et al.* [26] directly used the purified nitrate reductase from the organism *Fusarium oxysporum* for the synthesis of silver nanoparticles in the test tube. Their reaction mixture contained only the enzyme nitrate reductase, silver nitrate and NADPH. Slowly, the reaction mixture turned brown, indicating the formation of silver nanoparticles. This was the first direct evidence for the involvement of nitrate reductase in the synthesis of silver NPs.During the catalysis, nicotinamide adenine dinucleotide (NADH) and NADH-dependent nitrate reductase enzymes are considered to be most important in the biogenic synthesis of metallic NPs as nitrate is converted to nitrite, and an electron will be shuttled to the incoming silver ions. Some authors have suggested that the binding of substrate ions with fungal cell wall during synthesis, entrapment of nanoparticles on the cell wall of the fungus, the carboxyl group of amino acid residues and the amine of peptide chains, along with reducing groups like aldehyde and ketone could be the mechanism behind the bioreduction of precursor metal ions [27 - 29]. In a recent study by Molnar *et al.* [30] in which they selected filamentous fungi *Thermoascus thermophilus* ATCC 26413 for AuNPs synthesis. The culture supernatants from 29 fungal species were fractioned using Microsep™ Advance Centrifugal Devices to identify biomolecules responsible for gold NPs. The result indicated that low molecular weight biomolecules such as peptides, primary metabolites, mono- and oligosaccharides, cofactors are present in all fractions and responsible for reducing auricchloride ions. Furthermore, biomolecules over 3 kDa molecular weight secreted are not capable of reducing Au(III) to Au(0), whereas smaller size biomolecules under 3 kDa reduceAu(III) to Au(0) efficiently and produce AuNPs. Most proteins species are mainly reducing the substrate however no specific protein was identified for reduction.The fungal culture filtrate upon incubation with precursor solution *e.g.* silver nitrate ($AgNO_3$), auric chloride ($AuCl_2$), ferrous sulphate ($FeSO_4$), magnetite (Fe_3O_4) *etc.* turned yellow, purple, grey, brown or dark brown colour indicating the formation of NPs while the control without the substrate remains as such during the entire incubation period. Thechange of colour of the reaction mixture from colourless to dark brown colour is due to the surface plasmon resonance (SPR) exhibited by the nanoparticles.Further colour of the NPs solutions is dependent on the concentration of precursor ions and incubation period as increase the intensity of the colour varies when biogenic nanomaterial is exhibited.

Fig. (2). Fungal mediated synthesis of different metal Nanoparticles.

CHARACTERIZATION OF FUNGAL MEDIATED NANOPARTICLES

The methods for characterization of extracellularly synthesized NPs of various kinds under best-optimized conditions are by the following techniques:

- Ultra Violet-Visible (UV-Vis) spectrophotometer
- FTIR, XRD
- TEM
- SEM

The UV–Vis spectra at 450 nm confirmed the synthesis of NPs due to the excitation of free electrons which is responsible for surface plasmon resonance and might indicate the possible shape of generated particles. Various metals particles will have variable excitation and oscillation spectra and so-known excitation spectra for particular nanoparticles are normally used for identification of the specific nanoparticles using UV–Vis spectrophotometer. For instance, Tidke *et al*. [31] observed absorbance peak at 520 nm-550 nm corresponding to the surface plasmon absorbance of the gold nanoparticles using *Fusarium acuminatum*, Mohamad *et al*. [32] obtained a peak at 238nm for iron oxide NPs by *Alternaria alternata* and Kobashigawa *et al*. [33] noticed peak at 265 for AgNPs using fungus *Tramete strogii, etc.*

Fourier transform infrared (FT-IR) spectroscopy technique gives insight into the different functional groups present by measuring the vibrational frequencies of the chemical bonds involved. The vibrational excitation energy of molecules present in the range of 10^{13}–10^{14} Hz can be used for both quantitative and qualitative analysis of the molecule present. Surface adsorption of functional groups on nanoparticles and molecular data enables the investigators to deduce the structural and conformational changes of the coordinating self-assembled functional groups on the surface of nanoparticles. FTIR peak displayed different stretches of bonds at different peaks *e.g.* 1121.56—C=O stretch, 1637.58—C=C, 2071.75—C≡C, 2676.19—C–H; O–H, 3432.94—N–H stretch and 2777.28—single aldehyde. Priyadarshini *et al*. [34] synthesized AuNPs using *Aspergillus terreus* IF0 and obtained FTIR spectra with a broad contour in the range of 3600–3220 cm^{-1}that confirmed the presence of –OH groups whereas peak positioned at 2730 cm^{-1} pointed towards the –CH stretch of aldehydes. Similarly, Gudikandula *et al*. [35] used *Ganoderma enigmaticum* mediated AgNPs and observed peaks in the wavelength range of 3500–500 cm−1 that established amine vibration band at 3385 cm−1, which represents a primary amine (N-H) stretching and amide (N-H) bending vibration at 1638 cm−1. To further characterize the nanoparticles produced by fungal species, XRD technique is used based on Bragg's law and mainly gives valuable insight into the lattice structure of a crystalline substance such as chemical composition, bond angles, unit cell dimensions and crystallographic structure. Several authors used this technique to study the composition of the particles by comparing the intensity and position of the peaks with the standard one available from the International Centre for Diffraction Data (previously known as Joint Committee on Powder Diffraction Standards, JCPDS) database. Our previous studies explored stable silver nanoparticles synthesis with the size range of 5-25 nm using *Penicillium atramentosum* KM and *A. fumigatus* MA. The process parameters were optimized to produce stable AgNPs and showed antimicrobial activity against various enteric pathogens. XRD pattern depicted in Fig. (**3**) confirmed the crystalline nature of AgNPs [36, 37].

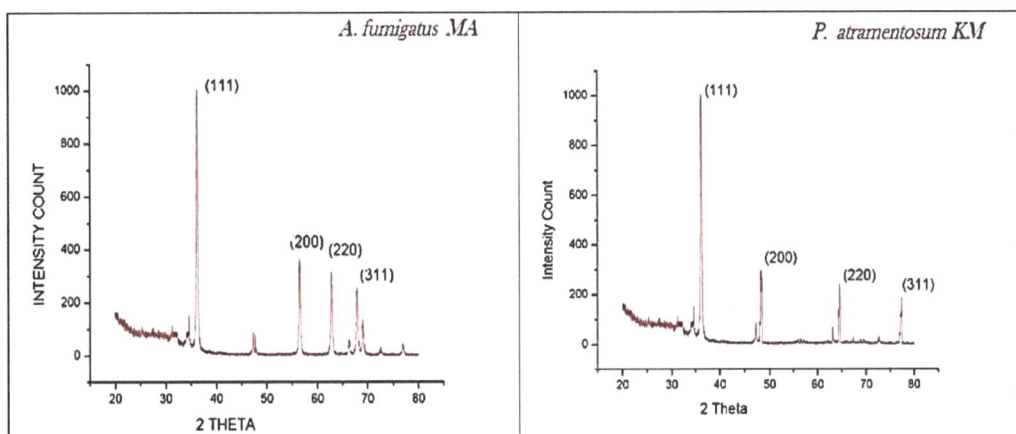

Fig. (3). XRD peaks of 111, 200, 220 and 311 nanoparticles synthesized by **a)** *Penicillium atramentosum* **b)** *Aspergillus fumigatus* MA implying crystalline nature of AgNPs.

The characterization of the size of nanomaterials is of utmost importance in nanotechnology as it exhibits size-dependent properties for example, the magnetic properties of iron nanoparticles, the optical properties of metallic nanoparticles, and optical properties of platinum particles or quantum dots, *etc.* which are different from their bulk constituent. Therefore, to elucidate the nanoparticles' structural size properties, fully electron microscopic techniques are used, namely atomic force microscopy (AFM), scanning electron microscopy (SEM), and transmission electron microscopy (TEM). All these above-mentioned techniques provided high-quality structural images apt for the measurement of sample dimensions. Various researchers applied these techniques for the measurement of the size of nanoparticles as the appropriate size has potential use in medical diagnostics, structural/material biology, and environmental applications. Feroze *et al*. [38] reported that 60-80 nm Ag-NPs were produced by *Penicillium oxalicum* fungus and spherical in size whereas Abdelrahim *et al*. [39] used *Rhizopus stolonifer* for AgNPs preparation and confirmed the presence of stable, spherical and mono-dispersed particles with a diameter around 9.47 nm and affirmed by high-resolution transmission electron microscopy (HR-TEM).SEM analysis of zinc oxide (ZnO) particles exhibited compactly arranged spherical shape within the size range of 40-75nm [40, 41]. Various other researchers successfully used SEM and TEM techniques to characterize the size and shape of finally synthesized NPs, as shown in Table **1**.

FACTORS AFFECTING FUNGAL MEDIATED NANOMATERIAL SYNTHESIS

There is always a continuous interaction between living organisms and the environment in which they live. The environmental conditions affect the growth and development of organisms and so the NPs production. The growth parameters such as incubation temperature, agitation, light, growth media and synthesis times vary depending on the fungal species used. The metabolite production by bacteria, fungi and plants is influenced by the conditions in which the organisms are cultivated. Therefore, optimization of conditions is very important to control as it will not only support good growth but also enhance desired NPs yield. Different studies have proved that fluctuation in temperature, precursor metal ion concentration, culture medium, pH, and amount of biomass affect the quality and quantity of the final product [89, 90]. Our previous studies also suggested that optimization of process parameters such as incubation temperature, pH, incubation time, the concentration of the metal ions, the concentration ratio of the extract and metal ions played an important role in increasing the productivity of AgNPs [37, 38, 91]. Shahzad *et al*. [92] used *Aspergillus fumigatus* BTCB10 strain and proved that 3:2 ratio of cell-free extract to silver nitrate solution (1mM $AgNO_3$) was appropriate and, when incubated at 25^0C produced AgNPs with spherical shape whereas Costa *et al*. [93] standardized temperature and pH conditions for AgNPs synthesis using *Duddingtonia flagans* and found that 60^0C and pH 10 is desired for spherical NPs production. In another study, it was shown that when 10g of *Trichoderma longibrachiatum* culture mixed with substrate metal ion and incubated at 28^0C temperature under without agitated condition produced suitable size silver NPs with the size of 24.43 nm [81]. Similarly, Mohamad *et al*. [32] exposed *A. alternata* fungal extract with standardized precursor salt iron (III) nitrate (1mM) solution at 28^0C in dark under shaking conditions synthesized cubic size particles with the size range of 5.5-12 nm. The amount of fungal biomass is reported to affect the synthesis of NPs as there is a relationship between the amount of biomass and the release of fungal proteins biomolecules. Rose *et al*. [94] reported higher production of AgNPs by employing higher biomass of *Penicillium oxalicum,* which was attributed to a higher release of nitrate reductase enzyme responsible for NPs synthesis whereas Balkumaran *et al.* [73] used 10, 20, 30g of *Guignardia mangiferae* fungal species biomass and obtained desired quality silver NPs with the lowest biomass. Tidke *et al.* [31] studied different parameters such as pH, temperature and salt concentration using *Fusarium acuminatum* and found that acidic pH, 1mM salt concentration and 37^0C incubation temperature was optimum for gold NPs synthesis.

APPLICATION OF NANOMATERIALS

Fungal synthesized NPs have been used in biomedical, agricultural, environmental and industrial fields due to their specific properties. Nanoparticles have been studied for their vital role in different fields, as shown in Fig. (**4**). A few emerging applications of nanoparticles are discussed below:

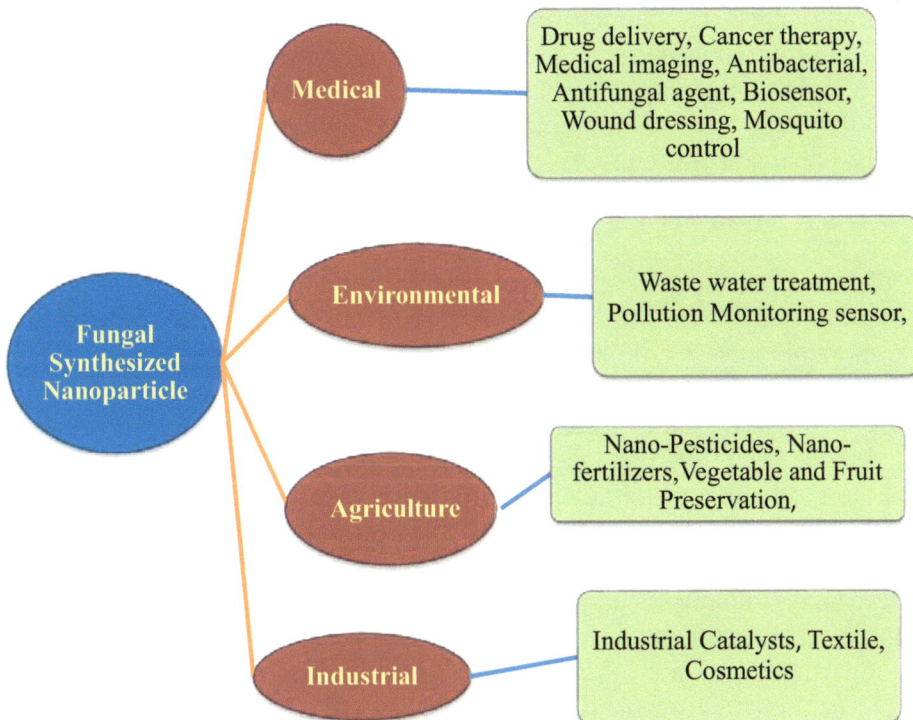

Fig. (4). Application of metal nanoparticles in various fields.

NANOPARTICLES AS DRUG DELIVERY AGENTS

Nanoparticles have been widely utilized in the pharmaceutical industry as a vehicle for drug delivery to the target organs [95]. Nanoparticles possess several advantages such as a small stable size than liposomes which are potential drug carriers. Moreover, nanoparticles have a longer useful life than liposomes and can entrap more drugs. Some studies showed that the number of nanoparticles that cross the intestinal epithelium is greater than that of the microspheres (>1μm) [96]. Polymer NPs from biodegradable and biocompatible polymers are the promising choice for delivery because they are expected to be adsorbed in an intact form in the gastrointestinal tract after oral administration [97, 98]. Biological approached nanoparticles can be employed as alternative drugs for the treatment of diabetes mellitus. Gold nanoparticles were synthesized by fungus *Trichoderma viride* with vancomycin that bound to the microbial surface by ionic interaction and suppressed the growth of vancomycin-resistant *Staphylococcus aureus* even at low concentration (approx. 8 μg/mL). The cell death of S. *aureus* was proven by TEM analysis, showing that the vancomycin-bound Au-NP had penetrated the bacterial membrane [99]. In another study, the researcher loaded doxorubicin into bacterial magnetosomes by using covalent attachment and which suppressed the tumour growth by 86.8% with the help of magnetosomes [100]. Brown *et al.* [101] investigated that gold nanoparticles functionalized with a thiolate polyethene glycol monolayer capped with a carboxylate group successfully enhanced the delivery of the anticancer drug oxaliplatin.

ANTICANCER ACTIVITY

Fungal NPs attract the special attention of researchers in the field of medicine and cancer therapy due to their specificity of cancerous cells. Drugs available for cancer treatment such as doxorubicin, cisplatin, vinblastine, bleomycin and daunorubicin are limited with poor specificity, high toxicity, high cost, and the emergence of drug-resistant diseases. Thus, new strategies for cancer treatment are needed to be discovered and nanotechnology plays a vital role in exploring a novel method of cancer therapy. AgNPs have been used in the screening, detection and imaging of tumours studded with biomarkers on nanoparticles for targeting tumours, drug delivery, and hyperthermia therapy [102]. PC3 cells, which are cancer-causing in the male reproduction system, were tested against silver nanoparticles synthesized by *P. djamor var. roseus*. The result showed that silver nanoparticles inhibited the PC3 cells proliferation by 10μg/ml IC_{50} in 24 hours. This occurs because nanoparticles may cause growth suppression, DNA condensation and apoptosis of PC3 cell line [103]. Several investigators also have used different cancerous cell lines such as B10F17 melanoma cell, MCF-7 human breast cancer cell line, SiHa cervical cancer cell line, A549 lung epithelial adenocarcinoma cell line, HeLa cell line to prove that biologically synthesized AgNPs revealed *in-vitro* as well as *in-vivo* antitumor activities [104, 105]. The action of nanomaterials on tumour cells may have occurred *via* DNA damage because AgNPs have been shown to negatively regulate the activity of DNA-dependent protein kinase, a key enzyme involved in DNA repair. In a recent study, Alalawy *et al.* [106] detected that *F. oxysporum* derived chitosan particles loaded with bee venom showed an increase in anticancer activity against HeLa cells (for cervix carcinoma) through induction of apoptosis and secondary necrosis markers. The activity of nanoconjugates is dose-dependent and most effective detected with 200µg/ml $IC_{50}.A. cladosporium$ synthesised gold nanoparticles showed activity against breast cancer cell line MCF-7 (IC50 38.23 µg/mL). Gold nanoparticles were tested on the Ehrlich Ascites Carcinoma mice model against tumour growth. They found nanoparticles induce apoptosis of Ehrlich Ascites Carcinoma cells. The body weight and ascites volume of mice decrease while life span increases [107].

ANTIBACTERIAL AND ANTIFUNGAL AGENTS

A silver nanoparticle is an effective antimicrobial agent against a broad spectrum of Gram-positive and negative bacterial pathogens, including antibiotic-resistant strains [108]. Metallic nanoparticles (*e.g.* silver, gold) synthesized using *Talaromyces purpureogenus, Phomopsis liquidambaris, Guignardia mangiferae, A.niger, Candida albicans, A. flavus, Trametes jubarsky, Ganoderma enigmaticum* are reported for antibacterial activity against gram positive and gram negative bacteria such *Klebsiella pneumonia, E. coli, Proteus vulgaris, P.aeroginosa, B. subtilis, Bacillus megaterium, Salmonella paratyphi* [16]. We have also tested antibacterial activity of AgNPs synthesized by *A. fumigatus* MA and *P. atramentosum* against pathogens such as *Staphylococcus aureus, Bacillus cereus, Micrococcus luteus, Salmonella typhimurium, Enterobacter aerogenes* and *Aeromonas hydrophilla*. The result indicated potential antimicrobial activity against mentioned pathogens [37].

The nanomaterials also exhibited effective fungicide activity against various fungal genera such as *Saccharomyces*, *Candida*, and *Aspergillus*. Nano formulations were tested to determine the inhibitory effect of fungal plant pathogens namely *B. cinerea, Curvularia lunata, Sclerotinia sclerotium, Macrophomina phaseolina, Rhizoctonia solani,* and *Alternaria alternata* [109]. The exact mechanism is not known to date, but it could be due to the large surface area for better contact with microorganisms. After attachment to the cell membrane, nanomaterial gets penetrated inside the bacteria. Silver nanoparticles release Ag+ ions inside the microbial cell, which may create free radicals and induce oxidative stress, thus further enhancing their bactericidal activity. Moreover, Sulphur containing proteins present in the cell membrane is the main target and nanoparticles attached with these could cause microbial cell disruption of protein and DNA [110 - 112].

MISCELLANEOUS APPLICATIONS

Metal nanoparticles are used in various ointments preparations such as skin gels for healing burns, cream for dermatophytes and antiseptic soap. These are used in electronic devices like LCDs, High-intensity LEDs and touch screens of various gadgets. Fungal mediated NPs showed wound healing activity. For instance, AgNPs from *Talaromyces purpureogenus* were found to be non-toxic to NIH3T3 normal cells and through microscopic analysis, it was confirmed that wound area treated with nanoparticles decreased after 48h when compared with the untreated group [113]. NPs have large number of applications in field of biological sensor [114], plasmonics [115], antimicrobial activities [116], catalysis [117], optoelectronics [118], surface-enhanced Raman scattering (SERS) [119] and DNA sequencing [120]. Nanoparticles are used in packaging material for foodstuff, treating contaminated water for absorption of pesticides, water purification and air disinfection.

CONCLUSION

It is a well known fact that fungal mediated synthesis of metal nanoparticles is a safe, non-hazardous and economical process as uniform shape, size and stable nanoparticles are produced. The fungal synthesis of nanoparticles has many advantages over bacteria, such as ease of culturing of fungal biomass for large scale production, high cell wall binding capacity, can grow on various inorganic substrates that lead to uniform distribution of nanoparticles. The fungal extract contains different enzymes and metabolites that facilitate the synthesis of NPs. However, the only drawback of this method is some remaining components of the fermentation media are present during the reaction that could destabilize the NPs. Different techniques such as FTIR, SEM, TEM are used for the characterization of the particles. The growth parameters like fungal species, growth medium, pH and fungal biomass play an important role in providing stability to the metal nanoparticles.

The NPs with appropriate size have a potential role in drug delivery, cancer therapy, gene treatment and DNA analysis, magnetic resonance imaging, catalysis, environmental sensing, textile engineering, food sectors, plant disease management, antibacterial factors, biosensors and MRI. Future work is required for the possible role in the control of pathogenic microorganisms as an alternative to antibiotic therapy.

CONSENT FOR PUBLICATION

Not applicable.

CONFLICT OF INTEREST

The authors declare no conflict of interest, financial or otherwise.

ACKNOWLEDGEMENT

Declared none.

REFERENCES

[1]   Zonooz NF, Salouti M. Extracellular biosynthesis of silver nanoparticles using cell filtrate of Streptomyces sp. ERI-3. Sci Iran 2011; 18(6): 1631-5.
[http://dx.doi.org/10.1016/j.scient.2011.11.029]

[2]   Faraday M. Experimental Relations o f Gold (and other Metals) to Light. Phil Trans R Soc 1857; 147(0): 145-81.

[3]   Drexler KE. Molecular engineering: An approach to the development of general capabilities for molecular manipulation. Proceedings of the National Academy of Sciences of the United States of America. 5275-8.
[http://dx.doi.org/10.1073/pnas.78.9.5275]

[4]   Rudge S, Peterson C, Vessely C, Koda J, Stevens S, Catterall L. Adsorption and desorption of chemotherapeutic drugs from a magnetically targeted carrier (MTC). J Control Release 2001; 74(1-3): 335-40.
[http://dx.doi.org/10.1016/S0168-3659(01)00344-3] [PMID: 11489515]

[5]   Kmis FE, Fissan H, Rellinghaus B. Sintering and evaporation characteristics of gas-phase synthesis of size-selected PbS nanoparticles. Materials Science and Engineering B. Solid-State Materials for Advanced Technology 2000; 69: 329-34.

[6]   Forster H, Wolfrum C, Peukert W. Experimental study of metal nanoparticle synthesis by an arc evaporation/condensation process. J Nanopart Res 2012; 14(7): 1-16.
[http://dx.doi.org/10.1007/s11051-012-0926-1]

[7]   Gurav AS, Duan Z, Wang L, Hampden-Smith MJ, Kodas TT. Synthesis of Fullerene-Rhodium Nanocomposites *via* Aerosol Decomposition. Chem Mater 1993; 5(2): 214-6.
[http://dx.doi.org/10.1021/cm00026a011]

[8]   Vodop yanov A V., Mansfeld DA, Samokhin A V., Alekseev N V., Tsvetkov Y V. Production of Nanopowders by the Evaporation–Condensation Method Using a Focused Microwave Radiation. Radiophys Quantum Electron 2017; 59(8–9): 698-705.

[9]   Gunther B, Kumpmann A. Ultrafine oxide powders prepared by inert gas evaporation. Nanostruct Mater 1992; 1(1): 27-30.
[http://dx.doi.org/10.1016/0965-9773(92)90047-2]

[10]   Chen Y, Yeh C. Ablation in salt, dispersed particles. 2002; 133-9.

[11]   Kabashin AV, Meunier M. Synthesis of colloidal nanoparticles during femtosecond laser ablation of gold in water. J Appl Phys 2003; 94(12): 7941-3.
[http://dx.doi.org/10.1063/1.1626793]

[12]   Mafune F, Kohno JY, Takeda Y, Kondow T, Sawabe H. Structure and stability of silver nanoparticles in aqueous solution produced by laser ablation. J Phys Chem B 2000; 104(35): 8336-7.
[http://dx.doi.org/10.1021/jp001803b]

[13]   Khan I, Saeed K, Khan I. Nanoparticles: Properties, applications and toxicities. Arab J Chem 2019; 12(7): 908-31.
[http://dx.doi.org/10.1016/j.arabjc.2017.05.011]

[14]   Mittal J, Batra A, Singh A, Sharma MM. Phytofabrication of nanoparticles through plant as nanofactories. Advances in Natural Sciences: Nanoscience and Nanotechnology 2014; 5(4)
[http://dx.doi.org/10.1088/2043-6262/5/4/043002]

[15]   Sharma D, Kanchi S, Bisetty K. Biogenic synthesis of nanoparticles: A review. Arab J Chem 2019; 12(8): 3576-600.
[http://dx.doi.org/10.1016/j.arabjc.2015.11.002]

[16]   Guilger-Casagrande M, de Lima R. Synthesis of Silver Nanoparticles Mediated by Fungi: A Review. Front Bioeng Biotechnol 2019; 7(October): 287.
[http://dx.doi.org/10.3389/fbioe.2019.00287] [PMID: 31696113]

[17]   Gholami-Shabani M, Akbarzadeh A, Norouzian D, *et al.* Antimicrobial activity and physical characterization of silver nanoparticles green synthesized using nitrate reductase from Fusarium oxysporum. Appl Biochem Biotechnol 2014; 172(8): 4084-98.
[http://dx.doi.org/10.1007/s12010-014-0809-2] [PMID: 24610039]

[18]   Azmath P, Baker S, Rakshith D, Satish S. Mycosynthesis of silver nanoparticles bearing antibacterial activity. Saudi Pharm J 2016; 24(2): 140-6.
[http://dx.doi.org/10.1016/j.jsps.2015.01.008] [PMID: 27013906]

[19]   Akther T, Vabeiryureilai Mathipi , Davoodbasha M, Srinivasan H, Srinivasan H. Fungal-mediated synthesis of pharmaceutically active silver nanoparticles and anticancer property against A549 cells through apoptosis. Environ Sci Pollut Res Int 2019; 26(13): 13649-57.
[http://dx.doi.org/10.1007/s11356-019-04718-w] [PMID: 30919178]

[20]   Rajput S, Werezuk R, Lange RM, McDermott MT. Fungal Isolate Optimized for Biogenesis of Silver Nanoparticles with Enhanced Colloidal Stability. Langmuir 2016; 32(34): 8688-97.
[http://dx.doi.org/10.1021/acs.langmuir.6b01813] [PMID: 27466012]

[21]   Gade AK, Bonde P, Ingle AP, Marcato PD, Durán N, Rai MK. Exploitation of Aspergillus niger for synthesis of silver nanoparticles. J Biobased Mater Bioenergy 2008; 2(3): 243-7.
[http://dx.doi.org/10.1166/jbmb.2008.401]

[22]   Velusamy P, Kumar GV, Jeyanthi V, Das J, Pachaiappan R. Bio-Inspired Green Nanoparticles: Synthesis, Mechanism, and Antibacterial Application. Toxicol Res 2016; 32(2): 95-102.
[http://dx.doi.org/10.5487/TR.2016.32.2.095] [PMID: 27123159]

[23]   Honary S, Barabadi H, Gharaei-fathabad E. Green Synthesis of Silver Nanoparticles Induced by the Fungus Penicillium citrinum. 2013; 12: 7-11.
[http://dx.doi.org/10.4314/tjpr.v12i1.2]

[24]   Leng Y, Fu L, Ye L, *et al.* Protein-directed synthesis of highly monodispersed, spherical gold nanoparticles and their applications in multidimensional sensing. 2016; 1-11.
[http://dx.doi.org/10.1038/srep28900]

[25]   Liu J, Peng Q. Protein-gold nanoparticle interactions and their possible impact on biomedical applications. Acta Biomater 2017; 55: 13-27.
[http://dx.doi.org/10.1016/j.actbio.2017.03.055] [PMID: 28377307]

[26]   Anil Kumar S, Abyaneh MK, Gosavi SW, *et al.* Nitrate reductase-mediated synthesis of silver nanoparticles from AgNO3. Biotechnol Lett 2007; 29(3): 439-45.
[http://dx.doi.org/10.1007/s10529-006-9256-7] [PMID: 17237973]

[27]   Durán N, Marcato PD, Alves OL, Souza GI, Esposito E. Mechanistic aspects of biosynthesis of silver nanoparticles by several Fusarium oxysporum strains. J Nanobiotechnology 2005; 3(1): 8.
[http://dx.doi.org/10.1186/1477-3155-3-8] [PMID: 16014167]

[28]   Velhal SG, Kulkarni SD, Latpate RV. Fungal mediated silver nanoparticle synthesis using robust experimental design and its application in cotton fabric. Int Nano Lett 2016; 6(4): 257-64.
[http://dx.doi.org/10.1007/s40089-016-0192-9]

[29] Mukherjee P, Ahmad A, Mandal D, *et al.* Fungus-mediated synthesis of silver nanoparticles and their immobilization in the mycelial matrix: a novel biological approach to nanoparticle synthesis. Nano Lett 2001; 1(10): 515-9.
[http://dx.doi.org/10.1021/nl0155274]

[30] Molnár Z, Bódai V, Szakacs G, *et al.* Green synthesis of gold nanoparticles by thermophilic filamentous fungi. Sci Rep 2018; 8(1): 3943.
[http://dx.doi.org/10.1038/s41598-018-22112-3] [PMID: 29500365]

[31] Tidke PR, Gupta I, Gade A, Rai M. Fungus- mediated synthesis of gold nanoparticles and standardization of parameters for its biosynthesis. 2014; 602

[32] Mohamed YM, Azzam AM, Amin BH, Safwat NA. Mycosynthesis of iron nanoparticles by Alternaria alternata and its antibacterial activity. Afr J Biotechnol 2015; 14(14): 1234-41.
[http://dx.doi.org/10.5897/AJB2014.14286]

[33] Kobashigawa JM, Robles CA, Martínez Ricci ML, Carmarán CC. Influence of strong bases on the synthesis of silver nanoparticles (AgNPs) using the ligninolytic fungi *Trametes trogii.* Saudi J Biol Sci 2019; 26(7): 1331-7.
[http://dx.doi.org/10.1016/j.sjbs.2018.09.006] [PMID: 31762592]

[34] Priyadarshini E, Pradhan N, Sukla LB, Panda PK. Controlled synthesis of gold nanoparticles using aspergillus terreus IF0 and its antibacterial potential against gram negative pathogenic bacteria. J Nanotechnol 2014; 2014: 1-9.
[http://dx.doi.org/10.1155/2014/653198]

[35] Gudikandula K, Vadapally P, Singara Charya MA. Biogenic synthesis of silver nanoparticles from white rot fungi: Their characterization and antibacterial studies. OpenNano 2017; 2: 64-78.
[http://dx.doi.org/10.1016/j.onano.2017.07.002]

[36] Sarsar V, Selwal MK, Selwal KK. Biofabrication, characterization and antibacterial efficacy of extracellular silver nanoparticles using novel fungal strain of Penicillium atramentosum KM. J Saudi Chem Soc 2015; 19(6): 682-8.
[http://dx.doi.org/10.1016/j.jscs.2014.07.001]

[37] Sarsar V, Selwal MK, Selwal KK. Biogenic synthesis, optimisation and antibacterial efficacy of extracellular silver nanoparticles using novel fungal isolate Aspergillus fumigatus MA 2015; 1-7.

[38] Feroze N, Arshad B, Younas M, Afridi MI, Saqib S, Ayaz A. Fungal mediated synthesis of silver nanoparticles and evaluation of antibacterial activity. Microsc Res Tech 2020; 83(1): 72-80.
[http://dx.doi.org/10.1002/jemt.23390] [PMID: 31617656]

[39] AbdelRahim K, Mahmoud SY, Ali AM, Almaary KS, Mustafa AE-ZMA, Husseiny SM. Extracellular biosynthesis of silver nanoparticles using *Rhizopus stolonifer.* Saudi J Biol Sci 2017; 24(1): 208-16.
[http://dx.doi.org/10.1016/j.sjbs.2016.02.025] [PMID: 28053592]

[40] Ibrahem EJ, Thalij KM, Saleh MK, Badawy AS. Biosynthesis of zinc oxide nanoparticles and assay of antibacterial activity. Am J Biochem Biotechnol 2017; 13(2): 63-9.
[http://dx.doi.org/10.3844/ajbbsp.2017.63.69]

[41] Kalpana VN, Kataru BAS, Sravani N, Vigneshwari T, Panneerselvam A, Devi Rajeswari V. Biosynthesis of zinc oxide nanoparticles using culture filtrates of Aspergillus niger: Antimicrobial textiles and dye degradation studies. OpenNano 2018; 3: 48-55.
[http://dx.doi.org/10.1016/j.onano.2018.06.001]

[42] Kowshik M, Ashtaputre S, Kharrazi S, *et al.* Extracellular synthesis of silver nanoparticles by a silver-tolerant yeast strain MKY3. Nanotechnology 2003; 14(1): 95-100.
[http://dx.doi.org/10.1088/0957-4484/14/1/321]

[43] Ahmad A, Mukherjee P, Mandal D, *et al.* Enzyme mediated extracellular synthesis of CdS nanoparticles by the fungus, Fusarium oxysporum. J Am Chem Soc 2002; 124(41): 12108-9.
[http://dx.doi.org/10.1021/ja027296o] [PMID: 12371846]

[44]　Shankar SS, Ahmad A, Pasricha R, Sastry M. Bioreduction of chloroaurate ions by geranium leaves and its endophytic fungus yields gold nanoparticles of different shapes. J Mater Chem 2003; 13(7): 1822.
[http://dx.doi.org/10.1039/b303808b]

[45]　Ahmad A, Senapati S, Khan MI, Kumar R, Sastry M. Extra-/intracellular biosynthesis of gold nanoparticles by an alkalotolerant fungus, trichothecium sp. J Biomed Nanotechnol 2005; 1(1): 47-53.
[http://dx.doi.org/10.1166/jbn.2005.012]

[46]　Bhainsa KC, D'Souza SF. Extracellular biosynthesis of silver nanoparticles using the fungus Aspergillus fumigatus. Colloids Surf B Biointerfaces 2006; 47(2): 160-4.
[http://dx.doi.org/10.1016/j.colsurfb.2005.11.026] [PMID: 16420977]

[47]　Vigneshwaran N, Nachane RP, Balasubramanya RH, Varadarajan PV. A novel one-pot 'green' synthesis of stable silver nanoparticles using soluble starch. Carbohydr Res 2006; 341(12): 2012-8.
[http://dx.doi.org/10.1016/j.carres.2006.04.042] [PMID: 16716274]

[48]　Gericke M, Pinches A. Microbial production of gold nanoparticles. Gold Bull 2006; 39(1): 22-8.
[http://dx.doi.org/10.1007/BF03215529]

[49]　Krumov N, Oder S, Perner-Nochta I, Angelov A, Posten C. Accumulation of CdS nanoparticles by yeasts in a fed-batch bioprocess. J Biotechnol 2007; 132(4): 481-6.
[http://dx.doi.org/10.1016/j.jbiotec.2007.08.016] [PMID: 17900736]

[50]　Xie J, Lee JY, Wang , Ting YP. Wang, Ting YP. High-yield synthesis of complex gold nanostructures in a fungal system. J Phys Chem C 2007; 111(45): 16858-65.
[http://dx.doi.org/10.1021/jp0752668]

[51]　Dasaratrao Sawle B, Salimath B, Deshpande R, Dhondojirao Bedre M, Krishnamurthy Prabhakar B, Venkataraman A. Biosynthesis and stabilization of Au and Au-Ag alloy nanoparticles by fungus, *Fusarium semitectum*. Sci Technol Adv Mater 2008; 9(3): 035012.
[http://dx.doi.org/10.1088/1468-6996/9/3/035012] [PMID: 27878009]

[52]　Sanghi R, Verma P. PH dependant fungal proteins in the 'green' synthesis of gold nanoparticles. Adv Mater Lett 2010; 1(3): 193-9.
[http://dx.doi.org/10.5185/amlett.2010.5124]

[53]　Salunkhe RB, Patil SV, Patil CD, Salunke BK. Larvicidal potential of silver nanoparticles synthesized using fungus Cochliobolus lunatus against Aedes aegypti (Linnaeus, 1762) and Anopheles stephensi Liston (Diptera; Culicidae). Parasitol Res 2011; 109(3): 823-31.
[http://dx.doi.org/10.1007/s00436-011-2328-1] [PMID: 21451993]

[54]　Castro-Longoria E, Vilchis-Nestor AR, Avalos-Borja M. Biosynthesis of silver, gold and bimetallic nanoparticles using the filamentous fungus Neurospora crassa. Colloids Surf B Biointerfaces 2011; 83(1): 42-8.
[http://dx.doi.org/10.1016/j.colsurfb.2010.10.035] [PMID: 21087843]

[55]　Verma VC, Singh SK, Solanki R, Prakash S. Biofabrication of anisotropic gold nanotriangles using extract of endophytic aspergillus clavatus as a dual functional reductant and stabilizer. Nanoscale Res Lett 2010.
[http://dx.doi.org/10.1007/s11671-010-9743-6] [PMID: 27502640]

[56]　Sheikhloo Z, Salouti M, Katiraee F. Biological synthesis of gold nanoparticles by fungus epicoccum nigrum. J Cluster Sci 2011; 22(4): 661-5.
[http://dx.doi.org/10.1007/s10876-011-0412-4]

[57]　Li G, He D, Qian Y, *et al.* Fungus-mediated green synthesis of silver nanoparticles using Aspergillus terreus. Int J Mol Sci 2012; 13(1): 466-76.
[http://dx.doi.org/10.3390/ijms13010466] [PMID: 22312264]

[58]　Das SK, Dickinson C, Lafir F, Brougham DF, Marsili E. Synthesis, characterization and catalytic activity of gold nanoparticles biosynthesized with Rhizopus oryzae protein extract. Green Chem 2012;

14(5): 1322.
[http://dx.doi.org/10.1039/c2gc16676c]

[59] Rajakumar G, Rahuman AA, Roopan SM, *et al.* Fungus-mediated biosynthesis and characterization of TiO$_2$ nanoparticles and their activity against pathogenic bacteria. Spectrochim Acta A Mol Biomol Spectrosc 2012; 91: 23-9.
[http://dx.doi.org/10.1016/j.saa.2012.01.011] [PMID: 22349888]

[60] Kaler A, Jain S, Banerjee UC. Green and rapid synthesis of anticancerous silver nanoparticles by Saccharomyces boulardii and insight into mechanism of nanoparticle synthesis. BioMed Res Int 2013; 2013: 872940.
[http://dx.doi.org/10.1155/2013/872940] [PMID: 24298556]

[61] Narayanan KB, Sakthivel N. Mycocrystallization of gold ions by the fungus Cylindrocladium floridanum. World J Microbiol Biotechnol 2013; 29(11): 2207-11.
[http://dx.doi.org/10.1007/s11274-013-1379-0] [PMID: 23736894]

[62] Qian Y, Yu H, He D, *et al.* Biosynthesis of silver nanoparticles by the endophytic fungus Epicoccum nigrum and their activity against pathogenic fungi. Bioprocess Biosyst Eng 2013; 36(11): 1613-9.
[http://dx.doi.org/10.1007/s00449-013-0937-z] [PMID: 23463299]

[63] Gupta S, Bector S. Biosynthesis of extracellular and intracellular gold nanoparticles by Aspergillus fumigatus and A. flavus. Antonie van Leeuwenhoek 2013; 103(5): 1113-23.
[http://dx.doi.org/10.1007/s10482-013-9892-6] [PMID: 23400423]

[64] Tarafdar JC, Raliya R. Rapid, low-cost, and ecofriendly approach for iron nanoparticle synthesis using aspergillus oryzae TFR9. Journal of Nanoparticles 2013; 2013: 1-4.
[http://dx.doi.org/10.1155/2013/141274]

[65] Raliya R, Tarafdar JC. ZnO nanoparticle biosynthesis and its effect on phosphorous-mobilizing enzyme secretion and gum contents in clusterbean (cyamopsis tetragonoloba L.). Agric Res 2013; 2(1): 48-57.
[http://dx.doi.org/10.1007/s40003-012-0049-z]

[66] Chowdhury S, Basu A, Kundu S. Green synthesis of protein capped silver nanoparticles from phytopathogenic fungus Macrophomina phaseolina (Tassi) Goid with antimicrobial properties against multidrug-resistant bacteria. Nanoscale Res Lett 2014; 9(1): 365.
[http://dx.doi.org/10.1186/1556-276X-9-365] [PMID: 25114655]

[67] Pereira L, Dias N, Carvalho J, Fernandes S, Santos C, Lima N. Synthesis, characterization and antifungal activity of chemically and fungal-produced silver nanoparticles against Trichophyton rubrum. J Appl Microbiol 2014; 117(6): 1601-13.
[http://dx.doi.org/10.1111/jam.12652] [PMID: 25234047]

[68] Sundaravadivelan C, Padmanabhan MN. Effect of mycosynthesized silver nanoparticles from filtrate of Trichoderma harzianum against larvae and pupa of dengue vector Aedes aegypti L. Environ Sci Pollut Res Int 2014; 21(6): 4624-33.
[http://dx.doi.org/10.1007/s11356-013-2358-6] [PMID: 24352539]

[69] Castro ME, Cottet L, Castillo A. Biosynthesis of gold nanoparticles by extracellular molecules produced by the phytopathogenic fungus Botrytis cinerea. Mater Lett 2014; 115: 42-4.
[http://dx.doi.org/10.1016/j.matlet.2013.10.020]

[70] Sarkar J, Ghosh M, Mukherjee A, Chattopadhyay D, Acharya K. Biosynthesis and safety evaluation of ZnO nanoparticles. Bioprocess Biosyst Eng 2014; 37(2): 165-71.
[http://dx.doi.org/10.1007/s00449-013-0982-7] [PMID: 23743731]

[71] Raliya R, Tarafdar JC, Choudhary K, *et al.* Synthesis of MgO nanoparticles using aspergillus tubingensis TFR-3. Journal of Bionanoscience 2014; 8(1): 34-8.
[http://dx.doi.org/10.1166/jbns.2014.1195]

[72] Chen G, Yi B, Zeng G, *et al.* Facile green extracellular biosynthesis of CdS quantum dots by white rot

fungus Phanerochaete chrysosporium. Colloids Surf B Biointerfaces 2014; 117: 199-205.
[http://dx.doi.org/10.1016/j.colsurfb.2014.02.027] [PMID: 24632392]

[73]   Balakumaran MD, Ramachandran R, Kalaichelvan PT. Exploitation of endophytic fungus, Guignardia
       mangiferae for extracellular synthesis of silver nanoparticles and their in vitro biological activities.
       Microbiol Res 2015; 178: 9-17.
       [http://dx.doi.org/10.1016/j.micres.2015.05.009] [PMID: 26302842]

[74]   El-Baz AF, El-Batal AI, Abomosalam FM, Tayel AA, Shetaia YM, Yang S-T. Extracellular
       biosynthesis of anti-Candida silver nanoparticles using Monascus purpureus. J Basic Microbiol 2016;
       56(5): 531-40.
       [http://dx.doi.org/10.1002/jobm.201500503] [PMID: 26515502]

[75]   Ammar HAM, El-Desouky TA. Green synthesis of nanosilver particles by Aspergillus terreus HA1N
       and Penicillium expansum HA2N and its antifungal activity against mycotoxigenic fungi. J Appl
       Microbiol 2016; 121(1): 89-100.
       [http://dx.doi.org/10.1111/jam.13140] [PMID: 27002915]

[76]   Balakumaran MD, Ramachandran R, Balashanmugam P, Mukeshkumar DJ, Kalaichelvan PT.
       Mycosynthesis of silver and gold nanoparticles: Optimization, characterization and antimicrobial
       activity against human pathogens. Microbiol Res 2016; 182: 8-20.
       [http://dx.doi.org/10.1016/j.micres.2015.09.009] [PMID: 26686609]

[77]   Kitching M, Choudhary P, Inguva S, et al. Fungal surface protein mediated one-pot synthesis of stable
       and hemocompatible gold nanoparticles. Enzyme Microb Technol 2016; 95: 76-84.
       [http://dx.doi.org/10.1016/j.enzmictec.2016.08.007] [PMID: 27866629]

[78]   Costa Silva LP, Oliveira JP, Keijok WJ, et al. Extracellular biosynthesis of silver nanoparticles using
       the cell-free filtrate of nematophagous fungus Duddingtonia flagrans. Int J Nanomedicine 2017; 12:
       6373-81.
       [http://dx.doi.org/10.2147/IJN.S137703] [PMID: 28919741]

[79]   Barabadi H, Kobarfard F, Vahidi H. Biosynthesis and characterization of biogenic tellurium
       nanoparticles by using Penicillium chrysogenum PTCC 5031: A novel approach in gold
       biotechnology. Iran J Pharm Res 2018; 17 (Suppl. 2): 87-97.
       [PMID: 31011345]

[80]   Vijayanandan AS, Balakrishnan RM. Biosynthesis of cobalt oxide nanoparticles using endophytic
       fungus Aspergillus nidulans. J Environ Manage 2018; 218: 442-50.
       [http://dx.doi.org/10.1016/j.jenvman.2018.04.032] [PMID: 29709813]

[81]   Elamawi RM, Al-Harbi RE, Hendi AA. Biosynthesis and characterization of silver nanoparticles using
       Trichoderma longibrachiatum and their effect on phytopathogenic fungi. Egypt J Biol Pest Control
       2018; 28(1): 28.
       [http://dx.doi.org/10.1186/s41938-018-0028-1]

[82]   Akther T, Vabeiryureilai Mathipi , Davoodbasha M, Srinivasan H. Fungal-mediated synthesis of
       pharmaceutically active silver nanoparticles and anticancer property against A549 cells through
       apoptosis. Environ Sci Pollut Res Int 2019; 26(13): 13649-57.
       [http://dx.doi.org/10.1007/s11356-019-04718-w] [PMID: 30919178]

[83]   Hefny M, El-Zamek F, Abd El-Fattah H, Mahgoub S. Biosynthesis of zinc nanoparticles using culture
       filtrates of Aspergillus, Fusarium and penicillium fungal species and their antibacterial properties
       against gram-positive and gram-negative bacteria. Zagazig Journal of Agricultural Research 2019;
       46(6): 2009-21.
       [http://dx.doi.org/10.21608/zjar.2019.51920]

[84]   Clarance P, Luvankar B, Sales J, et al. Green synthesis and characterization of gold nanoparticles
       using endophytic fungi Fusarium solani and its in-vitro anticancer and biomedical applications. Saudi
       J Biol Sci 2020; 27(2): 706-12.
       [http://dx.doi.org/10.1016/j.sjbs.2019.12.026] [PMID: 32210692]

[85] Ganesan V, Hariram M, Vivekanandhan S, Muthuramkumar S. Periconium sp. (endophytic fungi) extract mediated sol-gel synthesis of ZnO nanoparticles for antimicrobial and antioxidant applications. Mater Sci Semicond Process 2020; 105: 104739.
[http://dx.doi.org/10.1016/j.mssp.2019.104739]

[86] Abdelhakim HK, El-Sayed ER, Rashidi FB. Biosynthesis of zinc oxide nanoparticles with antimicrobial, anticancer, antioxidant and photocatalytic activities by the endophytic Alternaria tenuissima. J Appl Microbiol 2020; 128(6): 1634-46.
[http://dx.doi.org/10.1111/jam.14581] [PMID: 31954094]

[87] Al-limoun M, Qaralleh HN, Khleifat KM, *et al.* Culture media composition and reduction potential optimization of mycelia-free filtrate for the biosynthesis of silver nanoparticles using the fungus tritirachium oryzae W5H. Curr Nanosci 2020; 16(5): 757-69.
[http://dx.doi.org/10.2174/1573413715666190725111956]

[88] Iranmanesh S, Shahidi Bonjar GH, Baghizadeh A. Study of the biosynthesis of gold nanoparticles by using several saprophytic fungi. SN Applied Sciences 2020; 2(11): 1851.
[http://dx.doi.org/10.1007/s42452-020-03704-z]

[89] Liang P, Shi H, Zhu W, *et al.* Silver nanoparticles enhance the sensitivity of temozolomide on human glioma cells. Oncotarget 2017; 8(5): 7533-9.
[http://dx.doi.org/10.18632/oncotarget.13503] [PMID: 27893419]

[90] Saxena A, Tripathi RM, Singh RP. Biological synthesis of silver nanoparticles by using onion (Allium Cepa). Extract and their Antibacterial Activity 2010; 5(2): 427-32.

[91] Sarsar V, Selwal KK, Selwal MK. Nanosilver : Potent antimicrobial agent and its biosynthesis. 2014; 13(4): 546-.

[92] Shahzad A, Saeed H, Iqtedar M, *et al.* Size-controlled production of silver nanoparticles by aspergillus fumigatus BTCB10: likely antibacterial and cytotoxic effects. J Nanomater 2019; 2019: 1-14.
[http://dx.doi.org/10.1155/2019/5168698]

[93] Costa LP, Cunegundes MC, Ferraz CM, Araújo JV, Tobias FL. Extracellular biosynthesis of silver nanoparticles using the cell-free filtrate of nematophagous fungus Duddingtonia flagrans. 2017; 6373-81.
[http://dx.doi.org/10.2147/IJN.S137703]

[94] Rose GK, Soni R, Rishi P, Soni SK. Optimization of the biological synthesis of silver nanoparticles using Penicillium oxalicum GRS-1 and their antimicrobial effects against common food-borne pathogens. 2019; 144-56.
[http://dx.doi.org/10.1515/gps-2018-0042]

[95] Waldrop MM. After the fall published by : american association for the advancement of science stable http://www.jstor.org/stable/17000782016; 239(4843): 17-8.

[96] Desai MP, Labhasetwar V, Amidon GL, Levy RJ. Gastrointestinal uptake of biodegradable microparticles: effect of particle size. Pharm Res 1996; 13(12): 1838-45.
[http://dx.doi.org/10.1023/A:1016085108889] [PMID: 8987081]

[97] Florence AT, Hillery AM, Hussain N, Jani PU. Nanoparticles as carriers for oral peptide absorption: Studies on particle uptake and fate. J Control Release 1995; 36(1–2): 39-46.
[http://dx.doi.org/10.1016/0168-3659(95)00059-H]

[98] Dhillon GS, Brar SK, Kaur S, Verma M. Green approach for nanoparticle biosynthesis by fungi: current trends and applications. Crit Rev Biotechnol 2012; 32(1): 49-73.
[http://dx.doi.org/10.3109/07388551.2010.550568] [PMID: 21696293]

[99] Mohammed Fayaz A, Girilal M, Mahdy SA, Somsundar SS, Venkatesan R, Kalaichelvan PT. Vancomycin bound biogenic gold nanoparticles: A different perspective for development of anti VRSA agents. Process Biochem 2011; 46(3): 636-41.
[http://dx.doi.org/10.1016/j.procbio.2010.11.001]

[100] Sun J, Li Y, Liang XJ, Wang PC. Bacterial magnetosome: A novel biogenetic magnetic targeted drug carrier with potential multifunctions. Journal of Nanomaterials 2011.
[http://dx.doi.org/10.1155/2011/469031]

[101] Brown SD, Nativo P, Smith JA, *et al*. Gold nanoparticles for the improved anticancer drug delivery of the active component of oxaliplatin. J Am Chem Soc 2010; 132(13): 4678-84.
[http://dx.doi.org/10.1021/ja908117a] [PMID: 20225865]

[102] Ong C, Lim JZZ, Ng C-T, Li JJ, Yung L-YL, Bay B-H. Silver nanoparticles in cancer: therapeutic efficacy and toxicity. Curr Med Chem 2013; 20(6): 772-81.
[PMID: 23298139]

[103] Raman J, Reddy GR, Lakshmanan H, *et al*. Mycosynthesis and characterization of silver nanoparticles from Pleurotus djamor var. roseus and their in vitro cytotoxicity effect on PC3 cells. Process Biochem 2014.

[104] Selvi BCG, Madhavan J, Santhanam A. Cytotoxic effect of silver nanoparticles synthesized from Padina tetrastromatica on breast cancer cell line. Advances in Natural Sciences: Nanoscience and Nanotechnology 2016; 7(3): 035015.
[http://dx.doi.org/10.1088/2043-6262/7/3/035015]

[105] M JF, P L. Apoptotic efficacy of biogenic silver nanoparticles on human breast cancer MCF-7 cell lines. Prog Biomater 2015; 4(2-4): 113-21.
[http://dx.doi.org/10.1007/s40204-015-0042-2] [PMID: 26566469]

[106] Alalawy AI, Rabey HA, El , Almutairi FM, *et al*. Effectual anticancer potentiality of loaded bee venom onto fungal chitosan nanoparticles 2020.
[http://dx.doi.org/10.1155/2020/2785304]

[107] Munawer U, Raghavendra VB, Ningaraju S, *et al*. Biofabrication of gold nanoparticles mediated by the endophytic Cladosporium species: Photodegradation, in vitro anticancer activity and *in vivo* antitumor studies. Int J Pharm 2020; 588(May): 119729.
[http://dx.doi.org/10.1016/j.ijpharm.2020.119729] [PMID: 32768527]

[108] Percival SL, Bowler PG, Dolman J. Antimicrobial activity of silver-containing dressings on wound microorganisms using an in vitro biofilm model. Int Wound J 2007; 4(2): 186-91.
[http://dx.doi.org/10.1111/j.1742-481X.2007.00296.x] [PMID: 17651233]

[109] Krishnaraj C, Ramachandran R, Mohan K, Kalaichelvan PT. Optimization for rapid synthesis of silver nanoparticles and its effect on phytopathogenic fungi. Spectrochim Acta A Mol Biomol Spectrosc 2012; 93: 95-9.
[http://dx.doi.org/10.1016/j.saa.2012.03.002] [PMID: 22465774]

[110] Liau SY, Read DC, Pugh WJ, Furr JR, Russell AD. Interaction of silver nitrate with readily identifiable groups: relationship to the antibacterial action of silver ions. Lett Appl Microbiol 1997; 25(4): 279-83.
[http://dx.doi.org/10.1046/j.1472-765X.1997.00219.x] [PMID: 9351278]

[111] Matsumura Y, Yoshikata K, Kunisaki S, Tsuchido T. Mode of bactericidal action of silver zeolite and its comparison with that of silver nitrate. Appl Environ Microbiol 2003; 69(7): 4278-81.
[http://dx.doi.org/10.1128/AEM.69.7.4278-4281.2003] [PMID: 12839814]

[112] Selwal MK, Selwal KK. Biogenic synthesis of silver nanoparticles and their applications in medicine. 2017; pp. 171-87.
[http://dx.doi.org/10.1007/978-3-319-68424-6_9]

[113] Hu X, Saravanakumar K, Jin T, Wang MH. Mycosynthesis, characterization, anticancer and antibacterial activity of silver nanoparticles from endophytic fungus *Talaromyces purpureogenus*. Int J Nanomedicine 2019; 14: 3427-38.
[http://dx.doi.org/10.2147/IJN.S200817] [PMID: 31190801]

[114] Gracias DH, Tien J, Breen TL, Hsu C, Whitesides GM. Forming electrical networks in three

dimensions by self-assembly. Science 2000; 289(5482): 1170-2.
[http://dx.doi.org/10.1126/science.289.5482.1170] [PMID: 10947979]

[115] Maier SA, Brongersma ML, Kik PG, Meltzer S, Requicha AAG, Atwater HA. Plasmonics-A route to nanoscale optical devices. Adv Mater 2001; 13(19): 1501-5.
[http://dx.doi.org/10.1002/1521-4095(200110)13:19<1501::AID-ADMA1501>3.0.CO;2-Z]

[116] Shahverdi AR, Minaeian S, Shahverdi HR, Jamalifar H, Nohi A-A. Rapid synthesis of silver nanoparticles using culture supernatants of Enterobacteria: A novel biological approach. Process Biochem 2007; 42(5): 919-23.
[http://dx.doi.org/10.1016/j.procbio.2007.02.005]

[117] Kamat PV. Photophysical, photochemical and photocatalytic aspects of metal nanoparticles. J Phys Chem B 2002; 106(32): 7729-44.
[http://dx.doi.org/10.1021/jp0209289]

[118] Mirkin CA, Letsinger RL, Mucic RC, Storhoff JJ. A DNA-based method for rationally assembling nanoparticles into macroscopic materials. Nature 1996; 382(6592): 607-9.
[http://dx.doi.org/10.1038/382607a0] [PMID: 8757129]

[119] Matejka P, Vlckova B, Vohlidal J, Pancoska P, Baumruk V. The role of triton X-100 as an adsorbate and a molecular spacer on the surface of silver colloid: a surface-enhanced Raman scattering study. J Phys Chem 1992; 96(3): 1361-6.
[http://dx.doi.org/10.1021/j100182a063]

[120] Cao Y, Jin R, Mirkin CA. DNA-modified core-shell Ag/Au nanoparticles. J Am Chem Soc 2001; 123(32): 7961-2.
[http://dx.doi.org/10.1021/ja011342n] [PMID: 11493092]

CHAPTER 6

Extracellular Biosynthesis of Gold and Silver Nanoparticles using Fungal Extracts

Lham Dorjee[1,*], **Shweta Meshram**[1] and **Ankita Verma**[2]

[1] *Division of Plant Pathology, ICAR-Indian Agricultural Research Institute, New Delhi-110 012, India*

[2] *Division of Agricultural Chemicals, ICAR-Indian Agricultural Research Institute, New Delhi-110 012, India*

Abstract: Nanoscience has opened new vistas to manage phytopathogens, improve crop productivity by the development of new varieties, and control infectious diseases in humans. Silver nanoparticles (AgNPs) and gold nanoparticles (AuNPs) are highly acclaimed for their wide potential application in various fields. Chemical and physical methods of synthesis of AgNPs and AuNPs are widely used; however, such methods possess numerous setbacks, such as the production of toxic residues and indispensable need for high energy. Biosynthesis of nanoparticles is a cost-effective and environmentally friendly method. A plethora of species of plant, bacteria, fungi, *etc.* is available with potential biosynthesis ability. Fungi are a highly preferred organism owing to the ability to secrete a large number of extracellular enzymes, metal toxicity tolerance and bioaccumulation ability, and ease of handling of its biomass. Extracellular enzymes act both as reducing as well as capping agents. Two different methods are used by fungi for synthesis *viz.*, intercellular and extracellular synthesis. Extracellular synthesis is preferred over intercellular as it bypasses several down streaming processes. During the reduction process, the metal ions (Ag^{2+} and Au^{3+}) are converted to an elemental state (Ag^0 and Au^0) which is in the nano range. Due to their large surface-to-volume ratio and other properties, they become very effective against other pathogens. There is an excellent prospect of the use of nanoparticles in the field of agriculture and health and nanoparticles synthesized using a biological method involving fungi could be a boon.

Keywords: Bacteria, Biosynthesis, Extracellular synthesis, Fungi, Nanoparticles, Nanoscience, Phytopathogens.

INTRODUCTION

Since time immemorial, chemicals have been utilized to manage the phytopathogens directly or indirectly, and perhaps the use of chemical pesticides

* **Corresponding Author Lham Dorjee:** Division of Plant Pathology, ICAR-Indian Agricultural Research Institute, New Delhi-110 012, India; Tel:9821018341; Email: lamdorg12@gmail.com

is inevitable, taking into consideration the incessant increase of the human population. Unfortunately, pesticides have numerous setbacks which bring them to the limelight of criticism. Such conventional methods adversely affect the non-target organism and environment as a whole. Although systemic fungicides are known to have potential controlling effects, nevertheless, indiscriminate use of systemic fungicides leads to the development of resistance due to its specific site of action. Detrimental and inimical effects of pesticides can be attributed to their high toxicity, non-biodegradable nature, and residual activity. Therefore, there is an utmost need for an alternate avenue to combat the phytopathogens, and employing nanotechnology in agriculture could be a boon. Such one application is the use of nanoparticles (NPs) as nano-fungicides to manage phytopathogens. A nanoparticle by virtue of its high surface-to-volume ratio has high efficacy and durability and also delays the development of resistance in pathogens. Several metallic nanoparticles such as gold nanoparticles (AuNPs), silver nanoparticles (AgNPs), copper nanoparticles (CuNPs), Platinum nanoparticles (PtNPs), Zinc nanoparticles (ZnNPs), *etc.* have been synthesized that can be exploited astoundingly to manage plant Pathogens. Three important methods are widely used for the synthesis of metallic nanoparticles (AuNPs & AgNPs) *viz.*, chemical method, physical method, and biological method. Physical and chemical methods are capable of generating large quantities of NPs with a defined shape and size in a short duration but these methods also have some major setbacks such as complex and out of date protocol requiring high expenditure, inefficient and produce hazardous toxic wastes that are inimical to the environment as well as to the human beings [1]. Propitiously "*Biosynthesis method*" satisfies almost all the criteria to fit and is an eco-friendly and efficient method for the synthesis of nanoparticles. Biological methods of synthesis are carried out by using plant extract, fungi, bacteria, or their metabolites which act both as reducing as well as stabilizing agents [2]. Biosynthesis of metallic nanoparticles using microorganisms is attracting the interest of many, which is a blooming environmentally safe science of nanoparticles synthesis with well-defined sizes, shapes, and controlled mono-dispersity [3].

The Silver nanoparticles (AgNPs) are one of the widely studied and utilized nanoparticles. It is known to have remarkable inhibitory/antimicrobial effects against various plant pathogens by virtue of its high surface area to the volume ratio [4]. Silver (Ag) is the second-most malleable and ductile element after gold (Au). Just an ounce of silver can be drawn into 8,000 feet long wire. The silver was first mined in Anatolia (present Turkey) between 5,000 and 6,000 years ago. Around 700 B.C., early Mediterranean civilizations started using silver as currency. The silver has powerful antibacterial/antimicrobial properties, which have been well known to humans during the times of the ancient Phoenicians, who kept their wine and food items fresh, by keeping them into silver vessels.

Nowadays, silver is widely used in health care tools such as bandages, surgical instruments, stethoscopes, *etc.* Unlike other antibiotics, silver also prevents bacteria from developing resistance to it.

Silver now has been accepted as a replacement for agrochemicals and has the potential to manage microorganisms. It has been used as a foliar spray to stop fungi, moulds, rot, and many other plant diseases. Silver has also been reported to act as a potential plant growth stimulator [5].

Lately, gold nanoparticles (AuNPs) have attracted the interest of many due to their significant applications and astounding properties. AuNPs are important as drug carriers to treat cancer, heart diseases, *etc.* by virtue of their large bioavailable surface area, which confers effortless functionalization [6]. AuNPs are also found to have antimicrobial activity. It is used to remove environmental pollutants. AuNPs are also used as nanosensors to detect viruses and other pathogens [7]. Moreover, AuNPs are recalcitrant to chemical oxidation and are capable to remain stable in varied environmental conditions [8, 9].

Biosynthesis of Silver Nanoparticles

Silver nanoparticles (AgNPs) are highly acclaimed nanoparticles for their broad-spectrum antimicrobial properties [10, 11]. AgNPs are widely synthesized by using the chemical method and physical methods. In chemical methods, chemical reagents are used which act as the reducing agent and stabilizing agent. However, reagents such as sodium borohydride, potassium bitartrate, hydrazine, *etc.* are highly toxic and can lead to health hazards and various environmental consequences [1, 12]. Physical methods are also capable of generating large quantities of NPs with a defined size and shape in a relatively short period of time, but this method has some major setbacks which cannot be ignored, such as they are complicated, outdated, costly (as it requires energy in form of heat and pressure), therefore pose a major threat to the environment [1].

Biosynthesis of AgNPs provides advancement over chemical and physical methods as it is cost-effective, environment friendly, easily scaled up for large-scale synthesis, and importantly it doesn't involve high energy or any toxic chemicals [13]. The biosynthesis method generates nanoparticles of defined sizes, shapes and controlled mono-dispersity [3]. Several microorganisms are standing as impeccable candidates for the synthesis of NPs. Among numerous microorganisms, fungi are one of the highly preferred organisms by virtue of their high tolerance and bioaccumulation ability towards metals and are excellent efficient secretors of enzymes. Other advantages attributed to the biosynthesis of NPs are economic viability and ease of biomass handling [14]. Furthermore, it requires simple nutrients, possesses high wall binding capacity, and also has intracellular metal uptake capabilities. The extracellular synthesis method has an edge over the intracellular synthesis of nanoparticles; as such methods bypass the tedious down streaming process. Interestingly, the metabolism of fungi can be manipulated such that it would favour the synthesis of nanoparticles of desired morphology and monodispersity by adjusting temperature, pH, time quantity of biomass, *etc* [15]. Fungal culture has an edge over bacterial culture regarding nanoparticle synthesis for fungi provide good biomass; requires simple nutrients and ease of gene manipulation. Biosynthesis is conducted by using organisms (bacteria, fungi, and plants) themselves or the by-products of their metabolism, which act as reducing and stabilizing agents [2]. Moreover, bio-molecules produced as a result of metabolism are found to enhance the stability of the obtained nanoparticles.

In the case of the chemical method of nanoparticles synthesis, certain capping agents in form of polymers and surfactants such as Polyethylene Glycol (PEG), Polyvinylpyrrolidone (PVP), dodecyl benzyl sulfate, *etc.* are added to the reaction mixture for the stabilization of nanoparticles [12, 16]. The process is termed functionalization. In the biosynthesis method of nanoparticle generation, a certain functional group of biological origin or biomolecules from the organism employed in synthesis forms a cap around and hence stabilizing the nanoparticles generated in the process. The biomolecules derived from organisms have a high tendency to adhere and bond with metals, such as protein and amino acid-binding to nanoparticles results in high stability of nanoparticles preventing agglomeration. The stabilization of nanoparticles occurs due to the binding of proteins, amino acids, or cysteine residues at the surface. The nucleophilic OH- ions adsorbed at the surface confer stabilization to the AgNPs [17]. Certain mycelial cell wall enzymes secreted in the filtrate may also confer to the stabilization of the synthesized nanoparticles due to their negative carboxyl groups which provide electrostatic attraction [18]. Since the capping agent is of biological origin, therefore, it offers biocompatibility, which in turn increases the rate of internalization and retention of nanoparticles in the host system [19].

Biosynthesis can be carried out in two different ways *viz.,* intracellularly, or extracellularly (Fig. **1**). Intracellular synthesis involves the use of the metal precursor in mycelia, which is later internalized in the biomass. Further, the down streaming process needs to be carried out to extract the formed nanoparticles which include chemical treatment, centrifugation, and filtration that disrupt the biomass and aid in the release of the nanoparticles [20]. In the case of extracellular synthesis, fungal filtrate/supernatant is obtained through filtration and centrifugation process ensuring the final filtrate is devoid of spore or any cell. Then the desired metal precursor is added to the aqueous filtrate containing only the fungal biomolecules, which leads to the formation of nanoparticles in the dispersion. The second method is widely used as it bypasses several tedious cumbersome down streaming processes [21]. Further, nanoparticles can be purified and collected by filtration, centrifugation, and lyophilization [22].

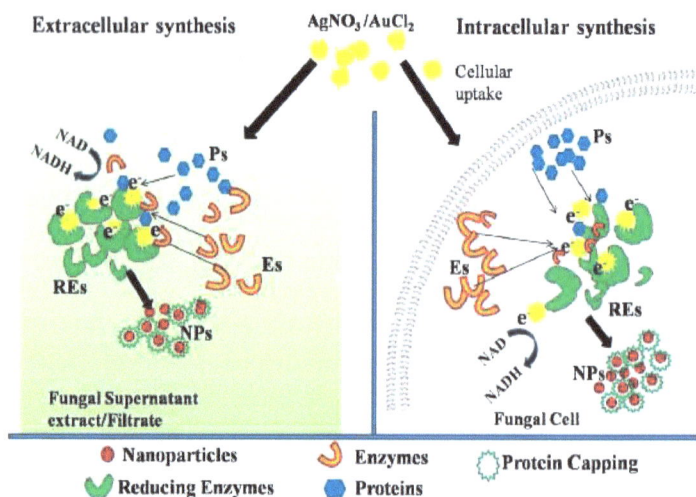

Fig. (1). Schematic representation of mechanism for the synthesis of nanoparticles extracellularly and intracellularly.

Fungi Mediated Synthesis of AgNPs

Nature has endowed us with plenty of microorganisms well capable of generating nanoparticles of defined size and monodispersity. Among them, fungi are considered an impeccable candidate for the synthesis of AgNPs owing to their capability of producing a large number of proteins and enzymes, metal tolerance, and bioaccumulation ability. The metabolites act as a reducing agent as well as a stabilizing agent. Moreover, fungi offer other benefits of ease of handling and cultivation on a large scale at a fairly low expense and eventually produce nanoparticles of controlled size and morphology [21]. Several proteins of fungal origin have been reported to actively participate in the biosynthesis of metallic nanoparticles as a potential reducing agent such as class I fungal hydrophobin from entomopathogenic fungus *Beauveria bassania* [23], tripeptide glutathione from certain yeast and fungi [24]. Moreover, fungal biomass has been extensively used to absorb cations from the water by virtue of its ability to large cationic biosorption sites (Das, 2010). A number of fungi have been utilized and found to possess the remarkable potential to synthesize AgNPs extracellularly (Table **1**). Biosynthesis of silver nanoparticles has been successfully achieved using the filamentous fungus *Neurospora crassa* wild type strain N150 [25]. The fungus *Fusarium oxysporum* has been exploited to synthesize AgNPs extracellularly by an enzymatic process by optimizing the temperature and pH at 50°C, respectively

[3, 18, 26, 27]. Devi and co-worker [28] used 75 different isolates of fungi *viz.*, *Trichoderma harzianum, T. asperellum, T. pseudokoningii, T. longibacterium,* and *T. virens.* Among them, VN-11 isolate belonging to *T. virens* was found to be the most efficient one. Total 25 different isolates of an entomopathogenic fungus *Beauveria bassania* were used by Kamil and co-worker [29] to synthesize AgNPs extracellularly using $AgNO_3$ as a precursor. The extracellular synthesis of AgNPs using *Alternaria alternata* was reported by Gajbhiye and co-worker [30] using 1mM $AgNO_3$ solution and biomass broth of aforesaid fungus. Nithya and team [31] biologically synthesized AgNPs using mushroom fungus *Pleurotus sajor*-caju by subjecting $AgNO_3$ to 20 ml cell filtrate and incubating it at 25 °C under dark conditions with constant agitation. Extracellular synthesis of AgNPs using *Trichoderma reesei* has been reported by Vahabi and co-worker [32]. For the synthesis of AgNPs by using *T. reesei* fungus wet biomass mixing with a 100 ml aqueous solution of 1 mM $AgNO_3$. The reaction was carried out at 28 °C for 5 days at 100 rpm in a rotating shaker. During the reaction process, the toxic Ag^+ ions get reduced to non-toxic metallic AgNPs by virtue of the catalytic effect of the extracellular enzyme and metabolites of the fungus. A very common fungus, *Aspergillus flavus* has been reported to synthesize AgNPs from the precursor 1mM $AgNO_3$ in dark conditions at 37°C at 200 rpm by incubating in shaking incubator [33]. Bhainsa and co-worker [34] also biosynthesized AgNPs using *A. fumigatus.* Silver nanoparticles were biologically synthesized using other species of *Aspergillus viz., A. terreus* [1], and *A. niger* [35]. Biological synthesis of AgNPs using an aqueous extract from the compactin-producing fungal strain (*Penicillium brevicompactum* WA 2315) was reported by Shaligram and co-worker [36]. Mycelium was rinsed with deionized water followed by resuspension in deionized water and incubated at 180 rpm for 96 hrs. The filtrate obtained was mixed with 100 mL of 1 mM $AgNO_3$ solution followed by incubation in a dark conditioned rotatory shaker at 180 rpm at 25°C. The pathogenic fungus *Phytophthora infestans* had also been successfully exploited for the synthesis of nanoparticles, particularly AgNPs. The solution of 1mM $AgNO_3$ was prepared and to which 50ml of cell filtrate was added followed by incubation in dark at 25°C with shaking. The colour change indicated the formation of NPs [37]. Ahluwalia and co-workers [38] reported the extracellular synthesis of AgNPs using *Trichoderma harzianum* by using 1mM $AgNO_3$ and incubating at 40 °C which resulted in the size of the nanoparticles 51.10 nm. Balakumaran and co-worker [39] used *Guignardia mangifera* for synthesizing $AgNO_3$ by optimizing temperature (30°C) and pH (7) and concentration (1mM). In a study conducted by Phanjo and Ahmed [40], spherical nanoparticles of size ranging 7-27 nm were generated using *Aspergillus niger* at relatively high temperature and pH of 90°C and 10, respectively. However, the irregular shape of nanoparticles was formed of size ranging 5-60 nm at high temperature (50-80 °C) and alkaline pH by using fungus *Colleotrichum sp* [21]. Nevertheless, a low temperature of 26 °C could also efficiently generate AgNPs (24.4nm) by exploiting *Trichoderma longibrachiatum* [41]. AbdelRahim an co-workers [42] used *Rhizopus stolonifer* for the synthesis of AgNPs by providing a temperature of 50 °C and alkaline conditions at optimized at 10 mM concentration of $AgNO_3$.

Table 1. Fungi used for the synthesis of AgNPs extracellularly.

Fungi	AgNPs characteristics	Optimized parameters	References
Fusarium oxysporum	Size=5-50 nm; spherical	Temperature pH	[43]
Fusarium oxysporum	Size:10-20 nm; spherical	Temp:40 °C; pH:9-11; conc:1.5 mM of $AgNO_3$	[26]
Fusarium oxysporum	Size:5-13 nm; spherical	Temp:50°C; pH:6; Conc:1 mM of $AgNO_3$	[18]
Fusarium oxysporum	Size: 25-50nm; spherical	Temp:28 °C; Conc: 5mM of $AgNO_3$	[27]
Fusarium oxysporum	Size:20-25 nm; spherical	Temp:28 °C	[44]
Alternaria alternata	Size=20-60 nm; spherical	Temp: 25°C; Conc:1mM of $AgNO_3$	[30]
Pleurotus sajor-caju	Size:5- 50nm; round & spherical	Temp: 25°C; Conc:1mM of $AgNO_3$	[31]
Trichoderma reesei	Size:5-50 nm	Temp:28 °C; Conc:1mM of $AgNO_3$	[32]
Trichoderma sp.	Size: 8-60 nm	Temp:40°C;pH:8; Conc:1nm of $AgNO_3$	[28]
Trichoderma longibrachiatum	Size:24.23 nm; spherical	Temp:28°C	[41]
Trichoderma harzianum	Size:51.10nm; spherical	Temp:40 °C; Conc:1 mM of $AgNO_3$	[38]
Trichoderma viride	Size: 2-4 nm; spherical	Temp: 27 °C	[45]
Verticillium sp.	Size:20±8	Temp25-28 °C; pH: 5.5 ± 6.0	[46]
Neurospora crassa	Size:3-50 nm	Temp:28 °C; pH:9; Conc: 1mM	[25]
Penicillium brevicompactum	Size:58.35 nm	Temp:25 °C; Conc: 1mM	[36]
Penicillium oxalicum	Size:36.53 nm; spherical	Temp:28 °C; Conc: 1mM	[47]
Penicillium purpurogenum	Size: 8-10 nm; spherical	pH:8; Conc:1mM $AgNO_3$	[48]
Phytophthora infestans		Temp: 25°C; Conc: 1 mM of $AgNO_3$	[37]
Aspergillus fumigatus	Size:5-25 nm	Temp: 25 °C; Conc: 1 mM of $AgNO_3$	[35]
Aspergillus oryzae	Size:7-27 nm; spherical	Temp:90°C; pH:10; Conc: 1mM	[40]
Aspergillus fumigatus	Size:328 nm;spherical	Temp:25°C;pH:6;Conc: 1mM of $AgNO_3$	[49]

(Table 1) cont.....

Fungi	AgNPs characteristics	Optimized parameters	References
Aspergillus niger	Size: 3-30 nm;Spherical	Temp:28 ± 4 °C; Conc: 1mM of AgNO$_3$	[35]
Aspergillus flavus	Size: 8.92±4.61 nm	Temp: 37°C; Conc:1 nm of AgNO$_3$	[33]
Aspergillus terreus	Size:1-20 nm;spherical	Temp: 28°C; Conc: 1 nm of AgNO$_3$	[1]
Aspergillus niger	Size:27°C; pH:8.5	Temp:25°C; pH:6	[50]
Colleotrichum sp.	Size:5-60 nm; irregular shapes	Temp: 50-80°C; Alkaline pH	[21]
Yeast strain MKY3	Size:2-5 nm; spherical	Temp:30°C; Conc: 2.0 mM of AgNO$_3$	[51]

Mechanisms of Fungal Mediated Synthesis of AgNPs

Despite numerous publications pertaining to the biological synthesis of AgNPs, the exact mechanism of formation of AgNPs is still uncertain. With the current data and information, it is accepted that the metabolites secreted by fungi *viz.,* enzymes, and proteins are responsible for the generation of AgNPs from the metallic precursors used. Basically, the enzymes present in the fungal filtrate reduce the silver ions (Ag$^+$) into elemental silver (Ag^0) (Fig. **2**). Change in the color of the reaction mixture is the first and foremost indication confirming the formation of nanoparticles which can be inspected by visual inspection. UV-spectroscopy can be used to observe the change in the surface plasmon resonance which gives an idea about the changes in the optical properties. Characteristics peaks or absorption bands are obtained around 413-420 nm [28, 30, 29, 32, 33]. The absorption peaks are due to the excitation of surface plasmon resonance (SPR) which depends on the particle size as well as the refractive index of the solvent, or in other words, it is due to the inter-band transitions excitation of metallic particles which are the characteristics properties of the metals [52]. The absorption band or peak at a longer wavelength perhaps indicates the formation of the larger nanoparticles [41].

Metabolites/biomolecules of fungal origin react with the silver precursor (AgNO$_3$), acting as a reducing agent and capping agent, stabilizing the generated the AgNPs. Biomolecules associated with pathways involved in electron transfer during the conversion of NADPH/NADH to NADP$^+$ /NAD$^+$ such as Nicotinamide adenine dinucleotide (NADH) and NADH-dependent nitrate reductase enzymes are believed to play a vital role in reducing the particles size in the biological synthesis of CuNPs [14, 28, 53, 54]. Presence of diverse biomolecules *viz.,* carboxyl group, aldehydes, primary secondary amines, the band I, II and III of amides, perhaps acting as a possible tool for the reduction of the particle size as well as capping agent [29, 55]. In some instances, nitrate reductase enzyme along with anthraquinones [44] and quinones [56] obtained from *Fusarium oxysporum* is accountable for the generation of AgNPs. However, recently it has been reported that NADPH alone without the aid of nitrate reductase enzyme is capable of synthesizing the AgNPs [57].

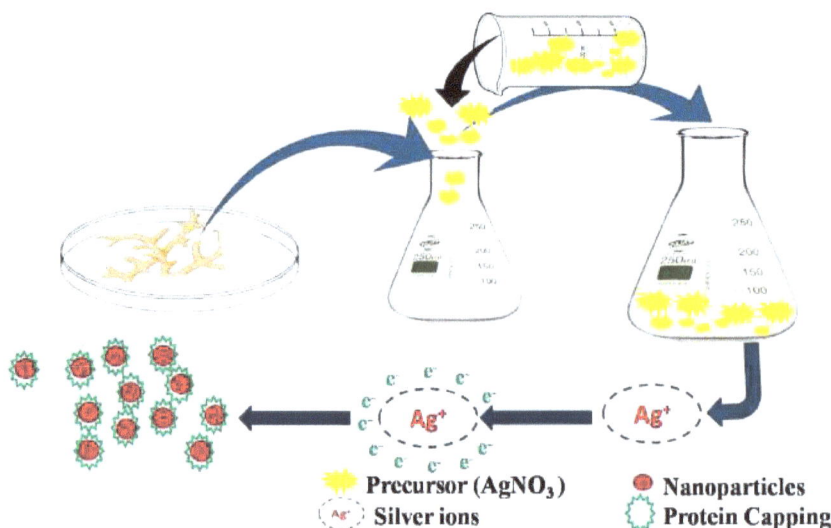

Fig. (2). Mechanisms of Biological synthesis of AgNPs.

Factors affecting the Synthesis of AgNPs

The morphology and size of the AgNPs are highly influenced by several factors or parameters *viz.,* temperature and pH of the medium, type or species or strain of fungi used, agitation provided in terms of revolution per minute (RPM), and light. Optimizations of aforesaid parameters are imperative to obtain the nanoparticles of desired size, shape, monodispersity and stability.

Temperature

Temperature is one of the most important factors that need to be taken into account. It influences the speed of synthesis, stability, and also size of the nanoparticles [41]. Several shreds of evidence suggest at higher temperature synthesis of nanoparticles is faster, comparatively. Higher temperature may also affect the nanoparticles' size, stability, and shorten the duration of the synthesis. Azmath and co-workers [21] observed that the reaction rate increased with the increase in temperature along with the shortening of the duration while using *Colletotrichum* sp. *Fusarium oxysporum* has been found to secrete more amount of metabolites at a higher temperature (60-80 °C) which in turn provides more efficiency in synthesizing AgNPs [26]. In a work carried out by AbdelRahim and co-workers [42] nanoparticles at 20, 40, and 80 °C were of sizes of 2.86, 25.89, and 48.43 nm, respectively. Another work came up with a similar conclusion using *Aspergillus oryzae*. A higher temperature of 30, 50, 70, and 90 °C led to the end of the synthesis with an increased rate in 6 h, 1 h, 45 min, and 20 min, respectively [40]. Sometimes higher temperature, in the case of some fungi may lead to the degeneration and inactivation of enzymes [42]. An increase in temperature also results in a decrease in the size of nanoparticles. The size of AgNPs was found to be decreased on using *Trichoderma viride* [45], *Fusarium oxysporum* [18] with the increase of temperature. The effect of temperature may also vary with the type of fungal species used. At higher temperatures, electrons can be transferred from free amino acids to silver ions favouring the formation of nanoparticles by some fungus. However, very high temperatures, between 80 and 100°C, may result in the denaturation and inactivation of the proteins required for capping and hence stabilization of the nanoparticles.

The pH of The Reaction Medium

The pH of the medium also influences the characteristics of the synthesized nanoparticles. With the increase in pH, the conformation of nitrate reductase enzymes changes according to the concentration of proton present in the reaction mixture which in turn influences the size and shape of the nanoparticles. Basically, at alkaline conditions, the competition between the protons and metal ions enhances establishment bonds with negatively charged regions and favours the synthesis of nanoparticles [58]. Moreover, the increase in pH resulted in shorter synthesis duration and nanoparticles of smaller size [47]. Devi and co-workers [28] observed that an alkaline pH resulted in AgNPs of desired shape and morphology using 75 different isolates of biocontrol agent *viz., Trichoderma harzianum, T. asperellum, T. pseudokoningii, T. longibacterium,* and *T.virens.* Kamil and co-workers [29] used the fungus *Beauveria bassania* for the synthesis of AgNPs and concluded with similar results of alkaline favouring swift synthesis. Synthesis rate was found to be increased at alkaline pH and was completed within 20 min by using *Colletotrichum* sp [21]. Lower pH usually renders aggregation of particles.

Culture Medium

The culture medium is an important parameter as microorganisms are cultivated in the culture medium and constituents of the medium leads to the synthesis of various metabolites and proteins by the fungi. Providing a particular substrate triggering an enzyme responsible for nanoparticle synthesis can be indeed helpful. The nanoparticles produced by using filtrate/supernatant extracted from the fungus (*F. oxysporum*) grown in an enzyme (nitrate reductase enzyme) induction medium (0.35%yeast extract, 1% peptone, 0.35% potassium nitrate, and 1.5% glucose) were of smaller sizes with higher concentrations [27].

Precursor Concentration

An optimum concentration of precursor should be used for the proper and successful synthesis of AgNPs. A higher concentration of precursor may result in agglomeration of nanoparticles due to nuclei- nuclei interactions. In most of the studies conducted, a concentration of $AgNO_3$ has been optimized to 1 mM as it favours smaller nanoparticles of the desired concentration and improved dispersion. It has been found that up to 8 mM $AgNO_3$ concentration, the size of AgNPs ranged between 7 to 17 nm, however, the size of AgNPs dramatically increased to 45.9 and 62.13 at 9 mM and 10 mM concentrations of $AgNO_3$ [40]. However, AbdelRahim and his team [42] achieved the smallest nanoparticles of 2.86 nm at an $AgNO_3$ concentration of 10 mM whereas 100 mM and 1 mM resulted in the nanoparticles of size 54.67 nm and 14.23 nm, respectively. Such effects possibly could be accounted to the lack of functional group in the reaction mixture to react with increased concentration of precursor.

Agitation and Light

A constant agitation at 100-150 RPM aids in the swift synthesis of nanoparticles. Agitation thwarts agglomeration of nanoparticles which could happen due to nuclei- nuclei interaction. It has been observed that providing dark conditions favours the better synthesis of nanoparticles [28]. Perhaps, under dark conditions, the enzymes become more active and work better.

Applications of AgNPs

Silver nanoparticles generated using fungal extract have several prospective applications in the field of agriculture and medical sciences. In agriculture, the AgNPs have shown promising result in managing pests (pathogens and insects). There are several studies on with the potential effectiveness of fungal-mediated AgNPs in controlling the pest in agriculture [59]. Colloidal nano-silver of average size 1.5 nm showed effective antifungal activity against rose powdery mildew (*Sphaerotheca pannosa* var *rosae*) [60]. Min and co-worker [61] observed that 7 ppm of AgNPs was effective in inhibiting *Rhizoctonia solani*, *Sclerotinia sclerotiorum* and *S.minor*. In a study conducted by Kim and co-worker [62], a concentration ranging from 10 to 100 ppm could effectively control fungi *viz.*, *Alternaria alternata*, *A. brassicicola*, *Botrytis cinerea*, *Cladosporium cucumerinum*, *Corynespora cassiicola*, *Cylindrocarpon destructans*, *Didymell bryoniae*, *Fusarium oxysporum* f. sp. *cucumerinum*, *F. oxysporum* f. sp. *lycopersici*, *F. solani*, *Glomerella cingulata*, *Monosporas cuscannonballus*,

Phythium aphanidermatum, P. spinosum and *Stemphylium lycopersici.* Balakumaran and co-workers [39] synthesized AgNPs using *Guignardia mangiferae* and results showed their effectiveness against *Colletotrichum* sp., *Rhizoctonia solani,* and *Curvularia lunata.* AgNPs have also been found effective in inhibiting the growth of fungus *Fusarium solani* [63]. The antibacterial property of the AgNPs is well documented. Shirley *et al.,* 2010 [64] reported the antibacterial effect of AgNPs against various gram-positive and gram-negative bacterial strains. AgNPs have been found very effective at 20 ppm against *Staphylococcus aureus* [65], at 0.3 ppm against *Bacillus subtilis* and *Escherichia coli* [66], at 10-30 ppm against *Pseudomonas aeruginosa* [67].

There are several reports pertaining to the investigation of AgNP's potential effects against insect pests. AgNPs synthesized using *Beauveria bassania,* exhibited insecticidal properties against *Lipaphis erysimi* [29]. A concentration of 0.2-1% of AgNPs synthesized using *Trichoderma harzianum* and Beauveria *bassania* was found against *Aedes aegypti* causing dengue and malaria [68].

In medical science, fungal-mediated AgNPs have been proven to control many bacteria and fungi. Gram negative bacteria were found to be inhibited when treated with AgNPs synthesized using *Guignardia mangiferae* due to disruption of membrane and DNA [39]. Bacteria such as *Staphylococcus aureus* and *Klebsiella pneumonia* were effectively controlled using AgNPs synthesized by using *Trichoderma harzianum* [38]. Fatima and co-workers [69] reported the effectiveness of AgNPs synthesized using *Aspergillus flavus* against *Bacillus cereus, Bacillus subtilis, Enterobacter aerogenes, Escherichia coli,* and *Staphylococcus aureus.*

Antifungal activity of AgNPs has been reported against *Candida* spp. synthesized using *Pleurotus* spp [70]. AgNPs synthesized using *Trichophyton rubrum* [71] and *Fusarium oxysporum* [72] was found effective against *Candida albicans* at a concentration of $4 \mu g/mL$.

AgNPs have also been found to possess potential antitumor activity. AgNPs at a concentration of 121.23 μgcm^3 synthesized using *Fusarium oxysporum* could effectively control tumor cell line (MCF-7-human breast adenocarcinoma) exhibiting cytotoxicity against tumor cell. High cytotoxicity was also observed against HeLa (Human cervical carcinoma) and MCF-7 tumor cells, when treated with AgNPs, which were synthesized using *Guignardia mangiferae* [39]. Such effects are possibly due to membrane damage, disruption of microbial respiratory chain, and production of reactive oxygen species.

Mechanism of Antimicrobial Activity of AgNPs

Antimicrobial activity can be attributed to the disruption of the cell wall, metabolic process, and production of reactive oxygen species. Nevertheless, the size of the nanoparticles is the main factor responsible for their antimicrobial property. By virtue of their small size, AgNPs can penetrate the membrane easily and affect the respiratory chain, altering membrane permeability, rendering DNA and RNA damage, disrupting cell division, and eventually leading to cell death. Intriguingly, nanoparticles are found to damage microorganism DNA and inhibition of the activity of DNases by interacting with the thiol groups of essential enzymes resulting in the release of Ag^+ ions that form complexes with nucleotides [73]. The binding of nanoparticles also leads to the inactivation of several indispensable enzymes which result in oxidative stress, which influences electron transport and protein oxidation [69, 74]. Because of the large surface area of silver, the antimicrobial activity is greatly enhanced. Due to the large surface of ions released, ions can easily bind with proteins containing sulphur which in turn affect the cell permeability and render disruption to the DNA replication process.

Fungal Mediated Extracellular Synthesis of Gold Nanoparticles (AuNPs)

So far around 30 fungal species have been exploited to synthesize AuNPs. Bacteria and fungi both are well capable of biosynthesizing AuNPs. It is a generally known idea that during the reduction process, Au^{3+} ions are reduced by microorganisms and form metal-protein conjugates as a response to the toxic stress of metal. Such reduction is carried out by the enzymes, which are secreted in large amounts in fungi. Fungi offer several advantages over other microorganisms such as large secretion of a large number of extracellular enzymes (NADH-reductases; phytochelatins) and proteins (melanin) responsible for the reduction and stabilization of AuNPs [46]. Several other advantages include very high metal toxicity resistance [75], very fast biosynthesis rate (less than an hour) [76], the likelihood of scalability to industrial process and largely extracellular mode of synthesis [2].

Several fungi have been used for the extracellular synthesis of AuNPs (Table **2**). However, there are reports of intercellular synthesis as well. Mukherjee and co-workers [46] reported the synthesis of AuNPs using *Verticillium* sp. using aqueous $AuCl_4$ as a precursor which generated nanoparticles of size 20 nm. The cell-free filtrate of *Alternaria alternata* and *Aspergillus clavatus* has been exploited to synthesize AuNPs rendering in AuNPs of 12±5 nm [77, 78]. Mishra and team [79] synthesized AuNPs (10-60 nm) extracellularly using the supernatant of *Penicillium brevicompactum*. Active biomass of *Fusarium* sp. is also found to be efficient in synthesizing AuNPs extracellularly of spherical and triangular shape and size of 10-80 nm such as *F. nigrum* [80] and *F.oxysporum* [81]. Pathogenic fungi *Helminthosporium solani was* also found capable of synthesizing AuNPs of spherical shape and size ranging between 2-70 nm [56].

Table 2. Fungi reported for the synthesis of AuNPs.

Fungi	AuNPs Characteristics	Optimized Parameters	References
Aspergillus clavatus	Size:24±11 nm; shape: spherical, triangular, hexagonal and elliptical	Temp: Room temperature (RT); pH:neutral	[78]
Aspergillus sydowii	Size:8.7-15.6 nm; Spherical	Temp:N.A.	[82]
Epicoccum nigrum	Size: 5–50 nm; spherical	Temp: 27-29 °C	[83]
Fusarium oxysporum	Size:8-40 nm; Spherical, triangular	Temp: NA	[80]
Fusarium semitectum	Size: 10-80 nm; Spherical	Temp: RT	[82]
Helminthsosporum solani	Size: 2–70 nm; Spheres, rods, triangles, pentagons,	Temp: 37±1 °C	[56]
Hormoconis resinae	Size:3-20 nm; Spherical	Temp: 30 °C	[84]
Penicillium brevicompactum	Size: 10-60 nm; spherical, triangular, hexaconal	Temp: 30 °C	[79]
Verticillium sp.	Size:20-50 nm; spherical	Temp: Temp25-28 °C; pH: 5.5 ± 6.0	[46]
Neurospora crassa	Size: 3-100 nm; spherical	Temp: Temp:28 °C; pH:9; Conc: 1mM	[86]

Mechanism of AuNPs Synthesis Extracellularly

In the case of intercellular synthesis, cystolic redox mediators are believed to play role in reducing Au^{3+} ions that entered the cell [87]. However, extracellular synthesis of AuNPs occurs when Au^{3+} ions are adsorbed to the cell wall and the protein native to the cell wall reduces it into smaller particles (Fig. 3). Adsorption to the cell wall and exposure to the enzymes possibly could occur due to electrostatic interaction [80]. Compounds of fungal origin like Nicotinamide adenine dinucleotide (NADH)/nicotinamide adenine dinucleotide phosphate (NADPH) oxido-reductase either in the cell surface or in the cytoplasm have been long speculated to participate in the reduction of Au ions. It is important to note that, generally two main precursors are used for the biosynthesis of AuNPs *viz.*, AuCl and $HAuCl_4$ which dissociate into Au^+ and Au^{3+}, respectively. Au^{3+} is largely investigated due to its higher solubility. It is apparent that in the case of Au^{3+}, a three-electron reduction process occurs while only one electron reduction occur in the case of Au^+ suggesting a number of chemical transformations in the former case [88].

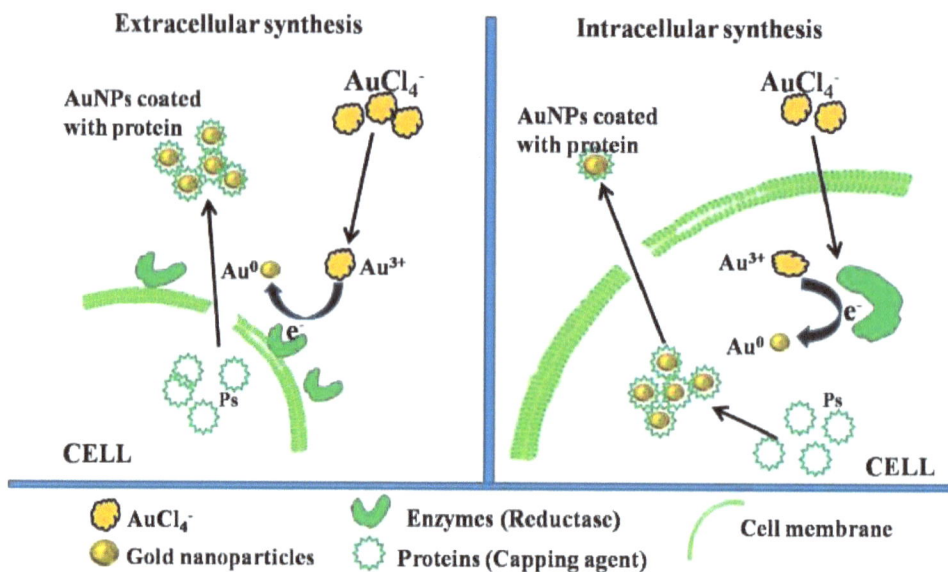

Fig. (3). Mechanism of AuNPs synthesis by fungi.

Mukherjee *et al.* (2002) [80] reported the involvement of NADH-dependent reductase from *Fusarium oxysporum* in the biosynthesis of AuNPs. A similar possible mechanism has been postulated by Narayan and Sakthivel, 2011 [54], where cell filtrate of *Sclerotium rolfsii* containing co-enzyme NADPH could efficiently reduce the Au^{3+} ions into AuNPs. Another study showed the possible role of tyrosine and tryptophan in the reduction of Au^{3+} ions in the filtrate obtained from *Phanerochaete chrysosporium* [89]. Certain functional groups such as hydroxyl, amine, and carboxyl groups are also responsible for the reduction of Au^{3+} to AuNPs in the cell free extract. More specifically carboxyl groups are the main player in the synthesis process [90]. Certain amides and carboxyl groups secreted by the cell wall of *Magnusiomyces ingens* are proven to be potentially reducing and capping for AuNPs [91]. Phytochelatins an oligomer produced by fungus *Candida albicans* which in the presence of glutathione metal ions (Au), renders synthesis of AuNPs by reducing Au^{3+} ions [92]. The glutathione molecules further cap the AuNPs conferring stability. One study demonstrates the participation of biosynthetic redox mediators in the biosynthesis of AuNPs in *Yarrowia lipolytica*. Such fungi produce melanin in large quantity which possibly could be linked to the reduction of Au^{3+} to AuNPs and therefore, are responsible to mediate the reduction of metal salts to their elemental forms as nanostructures [93].

Stabilization of AuNPs by Biomolecules Capping

Nanoparticles (Au^0) formed are highly unstable and consequently leads to aggregation of nanoparticles. Therefore, in any method, stabilization of synthesized nanoparticles is an important aspect of synthesis. In the case of biosynthesis of AuNPs, extracellular proteins act as a capping agent which prevents the aggregation. The FTIR interpretation is the most commonly used to analyze the capping/stabilizing agent. The FTIR determines the functional groups associated with the AuNPs, which could be the capping agent. However, numbers of organic molecules are produced; therefore, it becomes difficult to detect the exact capping agents. Nevertheless, a slight shift in the peaks of the functional groups to lower frequencies indicates the interactions among groups, thus confirming the capping mechanism. For instance, the likelihood of phosphate bonds in AuNPs capping was found, synthesized using *Rhizopus oryzae* in which the band was shifted from 1034 cm^{-1} to 1025 cm^{-1} corresponding to C-N stretching mode when compared with the pristine biomass control bands [94]. The experiments confirmed the presence of amide I, II, and III groups and the disappearance of then carboxyl groups in the mycelia indicating the role of polypeptides in the biosynthetic mechanism. Since the amide I, II, and III peaks shifted from 1652.9, 1550, and 1379 cm^{-1} to 1635.5, 1544, and 1371.2 cm^{-1}, respectively, this confirms the involvement of these amides in the capping process. Several other studies have shown the amide group as possible capping agent [77, 95]. Zhang and co-workers [96] identified three putative capping proteins *viz.,* plasma membrane ATPase, 3-glucan binding protein, and glyceraldehyde-3-phosphate dehydrogenase.

Factor/Parameters affecting AuNPs Biosynthesis

Different parameters such as temperature, gold ion concentration, and pH have a significant effect on the reduction and stabilization of the AuNPs.

Precursor Concentration

The formation of AuNPs is favoured by 2mM $HAuCl_4$. Generally, a lower concentration of precursor ($HAuCl_4$) decreases the formation of initial metal nuclei, resulting in the lower generation of AuNPs. But higher $HAuCl_4$ concentration may lead to polydispersity; multiple shapes ad large sizes [85]. The concentration of precursors has direct effects on the size and shape of AuNPs synthesized. Therefore, the optimization of the concentration is imperative.

The pH of the Medium

The acidic and alkaline condition affects the proteins which in turn disturbs the stability of AuNPs. Neutral conditions are found to be more suitable for the synthesis of AuNPs [85]. Low pH has resulted in increased aggregation of AuNPs [97].

Temperature

Optimization of temperature is critical as it may affect the reducing enzymes. Generally elevated temperatures (ranging from 50 °C to 90°C), results in the formation of uniform particles. In optimized conditions, the gold ions completely convert into gold nanoparticles with an increase in the rate of the reaction. However, the best temperature for synthesis may vary according to the concentration of precursors and type of fungal species. High temperature is usually conducive to nucleation while the low temperature is conducive to the growth of nanoparticle synthesis [98].

Application of AuNPs

The intrinsic optical properties of AuNPs make them impeccable for use as therapeutic agents in the clinic. AuNPs are widely used for the treatment of oncological diseases, skin, and other infectious diseases. It renders apoptosis or necrosis is induced in tumor cells by highly generating free radicals and single oxygen *via* the energy of photosensitizers [99]. AuNPs are also used for X-ray imaging as an X-ray contrast agent as it represents a high X-ray absorption coefficient, ease of synthetic manipulation, non-toxicity, surface functionalization for colloidal stability and targeted delivery [100]. AuNPs are widely used as a drug delivery agent. It is an effective carrier and is capable of transferring various drugs such as peptides [101], proteins [102], plasmid DNAs (cDNAs), small interfering RNAs (siRNAs), and chemotherapeutic agents [103].AuNPs are also used as a biosensor to detect biomolecules and molecules like nucleic acids of pathogens. Different types of Nanobiosensors employ different features of AuNPs. For instance, AuNPs Sensor based on Gold nanoislands (AuNIs) on glass substrate functionalized with artificially produced DNA probe receptor was developed to detect COVID-19, was found to be highly sensitive, fast, and reliable [7]. Au NP-based Bio-Barcode Assay is based on the strong binding affinity of AuNPs to thiols which further leads to the colour change due to aggregation of AuNPs [104].

AuNPs possess antibacterial/antifungal properties that can be exploited to protect plants from phytopathogens. Moreover, they are used for pesticide identification phytopathogens detection, and water purification.

Characterization in Nanoparticles

Characterization of synthesized nanoparticles is conducted to confirm the formation of nanoparticles and to determine their shape, size, particle charge, and stability, and hence characterization is mainly based on the size, shape, charge, and stability of NPs so formed. Various techniques are used for the characterization *viz.*, UV-Visible Spectra Analysis, X-ray Diffraction (XRD) analysis, Fourier Transform Infrared (FTIR) Spectroscopy, Particle size analyzer (PSA), Zeta size analyzer (ZSA) and microscopic techniques such as Transmission Electron Microscopy (TEM), Scanning Electron Microscopy (SEM) and Atomic Force Microscopy (AFM) (Table **3**).

Table 3. Characterisation of AgNPs using various instruments.

Fungi	UV-Vis (λ) nm	XRD (nm)	FTIR	TEM	References
Alternaria alternata	Peak at 413 nm	--	Amide II band: 1513 cm^{-1}; C–C stretching: 1629 cm^{-1}; C-N stretching: 1356 cm^{-1}	Shape:pol-ydisperse spherical; Size: 20 to 60 nm.	[30]
Trichoderma reesei	--	--	C=O:1650 cm^{-1}; N-H: 1450 cm^{-1}	Size: 5-50nm	[32]
Aspergillus flavus	Absorption peak at 420 nm	--	Primary amines: 3280 cm-1 and secondary	Size: 8.92±1.61 nm.	[33]
Penicillium brevicompactum WA 2315	Absorption band at 420 nm	The peaks at 2Θ values of 35.918, 45.548, 62.688 and 75.758 correspond to (1 1 1), (2 0 0), (2 2 0) and (3 1 1) planes of silver respectively	Stretching vibrations of primary: 3356 cm^{-1}; secondary amines: 2922 cm^{-1}C–N stretching vibrations of aromatic: 1412 cm^{-1}; aliphatic amines:1029 cm^{-1}	Size: 58.35 ±17.88 nm.	[36]
Aspergillus fumigatus	Peak at 420 nm	2θ values corresponding to the silver nanocrystal	--	Size: 5–25 nm.	[34]

Fungi	UV-Vis (λ) nm	XRD (nm)	FTIR	TEM	References
Yeast strain MKY3	Peak at 420 nm	Latter component is of fcc-type bulk silver	--	Size:2–5 nm	[51]
Aspergillus terreus	peak at 440 nm	Good agreement with the unit cell of the face centred cubic (fcc) structure; average size= 5.2 nm.	--	Shape:spherical Size:1- 20 nm	[1]
Alternaria alternata	Absorbance peak at 420 nm	--	Amide II band:1513 cm^{-1}; C–C stretching: 1629 cm^{-1}; C-N stretching:1356 cm^{-1}	Shape:pol-ydisperse spherical; Size:20 to 60 nm	[30]
Beauveria bassiana	Absorbance peak at 420 nm	Lattice plane value index at 524,520,296,279,266,210	Aldehyde:3420 cm^{-1};aromatic groups:1037 cm^{-1}; hydroxyl group: 2360 cm^{-1}	Size: 3- 25 nm	[29]
Pleurotus sajor-caju	Absorption band at 381nm	--	--	Size: 5-50nm	[32]
Verticillium sp.	Absorption band centred at ca. 450 nm	--	--	Size: 25-12 nm.	[46]
Fusarium oxysporum	Absorption at ca. 415–420 nm	--	--	Shape:spherical ;size:20–50 nm	[44]
Trichoderma asperellum, T. harzianum, T.longibrachiatum, T.pseudokoningii and *T. virens*	Absorption peak at 420 nm	--	--	Size:8-60 nm	[28]
Verticillium sp.	Absorption peak around 540 nm	Bragg reflections of gold agree with gold nano crystals; size:25 nm	--	Size:20±8 nm	[46]
Neurospora crassa	Peaks at 520 nm	--	--	Size: 3-100 nm	[25]

UV-Visible Spectral Analysis

Numerous works have demonstrated a progressive characteristic peak as the reaction time is prolonged, which could be mainly because of the increased surface area to volume ratio or in other words due to the reduction in the size of the particles. In fact, the characteristic peaks observed in the UV-Vis absorption spectrum are due to the Surface Plasmon Resonance (SPR) of NPs [105].

X-ray Diffraction (XRD) Analysis

XRD analysis is mainly used to determine the metallic or crystalline nature of NP [106]. It provides information regarding translational symmetry, the electron density of a unit cell, the shape, and size of a particle [107].

Fourier Transform Infrared (FTIR) Spectroscopy

FTIR spectroscopy determines the association of functional group from biological extract coated around NPs. This technique also helps to determine the structural features of biological extract associated with the NPs. It measures infrared intensity *vs* wavelength which is used to find out the functional groups associated with the synthesized nanoparticles.

Particle Size Analyser (PSA)

A PSA analyser is used to determine the size of an individual particle as well as a size distribution range for a given sample. It measures nanoparticle size by light scattering, laser diffraction, photon correlation spectroscopy, and sedimentation. The aforesaid three methods work by watching how light or lasers are shifted due to the particles and sedimentation determines particle size by measuring how fast the particles fall to the bottom of the sample. Some features to look for nano PSA includes the ability to process wet and/or dry dispersions; the particle size range that can be measured; whether it can determine shape as well as size; and whether it can measure concentration, zeta potential, and aggregation.

Transmission Electron Microscopy (TEM)

This Technique is mostly used for morphological studies of synthesized NPs. TEM studies helps in determining the distribution of NPs in various sizes ranges [108].

Scanning Electron Microscopy (SEM)

This technique is also used to study the morphology of the synthesized nanoparticles *viz.*, shape, size, and distribution.

CONCLUSION AND FUTURE PROSPECTS

Biosynthesis of nanoparticles has garnered much attention because of its several advantages. Several recent studies show the advantages of the biosynthesis method over the chemical and physical methods. Although chemical and physical methods are efficient in generating a large amount of NPs, unfortunately, several setbacks can be listed out. However, resorting to the biological method of nanoparticle synthesis can be one solution to many problems. Although, a plethora of organisms is at the disposal for the synthesis fungi have proven to be a better candidate in many aspects. Fungi win the race due to their ability to secrete a large amount of extracellular enzymes responsible for reducing the metal ions and stabilizing the same by coating with proteins. Such coatings based on the type of fungal species used may also act in synergy in some pathosystems. Viewing the diversity of fungi, certainly, there is an untapped potential waiting to be exploited for nanoparticle synthesis. Basically, an enzymatic process in extracellular synthesis is the same as metal detoxification.

To tap the true potential of fungi in the biosynthesis of nanoparticles, more research should be directed towards understanding the fungus growth and different parameters affecting it. The parameters affecting the reaction also need to be optimized like pH, temperature, precursor concentration, *etc.* to achieve NPs of defined size and shape. Generally, a difficulty arises in scaling up production which can be solved with research-oriented to fungi and biosynthesis. Moreover, there is a need to understand in detail the exact mechanism of nanoparticles formation using fungi and metals. Understanding the mechanisms of its inhibitory affects, on pathogens can definitely help us in better management of pathogens.

Among different fungi, filamentous fungi *Aspergillus* sp. and *Trichoderma* sp. produces a number of extracellular enzymes and their genomic composition are well known and hence can serve as a model organism to understand the genetic and biochemical process of bioreduction by fungi [109]. The certain gene encoding for the enzymes responsible for the NP synthesis can be up-regulated for the enhanced synthesis of nanoparticles. The use of fungi in Silver and gold nanoparticle synthesis can open new vistas.

CONSENT FOR PUBLICATION

Not applicable.

CONFLICT OF INTEREST

The authors declare no conflict of interest, financial or otherwise.

ACKNOWLEDGEMENT

Declared none.

REFERENCES

[1]   Li G, He D, Qian Y, *et al.* Fungus-mediated green synthesis of silver nanoparticles using *Aspergillus terreus.* Int J Mol Sci 2012; 13(1): 466-76.
[http://dx.doi.org/10.3390/ijms13010466] [PMID: 22312264]

[2]   Durán N, Marcato PD, Durán M, Yadav A, Gade A, Rai M. Mechanistic aspects in the biogenic synthesis of extracellular metal nanoparticles by peptides, bacteria, fungi, and plants. Appl Microbiol Biotechnol 2011; 90(5): 1609-24.
[http://dx.doi.org/10.1007/s00253-011-3249-8] [PMID: 21484205]

[3]   Ahmad A, Mukherjee P, Senapati S, *et al.* Extracellular biosynthesis of silver nanoparticles using the fungus *Fusarium oxysporum.* Colloids Surf B Biointerfaces 2003; 28(4): 313-8.
[http://dx.doi.org/10.1016/S0927-7765(02)00174-1]

[4]   Jo YK, Kim BH, Jung G. Antifungal activity of silver ions and nanoparticles on phytopathogenic fungi. Plant Dis 2009; 93(10): 1037-43.
[http://dx.doi.org/10.1094/PDIS-93-10-1037] [PMID: 30754381]

[5]   Sharon M, Choudhary AK, Kumar R. Nanotechnology in agricultural diseases and food safety. J Phytol 2010; 2(4)

[6]   Giljohann DA, Seferos DS, Daniel WL, Massich MD, Patel PC, Mirkin CA. Gold nanoparticles for biology and medicine. Angew Chem Int Ed Engl 2010; 49(19): 3280-94.
[http://dx.doi.org/10.1002/anie.200904359] [PMID: 20401880]

[7]   Qiu G, Gai Z, Tao Y, Schmitt J, Kullak-Ublick GA, Wang J. Dual-functional plasmonic photothermal biosensors for highly accurate severe acute respiratory syndrome coronavirus 2 detection. ACS Nano 2020; 14(5): 5268-77.
[http://dx.doi.org/10.1021/acsnano.0c02439] [PMID: 32281785]

[8]   Bhumkar DR, Joshi HM, Sastry M, Pokharkar VB. Chitosan reduced gold nanoparticles as novel carriers for transmucosal delivery of insulin. Pharm Res 2007; 24(8): 1415-26.
[http://dx.doi.org/10.1007/s11095-007-9257-9] [PMID: 17380266]

[9]   Shukla R, Bansal V, Chaudhary M, Basu A, Bhonde RR, Sastry M. Biocompatibility of gold nanoparticles and their endocytotic fate inside the cellular compartment: a microscopic overview. Langmuir 2005; 21(23): 10644-54.
[http://dx.doi.org/10.1021/la0513712] [PMID: 16262332]

[10]   Gupta RK, Kumar V, Gundampati RK, Malviya M, Hasan SH, Jagannadham MV. Biosynthesis of silver nanoparticles from the novel strain of *Streptomyces* Sp. BHUMBU-80 with highly efficient electroanalytical detection of hydrogen peroxide and antibacterial activity. J Environ Chem Eng 2017; 5(6): 5624-35.
[http://dx.doi.org/10.1016/j.jece.2017.09.029]

[11]   Loo YY, Rukayadi Y, Nor-Khaizura MA, *et al. In vitro* antimicrobial activity of green synthesized silver nanoparticles against selected gram-negative foodborne pathogens. Front Microbiol 2018; 9: 1555.
[http://dx.doi.org/10.3389/fmicb.2018.01555] [PMID: 30061871]

[12]   Zhu HT, Zhang CY, Yin YS. Rapid synthesis of copper nanoparticles by sodium hypophosphite reduction in ethylene glycol under microwave irradiation. J Cryst Growth 2004; 270(3-4): 722-8.
[http://dx.doi.org/10.1016/j.jcrysgro.2004.07.008]

[13]   Ramya M, Subapriya MS. Green synthesis of silver nanoparticles. Int J Pharm Med Biol Sci 2012; 1(1): 54-61.

[14]   Thakkar KN, Mhatre SS, Parikh RY. Biological synthesis of metallic nanoparticles. Nanomedicine 2010; 6(2): 257-62.
[http://dx.doi.org/10.1016/j.nano.2009.07.002] [PMID: 19616126]

[15]   Zielonka A, Klimek-Ochab M. Fungal synthesis of size-defined nanoparticles. Advances in Natural Sciences: Nanoscience and Nanotechnology 2017; 8(4): 043001.

[16]   Park HJ, Kim SH, Kim HJ, Choi SH. A new composition of nanosized silica-silver for control of various plant diseases. Plant Pathol J 2006; 22(3): 295-302.
[http://dx.doi.org/10.5423/PPJ.2006.22.3.295]

[17]   Gurunathan S, Kalishwaralal K, Vaidyanathan R, *et al.* Biosynthesis, purification and characterization of silver nanoparticles using Escherichia coli. Colloids Surf B Biointerfaces 2009; 74(1): 328-35.
[http://dx.doi.org/10.1016/j.colsurfb.2009.07.048] [PMID: 19716685]

[18]   Husseiny SM, Salah TA, Anter HA. Biosynthesis of size controlled silver nanoparticles by *Fusarium oxysporum*, their antibacterial and antitumor activities. Beni Suef Univ J Basic Appl Sci 2015; 4(3): 225-31.
[http://dx.doi.org/10.1016/j.bjbas.2015.07.004]

[19]   Mohanta YK, Nayak D, Biswas K, *et al.* Abd_Allah EF, Hashem A, Alqarawi AA, Yadav D, Mohanta TK. Silver nanoparticles synthesized using wild mushroom show potential antimicrobial activities against food borne pathogens. Molecules 2018; 23(3): 655.
[http://dx.doi.org/10.3390/molecules23030655] [PMID: 29538308]

[20]   Molnár Z, Bódai V, Szakacs G, *et al.* Green synthesis of gold nanoparticles by thermophilic filamentous fungi. Sci Rep 2018; 8(1): 3943.
[http://dx.doi.org/10.1038/s41598-018-22112-3] [PMID: 29500365]

[21]   Azmath P, Baker S, Rakshith D, Satish S. Mycosynthesis of silver nanoparticles bearing antibacterial activity. Saudi Pharm J 2016; 24(2): 140-6.
[http://dx.doi.org/10.1016/j.jsps.2015.01.008] [PMID: 27013906]

[22]   Qidwai A, Kumar R, Shukla SK, Dikshit A. Advances in biogenic nanoparticles and the mechanisms of antimicrobial effects. Indian J Pharm Sci 2018; 80(4): 592-603.
[http://dx.doi.org/10.4172/pharmaceutical-sciences.1000398]

[23]   Kirkland BH, Keyhani NO. Expression and purification of a functionally active class I fungal hydrophobin from the entomopathogenic fungus *Beauveria bassiana* in *E. coli.* J Ind Microbiol Biotechnol 2011; 38(2): 327-35.
[http://dx.doi.org/10.1007/s10295-010-0777-7] [PMID: 20640587]

[24]   Chen YL, Tuan HY, Tien CW, Lo WH, Liang HC, Hu YC. Augmented biosynthesis of cadmium sulfide nanoparticles by genetically engineered *Escherichia coli.* Biotechnol Prog 2009; 25(5): 1260-6.
[http://dx.doi.org/10.1002/btpr.199] [PMID: 19630084]

[25]   Castro-Longoria E, Moreno-Velázquez SD, Vilchis-Nestor AR, Arenas-Berumen E, Avalos-Borja M. Production of platinum nanoparticles and nanoaggregates using *Neurospora crassa.* J Microbiol Biotechnol 2012; 22(7): 1000-4.
[http://dx.doi.org/10.4014/jmb.1110.10085] [PMID: 22580320]

[26]   Birla SS, Gaikwad SC, Gade AK, Rai MK. Rapid synthesis of silver nanoparticles from *Fusarium oxysporum* by optimizing physicocultural conditions. The Scientific World Journal 2013; 2013: 796018.
[http://dx.doi.org/10.1155/2013/796018] [PMID: 24222751]

[27]   Hamedi S, Ghaseminezhad M, Shokrollahzadeh S, Shojaosadati SA. Controlled biosynthesis of silver nanoparticles using nitrate reductase enzyme induction of filamentous fungus and their antibacterial evaluation. Artif Cells Nanomed Biotechnol 2017; 45(8): 1588-96.
[http://dx.doi.org/10.1080/21691401.2016.1267011] [PMID: 27966375]

[28]   Devi TP, Kulanthaivel S, Kamil D, Borah JL, Prabhakaran N, Srinivasa N. Biosynthesis of silver

nanoparticles from *Trichoderma* species.

[29]   Kamil D, Prameeladevi T, Ganesh S, Prabhakaran N, Nareshkumar R, Thomas SP. Green synthesis of silver nanoparticles by entomopathogenic fungus Beauveria bassiana and their bioefficacy against mustard aphid (*Lipaphis erysimi* Kalt.).

[30]   Gajbhiye M, Kesharwani J, Ingle A, Gade A, Rai M. Fungus-mediated synthesis of silver nanoparticles and their activity against pathogenic fungi in combination with fluconazole. Nanomedicine 2009; 5(4): 382-6.
[http://dx.doi.org/10.1016/j.nano.2009.06.005] [PMID: 19616127]

[31]   Nithya R, Ragunathan R. Synthesis of silver nanoparticle using *Pleurotus sajorcaju* and its antimicrobial study. Dig J Nanomater Biostruct 2009; 4(4): 623-9.

[32]   Vahabi K, Mansoori GA, Karimi S. Biosynthesis of silver nanoparticles by fungus *Trichoderma reesei* (a route for large-scale production of AgNPs). Insciences J 2011; 1(1): 65-79.
[http://dx.doi.org/10.5640/insc.010165]

[33]   Vigneshwaran N, Ashtaputre NM, Varadarajan PV, Nachane RP, Paralikar KM, Balasubramanya RH. Biological synthesis of silver nanoparticles using the fungus Aspergillus flavus. Mater Lett 2007; 61(6): 1413-8.
[http://dx.doi.org/10.1016/j.matlet.2006.07.042]

[34]   Bhainsa KC, D'Souza SF. Extracellular biosynthesis of silver nanoparticles using the fungus *Aspergillus fumigatus.* Colloids Surf B Biointerfaces 2006; 47(2): 160-4.
[http://dx.doi.org/10.1016/j.colsurfb.2005.11.026] [PMID: 16420977]

[35]   Jaidev LR, Narasimha G. Fungal mediated biosynthesis of silver nanoparticles, characterization and antimicrobial activity. Colloids Surf B Biointerfaces 2010; 81(2): 430-3.
[http://dx.doi.org/10.1016/j.colsurfb.2010.07.033] [PMID: 20708910]

[36]   Shaligram NS, Bule M, Bhambure R, *et al.* Biosynthesis of silver nanoparticles using aqueous extract from the compactin producing fungal strain. Process Biochem 2009; 44(8): 939-43.
[http://dx.doi.org/10.1016/j.procbio.2009.04.009]

[37]   Thirumurugan G, Shaheedha SM, Dhanaraju MD. *in vitro* evaluation of antibacterial activity of silver nanoparticles synthesised by using *Phytophthora infestans.* Int J Chemtech Res 2009; 1(3): 714-6.

[38]   Ahluwalia V, Kumar J, Sisodia R, Shakil NA, Walia S. Green synthesis of silver nanoparticles by *Trichoderma harzianum* and their bio-efficacy evaluation against *Staphylococcus aureus* and *Klebsiella pneumonia.* Ind Crops Prod 2014; 55: 202-6.
[http://dx.doi.org/10.1016/j.indcrop.2014.01.026]

[39]   Balakumaran MD, Ramachandran R, Kalaichelvan PT. Exploitation of endophytic fungus, *Guignardia mangiferae* for extracellular synthesis of silver nanoparticles and their *in vitro* biological activities. Microbiol Res 2015; 178: 9-17.
[http://dx.doi.org/10.1016/j.micres.2015.05.009] [PMID: 26302842]

[40]   Phanjo P, Ahmed G. Effect of different physicochemical conditions on the synthesis of silver nanoparticles using fungal cell filtrate of *Aspergillus oryzae* (MTCC No. 1846) and their antibacterial effect. Advances in Natural Sciences: Nanoscience and Nanotechnology 2017; 8(4): : 045016..

[41]   Elamawi RM, Al-Harbi RE, Hendi AA. Biosynthesis and characterization of silver nanoparticles using *Trichoderma longibrachiatum* and their effect on phytopathogenic fungi. Egypt J Biol Pest Control 2018; 28(1): 1-1.
[http://dx.doi.org/10.1186/s41938-018-0028-1]

[42]   AbdelRahim K, Mahmoud SY, Ali AM, Almaary KS, Mustafa AE, Husseiny SM. Extracellular biosynthesis of silver nanoparticles using *Rhizopus stolonifer.* Saudi J Biol Sci 2017; 24(1): 208-16.
[http://dx.doi.org/10.1016/j.sjbs.2016.02.025] [PMID: 28053592]

[43]   Ahmad A, Mukherjee P, Mandal D, *et al.* Enzyme mediated extracellular synthesis of CdS nanoparticles by the fungus, *Fusarium oxysporum.* J Am Chem Soc 2002; 124(41): 12108-9.

[http://dx.doi.org/10.1021/ja027296o] [PMID: 12371846]

[44]   Durán N, Marcato PD, Alves OL, Souza GI, Esposito E. Mechanistic aspects of biosynthesis of silver nanoparticles by several *Fusarium oxysporum* strains. J Nanobiotechnology 2005; 3(1): 8.
       [http://dx.doi.org/10.1186/1477-3155-3-8] [PMID: 16014167]

[45]   Mohammed Fayaz A, Balaji K, Kalaichelvan PT, Venkatesan R. Fungal based synthesis of silver nanoparticles--an effect of temperature on the size of particles. Colloids Surf B Biointerfaces 2009; 74(1): 123-6.
       [http://dx.doi.org/10.1016/j.colsurfb.2009.07.002] [PMID: 19674875]

[46]   Mukherjee P, Ahmad A, Mandal D, *et al.* Bioreduction of AuCl4− ions by the fungus, *Verticillium* sp. and surface trapping of the gold nanoparticles formed. Angew Chem Int Ed 2001; 40(19): 3585-8.
       [http://dx.doi.org/10.1002/1521-3773(20011001)40:19<3585::AID-ANIE3585>3.0.CO;2-K]

[47]   Du L, Xu Q, Huang M, Xian L, Feng JX. Synthesis of small silver nanoparticles under light radiation by fungus *Penicillium oxalicum* and its application for the catalytic reduction of methylene blue. Mater Chem Phys 2015; 160: 40-7.
       [http://dx.doi.org/10.1016/j.matchemphys.2015.04.003]

[48]   Nayak BK, Nanda A, Prabhakar V. Biogenic synthesis of silver nanoparticle from wasp nest soil fungus, *Penicillium italicum* and its analysis against multi drug resistance pathogens. Biocatal Agric Biotechnol 2018; 16: 412-8.
       [http://dx.doi.org/10.1016/j.bcab.2018.09.014]

[49]   Shahzad A, Saeed H, Iqtedar M, *et al.* Size-controlled production of silver nanoparticles by *Aspergillus fumigatus* BTCB10: likely antibacterial and cytotoxic effects. J Nanomater 2019; 2019: 2019.
       [http://dx.doi.org/10.1155/2019/5168698]

[50]   Thokala PD, Kamil D, Toppo RS. Silver nanoparticles production by Aspergillus niger and their antibacterial efficacy against *Xanthomonas citri* and *Ralstonia solanacearum.* J Environ Biol 2018; 39(4): 493-9.
       [http://dx.doi.org/10.22438/jeb/39/4/MRN-489]

[51]   Kowshik M, Ashtaputre S, Kharrazi S, *et al.* Extracellular synthesis of silver nanoparticles by a silver-tolerant yeast strain MKY3. Nanotechnology 2002; 14(1): 95-100.
       [http://dx.doi.org/10.1088/0957-4484/14/1/321]

[52]   Creighton JA, Eadon DG. Ultraviolet–visible absorption spectra of the colloidal metallic elements. J Chem Soc, Faraday Trans 1991; 87(24): 3881-91.
       [http://dx.doi.org/10.1039/FT9918703881]

[53]   Baymiller M, Gobble A, Huang F. Rapid one-step synthesis of gold nanoparticles using the ubiquitous coenzyme NADH. Matters (Zur) 2017; 3(7): e201705000007.
       [http://dx.doi.org/10.19185/matters.201705000007]

[54]   Narayanan KB, Sakthivel N. Facile green synthesis of gold nanostructures by NADPH-dependent enzyme from the extract of *Sclerotium rolfsii.* Colloids Surf A Physicochem Eng Asp 2011; 380(1-3): 156-61.
       [http://dx.doi.org/10.1016/j.colsurfa.2011.02.042]

[55]   Hulkoti NI, Taranath TC. Biosynthesis of nanoparticles using microbes- a review. Colloids Surf B Biointerfaces 2014; 121: 474-83.
       [http://dx.doi.org/10.1016/j.colsurfb.2014.05.027] [PMID: 25001188]

[56]   Anil Kumar S, Abyaneh MK, Gosavi SW, *et al.* Nitrate reductase-mediated synthesis of silver nanoparticles from AgNO3. Biotechnol Lett 2007; 29(3): 439-45.
       [http://dx.doi.org/10.1007/s10529-006-9256-7] [PMID: 17237973]

[57]   Hietzschold S, Walter A, Davis C, Taylor AA, Sepunaru L. Does nitrate reductase play a role in silver nanoparticle synthesis? Evidence for NADPH as the sole reducing agent. ACS Sustain Chem& Eng

2019; 7(9): 8070-6.
[http://dx.doi.org/10.1021/acssuschemeng.9b00506]

[58]   Sintubin L, De Windt W, Dick J, *et al.* Lactic acid bacteria as reducing and capping agent for the fast and efficient production of silver nanoparticles. Appl Microbiol Biotechnol 2009; 84(4): 741-9.
[http://dx.doi.org/10.1007/s00253-009-2032-6] [PMID: 19488750]

[59]   Graily-Moradi F, Mallak AM, Ghorbanpour M. Biogenic Synthesis of Gold Nanoparticles and Their Potential Application in Agriculture. InBiogenic Nano-Particles and their Use in Agro-ecosystems. Singapore: Springer 2020; pp. 187-204.

[60]   Kim HS, Kang HS, Chu GJ, Byun HS. Antifungal effectiveness of nanosilver colloid against rose powdery mildew in greenhouses. InSolid State Phenomena. Trans Tech Publications Ltd. 2008; Vol. 135: pp. 15-8.

[61]   Min JS, Kim KS, Kim SW, *et al.* Effects of colloidal silver nanoparticles on sclerotium-forming phytopathogenic fungi. Plant Pathol J 2009; 25(4): 376-80.
[http://dx.doi.org/10.5423/PPJ.2009.25.4.376]

[62]   Kim SW, Jung JH, Lamsal K, Kim YS, Min JS, Lee YS. Antifungal effects of silver nanoparticles (AgNPs) against various plant pathogenic fungi. Mycobiology 2012; 40(1): 53-8.
[http://dx.doi.org/10.5941/MYCO.2012.40.1.053] [PMID: 22783135]

[63]   Abd El-Aziz AR, Al-Othman MR, Mahmoud MA, Metwaly HA. Biosynthesis of silver nanoparticles using *Fusarium solani* and its impact on grain borne fungi. Dig J Nanomater Biostruct 2015; 10(2): 655-62.

[64]   Shirley AD, Dayanand A, Sreedhar B, Dastager SG. Antimicrobial activity of silver nanoparticles synthesized from novel *Streptomyces*. Dig J Nanomater Biostructures 2010; 5(2): 447-51.
[http://dx.doi.org/10.1166/jbn.2007.022]

[65]   Durán N, Marcato PD, De Souza GI, Alves OL, Esposito E. Antibacterial effect of silver nanoparticles produced by fungal process on textile fabrics and their effluent treatment. J Biomed Nanotechnol 2007; 3(2): 203-8.
[http://dx.doi.org/10.1166/jbn.2007.022]

[66]   Valodkar M, Modi S, Pal A, Thakore S. Synthesis and anti-bacterial activity of Cu, Ag and Cu–Ag alloy nanoparticles: a green approach. Mater Res Bull 2011; 46(3): 384-9.
[http://dx.doi.org/10.1016/j.materresbull.2010.12.001]

[67]   Morones JR, Elechiguerra JL, Camacho A, *et al.* The bactericidal effect of silver nanoparticles. Nanotechnology 2005; 16(10): 2346-53.
[http://dx.doi.org/10.1088/0957-4484/16/10/059] [PMID: 20818017]

[68]   Banu AN, Balasubramanian C, Moorthi PV. Biosynthesis of silver nanoparticles using *Bacillus thuringiensis* against dengue vector, *Aedes aegypti* (Diptera: Culicidae). Parasitol Res 2014; 113(1): 311-6.
[http://dx.doi.org/10.1007/s00436-013-3656-0] [PMID: 24173811]

[69]   Fatima F, Verma SR, Pathak N, Bajpai P. Extracellular mycosynthesis of silver nanoparticles and their microbicidal activity. J Glob Antimicrob Resist 2016; 7: 88-92.
[http://dx.doi.org/10.1016/j.jgar.2016.07.013] [PMID: 27689341]

[70]   Owaid MN, Raman J, Lakshmanan H, Al-Saeedi SS, Sabaratnam V, Abed IA. Mycosynthesis of silver nanoparticles by *Pleurotus cornucopiae* var. *citrinopileatus* and its inhibitory effects against Candida sp. Mater Lett 2015; 153: 186-90.
[http://dx.doi.org/10.1016/j.matlet.2015.04.023]

[71]   Moazeni M, Rashidi N, Shahverdi AR, Noorbakhsh F, Rezaie S. Extracellular production of silver nanoparticles by using three common species of dermatophytes: *Trichophyton rubrum, Trichophyton mentagrophytes* and *Microsporum canis*. Iran Biomed J 2012; 16(1): 52-8.
[PMID: 22562033]

[72]   Ishida K, Cipriano TF, Rocha GM, *et al.* Silver nanoparticle production by the fungus *Fusarium oxysporum*: nanoparticle characterisation and analysis of antifungal activity against pathogenic yeasts. Mem Inst Oswaldo Cruz 2014; 109(2): 220-8.
[http://dx.doi.org/10.1590/0074-0276130269] [PMID: 24714966]

[73]   Li WR, Xie XB, Shi QS, Zeng HY, Ou-Yang YS, Chen YB. Antibacterial activity and mechanism of silver nanoparticles on *Escherichia coli*. Appl Microbiol Biotechnol 2010; 85(4): 1115-22.
[http://dx.doi.org/10.1007/s00253-009-2159-5] [PMID: 19669753]

[74]   Rai M, Kon K, Ingle A, Duran N, Galdiero S, Galdiero M. Broad-spectrum bioactivities of silver nanoparticles: the emerging trends and future prospects. Appl Microbiol Biotechnol 2014; 98(5): 1951-61.
[http://dx.doi.org/10.1007/s00253-013-5473-x] [PMID: 24407450]

[75]   Rajapaksha RM, Tobor-Kapłon MA, Bååth E. Metal toxicity affects fungal and bacterial activities in soil differently. Appl Environ Microbiol 2004; 70(5): 2966-73.
[http://dx.doi.org/10.1128/AEM.70.5.2966-2973.2004] [PMID: 15128558]

[76]   Du L, Xian L, Feng JX. Rapid extra-/intracellular biosynthesis of gold nanoparticles by the fungus *Penicillium* sp. J Nanopart Res 2011; 13(3): 921-30.
[http://dx.doi.org/10.1007/s11051-010-0165-2]

[77]   Sarkar J, Ray S, Chattopadhyay D, Laskar A, Acharya K. Mycogenesis of gold nanoparticles using a phytopathogen *Alternaria alternata*. Bioprocess Biosyst Eng 2012; 35(4): 637-43.
[http://dx.doi.org/10.1007/s00449-011-0646-4] [PMID: 22009439]

[78]   Verma VC, Singh SK, Solanki R, Prakash S. Biofabrication of anisotropic gold nanotriangles using extract of endophytic *Aspergillus clavatus* as a dual functional reductant and stabilizer. Nanoscale Res Lett 2011; 6(1): 1-7.
[http://dx.doi.org/10.1186/1556-276X-6-261]

[79]   Mishra A, Tripathy SK, Wahab R, *et al.* Microbial synthesis of gold nanoparticles using the fungus *Penicillium brevicompactum* and their cytotoxic effects against mouse mayo blast cancer C 2 C 12 cells. Appl Microbiol Biotechnol 2011; 92(3): 617-30.
[http://dx.doi.org/10.1007/s00253-011-3556-0] [PMID: 21894479]

[80]   Sawle BD, Salimath B, Deshpande R, Bedre MD, Prabhakar BK, Venkataraman A. Biosynthesis and stabilization of Au and Au–Ag alloy nanoparticles by fungus, Fusarium semitectum. Science and Technology of Advanced Materials 2008 Oct 8; 9(3): 035012.

[81]   Dasaratrao Sawle B, Salimath B, Deshpande R, Dhondojirao Bedre M, Krishnamurthy Prabhakar B, Venkataraman A. Biosynthesis and stabilization of Au and Au-Ag alloy nanoparticles by fungus, *Fusarium semitectum.* Sci Technol Adv Mater 2008; 9(3): 035012.
[http://dx.doi.org/10.1088/1468-6996/9/3/035012] [PMID: 27878009]

[82]   Vala AK. Exploration on green synthesis of gold nanoparticles by a marine☐derived fungus *Aspergillus sydowii.* Environ Prog Sustain Energy 2015; 34(1): 194-7.
[http://dx.doi.org/10.1002/ep.11949]

[83]   Sheikhloo Z, Salouti M, Katiraee F. Biological synthesis of gold nanoparticles by fungus *Epicoccum nigrum.* J Cluster Sci 2011; 22(4): 661-5.
[http://dx.doi.org/10.1007/s10876-011-0412-4]

[84]   Mishra AN, Bhadauria S, Gaur MS, Pasricha R. Extracellular microbial synthesis of gold nanoparticles using fungus *Hormoconis resinae.* J Miner Met Mater Soc 2010; 62(11): 45-8.
[http://dx.doi.org/10.1007/s11837-010-0168-6]

[85]   Mishra P, Ray S, Sinha S, *et al.* Facile bio-synthesis of gold nanoparticles by using extract of *Hibiscus sabdariffa* and evaluation of its cytotoxicity against U87 glioblastoma cells under hyperglycemic condition. Biochem Eng J 2016; 105: 264-72.
[http://dx.doi.org/10.1016/j.bej.2015.09.021]

[86] Castro-Longoria E, Vilchis-Nestor AR, Avalos-Borja M. Biosynthesis of silver, gold and bimetallic nanoparticles using the filamentous fungus *Neurospora crassa*. Colloids Surf B Biointerfaces 2011; 83(1): 42-8.
[http://dx.doi.org/10.1016/j.colsurfb.2010.10.035] [PMID: 21087843]

[87] Das SK, Dickinson C, Lafir F, Brougham DF, Marsili E. Synthesis, characterization and catalytic activity of gold nanoparticles biosynthesized with *Rhizopus oryzae* protein extract. Green Chem 2012; 14(5): 1322-34.
[http://dx.doi.org/10.1039/c2gc16676c]

[88] Das SK, Liang J, Schmidt M, Laffir F, Marsili E. Biomineralization mechanism of gold by zygomycete fungi Rhizopus oryzae. ACS Nano 2012; 6(7): 6165-73.
[http://dx.doi.org/10.1021/nn301502s] [PMID: 22708541]

[89] Sanghi R, Verma P, Puri S. Enzymatic formation of gold nanoparticles using *Phanerochaete chrysosporium*. Advances in Chemical Engineering and Science 2011; 1(03): 154-62.
[http://dx.doi.org/10.4236/aces.2011.13023]

[90] Pei X, Qu Y, Shen W, *et al.* Green synthesis of gold nanoparticles using fungus *Mariannaea* sp. HJ and their catalysis in reduction of 4-nitrophenol. Environ Sci Pollut Res Int 2017; 24(27): 21649-59.
[http://dx.doi.org/10.1007/s11356-017-9684-z] [PMID: 28752308]

[91] Zhang X, Qu Y, Shen W, *et al.* Biogenic synthesis of gold nanoparticles by yeast *Magnusiomyces ingens* LH-F1 for catalytic reduction of nitrophenols. Colloids Surf A Physicochem Eng Asp 2016; 497: 280-5.
[http://dx.doi.org/10.1016/j.colsurfa.2016.02.033]

[92] Chauhan A, Zubair S, Tufail S, *et al.* Fungus-mediated biological synthesis of gold nanoparticles: potential in detection of liver cancer. Int J Nanomedicine 2011; 6: 2305-19.
[PMID: 22072868]

[93] Apte M, Girme G, Bankar A, RaviKumar A, Zinjarde S. RaviKumar A, Zinjarde S. 3, 4-dihydroxy-L-phenylalanine-derived melanin from *Yarrowia lipolytica* mediates the synthesis of silver and gold nanostructures. J Nanobiotechnology 2013; 11(1): 1-9.
[http://dx.doi.org/10.1186/1477-3155-11-2]

[94] Das SK, Das AR, Guha AK. Gold nanoparticles: microbial synthesis and application in water hygiene management. Langmuir 2009; 25(14): 8192-9.
[http://dx.doi.org/10.1021/la900585p] [PMID: 19425601]

[95] Shankar SS, Ahmad A, Pasricha R, Sastry M. Bioreduction of chloroaurate ions by geranium leaves and its endophytic fungus yields gold nanoparticles of different shapes. J Mater Chem 2003; 13(7): 1822-6.
[http://dx.doi.org/10.1039/b303808b]

[96] Zhang X, He X, Wang K, Yang X. Different active biomolecules involved in biosynthesis of gold nanoparticles by three fungus species. J Biomed Nanotechnol 2011; 7(2): 245-54.
[http://dx.doi.org/10.1166/jbn.2011.1285] [PMID: 21702362]

[97] Sneha K, Sathishkumar M, Kim S, Yun YS. Counter ions and temperature incorporated tailoring of biogenic gold nanoparticles. Process Biochem 2010; 45(9): 1450-8.
[http://dx.doi.org/10.1016/j.procbio.2010.05.019]

[98] Liu H, Zhang H, Wang J, Wei J. Effect of temperature on the size of biosynthesized silver nanoparticle: deep insight into microscopic kinetics analysis. Arab J Chem 2020; 13(1): 1011-9.
[http://dx.doi.org/10.1016/j.arabjc.2017.09.004]

[99] Narang J, Malhotra N, Singh G, Pundir CS. Electrochemical impediometric detection of anti-HIV drug taking gold nanorods as a sensing interface. Biosens Bioelectron 2015; 66: 332-7.
[http://dx.doi.org/10.1016/j.bios.2014.11.038] [PMID: 25437372]

[100] Mackey MA, Ali MR, Austin LA, Near RD, El-Sayed MA. The most effective gold nanorod size for

plasmonic photothermal therapy: theory and *in vitro* experiments. J Phys Chem B 2014; 118(5): 1319-26.
[http://dx.doi.org/10.1021/jp409298f] [PMID: 24433049]

[101] Lu X, Dong X, Zhang K, Han X, Fang X, Zhang Y. A gold nanorods-based fluorescent biosensor for the detection of hepatitis B virus DNA based on fluorescence resonance energy transfer. Analyst (Lond) 2013; 138(2): 642-50.
[http://dx.doi.org/10.1039/C2AN36099C] [PMID: 23172079]

[102] Love AJ, Makarov VV, Sinitsyna OV, *et al.* A genetically modified tobacco mosaic virus that can produce gold nanoparticles from a metal salt precursor. Front Plant Sci 2015; 6: 984.
[http://dx.doi.org/10.3389/fpls.2015.00984] [PMID: 26617624]

[103] Liu J, Detrembleur C, De Pauw-Gillet MC, Mornet S, Jérôme C, Duguet E. Gold nanorods coated with mesoporous silica shell as drug delivery system for remote near infrared light-activated release and potential phototherapy. Small 2015; 11(19): 2323-32.
[http://dx.doi.org/10.1002/smll.201402145] [PMID: 25580816]

[104] Yeo ELL, Cheah JU, Neo DJH, *et al.* Exploiting the protein corona around gold nanorods for low-dose combined photothermal and photodynamic therapy. J Mater Chem B Mater Biol Med 2017; 5(2): 254-68.
[http://dx.doi.org/10.1039/C6TB02743A] [PMID: 32263544]

[105] Swarnkar RK, Singh SC, Gopal R. Effect of aging on copper nanoparticles synthesized by pulsed laser ablation in water: structural and optical characterizations. Bull Mater Sci 2011; 34(7): 1363-9.
[http://dx.doi.org/10.1007/s12034-011-0329-4]

[106] Varshney R, Bhadauria S, Gaur MS, Pasricha R. Characterization of copper nanoparticles synthesized by a novel microbiological method. J Miner Met Mater Soc 2010; 62(12): 102-4.
[http://dx.doi.org/10.1007/s11837-010-0171-y]

[107] Prema P. Chemical mediated synthesis of silver nanoparticles and its potential antibacterial application. In Progress in Molecular andEnvironmental Bioengineering –From Analysis and Modeling to TechnologyApplications (Ed Prof Angelo Carpi). In TechEurope 2011; pp. 151-66.
[http://dx.doi.org/10.5772/22114]

[108] Yeshchenko OA, Dmitruk IM, Dmytruk AM, Alexeenko AA. Influence of annealing conditions on size and optical properties of copper nanoparticles embedded in silica matrix. Mater Sci Eng B 2007; 137(1-3): 247-54.
[http://dx.doi.org/10.1016/j.mseb.2006.11.030]

[109] Kitching M, Ramani M, Marsili E. Fungal biosynthesis of gold nanoparticles: mechanism and scale up. Microb Biotechnol 2015; 8(6): 904-17.
[http://dx.doi.org/10.1111/1751-7915.12151] [PMID: 25154648]

CHAPTER 7

Role of Fungal Nanotechnology in Bioremediation of Heavy Metals

Sandra Pérez Álvarez[1,*], Lorena Patricia Licón Trillo[1], Eduardo Fidel Héctor Ardisana[2], Ana Elsi Ulloa Pérez[3] and María Esther González Vega[4]

[1] *Facultad de Ciencias Agrícolas y Forestales, Universidad Autónoma de Chihuahua, Km 2.5, carretera Delicias-Rosales, campus Delicias, CD. Delicias, Chihuahua, CP 33000, Mexico*

[2] *Instituto de Posgrado, Universidad Técnica de Manabí, Portoviejo, Ecuador, C. P. EC130105, Mexico*

[3] *Instituto Politécnico Nacional, CIIDIR Unidad Sinaloa, Depto. de Medio Ambiente, Blvd. Juan de Dios Bátiz Paredes 250 Guasave, Sinaloa, CP 81101, Mexico*

[4] *Instituto Nacional de Ciencias Agrícolas (INCA), Carretera a Tapaste, Km 3,5. San José de las Lajas, Mayabeque, CP 32700, Mexico*

Abstract: The main sources of soil contamination are anthropogenic activities, which result in the accumulation of contaminants that can reach levels considered toxic. One of the main soils contaminants these days is heavy metals. These metals are bioaccumulative and are not biodegradable, so many of them are toxic when they exceed certain limits. Heavy metals, when accumulated in the tissues of plants, animals and humans, induce severe symptoms that can even cause death. Bioremediation is a widely used technology to decrease the levels of these metals in soils and waters using microorganisms, among which fungi stand out. Nanotechnology currently applies the bases of bioremediation using fungi at the nanoparticle level to treat soils contaminated with heavy metals. This chapter will discuss novel aspects related to heavy metals in modern agriculture, bioremediation and nanotechnology using fungi with bioremediation purposes.

Keywords: Agriculture, Fungi, Heavy metals, Nanotechnology, Soil contamination.

INTRODUCTION

Intensification in agriculture combined with industrialization, wars and mining have left a legacy of contaminated soil throughout the world [1, 2]. Since the urbanization began, the soil has been used as a solid and liquid waste dump beca-

---
* **Corresponding Author Sandra Pérez Álvarez**: Facultad de Ciencias Agrícolas y Forestales, Universidad Autónoma de Chihuahua, Km 2.5, carretera Delicias-Rosales, campus Delicias, CD. Delicias, Chihuahua, CP; Tel: 00526871269840; Email: spalvarez@uach.mx

use it was thought that once buried, over time, the pollutants would not represent any risk to human health or the environment and, somehow, they would disappear [3]. The main sources of soil contamination are anthropogenic, which results in the accumulation of contaminants that can reach levels considered toxic [4].

Heavy metals are a group of metals and metalloids that have an atomic density above 4 g cm^3. This group include toxic plant metals such as lead (Pb), mercury (Hg), arsenic (As), cobalt (Co), cadmium (Cd), vanadium(V), chromium (Cr), nickel (Ni), selenium (Se) and some others that are essential to plants, like manganese (Mn), iron (Fe), zinc (Zn) and copper (Cu). These metals can affect animals, plants, and humans due to a long-term exposure [5].

In the actual agricultural soil context, bioremediation is a useful alternative to reduce the great accumulation of heavy metals [6]. This technique is an efficient, cost-effective and environmentally friendly strategy that uses microorganisms such as yeast, bacteria, algae and fungi to diminish heavy metal contamination [7]. At present, the use of microorganisms, particularly fungi, for the removal of heavy metals has been extensively investigated [8, 9].

Conventional technologies to decrease soil contamination are restricted by rigorous health policies and emergent contaminants. Fungi-based nanotechnology is an effective technology to treat soil contaminated with heavy metals. Nanotechnology studies materials and applications occurring at a very small scale. The technology use structures sizing less than 100 nanometers (nm) and includes developing materials or devices within that size [10].

Nanoparticles (NPs) have several special properties that include bulk material and also have exceptional visible properties, because they are small enough to confine their electrons and produce quantum effects [11].

The aim of this chapter is to discuss and present novel aspects about agricultural soil contaminated with heavy metals and the application of fungal nanotechnology to diminish the damage of these metals to the health of plants, animal and humans.

HEAVY METALS IN MODERN AGRICULTURE

The contamination of soils with heavy metals is a great problem around the world [12], created principally by human activities [13] such as agriculture, industrialization, technology, urbanization, and mining [12, 14] (Fig. **1**). Mineral processing and mining are considered the main sources of soil contamination [12, 15]. The contamination of soils with heavy metals could generate serious problems for plants, animals, human health and the normal balance of the ecosystems [16].

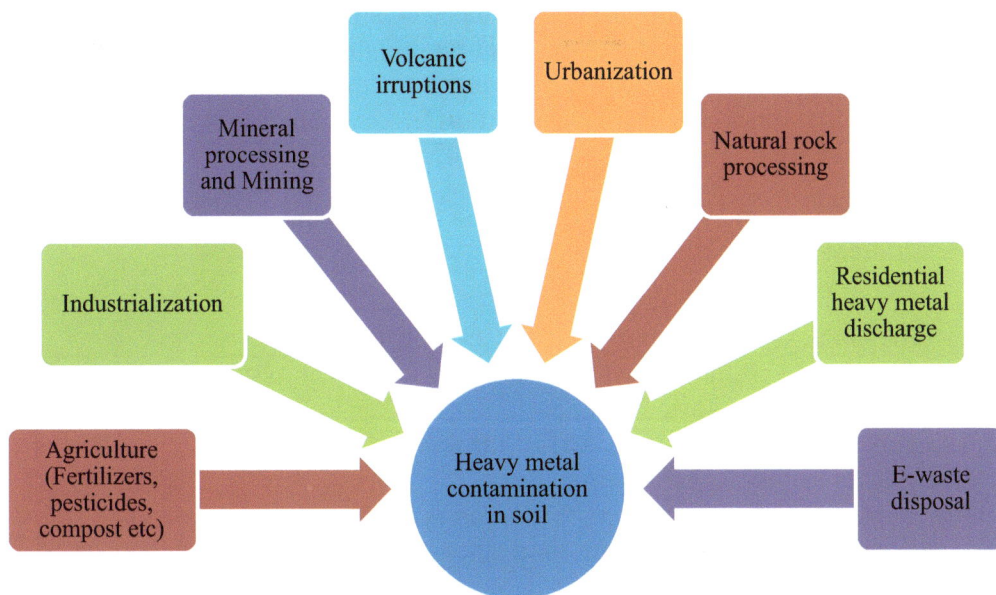

Fig. (1). Sources of heavy metals contamination in soil.

Several metals (Mn, Co, Zn and Cu) in low concentrations are essential for some metabolic functions in humans [17], but others do not have any positive effects on human health and they are considered carcinogens, such as As, Cr, Cd [17, 18]. Hg can produce ataxia in adults; in children, it affects language abilities as well as causes attention deficit [19], whereas Pb is considered possibly carcinogenic [18].

Agricultural production systems are one of the major sources of heavy metals because these metals can accumulate in the soil and be transferred to plants, entering the consumer chain, mostly in places where these processes are carried out intensively and without rest periods or rotation of crops [20].

According to Alloway [21], some heavy metals are directly related to pesticides (As, Cu, Mn, Pb, Hg, Zn), manure (Zn, Cu, As), fertilizers (Pb, Cr, Cd, Mo, Zn) and compost derived from waste conventional solids (Zn, Cu, Cd, Ni, Pb). Also, Laegreid *et al* [22]. state that phosphorous fertilizers are a Cd source due to apatite contents, which in addition to phosphorus (P) contains Cd in concentrations between 8 and 500 mg kg^{-1}. Similarly, the irrigation waters from treatment systems or contaminated sources increase the metal contents in the soil.

Mechanisms of Heavy Metals Toxicity

Heavy metals can be present in the soil as free or available ions, as part of soluble metal salts, or insoluble compounds or with partial solubility as oxides, carbonates and hydroxides [23]. One of the most important characteristics of trace elements in soils, in terms of availability and potential to leach out of soil profiles into groundwater, is their relative mobility and differs in whether its origin is natural or anthropic (the type of anthropic source) [24].

According to Garcíaand Dorronsoro [25], the heavy metals present in the soil can follow different routes; they can be (Fig. **2**):

a. Specifically adsorbed on inorganic constituents of the soil.
b. Caught on the ground, either dissolved in the aqueous phase of the soil or occupying exchange sites.
c. Associated with the organic matter of the soil.
d. Precipitated as pure solids or mixed.
e. Absorbed by plants and thus enter the trophic chains.
f. Volatilized to the atmosphere and then mobilized to surface waters or underground.

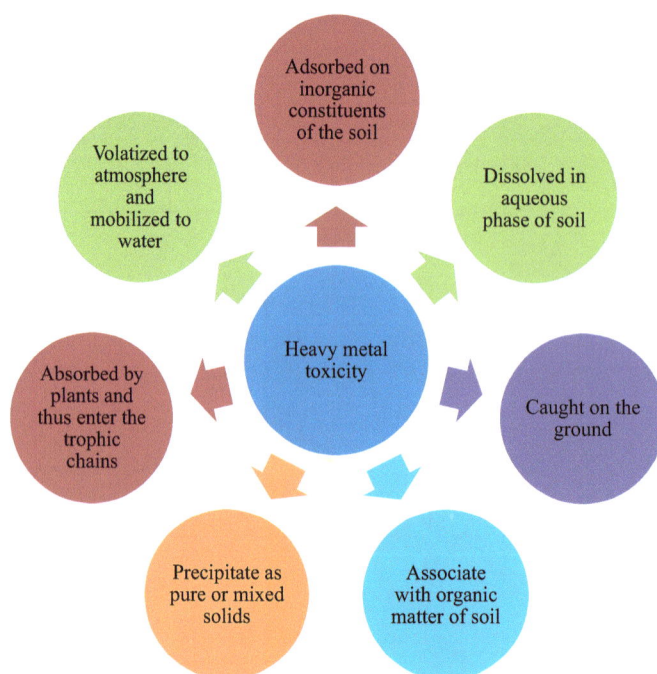

Fig. (2). Routes follow by heavy metals present into soil.

Most metals tend to be more available at acidic pH, except As, Mo, Se, and Cr, which tend to be more available at alkaline pH. The pH is an important parameter to define cation mobility, because in soils where the pH is moderately high, these metals can precipitate to hydroxides. In highly alkaline media, they can again pass into solution as hydroxides-complexes. Some metals may be in the soil dissolution as soluble anions [26].

Plants have several strategies against the presence of metals in their environment [27]. Many species tolerate high concentrations of metals in the soil because they are able to control their absorption and/or translocation to the leaves (exclusion strategy). Other species actively absorb and accumulate metals in their aerial biomass (accumulation strategy), which requires highly specialized physiology [28].

The main mechanisms of heavy metals toxicity at a molecular level are [29]:

a. Blocking essential functional groups in biomolecules, due to the high affinity of cation metals by the sulfhydryl groups of proteins, specifically to cysteine residues, which causes their denaturation [30, 31].
b. Displacement of cationic centers in important enzymes, causing function loss, such as ribulose 1-5 bisphosphate carboxylase-oxygenase (rubisco), which has a cationic center of Mg^{2+} that can be displaced in the presence of divalent cations such as Co^{2+}, Ni^{2+} and Zn^{2+} [30, 32].
c. Formation of reactive oxygen species (ROS) due to autoxidation of metals as Fe^{2+} or Cu^+, which results in the formation of H_2O_2 and $\cdot OH$ radicals *via* Haber-Weiss reaction (Fenton reaction when Fe^{2+} iron is the transition metal) [30, 33]. The radical $\cdot OH$ is one of the most reactive knowns for its ability to initiate chain reactions of free radicals that cause modifications and irreversible damage to cellular compounds such as carbohydrates, deoxyribonucleic acid (DNA), proteins and particularly lipids [34].

The Haber-Weiss reaction with metals also induces modification in free proteins and amino acids [35]. Normally, in a protein, only one amino acid residue is modified by oxidative process and the specific target is an amino acid residue at metal-binding sites. Proteins can be a target for degradation by proteases; this is the greatest consequence of oxygen-free radical damage to proteins [36].

Genotoxicity is another mechanism in heavy metals plant toxicity. This is mainly caused when a metal binds to the nucleus of the cell and induces DNA strand breaks, DNA base modifications, rearrangements, and depurination, inter- and intramolecular cross-linkage of DNA and proteins [37].

When plants are exposed to high concentrations of metal ions the first line of defense is to reduce metal ions uptake into root cells by restricting them to the apoplast or inhibiting long-distance transport [38]. If this is not enough, plants can use some detoxification strategies, such as synthesis of specific transporters for heavy metals, immobilization, trafficking, chelation, and sequestration of heavy metals by phytochelatins or metallothioneins (ligands). At last, to diminish the effects of ROS, plants trigger oxidative stress defense mechanisms, like upregulation of antioxidant and glyoxalase system, and the synthesis of several signaling molecules and stress-related proteins (hormones, heat shock proteins, biosynthesis of signaling molecules such as nitric oxide and salicylic acid [39].

Impact of Heavy Metals in Agriculture

Around the world, there are many contaminated soils because of wars, mining, industrialization, and intensification in agriculture [1]. Soils have been used as dump for solid and liquid waste since urban expansion. People think that if the contaminants are buried and out of sight, they do not have risk to human health or the environment and would somehow disappear [3].

Anthropogenic activities are considered the main sources of soil contamination, resulting in the accumulation of soil pollutants that can reach worrying levels [4]. The main anthropogenic sources of soil contamination are chemicals used or produced as derivatives of industrial activities, domestic and municipal wastes, including sewage, agrochemicals and petroleum products [40].

The main agricultural sources of soil pollutants are agrochemicals, such as fertilizers, animal manure, and pesticides. The trace metals contained in these agrochemicals, such as Cu, Cd, Pb and Hg, are also considered soil pollutants since they can harm the metabolism of plants and decrease the productivity of crops. Water sources used for irrigation can also cause soil contamination if they consist of agricultural, industrial or urban wastewater. Excess of nitrogen (N) and heavy metals are not only a source of soil contamination, but also pose a threat to food security, water quality and human health when they enter the food chain [41].

In plants, the most common effect of heavy metals is the damage to the photosynthetic apparatus. This is a characteristic for all heavy metals, not for one in particular [42]. The photosynthetic apparatus is damaged because of the degradation of chlorophyll when the chloroplast is the first organelle attack by air pollutants [43].

Organic and inorganic pesticides contain trace metals (Cu-based) that represent a major environmental and toxicological problem [44]. Soil organic matter, Fe- and

Mn- (hydro) oxides can easily immobilize Cu, remaining in high concentrations in the upper layers of soils [45].

According to Bhattacharya *et al* [46], As is one of the most widespread and worrying metalloid pollutants in agricultural soils. If being present in the environment, it is toxic to man, plants, animals, and other living organisms [47]. The flooded systems for rice cultivation favor the contamination of the food chain by As, because in the low oxide-reduction conditions in flooded soils, the As is mobilized by solubilizing the iron oxide minerals bond to it, and also reduce the arsenate ion to convert it into arsenite, which is more mobile in soil than arsenate [48].

The uptake of As in plants is in the form of As^{5+} by the competition with phosphate (PO_4^{3-}) and this mechanism has been analyzed in *Oryza sativa* L [49]. and *Brassica juncea* (L.) [50] among others. On the other hand, the ions As^{5+}, As^{3+} use aquaglyceroporin multifunctional channels (aquaporin family) for transportation. These channels have been studied in *O. sativa* [51].

Cr toxicity is associated with its oxidation status and the most toxic ion is Cr^{6+} because it goes into cells and can be reduced to reactive intermediated (thiylradicals, hydroxyl radicals, Cr^{5+}, Cr^{4+}) and finally, Cr^{3+}. Cr in its different oxidative status affect plants by attacking DNA, membrane lipids, proteins, resulting in disruption of cellular integrity [52]. Additionally, when plants take high concentrations of Cr, the quantity of macronutrients such as potassium (K), P, Fe and magnesium (Mg) decreases [53].

Pb can enter the plant passively through the roots or leaves (waxy cuticles) and it can damage enzymes, membranes and several components of the proteins. Similarly, Pb, in small quantities, may affect respiration and photosynthesis because the electron transfer chain reaction is disrupted [54]. The most serious problem that Pb can cause in plants is the plasma lemma due to its destruction that induces difficulty in the permeability of the water membrane; it also imitates the physiological activities of calcium (Ca) causing restriction of some enzymatic reactions.

Cd is a very mobile metal and can enter the plant through absorption by roots and leaves. The damage that this element can produce to crops is related to enzyme activities and the inhibition of the function of several groups such as phosphate, sulfhydryl, and side chains proteins because of its affinity with them [53].

Hg affects plant growth, root development, synthesis of chlorophyll, respiration, gas exchange and induces the inhibition of the photosynthetic activity. The excessive accumulation of Hg in plant roots inhibits the absorption of K^+ and lower quantities stimulate K^+ uptake. This metal has a high affinity with several proteins and enzymes. Also Hg affinity with sulfhydryl groups may explain the disorders in plant metabolism [55].

Remediation of Soils Contaminated with Heavy Metals

Agricultural sustainable development is needed for the production of quality food and to accomplish this goal soil remediation is necessary, taking in account the high accumulation rate of heavy metals [56]. These metals in soils are present in form of cations and soil particles retain them, or they are forming bonds with ligand ions (organic or inorganic) [57]. Several remediation techniques are used (Fig. **3**), such as:

a. Washing the contaminated soil with strong chelating agents [58] to release heavy metal on the surface of soil particles, which confronts three disadvantages: a high intake of chelating agents occurs [59]; the washing effluent must be treated and there is no strategy available for it [60], and several nutrients are washed from the soil [61].
b. Using high-surface-area sorbents to reduce the mobility and bioavailability of heavy metal cations [62]; however, the speed of metals capture is slow [63]. The stability of heavy metals immobilization similarly requires long-term monitoring [64].
c. Phytoremediation is used as a high energy efficient method [65] and it is based on the use of plants [66].
d. Bioremediation is fundamentally a metabolic process of electron transfer, necessary for microbial growth [67], obtained during the oxidation process of reduced materials, where microbial enzymes catalyze this transfer [68].

Phytoremediation

Phytoremediation is a technique where plant species are used for decontamination of sites (natural and anthropic contamination); it is a sustainable, low-cost alternative, respectful of the ecological processes of the edaphic ecosystem, which has led it to be considered a more socially, aesthetically and environmentally accepted technology. This technique was originated in the last decades of the 20th century due to the increase in pollution [69, 70].

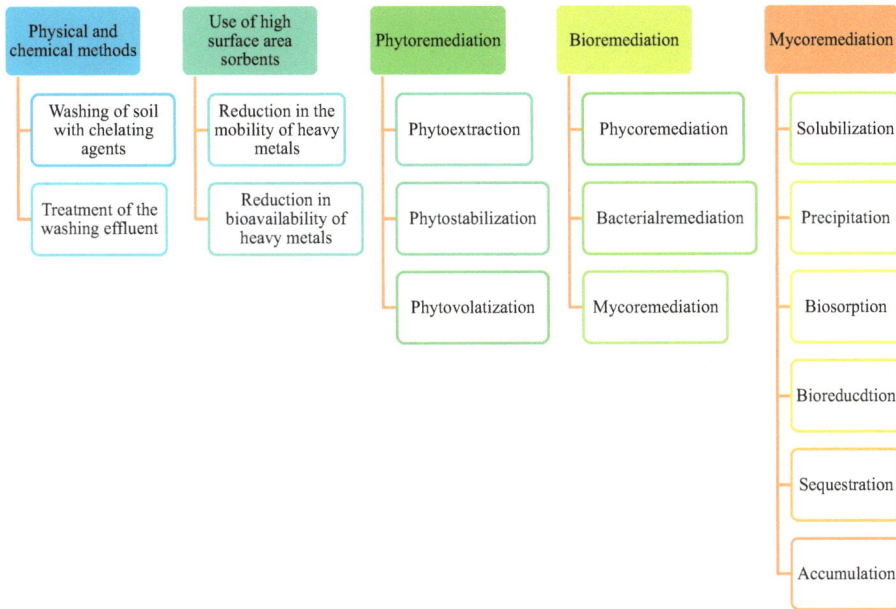

Fig. (3). Remediation techniques uses to remove heavy metals from soil.

Plants can use three mechanisms for the recuperation of contaminated sites called phytotechnologies, namely:

- Phytoextraction: Metals are absorbed by roots and transported to the aerial part of the plants (shoots, leaves) where they are accumulated. When plants have grown enough, they are harvested and the soil is cleared of the plant biomass [71].

There are several plant species reported as hyperaccumulators of heavy metals that can be used for the phytoextraction process [72]. For example, *Brassica juncea* L [73]. and *Brassica napus* L [74]. can accumulate Pb, Cu ad Zn; *Cardaminopsis halleri* [75] Zn, Pb, Cu and Cd; *Cicer aeritinum* L [76], Cd, Pb; *Jatropha curcas* L [77], *Pisum sativum* L [78]. Cr, and so on.

According to Mahar *et al* [79]. hyperaccumulator plants are those that can grow on soils with heavy metals and accumulate in shoots, leaves (aerial parts) in relation to the dry matter, >1000 mg kg^{-1} of Ni, Pb, As and Cu; >100 mg kg^{-1} of Se and Cd, or >10,000 mg kg^{-1} of Zn and Mn.

- Phytostabilisation: Contaminants such as heavy metals are immobilized in root tissues [80]. Several plant species like *Agrostis tenuifolia* M. Bieb. and *Festuca rubra* L. can be used for the phytostabilization of Pb, Zn and Cu [81].
- Phytovolatilisation: Some plants absorb contaminants (heavy metals like Hg and metalloids like Se and As) present in the soil, and they are processed in less toxic compounds [75].

Bioremediation

By definition, bioremediation is a technique where contaminants are processed and the natural environment is returned to its original condition [82]. Contaminants are transformed or degraded by microorganisms (fungi, bacteria) that are low-cost, low-technology techniques, which mostly have great public approval [68].

According to Tonini *et al* [83]. and Oliveira & Alves [84] a great number of bacteria can participate in this process, among them *Pseudomonas aeruginosa* stands out due to its great versatility in the use of carbon sources.

Liu *et al* [85]. investigated the biodegradation of phenanthrene (PHE, common contaminant) and heavy metals by the bacteria *Burkholderia fungorum* FM-2 (GenBank accession no. KM263605). Some of the results show that the most toxic heavy metal was Pb (II) inducing the non-degradation of PHE by strain FM-2. Resistance to heavy metals degradation by strain FM-2 was, first Zn (II), followed by Cd (II) and finally Pb (II) being Pb (II) the most resistant.

Chen *et al* [86]. demonstrated that the use of arbuscular mycorrhizal fungi with several local plant species in China was useful for the restoration of copper mine sites.

MYCOREMEDIATION OF HEAVY METALS IN AGRICULTURE

There are numerous fungi species recognized at present as able to remove heavy metals from polluted sites. Some examples are *Hypholoma capnoides* [Mn, Ti (titanium) and Sr (strontium)], *Marasmius oreades* (bismuth (Bi) and Ti), *Galerina vittiformis* (Zn, Cu, Pb, Cr and Cd), *Lentinus squarrosulus, Pleurotus ostreatus, Agaricus bisporus, Pleurotus tuber-regium, P. pulmonarius, Trametes versicolor* and *Phanerochaete chrysosporium* [87].

Fungi play a great role in the good functioning of medicine, biosphere, and research on earth [88]. The majority of the fungi are microscopic and live in soils, using multiple enzymes that can break down some organic compounds that are complex to simpler ones, which are used by themselves or other organisms [89].

They can decompose plant polymers (lignin and cellulose) in the environment [90], and also be used for bioremediation of several pollutants because they are capable of tolerate dangerous levels of metal concentration, fluctuating pH and display temperature adaptability or nutrient availability [91, 92].

Mycoremediation has several benefits, for example, it decreases the use of chemicals producing fewer toxic elements, high efficiency, and low cost of operation [93, 94]. This technology could be used for the *in situ* remediation of some contaminants (pharmaceutical drugs and herbicides) or bioreactors (promote microbial growth with the control of physical and chemical conditions) [95].

The most appropriate fungi used in soil remediation are basidiomycetes, specifically, the ecological groups of biotrophic and saprotrophic fungi [96].

The wood-degrading fungi are an example of saprotrophic basidiomycetes (they use dead organic matter as a carbon source). The most representative group of them for biodegradation are white-rot fungi (WRF) [97], because they have the ability to degrade until mineralization of both cellulose and lignin biopolymers [98]. Some examples of these fungi are *Pleurotus pulmonarius, P. ostreatus, P. tuber-regiumLentinula edodes, Agaricus bisporus, Phanerochaete chrysosporium, Irpex lacteus* and *Bjerkandera adusta* [99, 100].

Biodegradation like any other process in nature is affected by several factors such as nutrient availability, temperature, pH and any other that influence the microbial metabolism, for example, the pollutant type and concentration, toxicity, bioavailability and mobility [101].

Bioreactors can be used *ex-situ* for the remediation of herbicides, polycyclic aromatic hydrocarbons, pesticides, chlorinated solvents and other contaminants in soil [102]. Qayyum *et al* [103]. reported the use of filamentous fungi, specifically *Aspergillus flavus*, in the process of bioleaching heavy metals from polluted soil with an efficiency in one step 58.22% for Zn^{+2}, 18.16% for Pb and 39.77% for Cd, whereas in two-step the efficiency was 65.73% for Zn^{+2}, 16.91% for Pb and 49.66% for Cd. Bioreactors can also be used for the management of waste from industries such as pharmaceutical and sugar [95, 104]

Several fungi can be used to bioremediate heavy metals, for example, diverse *Aspergillus species* were isolated from industrial soil in Pakistan and they can remove Pb and Hg [105]. Also *Penicillium rubens* and *Aspergillus* species from the same place removed Cd and Cr [92].

According to Da Silva *et al* [106], in about 2 months, the free Pb concentration in soil was reduced by *Chaetomium aureum* isolate unrelatedly of its association with native microorganisms.

Agricultural soils contaminated with several pollutants are a source for fungal species isolation, for example, *Emericella, Fusarium* and *Rhizomucor* species were isolated from soils polluted with As and they were found resistant to elevated concentrations of this metal. Some of these species were capable of increasing the growth and yield in plants when they were irrigated with sterile water containing As [108].

Siderophores can promote transport of metals from fungal cells [109]. A siderophore (Greek: "iron carrier") is an iron chelating compound secreted by microorganisms which function is to dissolve the iron ion Fe^{3+} (it has very little solubility at neutral pH and therefore cannot be used by organisms) to Fe^{2+} complexes, which can be assimilated by active transport mechanisms [110].

The solubilization and mobility of heavy metals such as Cu, Cd, Pb, Ni and Zn were very effective with siderophores [111].

The overexpression of GiPT (phosphate transporter) in the fungus *Glomus intraradices* is produced by As (V); this enhancement in transport contributes to the tolerance of this metal by *Medicago truncatula* L [112].

Su *et al* [113]. studied the intracellular uptake of As by *Trichoderma asperellum, Fusarium oxysporum* and *Pencillium janthinellum* cultivated with As (V), and found the accumulation of 82.2, 63.4%; and 50%, respectively.

Toxic Cr (VI) can be converted to Cr (III) (less toxic or nontoxic form) by using dead fungal biomass of several fungi (*Rhizopus oryzae, Penicillium chrysogenum, Aspergillus niger, Saccharomyces cerevisiae*) [114].

Biosurfactants are biological surfactants that are formed extracellularly or as part of the cell membrane by bacteria, yeast, marine microorganisms, or fungi living in several substrates including oils, waste sugars and alkanes [115].

Several biosurfactants produced by *Candida sphaerica* have the characteristic to remove Zn (90%), Pb (79%) and Fe (95%) [116]. According to Juwarkar *et al* [117]. biosurfactant produced by *Pseudomonas aeruginosa* strain BS2 can remove Pb(II) (91%) and Cd(II) (87%) from artificially contaminated soil.

The use of fungi for the decontamination of agricultural soils polluted with heavy metals has been seen as a useful tool through the examples and studies carried out by different researchers. Nevertheless, in the proposal of a mycoremediation process is important to take an account the fungal species and the interaction with other microorganisms present in the soil.

NANOTECHNOLOGY

Impact of Nanotechnology in Agriculture

Nowadays, nanotechnology is increasingly present in daily life because it contributes to sustainability and development in several areas of industries and agricultural applications [118]. This can be defined as a science innovative and interdisciplinary technology that can control and understand the matter at nanometric scale (1-100 nm) and make possible novel applications [119, 120].

Facing problems like the increase in global population, soil and water contamination, salinity, drought, diseases, and others, agriculture today uses new technologies such as molecular techniques for rapid detection and management of diseases, improve plants to absorb nutrients and increase efficiency and yield with less cost and waste [121, 122]. Nanotechnology improves biomass-to-fuel production [123]; the use of nanomaterials as insecticides reduces the spraying of chemical products for plant protection [124].

Nanoparticles (NPs) could be synthesized using plant extracts, being a convenient, simple and eco-friendly method that can increase crop efficiency by improving the application of pesticides, fertilizers and phytohormones (plant growth regulators) [124]. Another important aspect is the reduction of pollutants in the environment because of the decrease in the sprayed amounts of dangerous chemicals [125].

Nanotechnology has several applications that can explain the increase in agricultural production, such as [123]:

- Application of fertilizers and pesticides using nanoformulations;
- Genetic engineering of plants using nanodevices;
- Identification of plant diseases and agrochemicals wastes with nanosensors;
- Effective diagnosis of plant diseases;
- Improvement of postharvest and,
- Animal health, breeding, poultry production.

The use of NPs in agriculture makes more harmless and effective the use of chemicals by crops. Some examples illustrate the use of several NPs on plant growth:

- The effect of Zn, Al, ZnO on root growth and seed germination of several plant species like rape (*Brassica napus* L.), radish (*Raphanus sativus* L.), corn (*Zea mays* L.), cucumber (*Cucumis sativus* L.) and lettuce (*Lactuca sativa* L.) was analyzed [126];
- In wheat (*Triticum spp*) the effect of silver NPs (Ag NPs) was studied in seedling growth [126];
- In tomato plants (*Solanum lycopersicum* L.) sulfur NPs were studied [127];
- In mungbean [*Vigna radiate* (L.) R. Wilczek] ZnO NPs were investigated [128].
- Aphidicidal potential of green synthesized magnesium hydroxide nanoparticles using *Olea europaea* leaves extract [129].

Nanotechnology applied to soil decontamination has not been studied much; efforts in this field have been applied more to water or some other aqueous solutions [130].

New technologies for remediation include the application of NPs to restore contaminated soils [107]. The most widely recognized nanotechnology in soil remediation is the application of nano-zero-valent iron (nZVI) to reduce the impact of both organic and inorganic contaminants. For example, nZVI can effectively degrade chlorinated hydrocarbons and organochlorine pesticides [80, 131]. Carbon nanotubes have been shown to be a feasible remediation material due to their high adsorption capacity for metal ions [132], radionuclides [133] and organic compounds [134].

Another example of nZVI NPs and the same NPs using bentonite (B-nZVI) was studied by Shi *et al* [130], to remove Cr(VI) not only in water but also in soil solution; a better result with nZVI NPs with bentonite (B-nZVI) was found.

Cabbage mustard was cultivated with Cr(VI) and Fe NPs supported on bio-carbon; the growth of the crop was better with the use of NPs and also the quantity of Cr(VI) decreased [135].

The effect of TiO_2 NPs on *Daphnia magna* with Cu toxicity was studied by Fan *et al* [136], finding that TiO_2 NPs together with organic matter may influence the toxicity of other pollutants; also, with humic acid, the effect of TiO_2 NPs on Cu toxicity decreased considerably.

Baragaño *et al* [137]. used graphene oxide nanoparticles (nGOx) compared with nZVI to study mobilization or immobilization of As and some other pollutants in soils. The doses of nZVI and nGOx analyzed were 0.2%, 1% and 5% and they found that nGOx can immobilized Cd, Cu, and Pb; however, mobilized P and As. On the other hand, nZVI induced remarkable immobilization of Pb and As, less immobilization for Cd, and better availability for Cu.

It is very important to also note that metal nanomaterials can induce certain toxicity to biology and the environment [138]. Some examples are shown in Table **1**.

Table 1. Example of metal nanomaterials that can induce certain toxicity to biology and environment.

Nanomaterials	Damage	References
SiO$_2$ NPs	Produce damage to several immune organs of the organism and they also affect the biological functions of macrophages, forming some immune system toxicity.	[139]
Polystyrene	Cell membrane, mitochondria and some other organelles of the killifish can be damaged.	[140]
ZnO NPs	The damage on *Staphylococcus aureus* and *Escherichia coli* depends on the particle size: when the particle size of ZnO NPs decreases its toxicity increases.	[141]
Elevated concentrations of NPs	In soil, can be affected the activity of the dehydrogenase enzymes.	[142]
The intensive use of NPs	They can produce alterations of nutrient balance in soil, which will affect plant nutrition and soil fertility.	[143, 144]
Ag at concentrations of 1-10 mg L^{-1}.	Seedling growth inhibition in *Triticum aestivum* L.	[145]
Ag at concentrations of 1000-3000 mM	Affect negatively chlorophyll fluorescence in *Pisum sativum* L.	[146]
Gold (Au) at concentrations of 25-100 mg L^{-1}	Reduce the length of root by 75%, and can be accumulated in *Arabidopsis thaliana* L.	[147]
ZnO at concentrations of 0.05-0.5 g kg^{-1} in soil	In *Glycine max* L. may increase ROS and lipid peroxidation	[148]

Fungal Nanotechnology

Fungal Nanotechnology for Soil Bioremediation of Heavy Metals

Nanobioremediation is the name given to a process in which plants or microorganisms (fungi, bacteria) produce NPs or nanomaterials and they are used for the decontamination (heavy metals or some other pollutants) of soils and waters [149].

Several researchers have studied the use of fungi to eliminate or remove polluted materials, like toxic metals, from contaminated sites. The biosynthesis of Cu NPs has been achieved with *Trichoderma koningiopsis* [150] and *Rhodotorula mucilaginosa* [151]. Some fungi isolated from polluted sites are able to produce NPs, such as *Fusarium* sp. isolated from a mine with Zn in South Korea, that according to Velmurugan *et al* [152], bioabsorbs up to 320 mg L^{-1} of Zn and produces ZnO NPs.

Remarkably, fungi are being applied in an effective way to eliminate NPs from an aqueous medium. Salvadori *et al* [153, 154]. demonstrated that dead biomass of *Hypocrea lixii* (the teleomorph of *Trichoderma harzianum*) can be used with success to transform Ni and Cu ions into NiO and CuO NPs in aqueous solution. *Pleurotus eryngii* and *Trametes versicolor* can remove Al$_2$O$_3$ NPs in quantities from 61 and 86%, also *P. eryngii* removes 58% of platinum NPs (Pt NPs) [155].

The bioremediation by fungi can be supplemented with the applications of NPs. The removal possibility of Cd(II) by *Phanerochaete chrysosporium* increases with Ag NPs at the concentrations of 1 mg L^{-1} [156]. Tan *et al* [157]. studied the biorediation capacity of immobilised *P. chrysosporium* with titanium oxide NPs for Cd adsorption and 2,4-dichlorophenol degradation.

Aguilar-Méndez *et al* [158]. used Ag NPs against the phytopathogen *Colletotrichum gloeosporioides*. Besides the antimicrobial capacity of several NPs such as Al$_2$O$_3$, ZnO, Fe, Si, TiO$_2$, Al, Zn, CeO$_2$, they may have several dangerous effects on plant development [159] or some of them have a positive influence in soil bacterial growth, for example, *Pseudomonas putida* KT2440 [160].

Great interest is displayed in diminishing the applications of pesticides and herbicides and changing them with eco-friendly products like NPs, so the intensive use of chemicals in agriculture may be eliminated [161]. Some examples include *Trichoderma longibrachiatum* for the synthesis of Ag NPs with a spherical shape, 1-25 nm and its application as antifungal agent to control *Pyricularia grisea, Penicillium glabrum, Alternaria alternata, Helminthosporium oryzae* and *Fusarium verticillioides* in multiple crops [162]; *Aspergillus versicolor* isolated from soil used to synthetize Ag (5-39 m) with spherical shape and its use as antifungal agent for the control of *Sclerotinia sclerotiorum* and *Botrytis cinerea* in strawberry (*Fragaria sp.*) [163]; *Aspergillus niger* isolated from grape (*Vitis vinifera* L.) synthetized Ag NPs (10-100 m) with spherical shape and the NPs were used as antifungal agent against *Penicillium digitatum, Aspergillus flavus*, and *Fusarium oxysporum* in multiple crops [164].

One of the most studied NPs is Ag NPs because of their abundant benefits on other nanoparticles. The filtrate of *Rhizopus stolonifer* was used for the synthesis of Ag NPs and different NPs sizes were obtained depending on the temperature (2.86 nm at 40°C; 25.89 nm at 20°C and 48.43 nm at 60°C) [165]. *Fusarium oxysporum* was used with the same purpose and the temperature used for incubation and the concentration of the substrate influenced the size of the Ag NPs [166]. According to Banu and Rathod [167], when the temperature is higher than 50°C, the size of the Ag NPs increases. Also, in the synthesis of these NPs using *F. oxysporum* the pH was an important factor, where at lower pH (acid) NPs size was large [167]. Some other fungi can not only form Ag NPs but also accumulate them, and examples are: *Amanita submembranacea, Aspergillus fumigatus, Cochliobolus lunatus* and *Trichoderma harzianum* [168 - 171].

Compared to bacteria at low pH metals, biosorption is higher on fungal biomass. Several bacteria (Gram-negative) are capable of immobilizing Au^{3+} at pH 3 (0.35 mM g^{-1} dry cells) [172] and *Aspergillus* sp. at pH 2.5 is able to immobilize around 1 mM g^{-1} dry cells [173].

Pulit *et al* [174]. investigated the effect of Ni NPs at 100 ppm over the mycelial growth of *F. oxysporum* f. sp. Lycopersici and *F. oxysporum* f. sp. lactucae. They found that at the concentration used, the mycelial growth was inhibited by 59.77 and 60.23%, respectively, compared to the control. Table **2** summarizes the production and remotion of nanoparticles by some fungi.

Table 2. Production and remotion of nanoparticles by some fungi.

Fungi	NPs	References
Trichoderma koningiopsis and *Rhodotorula mucilaginosa*	Biosynthesis of Cu NPs	[150, 151]
Fusarium sp.	Produces ZnO NPs.	[152]
Trichoderma harzianum	Transform Ni and Cu ions into NiO and CuO NPs	[153, 154]
Pleurotus eryngii and *Trametes versicolor*	Remove Al_2O_3 NPs and Pt NPs	[155]
Trichoderma longibrachiatum	Synthesis of Ag NPs	[162]
Aspergillus versicolor	Synthesis of Ag NPs	[163]
Aspergillus niger	Synthesis of Ag NPs	[164]
Rhizopus stolonifer	Synthesis of Ag NPs	[165]
Fusarium oxysporum	Synthesis of Ag NPs	[166]
Amanita submembranacea, Aspergillus fumigatus, Cochliobolus lunatus and *Trichoderma harzianum*	Synthesis and accumulate Ag NPs	[168 - 171]

CONCLUSION

The contamination of soils and waters with heavy metals, resulting from anthropogenic activity, is causing great interest among researchers due to the growing demands of producers and environmentalists. The remediation of contaminated soils can be done through plants (phytoremediation) and microorganisms (bioremediation).

Fungi begin to play an important role in heavy metal bioremediation processes through mechanisms such as biodegradation, the use of siderophores and the synthesis of biosurfactants. On the other hand, fungi have been used for the synthesis of nanoparticles that are used in the elimination of heavy metals and other polluting agents. The possibilities of using mycoremediation combined with nanotechnology are very wide, although they must be explored carefully, as some nanomaterials have shown toxic effects on various organisms.

CONSENT FOR PUBLICATION

Not applicable.

CONFLICT OF INTEREST

The authors declare no conflict of interest, financial or otherwise.

ACKNOWLEDGEMENT

Declared none.

REFERENCES

[1]   Bundschuh J, Litter MI, Parvez F, *et al.* One century of arsenic exposure in Latin America: a review of history and occurrence from 14 countries. Sci Total Environ 2012; 429: 2-35.
        [http://dx.doi.org/10.1016/j.scitotenv.2011.06.024] [PMID: 21959248]

[2]   EEA (European Environment Agency) Progress in management of contaminated sites 2014.
        https://www.eea.europa.eu/data-and-maps/indicators/progress-in-management-of-contam-nated-sites/progress-in-management-of-contaminated-1

[3]   Swartjes FA. Dealing with contaminated sites. Dordrecht: Springer Netherlands 2011; p. 1104.
        [http://dx.doi.org/10.1007/978-90-481-9757-6]

[4]   Cachada A, Rocha-Santos T, Duarte AC. Soil and pollution: An introduction to the main issues Soil pollution. Elsevier: Academic Press 2018; pp. 1-28.
        [http://dx.doi.org/10.1016/B978-0-12-849873-6.00001-7]

[5]   Hawkes JS. Heavy metals. J Chem Educ 1997; 74(11): 1369-74.
        [http://dx.doi.org/10.1021/ed074p1374]

[6]   Bhatnagar S, Kumari R. Bioremediation: A sustainable tool for environmental management - A review. Annu Res Rev Biol 2013; 3: 974-93.

[7]   Aryal M, Liakopoulou-Kyriakides M. Bioremoval of heavy metals by bacterial biomass. Environ Monit Assess 2015; 187(1): 4173.
        [http://dx.doi.org/10.1007/s10661-014-4173-z] [PMID: 25471624]

[8]   S MS, M EA, Chidambaram R. Isotherm modelling, kinetic study and optimization of batch parameters using response surface methodology for effective removal of Cr(VI) using fungal biomass. PLoS One 2015; 10(3): e0116884.
        [http://dx.doi.org/10.1371/journal.pone.0116884] [PMID: 25786227]

[9]   Samuel MS, E A Abigail M, Ramalingam C. Biosorption of Cr(VI) by *Ceratocystis paradoxa* MSR2 using isotherm modelling, kinetic study and optimization of batch parameters using response surface methodology. PLoS One 2015; 10(3): e0118999.

[http://dx.doi.org/10.1371/journal.pone.0118999] [PMID: 25822726]

[10]   Yadav KK, Gupta N, Kumar V, Khan SA, Kumar A. A review of emerging adsorbents and current demand for defluoridation of water: Bright future in water sustainability. Environ Int 2018; 111: 80-108.
[http://dx.doi.org/10.1016/j.envint.2017.11.014] [PMID: 29190529]

[11]   Yadav KK, Singh JK, Gupta N, Kumar V. A review of nanobioremediation technologies for environmental cleanup: a novel biological approach. J Mater Environ Sci 2017; 8(2): 740-57.

[12]   Valentim dos Santos J, Varón-López M, Fonsêca Sousa Soares CR, Lopes Leal P, Siqueira JO, de Souza Moreira FM. Biological attributes of rehabilitated soils contaminated with heavy metals. Environ Sci Pollut Res Int 2016; 23(7): 6735-48.
[http://dx.doi.org/10.1007/s11356-015-5904-6] [PMID: 26662102]

[13]   Zhang Q, Yu R, Fu S, Wu Z, Chen HYH, Liu H. Spatial heterogeneity of heavy metal contamination in soils and plants in Hefei, China. Sci Rep 2019; 9(1): 1049.
[http://dx.doi.org/10.1038/s41598-018-36582-y] [PMID: 30705298]

[14]   Londoño-Franco L, Londoño-Muñoz P, Muñoz-Garcia F. Risk of heavy metals in human and animal health. Biotecnol Sector Agropecu Agroind 2016; 14(2): 145-53.

[15]   Pérez-Sirvent C, Hernández-Pérez C, Martínez-Sánchez M, García-Lorenzo M, Bech J. Metal uptake by wetland plants: implications for phytoremediation and restoration. J Soils Sediments 2017; 17(5): 1384-93.
[http://dx.doi.org/10.1007/s11368-016-1520-4]

[16]   Doležalová Weissmannová H, Mihočová S, Chovanec P, Pavlovský J. Potential Ecological Risk and Human Health Risk Assessment of Heavy Metal Pollution in Industrial Affected Soils by Coal Mining and Metallurgy in Ostrava, Czech Republic. Int J Environ Res Public Health 2019; 16(22): 4495.
[http://dx.doi.org/10.3390/ijerph16224495] [PMID: 31739633]

[17]   Barraza F, Maurice L, Uzu G, *et al.* Distribution, contents and health risk assessment of metal(loid)s in small-scale farms in the Ecuadorian Amazon: An insight into impacts of oil activities. Sci Total Environ 2018; 622-623: 106-20.
[http://dx.doi.org/10.1016/j.scitotenv.2017.11.246] [PMID: 29212049]

[18]   Kim HS, Kim YJ, Seo YR. An Overview of Carcinogenic Heavy Metal: Molecular Toxicity Mechanism and Prevention. J Cancer Prev 2015; 20(4): 232-40.
[http://dx.doi.org/10.15430/JCP.2015.20.4.232] [PMID: 26734585]

[19]   Tshala-Katumbay D, Mwanza JC, Rohlman DS, Maestre G, Oriá RB. A global perspective on the influence of environmental exposures on the nervous system. Nature 2015; 527(7578): S187-92.
[http://dx.doi.org/10.1038/nature16034] [PMID: 26580326]

[20]   Kabata-Pendias A. Soil-plant transfer of trace elements-an environmental issue. Geoderma 2004; 122(2): 143-9.
[http://dx.doi.org/10.1016/j.geoderma.2004.01.004]

[21]   Alloway BJ. Heavy metals in soils, trace metals and metalloids in soils and their bioavailability. United Kingdom 2013; p. 614.

[22]   Laegreid M, Bockman OC, Kaarstad O. Agriculture, fertilizers and the environment. CABI publishing 1999; p. 320.

[23]   Pineda HR. Presencia de hongos micorrízicos arbusculares y contribución de Glomus Intraradices en la absorción y translocación de cinc y cobre en girasol (Helianthus Annuus L) crecido en un suelo contaminado con residuos de mina. Tecoman, Colima: Tesis para Obtener el Grado de Doctor en Ciencias Universidad de Colima 2004.

[24]   Burt R, Wilson MA, Keck TJ, Dougherty BD, Strom DE, Lindahl JA. Trace element speciation in selected smelter-contaminated soils in Anaconda and Deer Lodge Valley, Montana, USA. Adv Environ Res 2003; 8(1): 51-67.

[http://dx.doi.org/10.1016/S1093-0191(02)00140-5]

[25]   García I, Dorronsoro C. Contaminación por metales pesados. Tecnología de suelos. España: Universidad de Granada. Departamento de Edafología y Química Agrícola 2005.http://edafologia.ugr.es

[26]   Kabata-Pendias A. Trace elements in soils and plants. 3rd ed. Boca Raton, USA: CRC Press, Inc. 2000; pp. 365-413.
       [http://dx.doi.org/10.1201/9781420039900]

[27]   Barceló J, Poschenrieder C. Phytoremediation: principles and perspectives. Contributions to Science, Institut dEstudis Catalans, Barcelona 2003; 333-44.

[28]   Baker AJM. Accumulators and excluders-strategies in the response of plants to heavy metals. J Plant Nutrition, Department of Botany, University of Sheffield, S Yorkshire, ZTN, UK 1981; 643-54.

[29]   Covarrubias SA, Peña JJC. Contaminación ambiental por metales pesados en México: Problemática y estrategias de fitorremediación. Rev Int Contam Ambient 2017; 33(esp01): 7-21.
       [http://dx.doi.org/10.20937/RICA.2017.33.esp01.01]

[30]   Schützendübel A, Polle A. Plant responses to abiotic stresses: heavy metal-induced oxidative stress and protection by mycorrhization. J Exp Bot 2002; 53(372): 1351-65.
       [http://dx.doi.org/10.1093/jexbot/53.372.1351] [PMID: 11997381]

[31]   Peralta-Videa JR, López ML, Narayan M, Saupe G, Gardea-Torresdey J. The biochemistry of environmental heavy metal uptake by plants: implications for the food chain. Int J Biochem Cell Biol 2009; 41(8-9): 1665-77.
       [http://dx.doi.org/10.1016/j.biocel.2009.03.005] [PMID: 19433308]

[32]   Smeets K, Cuypers A, Lambrechts A, *et al.* Induction of oxidative stress and antioxidative mechanisms in *Phaseolus vulgaris* after Cd application. Plant Physiol Biochem 2005; 43(5): 437-44.
       [http://dx.doi.org/10.1016/j.plaphy.2005.03.007] [PMID: 15890519]

[33]   Pryor WA. Why is the hydroxyl radical the only radical that commonly adds to DNA? Hypothesis: it has a rare combination of high electrophilicity, high thermochemical reactivity, and a mode of production that can occur near DNA. Free Radic Biol Med 1988; 4(4): 219-23.
       [http://dx.doi.org/10.1016/0891-5849(88)90043-3] [PMID: 2834274]

[34]   Mithöfer A, Schulze B, Boland W. Biotic and heavy metal stress response in plants: evidence for common signals. FEBS Lett 2004; 566(1-3): 1-5.
       [http://dx.doi.org/10.1016/j.febslet.2004.04.011] [PMID: 15147858]

[35]   Stadtman ER. Oxidation of free amino acids and amino acid residues in proteins by radiolysis and by metal-catalyzed reactions. Annu Rev Biochem 1993; 62(1): 797-821.
       [http://dx.doi.org/10.1146/annurev.bi.62.070193.004053] [PMID: 8352601]

[36]   Roseman JE, Levine RL. Purification of a protease from *Escherichia coli* with specificity for oxidized glutamine synthetase. J Biol Chem 1987; 262(5): 2101-10.
       [http://dx.doi.org/10.1016/S0021-9258(18)61623-0] [PMID: 2880842]

[37]   Kasprzak KS. Possible role of oxidative damage in metal-induced carcinogenesis. Cancer Invest 1995; 13(4): 411-30.
       [http://dx.doi.org/10.3109/07357909509031921] [PMID: 7627727]

[38]   Manara A. Plant responses to heavy metal toxicity. In: Furini A, Ed. Plants and heavy metals SpringerBriefs in Biometals Springer Netherlands, 2012. 2012; pp. 27-53.

[39]   Hossain MA, Piyatida P, Jaime A, da Silva T, Fujita M. Molecular mechanism of heavy metal toxicity and tolerance in plants: central role of glutathione in detoxification of reactive oxygen species and methylglyoxal and in heavy metal chelation. J Bot (Egypt) 2012; 2012: 1-37.
       [http://dx.doi.org/10.1155/2012/872875]

[40]   Rodríguez-Eugenio N, McLaughlin M, Pennock D. La contaminación del suelo: una realidad oculta.

Roma: FAO 2012; pp. 1-144.

[41]   FAO 2015.http://www.fao.org/3/a-i4324e.pdf

[42]   Appenroth K-J. What are "heavy metals" in Plant Sciences? Acta Physiol Plant 2010; 32(4): 615-9.
         [http://dx.doi.org/10.1007/s11738-009-0455-4]

[43]   Sewelam N, Kazan K, Schenk PM. Global plant stress signaling: reactive oxygen species at the cross-road. Front Plant Sci 2016; 7: 187.
         [http://dx.doi.org/10.3389/fpls.2016.00187] [PMID: 26941757]

[44]   Komárek M, Čadková E, Chrastný V, Bordas F, Bollinger J-C. Contamination of vineyard soils with fungicides: a review of environmental and toxicological aspects. Environ Int 2010; 36(1): 138-51.
         [http://dx.doi.org/10.1016/j.envint.2009.10.005] [PMID: 19913914]

[45]   Pietrzak U, McPhail DC. Copper accumulation, distribution and fractionation in vineyard soils of Victoria, Australia. Geoderma 2004; 122(2-4): 151-66.
         [http://dx.doi.org/10.1016/j.geoderma.2004.01.005]

[46]   Bhattacharya S, Bhattacharyya A, Roy S. Arsenic-induced responses in freshwater teleosts. Fish Physiol Biochem 2007; 33(4): 463-73.
         [http://dx.doi.org/10.1007/s10695-007-9173-2]

[47]   Brima EI. Toxic elements in different medicinal plants and the impact on human health. Int J Environ Res Public Health 2017; 14(10): 1-9.
         [http://dx.doi.org/10.3390/ijerph14101209] [PMID: 29019913]

[48]   Hamon RE, Lombi E, Fortunati P, Nolan AL, McLaughlin MJ. Coupling speciation and isotope dilution techniques to study arsenic mobilization in the environment. Environ Sci Technol 2004; 38(6): 1794-8.
         [http://dx.doi.org/10.1021/es034931x] [PMID: 15074691]

[49]   Abedin MJ, Feldmann J, Meharg AA. Uptake kinetics of arsenic species in rice plants. Plant Physiol 2002; 128(3): 1120-8.
         [http://dx.doi.org/10.1104/pp.010733] [PMID: 11891266]

[50]   Pickering IJ, Prince RC, George MJ, Smith RD, George GN, Salt DE. Reduction and coordination of arsenic in Indian mustard. Plant Physiol 2000; 122(4): 1171-7.
         [http://dx.doi.org/10.1104/pp.122.4.1171] [PMID: 10759512]

[51]   Meharg AA, Jardine L. Arsenite transport into paddy rice (*Oryza sativa*) roots. New Phytol 2003; 157(1): 39-44.
         [http://dx.doi.org/10.1046/j.1469-8137.2003.00655.x] [PMID: 33873701]

[52]   Tchounwou PB, Yedjou CG, Patlolla AK, Sutton DJ. Heavy metal toxicity and the environment. Experientia Suppl 2012; 101: 133-64.
         [PMID: 22945569]

[53]   Kabata-Pendias A, Pendias H. Trace elements in soils and plants. New York 2001; p. 331.

[54]   Doğanlar ZB, Atmaca M. Influence of airborne pollution on Cd, Zn, Pb, Cu, and Al accumulation and physiological parameters of plant leaves in Antakya (Turkey). Water Air Soil Pollut 2011; 214(1-4): 509-23.
         [http://dx.doi.org/10.1007/s11270-010-0442-9]

[55]   McNear DH Jr, Afton SE, Caruso JA. Exploring the structural basis for selenium/mercury antagonism in *Allium fistulosum*. Metallomics 2012; 4(3): 267-76.
         [http://dx.doi.org/10.1039/c2mt00158f] [PMID: 22278221]

[56]   Nriagu JO. A history of global metal pollution. Science 1996; 272(5259): 223-4.
         [http://dx.doi.org/10.1126/science.272.5259.223]

[57]   Young SD. Heavy Metals in Soils. Springer, Netherlands 2013; pp. 51-95.
         [http://dx.doi.org/10.1007/978-94-007-4470-7_3]

[58]   Dermont G, Bergeron M, Mercier G, Richer-Laflèche M. Soil washing for metal removal: a review of physical/chemical technologies and field applications. J Hazard Mater 2008; 152(1): 1-31.
[http://dx.doi.org/10.1016/j.jhazmat.2007.10.043] [PMID: 18036735]

[59]   Lestan D, Luo CL, Li XD. The use of chelating agents in the remediation of metal-contaminated soils: a review. Environ Pollut 2008; 153(1): 3-13.
[http://dx.doi.org/10.1016/j.envpol.2007.11.015] [PMID: 18155817]

[60]   Di Palma L, Ferrantelli P, Merli C, Biancifiori F. Recovery of EDTA and metal precipitation from soil flushing solutions. J Hazard Mater 2003; 103(1-2): 153-68.
[http://dx.doi.org/10.1016/S0304-3894(03)00268-1] [PMID: 14568703]

[61]   Jelusic M, Vodnik D, Macek I, Lestan D. Effect of EDTA washing of metal polluted garden soils. Part II: Can remediated soil be used as a plant substrate? Sci Total Environ 2014; 475: 142-52.
[http://dx.doi.org/10.1016/j.scitotenv.2013.11.111] [PMID: 24342493]

[62]   Guo G, Zhou Q, Ma LQ. Availability and assessment of fixing additives for the *in situ* remediation of heavy metal contaminated soils: a review. Environ Monit Assess 2006; 116(1-3): 513-28.
[http://dx.doi.org/10.1007/s10661-006-7668-4] [PMID: 16779609]

[63]   Bolan N, Kunhikrishnan A, Thangarajan R, *et al.* Remediation of heavy metal(loid)s contaminated soils--to mobilize or to immobilize? J Hazard Mater 2014; 266: 141-66.
[http://dx.doi.org/10.1016/j.jhazmat.2013.12.018] [PMID: 24394669]

[64]   Brandl F, Bertrand N, Lima EM, Langer R. Nanoparticles with photoinduced precipitation for the extraction of pollutants from water and soil. Nat Commun 2015; 6(7765): 7765.
[http://dx.doi.org/10.1038/ncomms8765] [PMID: 26196119]

[65]   Salt DE, Blaylock M, Kumar NP, *et al.* Phytoremediation: a novel strategy for the removal of toxic metals from the environment using plants. Biotechnology (N Y) 1995; 13(5): 468-74.
[PMID: 9634787]

[66]   Salt DE, Smith RD, Raskin I. Phytoremediation. Annu Rev Plant Physiol Plant Mol Biol 1998; 49(1): 643-68.
[http://dx.doi.org/10.1146/annurev.arplant.49.1.643] [PMID: 15012249]

[67]   Gadd GM. Metals, minerals and microbes: geomicrobiology and bioremediation. Microbiology 2010; 156(Pt 3): 609-43.
[http://dx.doi.org/10.1099/mic.0.037143-0] [PMID: 20019082]

[68]   Vidali M. Bioremediation. An overview. Pure Appl Chem 2001; 73(7): 1163-72.
[http://dx.doi.org/10.1351/pac200173071163]

[69]   Gisbert C, Ros R, De Haro A, *et al.* A plant genetically modified that accumulates Pb is especially promising for phytoremediation. Biochem Biophys Res Commun 2003; 303(2): 440-5.
[http://dx.doi.org/10.1016/S0006-291X(03)00349-8] [PMID: 12659836]

[70]   Carpena RO, Bernal MP. Claves de la fitorremediación: Fitotecnologías para la recuperación de suelos. Ecosistemas (Madr) 2007; 16(2): 1-3.

[71]   Cioica N, Tudora C, Iuga D, *et al.* A review on phytoremediation as an ecological method for *in situ* clean up of heavy metals contaminated soils. E3S Web of Conferences 2019; 112: 1-10.

[72]   Jakovljević T, Redovniković IR, Laslo A. Phytoremediation of heavy metals: Applications and experiences in Croatia. Zaštita materijala 2016; 57(3): 496-501.

[73]   Sharma H. Phytoremediation of lead using *Brasica juncea* and *Vetiveria zizanoides.* Int J Life Sci Res 2016; 4(1): 91.

[74]   Park J, Kim J-Y, Kim K-W. Phytoremediation of soil contaminated with heavy metals using *Brassica napus.* Geosystem Eng 2012; 15(1): 10-8.
[http://dx.doi.org/10.1080/12269328.2012.674428]

[75] Sharma S, Tiwari S, Hasan A, Saxena V, Pandey LM. Recent advances in conventional and contemporary methods for remediation of heavy metal-contaminated soils. 3 Biotech 2018; 8(4): 1-18.

[76] Dasgupta S, Satvat PS, Mahinrakar AB. Ability of *Cicer arientinum* (L.) for bioremoval of lead and chromium from soil IJTES. 2011; 2(3): 338-41.

[77] Yadav SK, Juwarkar AS, Kumar P, Thawale PR, Singh SK, Chakrabarti T. Bioaccumulation and phyto-translocation of arsenic, chromium and zinc by *Jatropha curcas* L.: impact of dairy sludge and biofertilizer. Biosource Tech 2009; 100(20): 4616-22.

[78] Sharma S, Sharma P, Mehrotra P. Bioaccumulation of Heavy Metals in *Pisum sativum* L. Growing in Fly Ash Amended Soil. J Am Sci 2010; 6(6): 43-50.

[79] Mahar A, Wang P, Ali A, *et al.* Challenges and opportunities in the phytoremediation of heavy metals contaminated soils: A review. Ecotoxicol Environ Saf 2016; 126: 111-21.
[http://dx.doi.org/10.1016/j.ecoenv.2015.12.023] [PMID: 26741880]

[80] Singh R, Singh A, Misra V, Singh RP. Degradation of lindane contaminated soil using zero-valent iron nanoparticles. J Biomed Nanotechnol 2011; 7(1): 175-6.
[http://dx.doi.org/10.1166/jbn.2011.1256] [PMID: 21485858]

[81] Galende MA, Becerril JM, Barrutia O, *et al.* Field assessment of the effectiveness of organic amendments for aided phytostabilization of a Pb-Zn contaminated mine soil. J Geochem Explor 2014; 145: 181-9.
[http://dx.doi.org/10.1016/j.gexplo.2014.06.006]

[82] Garbisu C, Alkorta I. Basic concepts on heavy metal soil bioremediation. Eur J Min Process Env Protec 2003; 3: 58-66.

[83] Tonini RMCW, de Rezende CE, Grativol AD. Degradação e biorremediacão de compostos do petróleo por bactérias: revisão. Oecol Aust 2010; 14(4): 1025-35.
[http://dx.doi.org/10.4257/oeco.2010.1404.11]

[84] Oliveira RM, Alves F. Diversidade microbiana utilizada na biorremediação de solos contaminados por petróleo e derivados. Acervo da Iniciação Científica 2013; 3(5): 1-14.
[http://dx.doi.org/10.15601/2238-1945/pcnb.v3n5p1-14]

[85] Liu X-X, Hu X, Cao Y, *et al.* Biodegradation of phenanthrene and heavy metal removal by acid-tolerant *Burkholderia fungorum* FM-2. Front Microbiol 2019; 10(408): 408.
[http://dx.doi.org/10.3389/fmicb.2019.00408] [PMID: 30930861]

[86] Chen M, Xu P, Zeng G, Yang C, Huang D, Zhang J. Bioremediation of soils contaminated with polycyclic aromatic hydrocarbons, petroleum, pesticides, chlorophenols and heavy metals by composting: Applications, microbes and future research needs. Biotechnol Adv 2015; 33(6 Pt 1): 745-55.
[http://dx.doi.org/10.1016/j.biotechadv.2015.05.003] [PMID: 26008965]

[87] D'Annibale A, Ricci M, Leonardi V, Quaratino D, Mincione E, Petruccioli M. Degradation of aromatic hydrocarbons by white-rot fungi in a historically contaminated soil. Biotechnol Bioeng 2005; 90(6): 723-31.
[http://dx.doi.org/10.1002/bit.20461] [PMID: 15858792]

[88] Scholtmeijer K, Wessels JG, Wösten HA. Fungal hydrophobins in medical and technical applications. Appl Microbiol Biotechnol 2001; 56(1-2): 1-8.
[http://dx.doi.org/10.1007/s002530100632] [PMID: 11499914]

[89] Stajich JE, Berbee ML, Blackwell M, *et al.* The fungi. Curr Biol 2009; 19(18): R840-5.
[http://dx.doi.org/10.1016/j.cub.2009.07.004] [PMID: 19788875]

[90] Romaní AM, Fischer H, Mille-Lindblom C, Tranvik LJ. Interactions of bacteria and fungi on decomposing litter: differential extracellular enzyme activities. Ecology 2006; 87(10): 2559-69.
[http://dx.doi.org/10.1890/0012-9658(2006)87[2559:IOBAFO]2.0.CO;2] [PMID: 17089664]

[91] Dhankhar R, Hooda A. Fungal biosorption--an alternative to meet the challenges of heavy metal pollution in aqueous solutions. Environ Technol 2011; 32(5-6): 467-91.
[http://dx.doi.org/10.1080/09593330.2011.572922] [PMID: 21877528]

[92] Khan I, Aftab M, Shakir S, *et al.* Mycoremediation of heavy metal (Cd and Cr)-polluted soil through indigenous metallotolerant fungal isolates. Environ Monit Assess 2019; 191(9): 585.
[http://dx.doi.org/10.1007/s10661-019-7769-5] [PMID: 31440913]

[93] Gadd GM. Biosorption: critical review of scientific rationale, environmental importance andsignificance for pollution treatment. J Chem Technol Biotechnol 2009; 84(1): 13-28.
[http://dx.doi.org/10.1002/jctb.1999]

[94] Harms H, Schlosser D, Wick LY. Untapped potential: exploiting fungi in bioremediation of hazardous chemicals. Nat Rev Microbiol 2011; 9(3): 177-92.
[http://dx.doi.org/10.1038/nrmicro2519] [PMID: 21297669]

[95] Aragão MS, Menezes DB, Ramos LC, *et al.* Mycoremediation of vinasse by surface response methodology and preliminary studies in air-lift bioreactors. Chemosphere 2020; 244: 125432.
[http://dx.doi.org/10.1016/j.chemosphere.2019.125432] [PMID: 31812763]

[96] Treu R, Falandysz J. Mycoremediation of hydrocarbons with basidiomycetes-a review. J Environ Sci Health B 2017; 52(3): 148-55.
[http://dx.doi.org/10.1080/03601234.2017.1261536] [PMID: 28121269]

[97] Ellouze M, Sayadi S. White-rot fungi and their enzymes as a biotechnological tool for Xenobiotic bioremediation. In: El-Din M, Saleh H, Eds. Management of Hazardous Wastes. London, UK: IntechOpen Limited 2016; pp. 103-20.
[http://dx.doi.org/10.5772/64145]

[98] Abdel-Hamid AM, Solbiati JO, Cann IKO. Chapter one: Insights into lignin degradation and its potential industrial applications. In: Sariaslani S, Gadd GM, Eds. Advances in Applied Microbiology. San Diego, CA: Academic Press/Elsevier 2013; pp. 1-28.

[99] Voběrková S, Solčány V, Vršanská M, Adam V. Immobilization of ligninolytic enzymes from white-rot fungi in cross-linked aggregates. Chemosphere 2018; 202: 694-707.
[http://dx.doi.org/10.1016/j.chemosphere.2018.03.088] [PMID: 29602102]

[100] Manavalan T, Manavalan A, Heese K. Characterization of lignocellulolytic enzymes from white-rot fungi. Curr Microbiol 2015; 70(4): 485-98.
[http://dx.doi.org/10.1007/s00284-014-0743-0] [PMID: 25487116]

[101] Lukić B, Panico A, Huguenot D, Fabbricino M, van Hullebusch ED, Esposito G. A review on the efficiency of landfarming integrated with composting as a soil remediation treatment. Environ Technol Rev 2017; 6(1): 94-116.
[http://dx.doi.org/10.1080/21622515.2017.1310310]

[102] Tekere M. Microbial bioremediation and different bioreactors designs applied. Biotechnology and Bioengineering. IntechOpen 2019; pp. 1-19.
[http://dx.doi.org/10.5772/intechopen.83661]

[103] Qayyum S, Meng K, Pervez S, Nawaz F, Peng Ch. Optimization of pH, temperature and carbon source for bioleaching of heavy metals by *Aspergillus flavus* isolated from contaminated soil. Main Group Met Chem 2019; 42(1): 1-7.
[http://dx.doi.org/10.1515/mgmc-2018-0038]

[104] Rodarte-Morales AI, Feijoo G, Moreira MT, Lema JM. Biotransformation of three pharmaceutical active compounds by the fungus *Phanerochaete chrysosporium* in a fed batch stirred reactor under air and oxygen supply. Biodegradation 2012; 23(1): 145-56.
[http://dx.doi.org/10.1007/s10532-011-9494-9] [PMID: 21695453]

[105] Khan I, Ali M, Aftab M, *et al.* Mycoremediation: a treatment for heavy metal-polluted soil using indigenous metallotolerant fungi. Environ Monit Assess 2019; 191(10): 622.

[http://dx.doi.org/10.1007/s10661-019-7781-9] [PMID: 31494726]

[106] Da Silva FMR, Volcão ML, Hoscha CL, Pereira SV. Growth of the fungus *Chaetomium aureum* in the presence of lead: Implications in bioremediation. Environ Earth Sci 2018; 77(7): 275.
[http://dx.doi.org/10.1007/s12665-018-7447-x]

[107] Pan B, Xing B. Applications and implications of manufactured nanoparticles in soils: a review. Eur J Soil Sci 2012; 63(4): 437-56.
[http://dx.doi.org/10.1111/j.1365-2389.2012.01475.x]

[108] Singh M, Srivastava PK, Verma PC, Kharwar RN, Singh N, Tripathi RD. Soil fungi for mycoremediation of arsenic pollution in agriculture soils. J Appl Microbiol 2015; 119(5): 1278-90.
[http://dx.doi.org/10.1111/jam.12948] [PMID: 26348882]

[109] Ahmed E, Holmström SJM. Siderophores in environmental research: roles and applications. Microb Biotechnol 2014; 7(3): 196-208.
[http://dx.doi.org/10.1111/1751-7915.12117] [PMID: 24576157]

[110] Winkelmann G. Microbial siderophore-mediated transport. Biochem Soc Trans 2002; 30(4): 691-6.
[http://dx.doi.org/10.1042/bst0300691] [PMID: 12196166]

[111] Schalk IJ, Hannauer M, Braud A. New roles for bacterial siderophores in metal transport and tolerance. Environ Microbiol 2011; 13(11): 2844-54.
[http://dx.doi.org/10.1111/j.1462-2920.2011.02556.x] [PMID: 21883800]

[112] González-Chávez MC, Ortega-Larrocea M del P, Carrillo-González R, *et al.* Arsenate induces the expression of fungal genes involved in As transport in arbuscular mycorrhiza. Fungal Biol 2011; 115(12): 1197-209.
[http://dx.doi.org/10.1016/j.funbio.2011.08.005] [PMID: 22115439]

[113] Su S, Zeng X, Bai L, Jiang X, Li L. Bioaccumulation and biovolatilisation of pentavalent arsenic by *Penicillin janthinellum, Fusarium oxysporum* and *Trichoderma asperellum* under laboratory conditions. Curr Microbiol 2010; 61(4): 261-6.
[http://dx.doi.org/10.1007/s00284-010-9605-6] [PMID: 20155358]

[114] Park D, Yun YS, Jo JH, Park JM. Mechanism of hexavalent chromium removal by dead fungal biomass of *Aspergillus niger.* Water Res 2005; 39(4): 533-40.
[http://dx.doi.org/10.1016/j.watres.2004.11.002] [PMID: 15707625]

[115] Açıkel YS. Use of biosurfactants in the removal of heavy metal ions from soils. Biomanagement of metal contaminated soils. Dordrecht, The Netherlands: Springer 2011; pp. 183-223.
[http://dx.doi.org/10.1007/978-94-007-1914-9_8]

[116] Luna JM, Rufino RD, Sarubbo LA. Biosurfactant from *Candida sphaerica* UCP0995 exhibiting heavy metal remediation properties. Process Saf Environ Prot 2016; 102: 558-66.
[http://dx.doi.org/10.1016/j.psep.2016.05.010]

[117] Juwarkar AA, Nair A, Dubey KV, Singh SK, Devotta S. Biosurfactant technology for remediation of cadmium and lead contaminated soils. Chemosphere 2007; 68(10): 1996-2002.
[http://dx.doi.org/10.1016/j.chemosphere.2007.02.027] [PMID: 17399765]

[118] Ali MA, Rehman I, Iqbal A, Din S, Rao AQ, Latif A, *et al.* Nanotechnology, a new frontier in agriculture. Adv Life Sci 2014; 1: 129-38.

[119] US Environmental Protection Agency Nanotechnology White Paper 2007. http://www.epa.gov/osainter/pdfs/nanotech/epa-nanotechnology-whitepaper-0207.pdf

[120] Parisi C, Vigani M, Rodríguez-Cerezo E. Agricultural nanotechnologies: What are the current possibilities? Nano Today 2015; 10(2): 124-7.
[http://dx.doi.org/10.1016/j.nantod.2014.09.009]

[121] Sharon M, Choudhary AK, Kumar R. Nanotechnology in agricultural diseases and food safety. J Phytol 2010; 2: 83-92.

[122] Servin AD, White JC. Nanotechnology in Agriculture: Next steps for understanding engineered nanoparticle exposure and risk. NanoImpact 2016; 1: 9-12.
[http://dx.doi.org/10.1016/j.impact.2015.12.002]

[123] Sekhon BS. Nanotechnology in agri-food production: an overview. Nanotechnol Sci Appl 2014; 7: 31-53.
[http://dx.doi.org/10.2147/NSA.S39406] [PMID: 24966671]

[124] Ghidan AY, Al-Antary TM, Salem NM, Awwad AM. Facile green synthetic route to the zinc oxide (ZnO NPs) nanoparticles: Effect on green peach aphid and antibacterial activity. J Agric Sci 2017; 9(2): 131-8.

[125] Huang S, Wang L, Liu L, Hou Y, Li L. Nanotechnology in agriculture, livestock, and aquaculture in China. A review. Agron Sustain Dev 2015; 35(2): 369-400.
[http://dx.doi.org/10.1007/s13593-014-0274-x]

[126] Lin D, Xing B. Phytotoxicity of nanoparticles: inhibition of seed germination and root growth. Environ Pollut 2007; 150(2): 243-50.
[http://dx.doi.org/10.1016/j.envpol.2007.01.016] [PMID: 17374428]

[127] Salem NM, Albanna LS, Abdeen A, Ibrahim OQ, Awwad AI. Sulfur nanoparticles improves root and shoot growth of tomato. J Agric Sci 2016; 8(4): 179 85.

[128] Ghidan AY, Al-Antary TM, Awwad AM, Ghidan OY, Al Araj SE, Ateyyat MA. Comparison of different green synthesized nanomaterials on green peach aphid as aphicidal potential. Fresenius Environ Bull 2018; 27(10): 7009-16.

[129] Ghidan AY, Al Antary TM, Awwad AM. Aphidicidal potential of green synthesized magnesium hydroxide nanoparticles using *Olea europaea* leaves extract. ARPN J Agr Biol Sci 2017; 12(10): 293-301.

[130] Shi LN, Zhang X, Chen ZL. Removal of chromium (VI) from wastewater using bentonite-supported nanoscale zero-valent iron. Water Res 2011; 45(2): 886-92.
[http://dx.doi.org/10.1016/j.watres.2010.09.025] [PMID: 20950833]

[131] Zhanqiang QXF. Degradation of halogenated organic compounds by modified nano zero-valent iron. Huaxue Jinzhan 2010; 22(2): 291-7.

[132] Rao G, Lu C, Su F. Sorption of divalent metal ions from aqueous solution by carbon nanotubes: A review. Separ Purif Tech 2007; 58(1): 224-31.
[http://dx.doi.org/10.1016/j.seppur.2006.12.006]

[133] Ren X, Chen C, Nagatsu M, Wang X. Carbon nanotubes as adsorbents in environmental pollution management: A review. Chem Eng J 2011; 170(2-3): 395-410.
[http://dx.doi.org/10.1016/j.cej.2010.08.045]

[134] Pan B, Xing B. Adsorption mechanisms of organic chemicals on carbon nanotubes. Environ Sci Technol 2008; 42(24): 9005-13.
[http://dx.doi.org/10.1021/es801777n] [PMID: 19174865]

[135] Su H, Fang Z, Tsang PE, *et al.* Remediation of hexavalent chromium contaminated soil by biochar-supported zero-valent iron nanoparticles. J Hazard Mater 2016; 318: 533-40.
[http://dx.doi.org/10.1016/j.jhazmat.2016.07.039] [PMID: 27469041]

[136] Fan W, Peng R, Li X, Ren J, Liu T, Wang X. Effect of titanium dioxide nanoparticles on copper toxicity to *Daphnia magna* in water: Role of organic matter. Water Res 2016; 105: 129-37.
[http://dx.doi.org/10.1016/j.watres.2016.08.060] [PMID: 27611640]

[137] Baragaño D, Forján R, Welte L, Gallego JLR. Nanoremediation of As and metals polluted soils by means of graphene oxide nanoparticles. Sci Rep 2020; 10(1): 1896-906.
[http://dx.doi.org/10.1038/s41598-020-58852-4] [PMID: 32024880]

[138] Tang WW, Zeng GM, Gong JL, *et al.* Impact of humic/fulvic acid on the removal of heavy metals

from aqueous solutions using nanomaterials: a review. Sci Total Environ 2014; 468-469: 1014-27.
[http://dx.doi.org/10.1016/j.scitotenv.2013.09.044] [PMID: 24095965]

[139] Zhang H, Huang XJ, Liu XY, Zhu QM. Study on biological and environmental effect of phase change composite nanomaterials. New Chem Mater 2015; 43(04): 236-8.

[140] Kashiwada S. Distribution of nanoparticles in the see-through medaka (*Oryzias latipes*). Environ Health Perspect 2006; 114(11): 1697-702.
[http://dx.doi.org/10.1289/ehp.9209] [PMID: 17107855]

[141] Yamamoto O. Influence of particle size on the antibacterial activity of zinc oxide. Int J Inorg Mater 2001; 3(7): 643-6.
[http://dx.doi.org/10.1016/S1466-6049(01)00197-0]

[142] Jośko I, Oleszczuk P, Futa B. The effect of inorganic nanoparticles (ZnO, Cr2O3, CuO and Ni) and their bulk counterparts on enzyme activities in different soils. Geoderma 2014; 232: 528-37.
[http://dx.doi.org/10.1016/j.geoderma.2014.06.012]

[143] Janvier C, Villeneuve F, Alabouvette C, Edel-Hermann V, Mateille T, Steinberg C. Soil health through soil disease suppression: which strategy from descriptors to indicators. Soil Biol Biochem 2007; 39(1): 1-23.
[http://dx.doi.org/10.1016/j.soilbio.2006.07.001]

[144] Suresh Y, Annapurna S, Bhikshamaiah G, Singh AK. Characterization of green synthesized copper nanoparticles: A novel approach. International Conference on Advanced Nanomaterials Emerging Engineering Technologies. 63-7.
[http://dx.doi.org/10.1109/ICANMEET.2013.6609236]

[145] Vannini C, Domingo G, Onelli E, *et al.* Phytotoxic and genotoxic effects of silver nanoparticles exposure on germinating wheat seedlings. J Plant Physiol 2014; 171(13): 1142-8.
[http://dx.doi.org/10.1016/j.jplph.2014.05.002] [PMID: 24973586]

[146] Tripathi DK, Singh S, Singh S, *et al.* Nitric oxide alleviates silver nanoparticles (AgNps)-induced phytotoxicity in *Pisum sativum* seedlings. Plant Physiol Biochem 2017; 110: 167-77.
[http://dx.doi.org/10.1016/j.plaphy.2016.06.015] [PMID: 27449300]

[147] Taylor AF, Rylott EL, Anderson CW, Bruce NC. Investigating the toxicity, uptake, nanoparticle formation and genetic response of plants to gold. PLoS One 2014; 9(4): e93793.
[http://dx.doi.org/10.1371/journal.pone.0093793] [PMID: 24736522]

[148] Priester JH, Moritz SC, Espinosa K, *et al.* Damage assessment for soybean cultivated in soil with either CeO2 or ZnO manufactured nanomaterials. Sci Total Environ 2017; 579: 1756-68.
[http://dx.doi.org/10.1016/j.scitotenv.2016.11.149] [PMID: 27939199]

[149] Koul B, Taak P. Biotechnological Strategies for Effective Remediation of Polluted Soils. Singapore: Springer Nature 2018; p. 240.
[http://dx.doi.org/10.1007/978-981-13-2420-8]

[150] Salvadori MR, Ando RA, Nascimento CA, Corrêa B. Bioremediation from wastewater and extracellular synthesis of copper nanoparticles by the fungus *Trichoderma koningiopsis*. J Environ Sci Heal A Tox Hazard Subst Environ Eng 2014; 49: 1286-95.

[151] Salvadori MR, Nascimento CA, Corrêa B. Nickel oxide nanoparticles film produced by dead biomass of filamentous fungus. Sci Rep 2014; 4(1): 6404.
[http://dx.doi.org/10.1038/srep06404] [PMID: 25228324]

[152] Velmurugan P, Shim J, You Y, *et al.* Removal of zinc by live, dead, and dried biomass of *Fusarium spp.* isolated from the abandoned-metal mine in South Korea and its perspective of producing nanocrystals. J Hazard Mater 2010; 182(1-3): 317-24.
[http://dx.doi.org/10.1016/j.jhazmat.2010.06.032] [PMID: 20599320]

[153] Salvadori MR, Ando RA, Oller do Nascimento CA, Corrêa B. Intracellular biosynthesis and removal of copper nanoparticles by dead biomass of yeast isolated from the wastewater of a mine in the

Brazilian Amazonia. PLoS One 2014; 9(1): e87968.
[http://dx.doi.org/10.1371/journal.pone.0087968] [PMID: 24489975]

[154] Salvadori MR, Ando RA, Nascimento CA, Corrêa B. Extra and intracellular synthesis of nickel oxide nanoparticles mediated by dead fungal biomass. PLoS One 2015; 10(6): e0129799.
[http://dx.doi.org/10.1371/journal.pone.0129799] [PMID: 26043111]

[155] Jakubiak M, Giska I, Asztemborska M, Bystrzejewska-Piotrowska G. Bioaccumulation and biosorption of inorganic nanoparticles: factors affecting the efficiency of nanoparticle mycoextraction by liquid-grown mycelia of *Pleurotus eryngii* and *Trametes versicolor*. Mycol Prog 2014; 13(3): 525-32.
[http://dx.doi.org/10.1007/s11557-013-0933-3]

[156] Zuo Y, Chen G, Zeng G, *et al.* Transport, fate, and stimulating impact of silver nanoparticles on the removal of Cd(II) by *Phanerochaete chrysosporium* in aqueous solutions. J Hazard Mater 2015; 285: 236-44.
[http://dx.doi.org/10.1016/j.jhazmat.2014.12.003] [PMID: 25497315]

[157] Tan Q, Chen G, Zeng G, *et al.* Physiological fluxes and antioxidative enzymes activities of immobilized *Phanerochaete chrysosporium* loaded with TiO_2 nanoparticles after exposure to toxic pollutants in solution. Chemosphere 2015; 128(128): 21-7.
[http://dx.doi.org/10.1016/j.chemosphere.2014.12.088] [PMID: 25638529]

[158] Madan HR, Sharma SC, Udayabhanu , *et al.* Facile green fabrication of nanostructure ZnO plates, bullets, flower, prismatic tip, closed pine cone: Their antibacterial, antioxidant, photoluminescent and photocatalytic properties. Spectrochim Acta A Mol Biomol Spectrosc 2016; 152: 404-16.
[http://dx.doi.org/10.1016/j.saa.2015.07.067] [PMID: 26241826]

[159] Rico CM, Majumdar S, Duarte-Gardea M, Peralta-Catalina JR, Marambio-Jones E, Hoek MV. A review of the antibacterial effects of silver nanomaterials and potential implications for human health and the environment. J Nanopart Res 2010; 12(5): 1531-51.
[http://dx.doi.org/10.1007/s11051-010-9900-y]

[160] Gajjar P, Pettee B, Britt DW, Huang W, Johnson WP, Anderson AJ. Antimicrobial activities of commercial nanoparticles against an environmental soil microbe, *Pseudomonas putida* KT2440. J Biol Eng 2009; 3(1): 9-10.
[http://dx.doi.org/10.1186/1754-1611-3-9] [PMID: 19558688]

[161] Baruah S, Dutta J. Nanotechnology applications in pollution sensing and degradation in agriculture: A review. Environ Chem Lett 2009; 7(3): 191-204.
[http://dx.doi.org/10.1007/s10311-009-0228-8]

[162] Elamawi RM, Al-Harbi RE, Hendi AA. Biosynthesis and characterization of silver nanoparticles using *Trichoderma longibrachiatum* and their effect on phytopathogenic fungi. Egypt J Biol Pest Control 2018; 28(28): 1-11.
[http://dx.doi.org/10.1186/s41938-018-0028-1]

[163] Elgorban AM, Aref SM, Seham SM, Elhindi KM, Bahkali AH, Sayed SR, *et al.* Extracellular synthesis of silver nanoparticles using *Aspergillus versicolor* and evaluation of their activity on plant pathogenic fungi. Mycosphere 2016; 7(6): 844-52.
[http://dx.doi.org/10.5943/mycosphere/7/6/15]

[164] Al-Zubaidi S, Alayafi AA, Abdelkader HS. Biosynthesis, Characterization and antifungal activity of silver nanoparticles by *Aspergillus niger* isolate. J Nanotechnol Res 2019; 1(1): 23-36.
[http://dx.doi.org/10.26502/jnr.2688-8521002]

[165] AbdelRahim K, Mahmoud SY, Ali AM, Almaary KS, Mustafa AE-ZMA, Husseiny SM. Extracellular biosynthesis of silver nanoparticles using *Rhizopus stolonifer*. Saudi J Biol Sci 2017; 24(1): 208-16.
[http://dx.doi.org/10.1016/j.sjbs.2016.02.025] [PMID: 28053592]

[166] Husseiny SM, Salah TA, Anter HA. Biosynthesis of size controlled silver nanoparticles by *Fusarium oxysporum*, their antibacterial and antitumor activities. Beni Suef Univ J Basic Appl Sci 2015; 4(3):

225-31.
[http://dx.doi.org/10.1016/j.bjbas.2015.07.004]

[167] Banu A, Rathod V. Synthesis and characterization of silver nanoparticles by *Rhizopus stolonier*. Int. J Biomed Adv Res 2011; 2: 148-58.

[168] Sabatini L, Battistelli M, Giorgi L, *et al.* Tolerance to silver of an *Aspergillus fumigatus* strain able to grow on cyanide containing wastes. J Hazard Mater 2016; 306: 115-23.
[http://dx.doi.org/10.1016/j.jhazmat.2015.12.014] [PMID: 26705888]

[169] El-Sayed ASA, Akbar A, Iqrar I, *et al.* A glucanolytic *Pseudomonas sp.* associated with *Smilax bona-nox* L. displays strong activity against *Phytophthora parasitica*. Microbiol Res 2018; 207: 140-52.
[http://dx.doi.org/10.1016/j.micres.2017.11.018] [PMID: 29458848]

[170] Cecchi G, Marescotti P, Di Piazza S, Zotti M. Native fungi as metal remediators: Silver myco-accumulation from metal contaminated waste-rock dumps (Libiola Mine, Italy). J Environ Sci Health B 2017; 52(3): 191-5.
[http://dx.doi.org/10.1080/03601234.2017.1261549] [PMID: 28121268]

[171] El-Sayed ASA, Ali GA. *Aspergillus flavipes* is a novel efficient biocontrol agent of *Phytophthora parasitica*. Biol Control 2020; 140: 104072.
[http://dx.doi.org/10.1016/j.biocontrol.2019.104072]

[172] Tsuruta T. Biosorption and recycling of gold using various microorganisms. J Gen Appl Microbiol 2004; 50(4): 221-8.
[http://dx.doi.org/10.2323/jgam.50.221] [PMID: 15754248]

[173] Kuyucak N, Volesky B. Biosorbents for recovery of metals from industrial solutions. Biotechnol Lett 1988; 10(2): 137-42.
[http://dx.doi.org/10.1007/BF01024641]

[174] Pulit J, Banach M, Szczygłowska R, Bryk M. Nanosilver against fungi. Silver nanoparticles as an effective biocidal factor. Acta Biochim Pol 2013; 60(4): 795-8.
[PMID: 24432334]

CHAPTER 8

Plant Fungal Disease Management by Nanotechnology

Sunaina Bisht[1], Anita Puyam[1,*] and Prem Lal Kashyap[2]

[1] *Department of Plant Pathology, College of Agriculture, Rani Lakshmi Bai Central Agricultural University, Jhansi, Uttar Pradesh-284003, India*

[2] *ICAR-Indian Institute of Wheat and Barley Research Karnal, Haryana-132001, India*

Abstract: With the enormous increase in global population, there is an increasing number of individuals to feed. Crop loss has become the biggest issue worldwide. Insects (14%), weeds (13%) and various plant diseases (13%) play a very important role in crop losses. The loss caused by plant diseases single-handedly causes an estimated loss of 2 trillion dollars per year. Due to the increasing demand of food, the use of synthetic chemicals has become today's fastest, easiest and cheapest way to control loss causing agents. But due to the immense use of these chemicals, it induces adverse effect on the environment, human beings, animals and also depleting natural resources. In the current scenario, there is a need to introduce control measures which are effective and increase crop production but on the other hand, they must be less harmful for the ecosystem. After the introduction of irrational use of fungicides, there is always a posed threat to the living system, killing not only the target fungi but also affecting beneficial living systems. Besides, there is an increase in resistance against fungicides in the fungal pathogen. It is becoming necessary to reassess our strategies and achieve disease management by alternate approaches such as nanotechnology. Nanofungicides based on metals like silver (Ag), copper (Cu), *etc.* and nano-emulsion has been becoming an important technology to tackle fungal pathogen problems in agriculture, having immense potential to cope with the fungal pathogen in the future. However, very little work has been done to bring this technology to field level. Nanotechnology has substantially advanced in medicine and pharmacology, but has received comparatively less interest for agricultural applications. They aim at acting directly into the plant's part where the pest or disease attacks, which means that only the required amount of chemical is delivered to the plant tissue as medication. Nanoparticles may act upon pathogens in a way similar to chemical pesticides or the nanomaterials can be used as a carrier of active ingredients of pesticides, host defense inducing chemicals, *etc.*, to the target pathogens. It is a more appropriate and suitable solution for crop protection and is also safer for the environment. It will improve agricultural output in the coming years by solving the above-mentioned problems in crop production therefore, extensive research work is needed. Nanotechnology

* **Corresponding Author Anita Puyam:** Department of Plant Pathology, College of Agriculture, Rani Lakshmi Bai Central Agricultural University, Jhansi, Uttar Pradesh-284003, India; Tel: +91 8283010061; Email: anitapau@gmail.com

Savita, Anju Srivastava, Reena Jain & Pratap Kumar Pati (Eds.)

may bring an evolution in industry as well as in the field of dealing with fungal pathogens.

Keywords: Agriculture, Delivery systems, Fungicides, Myconanotechnology nanodiagnostics, Nanofungicides, Nanomaterials.

INTRODUCTION

With the climate change scenario, the plants are becoming more venerable to many diseases. Among all the pathogen causing plant diseases, fungi are an important group of organisms that cause approximately $45 billion loss in crops all over the world every year. The demand for food is increasing day by day putting pressure on agriculturists for increasing the food supply. Fungi not only affect the seedling emergence by attacking the seeds but are also able to attack the plant tissues at different growth stages and lowering the yield ultimately [1]. Most of the disease management techniques involve the use of chemicals which we can achieve the necessary yield but these practices cost ecological damage. In spite of many advantages, like high availability, fast action, and reliability, a large portion of these chemicals does not reach their target and ultimately contaminate the ecosystem. Applied fungicides can be lost (90%) as a result of overflow and enter in the ecosystem disturbing both the agriculture and environment during its application in the open field. Therefore, farmers need to apply these fungicides in huge amount, and it becomes costly for them. In this situation, there is an increased motivation to develop cost-efficient, high-performing pesticides that are less harmful to the environment. Nanotechnology comes to focus, as it is the better way to manage plant diseases. Innovative tools are provided by nanotechnology to deliver agrochemicals at targeted area safely without disturbing the ecosystem. It has developed such carrier systems that cause the controlled release of compounds when needed; that is how the concentration of agrochemicals in the environment can be reduced [2, 3].

WHAT IS NANOTECHNOLOGY?

Nanotechnology is an advanced field of science, deals with production, manipulation and implication of matter at nanoscale. The concept of nanotechnology was seeded at first by a physicist "Richard Phillips Feynman" in 1959 and the term "nanotechnology" was coined by Professor Norio Taniguchi in 1969. The name is based on a Greek letter "nano" meaning dwarf. It is one billionth part of a meter or 10^{-9} m [4, 5]. Nanomaterials are very minute structures which range from 0.1 to 100 nm. Properties of these nanomaterials such as electrical conductance, magnetism, chemical reactivity, optical effects and physical strength vary from bulk materials due to their smaller size [6].

Nanomaterials are classified on the basis of dimension and on the basis of structural configuration. On dimensional basis nanomaterial are divided into three groups [7] and includes one, two and three dimensional nanoparticles.

Nanomaterials having less than 100 nm size with one dimension are grouped into one-dimensional nanoparticles, these are used for build-up of various chemical and biological sensors, solar cells, IT systems and optical devices. Nanomaterials having size less than 100 nm along two dimensions at least are grouped under two dimensional nanoparticles *e.g.* carbon nanotubes, fibers and platelets. Metallic nanomaterials having <100 nm in all dimensions *i.e.* quantum dots, dendrimers and hollow spheres, are three dimensional nanoparticles. On the basis of structural configuration, nanomaterials are grouped as Metallic nanoparticles, Nanocrystals Quantum dots, Carbon Nanotubes, Liposome, Dendrimer, Polymeric micelles and Polymeric nanoparticles [8, 9]. Two approaches *viz.*, Top to bottom approach and Bottom to top approach [10] are used to synthesize nanoparticles (Fig. **1**).

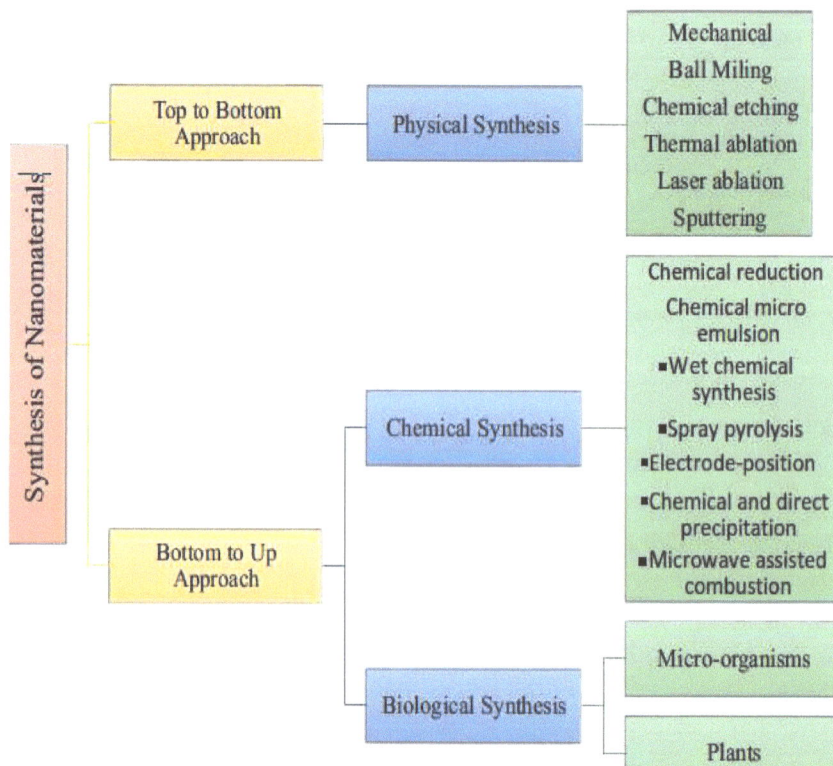

Fig. (1). Methods of synthesis of nanomaterials [10, 14, 15].

GREEN SYNTHESIS OF NANOPARTICLES

There are different methods for the synthesis of nanoparticles (NPs) but mostly, they are depended on chemical synthesis, which is expensive and usually requires toxic compounds. These methods have harmful impacts on human and environmental health due to detrimental radiations, synthetic reductants and stabilizing agents [11 - 13]. To avoid the detrimental effects, a growing need of research in this field has been stressed on different synthesis procedures of nanomaterials along with their characterization and application potential. This situation sensitizes scientists to develop and adopt an alternative "Green/biosynthesis" strategies and methods where, nontoxic and eco-friendly resources are used [16, 17] to produce biocompatible nanoparticles in a cost-effective way [18 - 20]. Plants and microbe-based synthesis of NPs are catching more attention and popularity. Microorganisms play an important role in green synthesis of nanoparticles. They produce several reductase enzymes which can reduce the metals into metallic nanoparticles and are importantly being used for synthesis of nanoparticles [21 - 29]. Using green synthesis, we can synthesize different varieties of nanoparticles that can further be applied in a diversity of applications. Synthesis of nanoparticles using greener methods is difficult to achieve, it is affected by various factors such as pH, temperature, incubation time and particle size and shape, which must be taken into consideration to obtain an optimal result [10, 30 - 33]. To overcome these constraints, a combination of appropriate strain selection and proper optimization of the incubation parameters (pH, temperature, time, concentration of metal ions, and amount of biological material) is useful. Particles like gold, silver, and platinum have drawn considerable technological and scientific attention [9, 16, 22, 34 - 39].

FUNGI MEDIATED BIOSYNTHESIS OF NANOPARTICLES

Green synthesis of nanoparticles by using microorganisms successfully interconnects nanotechnology and microbial biotechnology. Proteins, organic acids, and polysaccharides released by these biological resources acted as the key of the bio-reduction method leading toward sustainable production of nanoparticles (Table 1). With the use of green synthesis, biosynthesis of gold, silver gold–silver alloy, selenium, tellurium, platinum, palladium, silica, titania, zirconia, quantum dots, magnetite, and uraninite nanoparticles by several plant, bacteria, fungi, yeasts, and viruses have been established. This single-step green-synthesis process is rapid, can be conducted at ambient temperature and pressure, and is readily scalable [40]. Mycology conjugating with nano-technology has built up a new approach known as myco-nanotechnology, where fungi play an important role in synthesis of nanoparticles [13, 41]. Compared to the other microorganisms, fungi-mediated biosynthesis of nanomaterials, fast-growing fungi have more potential, especially filamentous fungi are exploited because of their quick growth on substrate and metabolites production [42]. Proteins, organic acids, and polysaccharides released by the fungi make them capable of sustaining varied physicochemical conditions in the bioreduction process thus leading to a contemporary and *via*ble approach for nanomaterial synthesis [43 - 46]. Several species in the plant-pathogenic genera of *Fusarium, Aspergillus, Verticillium*, and *Penicillium* have been employed to biosynthesize nanoparticles [13, 21, 40, 43, 47].

Table 1. Principal fungi used in the biosynthesis of nanoparticles [21, 48].

Fungi	Phylum	Type of NPs Size (nm)	Location Morphology	References
Agaricus bisporus	Basidiomycetes	Ag (5–50)	Extracellular Spherical	[49]
Aspergillus oryzae	Ascomycetes	Fe (10–24.6)	Intracellular	[50]
Calocybe indica	Basidiomycetes	Ag (5–50)	Extracellular Spherical	[49]
Lentinus edodes	Basidiomycetes	Au (5–50)	Extracellular Spherical	[51]
Penicillium spp.	Ascomycetes	Ag (25)	Spherical	[39]
Trichoderma spp.	Ascomycetes	Ag (8–60)	Extracellular Spherical	[52]
Rhizopus stolonifer	Phycomycetes	Ag (5–50)	Extracellular Spherical	[38]
Trichoderma viride	Ascomycetes	Ag (5–40)	Extracellular Spherical rod like	[53]

NANOTECHNOLOGY AND AGRICULTURE

Nanotechnology has received comparatively less interest for agricultural applications relative to the medicine and pharmacology field. Applications of nanotechnology have a promising future in agricultural science, and they can be a great source of innovation to improve yields and significantly contribute to precision agriculture farming practices [54]. The major reason for the adoption of nanotechnology in the field of agriculture is the production of food, feed and fiber. As the population is increasing tremendously, there is an increase in the cost of production of crops at an alarming rate due to limited resources. To overcome these limitations and to maximize agricultural production, precision farming is a much better solution. Various advanced techniques of nanotechnology are available which can be helpful in improving the practices of precision farming. Nanotechnology is basically the use of nanomaterials with exceptional properties, which will enhance the productivity of the crops and help in revolutionizing agricultural industry [55]. Nanotechnology focusing on various factors *viz.*, improves the quality of food, protect crop, monitor the growth of plant, and identify the disease-causing pathogens which will ultimately enhances the productivity. Nanotechnology has led to the development of new concepts and agricultural products with immense potential to manage plant losses.

In agricultural industry, nanotechnology is mainly used for food production and crop protection. Smart delivery of fertilizers and pesticides, detection of the contaminants and pathogens can be done by using suitable nanotools and sensors (Fig. **2**) [56].

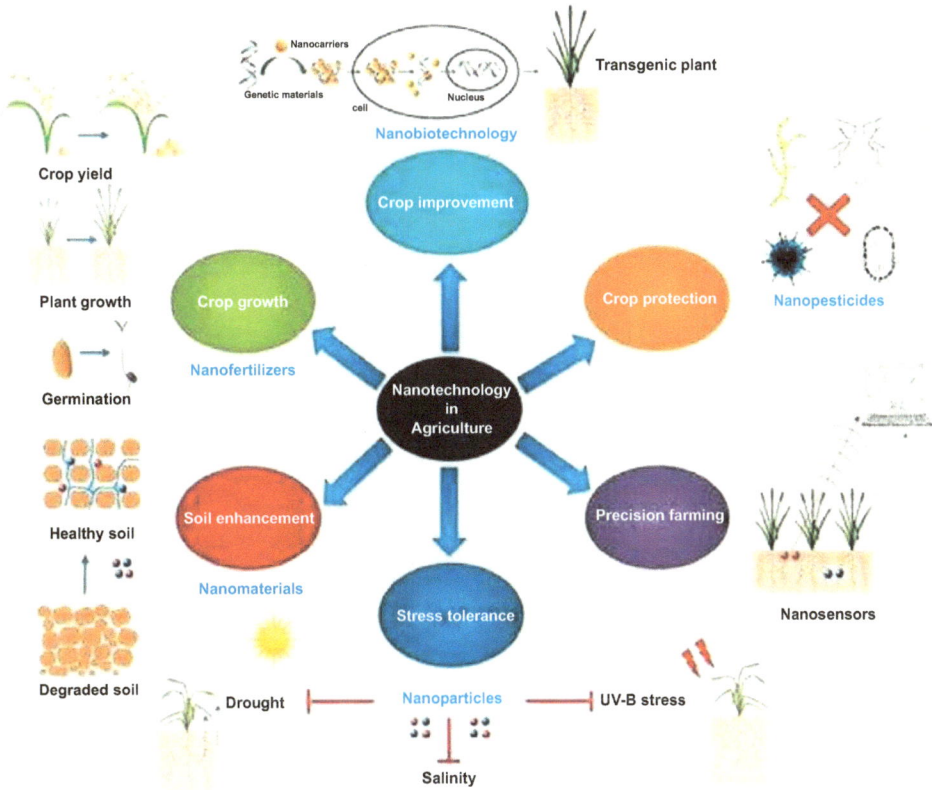

Fig. (2). Role of Nanotechnology in Agriculture.

NANOTECHNOLOGY FOR PLANT DISEASE MANAGEMENT

Crop loss, mainly caused by insects (14%), weeds (13%) and various plant diseases (13%), is the biggest issue worldwide. The value of loss caused by plant diseases is estimated 2000 billion dollars per year [57]. Current pest management relies heavily on the application of pesticides, such as insecticides, fungicides, and herbicides as this is the fastest, easiest and cheapest way to control loss causing agents. But these chemicals are harmful to the environment, humans and animals. These chemicals may enter the soil and water and causes direct and indirect effects on ecosystems. The aforementioned problems can be solved by nanotechnology, which has led to the development of new concepts and agricultural products. The use of nanotechnology in agriculture is currently being explored in plant hormone delivery, seed germination, water management, transfer of target genes, nanobarcoding, nanosensors, and controlled release of agrochemicals [58, 59]. Many scientists have engineered nanoparticles with desired characteristics to develop a new generation pesticide for plant disease management. They can then be used as protectants or for precise and targeted delivery *via* adsorption, encapsulation, and/or conjugation [60]. The use of nanoparticles to protect plants can occur *via* two different mechanisms: (a) nanoparticles themselves providing crop protection, or (b) nanoparticles as carriers for existing pesticides or other actives, such as double-stranded RNA (dsRNA), and can be applied by spray application or drenching/soaking onto seeds, foliar tissue, or roots (Fig. **3**).

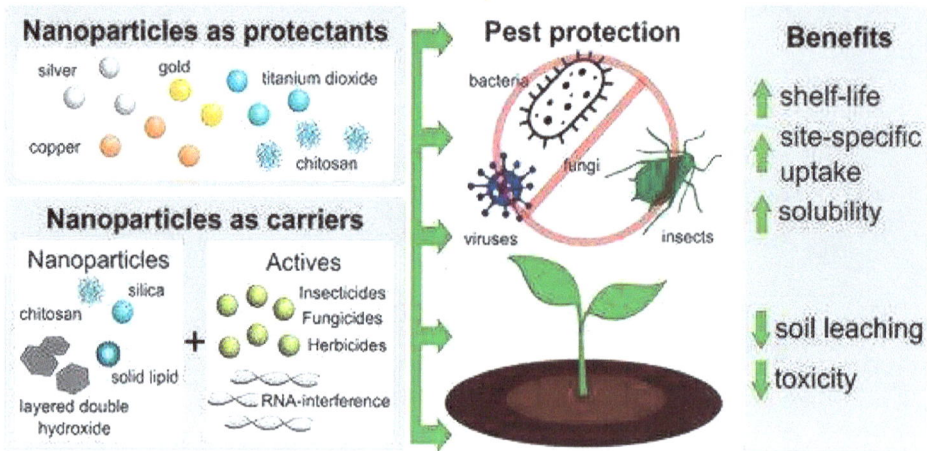

Fig. (3). Potential benefits of nanomaterial application [61].

Nanofungicides, Nanoherbicides, and insect repellent can be prepared by using nanoparticles, either prepared in combination of nanoparticles with pesticidal active ingredients or nanoparticles with high pesticidal activity [62, 63]. For the success of any plant disease management strategy, rapid detection of pathogens is essential [64]. In this area of diagnostics, nanotechnology offers major advances through faster and more-sensitive pathogen probes, as nanoparticles can be used as rapid diagnostic tools [9, 65 - 68]. The use of super paramagnetic iron oxide nanoparticles to aid in pathogen detection has only recently been explored where magnetic nanoparticles attach to biological tissue and to DNA, subsequently facilitating detection and/or extraction of *Fusarium oxysporum* and *Meloidogyne hapla* [69 - 72]. Biosensors are nanoanalytical devices consisting of metalloid or metal oxide nanoparticles that use a biological sensing element integrated into a physicochemical transducer to produce an electronic signal when in contact with the analyte of interest (pathogen) (Fig. **4**).

The biosensor can be loaded in sufficient quantity so that the electrical signal increases as a function of pathogen density. Several publications have demonstrated the capability of antibody-based biosensors for detection of *Cowpea mosaic virus*, *Tobacco mosaic virus*, and *Lettuce mosaic virus* [73, 74], nanobiosensor used nano-Au functionalized for the detection of *Ralstonia solanacearum* [75], quantum dot biosensors were useful in detecting rhizomania (*Beet necrotic yellow vein virus*) in *Polymyxa betae* [76]. The biosensors can be coupled with robotics and GPS systems to create smart delivery systems that detect, map, and treat specific areas in a field prior to or during the onset of symptoms. This technology could reduce agrochemical inputs and increase yield and profits [77, 78].

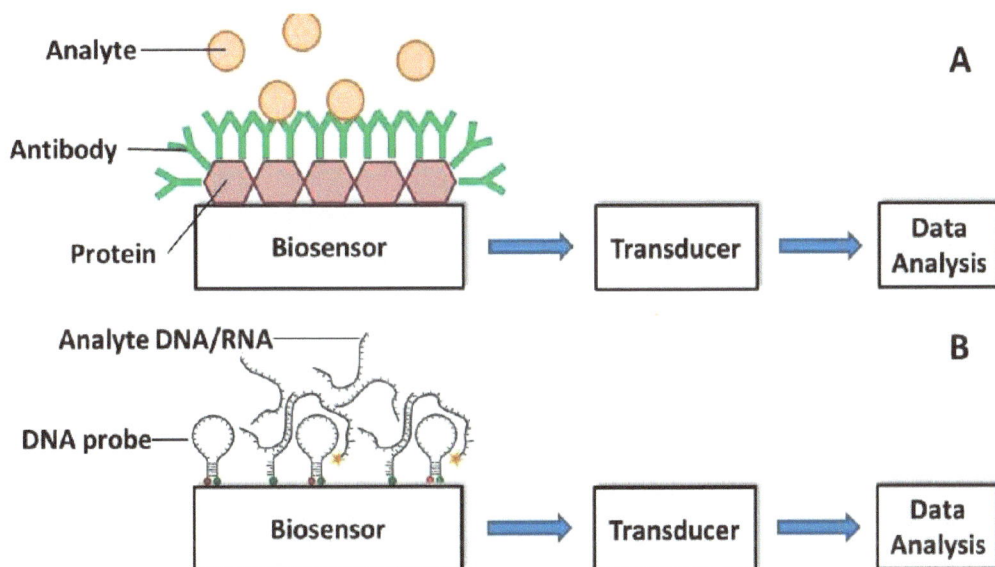

Fig. (4). Schematic illustration of (a) antibody-based and (b) DNA/RNA-based biosensor for pathogen detection [73].

TYPES OF NANOPARTICLES FOR PLANT DISEASE MANAGEMENT

Nanoparticles (NPs) have received greater interest in multidisciplinary research fields because of their extraordinary attributes. Physical and chemical properties are influenced by the NPs size, which is a key factor [79]. Nanomaterials show chemical, biological, and physical properties that are wholly diverse and distinct in their corresponding macro state [80].

Physicochemical properties are associated with changes in size and shape of NPs [81]. NPs are grouped into different classes on the basis of their chemical and physical nature, such as: (1) lipid, (2) metallic, (3) ceramics, (4) semiconductor, (5) polymer, and (6) carbon NPs.

Nanotechnology is currently being explored in nanomedicine against human pathogens making their way into the plant disease management arena. Among all the nanoparticles that exist, at present, nanoparticles of metalloids, metallic oxides, nonmetals (single and composites), carbon nanomaterials (single- and multiwalled carbon nanotubes, graphene oxides, and fullerenes), and functionalized forms of dendrimers, liposomes, and quantum dots (Table **2**) are most currently used for plant disease management. Mostly metalloids (B and Si), metallic oxides (Ag, Al, Au, Ce, Cu, Fe, Mg, Mn, Ni, Ti, and Zn), or nonmetals (S) nanoparticles are used as either bactericide/fungicides or nanofertilizers to affect disease resistance and suppress foliar, stem, fruit, and root rot pathogens [82]. Many researchers found nanoparticles effective against many foliar and soil borne pathogens *viz., Podosphaera xanthii, Sphaerotheca pannosa, Colletotrichum musae, Serratia* spp.*, Bipolaris sorokiniana, Phytophthora parasitica, Fusarium* spp.*, and Meloidogyne* spp.*, Xanthomonas perforans* [83 - 88]. The nano-Ag was found effective on suppressing plant-parasitic nematodes *Meloidogyne graminis, Meloidogyne incognita* and *Meloidogyne javanica* [89 - 91]. Cu nanoparticles are used as nanofertilizers/supplements against *Fusarium oxysporum f. sp. Niveum, Poria hypolateritia, Botrytis cinerea, Curvularia lunata Xiphinema index* by enhancement of disease resistance [92 - 94]. Similar to Ag and Cu, the antimicrobial activity of Zn nanoparticles to plant pathogens range of fungal pathogens, including *Alternaria alternata, Botrytis cinerea, Fusarium oxysporum, Mucor plumbeus, Penicillium expansum, Rhizoctonia solani, Rhizopus stolonifera, and Sclerotinia sclerotiorum* [95, 96], has been examined by several laboratories [97, 98]. Nanoparticles help in disease management by activation of disease resistance by expression of biomolecules such as total phenols, phenylalanine ammonia lyase, peroxidase and polyphenol oxidase [82, 94]. Imada *et al* [99]. reported that nano-MgO treated roots produce a rapid generation of reactive oxygen species along with upregulation of PR1, jasmonic acid, ethylene, and systemic resistance–related genes which were effective against *Alternaria alternata, Fusarium oxysporum, Rhizopus stolonifera*, and *Mucor plumbeus*. Nano-CeO was found effective against *Fusarium oxysporum* f. sp. *lycopersici* by stimulating the generation of chlorophyll, lycopene, catalase, peroxidase and polyphenol oxidase [100, 101]. Like in the formulation of other pesticides, the aims of nanoformulations are:

Table 2. Types of nanoparticles used for plant disease management [61].

Types	Definitions	Potential uses in Plant pathology
Metalloids, metallic oxides, nonmetals, and their composites	Engineered metals at nanoscale in cubes, spheres, bars and sheets	Bactericides/fungicides; Nanofertilizers; Delivery vehicle for antimicrobials and genetic material
Carbon nanomaterials	Allotropes of carbon designed at the nanoscale	Multiple uses
Single-walled or multiwalled nanotubes	Graphenes sheets rolled into single or multiple tubes	Antimicrobial agents; Delivery vehicle for antimicrobials and genetic materials
Fullerenes (buckyballs)	60 carbon atoms in a specific soccer-ball arrangement	Antimicrobial agents; Delivery vehicle for antimicrobials and genetic materials
Graphene oxide sheet (reduced or oxide forms)	Graphene oxide sheet	Antimicrobial agents; Delivery vehicle for antimicrobials and genetic materials
Liposomes	A lipid enclosing a water core	Delivery vehicle for genetic or antimicrobial products
Dendrimers	Nanomaterial with tree-like appendages that radiate from a central core	Delivery vehicle for genetic or antimicrobial products
Nanobiosensor	A nanoparticle that combines a biological component for detection	Diagnostics, research tool
Nanoshell	Nanoparticles composed of a gold shell surrounding a semiconductor	Diagnostics, research tool
Quantum dots	Inorganic fluorescent, crystalline semiconductor nanoparticles used in biosensors	Diagnostics, research tool

1. To enhance the solubility of the main ingredient.
2. To protect active ingredient against early degradation.
3. To release active ingredients in a slow manner.

The formulation of any pesticide, which includes nanoparticles has innovative properties such as optical properties, chemical reactivity, mechanical strength, and surface area. These properties are of great variations in different products which pose uniqueness to nanomaterials and enable them for a wide range of applications [102]. Different techniques are used to explore the structural characteristics of NPs [103]. Among characterization techniques, the most common are:

- Scanning electron microscopy (SEM)
- X-rays diffraction (XRD)
- Transmission electron microscopy (TEM)

- Infrared spectroscopy (IR)
- Zeta sizing
- Energy dispersive X-rays (EDX)
- X-rays photoelectron spectroscopy (XPS)

NANOPARTICLES AS CARRIERS FOR FUNGICIDES

Nanofungicide formulations enhance the active compounds solubility through a gradual and targeted release at slow rates, increasing the bioavailability of agrochemicals [104, 105]. Nanofungicide developed and tested so far proved effective in plant protection strategies [77, 106]. Several attempts have been made so far for the preparation of different nanofungicides in different ways [107]. Using nanoparticles as carriers might facilitate delivery of efficient nutrients, pesticides, herbicides, and other plant growth promoting inputs (Table **3**). Nanoencapsulation is one of the mechanisms through which fungicides are efficiently released in the host plants at target site for disease control [2]. Polymeric and solid-lipid nanocaspsules, which are loaded with tebuconazole and carbendazim, have been produced as nanofungicides [108, 109]. This mechanism also increases the pesticides and other chemicals stability, reducing degradation, and the amounts used for application as well. Nanogels of chitosan, evaluated in combination with copper and pheromones, were proved effective against *Fusarium graminearum* [20, 110, 111]. Nanofabrication tools have been helpful in improving plant disease management strategies by understanding the mechanism of physico-chemical and biological interactions between host cells and pathogens [112]. Nanoemulsions (NEs) with small size, lower viscosity and higher stability are the better option for nanofungicides [113]. Researchers claimed that the development of clay nanotubes as pesticide carriers helps in reducing the cost of pesticide up to 80% [114], by the development of microfabricated nano sized xylem vessels helps in study of mechanism of pathogenicity of xylem-inhabiting bacteria [115]. Various natural and synthetic polymers and inorganic compounds have been tested to explore their potential in nanofungicide formulations for crop protection [116].

Table 3. Nanoparticles as carrier of fungicides against target pathogen [117].

Fungicides with FRAC code	Carrier(s)	Pathogens	References
Validamycin	Calcium carbonate	*Rhizoctonia solani*	[118]
Pyraclostrobin	Chitosan–PLA graft copolymer -	*Colletotrichum gossypii*	[119]
Kaempferol	Lecithin/Chitosan	*Fusarium oxysporum*	[120]

(Table 3) cont.....

Fungicides with FRAC code	Carrier(s)	Pathogens	References
Zataria multiflora essential oil	Solid lipid nanoparticles	*Aspergillus ochraceus, Aspergillus niger, Aspergillus flavus, Alternaria solani, Rhizoctonia solani* and *Rhizopus stolonifera*	[121]
Pyraclostrobin	Chitosan/MSN -	*Puccinia asparagi*	[122]
Carbendazim	Chitosan/Pectin	*Fusarium oxysporum Aspergillus parasiticus,* and *Escherichia coli*	[123]
Prochloraz	PHSN	*Botrytis cinerea*	[124]
Clove essential oil	Chitosan -	*Aspergillus niger*	[125]

CONCLUSION

Each year, plant pests and pathogens cause 20%–40% of crops losses, the plant management strategies rely on toxic pesticides that are potentially harmful to humans and the environment. Nanotechnology can offer advantages to pesticides, like reducing toxicity, improving the shelf-life, and increasing the solubility of poorly water-soluble pesticides, all of which could have positive environmental impacts. Nanotechnology has a great breakthrough in the field of agricultural sciences/agronomy as nanomaterials can be applied to fasten the germination/ production of plants. Nanotechnology can provide solutions for agricultural applications and plant protection is another environment friendly application of effective nanoparticles in comparison to traditional approaches. Development of nano-pesticides can offer unprecedented advantages like improved solubility of poorly water-soluble pesticides, increased bioavailability and efficacy of pesticides when loaded onto nanoparticles and reduced pesticide toxicity, enhanced shelf-life and controlled delivery of actives by nanoscale carriers, target-specific delivery of the active molecules and pH dependent release, smart delivery of RNAi molecules for disease management, nanoparticles as carriers to slow down the degradation of active molecules and improve the formulations' UV stability and rain-fastness, nanopesticides to improve the selective toxicity and overcome pesticide resistance, nanosensors for disease detection. Nanotechnology will enable the development of multiple new methods for suppressing disease in the greenhouse and field, enhance disease diagnostics, and create many new tools for molecular manipulations of plants and pathogens. Engineered nanoparticles that have demonstrated activity in suppressing plant diseases are metalloids, metallic oxides, nonmetals and carbon nanoparticles. So far, nano-Ag, nano-Cu, and nano-Zn make up most reports that pertain to disease suppression. They can be integrated into disease management strategies as bactericides/fungicides and as nano-fertilizers to enhance plant health. However, there are limitations that need to be addressed. Further research is needed to be done for practical implementations of this technology. The health hazards due to the application of nanomaterials and their fate in the environment and food chain, the use of nanoscale in organic cropping system must be answered. Formulation chemistry will also need to be advanced to overcome the problems of losing efficacy over time. The struggle to integrate nanotechnology into phytopathology is currently being pursued by only a small number of laboratories, but by the increase in discovery, adaptation and application of nanotechnology the challenges to global food production can be lessened.

CONSENT FOR PUBLICATION

Not applicable.

CONFLICT OF INTEREST

The authors declare no conflict of interest, financial or otherwise.

ACKNOWLEDGEMENT

Declared none.

REFERENCES

[1]   Fernandez-Acero FJ, Carbú M, Garrido C, Vallejo I, Cantoral JM. Proteomic advances in phytopathogenic fungi. Curr Proteomics 2007; 4(2): 79-88.
        [http://dx.doi.org/10.2174/157016407782194620]

[2]   Ghormade V, Deshpande MV, Paknikar KM. Perspectives for nano-biotechnology enabled protection and nutrition of plants. Biotechnol Adv 2011; 29(6): 792-803.
        [http://dx.doi.org/10.1016/j.biotechadv.2011.06.007] [PMID: 21729746]

[3]   Werdin González JO, Gutiérrez MM, Ferrero AA, Fernández Band B. Essential oils nanoformulations for stored-product pest control - characterization and biological properties. Chemosphere 2014; 100: 130-8.
        [http://dx.doi.org/10.1016/j.chemosphere.2013.11.056] [PMID: 24359912]

[4]   Holdren JP. The national nanotechnology initiative strategic plan report at subcommittee on nanoscale science, engineering and technology of committee on technology. National Science Technology Council. Arlington: NSTC 2011.

[5]   Rai M, Ingle A. Role of nanotechnology in agriculture with special reference to management of insect pests. Appl Microbiol Biotechnol 2012; 94(2): 287-93.
        [http://dx.doi.org/10.1007/s00253-012-3969-4] [PMID: 22388570]

[6]   Boisseau P, Loubaton B. Nanomedicine, nanotechnology in medicine. C R Phys 2011; 12(7): 620-36.
        [http://dx.doi.org/10.1016/j.crhy.2011.06.001]

[7]   Hett A. Nanotechnology: small matter, many unknowns. Zurich: Swiss Reinsurance Company 2004.

[8]   Jasrotia P, Kashyap PL, Bhardwaj AK, Kumar S, Singh GP. Scope and applications of nanotechnology for wheat production: A review of recent advances. Wheat and Barley Research 2018; 10(1): 1-14.
        [http://dx.doi.org/10.25174/2249-4065/2018/76672]

[9] Kashyap PL, Rai P, Sharma S, *et al.* Nanotechnology for the detection and diagnosis of plant pathogens. In: Ranjan S, Ed. Nanoscience in food and agriculture 2, sustainable agriculture reviews 21. Basel: Springer 2016; pp. 253-76.
[http://dx.doi.org/10.1007/978-3-319-39306-3_8]

[10] Kashyap PL, Kumar S, Srivastava AK, Sharma AK. Myconanotechnology in agriculture: a perspective. World J Microbiol Biotechnol 2013; 29(2): 191-207.
[http://dx.doi.org/10.1007/s11274-012-1171-6] [PMID: 23001741]

[11] Joerger R, Klaus T, Granqvist CG. Biologically produced silver–carbon composite materials for optically functional thin-film coatings. Adv Mater 2000; 12(6): 407-9.
[http://dx.doi.org/10.1002/(SICI)1521-4095(200003)12:6<407::AID-ADMA407>3.0.CO;2-O]

[12] Panigrahi S, Kundu S, Ghosh S, Nath S, Pal T. General method of synthesis for metal nanoparticles. J Nanopart Res 2004; 6(4): 411-4.
[http://dx.doi.org/10.1007/s11051-004-6575-2]

[13] Oliveira MM, Ugarte D, Zanchet D, Zarbin AJ. Influence of synthetic parameters on the size, structure, and stability of dodecanethiol-stabilized silver nanoparticles. J Colloid Interface Sci 2005; 292(2): 429-35.
[http://dx.doi.org/10.1016/j.jcis.2005.05.068] [PMID: 16055140]

[14] Meyers MA, Mishra A, Benson DJ. [Meyers MA, Mishra A, Benson DJ. Mechanical properties of nanocrystalline materials. Prog Mater Sci 2006; 51(4): 427-556.
[http://dx.doi.org/10.1016/j.pmatsci.2005.08.003]

[15] Thakkar KN, Mhatre SS, Parikh RY. Biological synthesis of metallic nanoparticles. Nanomedicine 2010; 6(2): 257-62.
[http://dx.doi.org/10.1016/j.nano.2009.07.002] [PMID: 19616126]

[16] Vigneshwaran N, Ashtaputre NM, Varadarajan PV, Nachane RP, Paralikar KM, Balasubramanya RH. Biological synthesis of silver nanoparticles using the fungus Aspergillus flavus. Mater Lett 2007; 61(6): 1413-8.
[http://dx.doi.org/10.1016/j.matlet.2006.07.042]

[17] Ul Haq I, Ijaz S, Khan NA. Application of nanotechnology for integrated plant disease management. In: Plant Disease Management Strategies for Sustainable Agriculture Through Traditional and Modern Approaches (Eds ul Haq I, Ijaz S). Springer 2020; pp. 173-86.
[http://dx.doi.org/10.1007/978-3-030-35955-3_8]

[18] Iravani S. Green synthesis of metal nanoparticles using plants. Green Chem 2011; 13(10): 2638-50.
[http://dx.doi.org/10.1039/c1gc15386b]

[19] Mani A, Lakshmi GV. Bio-mimetic synthesis of silver nanoparticles and evaluation of its free radical scavenging activity. Int J Biol Pharm Res 2012; 3(4): 631-3.

[20] Kashyap PL, Xiang X, Heiden P. Chitosan nanoparticle based delivery systems for sustainable agriculture. Int J Biol Macromol 2015; 77: 36-51.
[http://dx.doi.org/10.1016/j.ijbiomac.2015.02.039] [PMID: 25748851]

[21] Kashyap PL, Rai P, Kumar R, *et al.* Microbial nanotechnology for climate resilient agriculture. In: Kashyap PL, Ed. Microbes for climate resilient agriculture. Hoboken: Wiley 2018; pp. 279-44.
[http://dx.doi.org/10.1002/9781119276050.ch13]

[22] Ahmad A, Mukherjee P, Senapati S, *et al.* Extracellular biosynthesis of silver nanoparticles using the fungus *Fusarium oxysporum.* Colloids Surf B Biointerfaces 2003; 27(4): 313-8.
[http://dx.doi.org/10.1016/S0927-7765(02)00174-1]

[23] Shankar SS, Rai A, Ahmad A, Sastry M. Rapid synthesis of Au, Ag, and bimetallic Au core-Ag shell nanoparticles using Neem (*Azadirachta indica*) leaf broth. J Colloid Interface Sci 2004; 275(2): 496-502.
[http://dx.doi.org/10.1016/j.jcis.2004.03.003] [PMID: 15178278]

[24]   Balaji DS, Basavaraja S, Deshpande R, Mahesh DB, Prabhakar BK, Venkataraman A. Extracellular biosynthesis of functionalized silver nanoparticles by strains of *Cladosporium cladosporioides* fungus. Colloids Surf B Biointerfaces 2009; 68(1): 88-92.
[http://dx.doi.org/10.1016/j.colsurfb.2008.09.022] [PMID: 18995994]

[25]   Kathiresan K, Manivannan S, Nabeel MA, Dhivya B. Studies on silver nanoparticles synthesized by a marine fungus, *Penicillium fellutanum* isolated from coastal mangrove sediment. Colloids Surf B Biointerfaces 2009; 71(1): 133-7.
[http://dx.doi.org/10.1016/j.colsurfb.2009.01.016] [PMID: 19269142]

[26]   Nanda A, Saravanan M. Biosynthesis of silver nanoparticles from Staphylococcus aureus and its antimicrobial activity against MRSA and MRSE. Nanomedicine 2009; 5(4): 452-6.
[http://dx.doi.org/10.1016/j.nano.2009.01.012] [PMID: 19523420]

[27]   Nabikhan A, Kandasamy K, Raj A, Alikunhi NM. Synthesis of antimicrobial silver nanoparticles by callus and leaf extracts from saltmarsh plant, Sesuvium portulacastrum L. Colloids Surf B Biointerfaces 2010; 79(2): 488-93.
[http://dx.doi.org/10.1016/j.colsurfb.2010.05.018] [PMID: 20627485]

[28]   Ponarulselvam S, Panneerselvam C, Murugan K, Aarthi N, Kalimuthu K, Thangamani S. Synthesis of silver nanoparticles using leaves of *Catharanthus roseus* Linn. G. Don and their antiplasmodial activities. Asian Pac J Trop Biomed 2012; 2(7): 574-80.
[http://dx.doi.org/10.1016/S2221-1691(12)60100-2] [PMID: 23569974]

[29]   Bonde SR, Rathod DP, Ingle AP, Ade RB, Gade AK, Rai MK. Murraya koenigii-mediated synthesis of silver nanoparticles and its activity against three human pathogenic bacteria. Nanoscience Methods 2012; 1(1): 25-36.
[http://dx.doi.org/10.1080/17458080.2010.529172]

[30]   Li S, Shen Y, Xie A, *et al.* Green synthesis of silver nanoparticles using *Capsicum annuum* L. extract. Green Chem 2007; 9(8): 852-8.
[http://dx.doi.org/10.1039/b615357g]

[31]   Nayak R, Pradhan N, Behera D, *et al.* Green synthesis of silver nanoparticle by *Penicillium purpurogenum* NPMF: The process and optimization. J Nanopart Res 2011; 13(8): 3129-37.
[http://dx.doi.org/10.1007/s11051-010-0208-8]

[32]   Sofi W, Gowri M, Shruthilaya M, Rayala S, Venkatraman G. Silver nanoparticles as an antibacterial agent for endodontic infections. BMC Infect Dis 2012; 12(S1): 60.
[http://dx.doi.org/10.1186/1471-2334-12-S1-P60]

[33]   Rai M, Yadav A. Plants as potential synthesiser of precious metal nanoparticles: progress and prospects. IET Nanobiotechnol 2013; 7(3): 117-24.
[http://dx.doi.org/10.1049/iet-nbt.2012.0031] [PMID: 24028810]

[34]   Mukherjee P, Ahmad A, Mandal D, *et al.* Fungus-mediated synthesis of silver nanoparticles and their immobilization in the mycelial matrix: A novel biological approach to nanoparticle synthesis. Nano Lett 2001; 1(10): 515-9.
[http://dx.doi.org/10.1021/nl0155274]

[35]   Ahmad A, Senapati S, Khan MI, Kumar R, Sastry M. Extra-/intracellular biosynthesis of gold nanoparticles by an alkalotolerant fungus, *Trichothecium* sp. J Biomed Nanotechnol 2005; 1(1): 47-53.
[http://dx.doi.org/10.1166/jbn.2005.012]

[36]   Riddin TL, Gericke M, Whiteley CG. Analysis of the inter- and extracellular formation of platinum nanoparticles by *Fusarium oxysporum* f. sp. *lycopersici* using response surface methodology. Nanotechnology 2006; 17(14): 3482-9.
[http://dx.doi.org/10.1088/0957-4484/17/14/021] [PMID: 19661593]

[37]   Gericke M, Pinches A. Microbial production of gold nanoparticles. Gold Bull 2006; 39(1): 22-8.
[http://dx.doi.org/10.1007/BF03215529]

[38] Afreen VR, Ranganath E. Synthesis of monodispersed silver nanoparticles by *Rhizopus stolonifer* and its antibacterial activity against MDR strains of *Pseudomonas aeruginosa* from burnt patients. Int J Environ Sci 2011; 1: 1582-92.

[39] Singh D, Rathod V, Ninganagouda S, Herimath J, Kulkarni P. Biosynthesis of silver nanoparticle by endophytic fungi *Pencillium* sp. isolated from *Curcuma longa* (turmeric) and its antibacterial activity against pathogenic gram negative bacteria. J Pharm Res 2013; 7(5): 448-53.
[http://dx.doi.org/10.1016/j.jopr.2013.06.003]

[40] Elmer W, White JC. The future of nanotechnology in plant pathology. Annu Rev Phytopathol 2018; 56(1): 111-33.
[http://dx.doi.org/10.1146/annurev-phyto-080417-050108] [PMID: 30149792]

[41] Rai M, Yadav A, Bridge P, Gade A. Myconanotechnology: A new and emerging science. In: Rai M, Bridge P, Eds. Applied Mycology. New York: CAB International 2009; Vol. 14: pp. 258-67.
[http://dx.doi.org/10.1079/9781845935344.0000]

[42] Alghuthaymi MA, Almoammar H, Rai M, Said-Galiev E, Abd-Elsalam KA. Myconanoparticles: synthesis and their role in phytopathogens management. Biotechnol Biotechnol Equip 2015; 29(2): 221-36.
[http://dx.doi.org/10.1080/13102818.2015.1008194] [PMID: 26019636]

[43] Sastry M, Ahmad A, Khan MI, Kumar R. Biosynthesis of metal nanoparticles using fungi and actinomycete. Curr Sci 2003; 85: 162-70.

[44] Mandal D, Bolander ME, Mukhopadhyay D, Sarkar G, Mukherjee P. The use of microorganisms for the formation of metal nanoparticles and their application. Appl Microbiol Biotechnol 2006; 69(5): 485-92.
[http://dx.doi.org/10.1007/s00253-005-0179-3] [PMID: 16317546]

[45] Gade AK, Bonde P, Ingle AP, Marcato PD, Duran N, Rai MK. Exploitation of *Aspergillus niger* for synthesis of silver nanoparticles. J Biobased Mater Bioenergy 2008; 2(3): 243-7.
[http://dx.doi.org/10.1166/jbmb.2008.401]

[46] Narayanan KB, Sakthivel N. Biological synthesis of metal nanoparticles by microbes. Adv Colloid Interface Sci 2010; 156(1-2): 1-13.
[http://dx.doi.org/10.1016/j.cis.2010.02.001] [PMID: 20181326]

[47] Bakshi M, Mahanty S, Chaudhuri P. Handbook of metal-microbe interactions and bioremediation. In: Surajit D, Dash HR, Eds. Fungi-mediated biosynthesis of nanoparticles and application in metal sequestration. Boca Raton, USA: CRC Press 2017; pp. 423-34.

[48] Pal G, Rai P, Pandey A. Green synthesis of nanoparticles: A greener approach for a cleaner future. Green synthesis, characterization and applications of nanoparticles. Elsevier 2019; pp. 1-26.
[http://dx.doi.org/10.1016/B978-0-08-102579-6.00001-0]

[49] Sujatha S, Tamilselvi S, Subha K, Panneerselvam A. Studies on biosynthesis of silver nanoparticles using mushroom and its antibacterial activities. Int J Curr Microbiol Appl Sci 2013; 2(12): 605-14.

[50] Tarafdar JC, Raliya R. Rapid, low-cost, and ecofriendly approach for iron nanoparticle synthesis using Aspergillus oryzae TFR9. J Nanoparticles Article ID 141274. 2013.
[http://dx.doi.org/10.1155/2013/141274]

[51] Vetchinkina EP, Loshchinina EA, Burov AM, Dykman LA, Nikitina VE. Enzymatic formation of gold nanoparticles by submerged culture of the basidiomycete *Lentinus edodes*. J Biotechnol 2014; 182-183: 37-45.
[http://dx.doi.org/10.1016/j.jbiotec.2014.04.018] [PMID: 24800960]

[52] Devi TP, Kulanthaivel S, Kamil D, Borah JL, Prabhakaran N, Srinivasa N. Biosynthesis of silver nanoparticles from *Trichoderma* species. Indian J Exp Biol 2013; 51(7): 543-7.
[PMID: 23898553]

[53]   Fayaz AM, Balaji K, Girilal M, Yadav R, Kalaichelvan PT, Venketesan R. Biogenic synthesis of silver nanoparticles and their synergistic effect with antibiotics: a study against gram-positive and gram-negative bacteria. Nanomedicine 2010; 6(1): 103-9.
      [http://dx.doi.org/10.1016/j.nano.2009.04.006] [PMID: 19447203]

[54]   Ying JY. Nanobiomaterials. Nano Today 2009; 4(1): 1-2.
      [http://dx.doi.org/10.1016/j.nantod.2008.10.016]

[55]   Batsmanova LM, Gonchar LM, Taran NY, Okanenko AA. 2013. https://nap.sumdu.edu.ua/ index.php/nap/nap2013/paper/view/1097/504

[56]   Ahmed F, Arshi N, Kumar S. Nanobiotechnology: scope and potential crop improvement. Crop improvement under adverse conditions. New York, NY: Springer 2013; pp. 245-69.
      [http://dx.doi.org/10.1007/978-1-4614-4633-0_11]

[57]   Pimentel D. Pesticides and pest control. Integrated pest management: innovation-development process. Dordrecht: Springer 2009; pp. 83-7.
      [http://dx.doi.org/10.1007/978-1-4020-8992-3_3]

[58]   Hayles J, Johnson L, Worthley C, Losic D. Nanopesticides: A review of current research and perspectives. New Pestic Soil Sens 2017; pp. 193-225.

[59]   Kashyap PL, Kumar S, Jasrotia P, Singh DP, Singh GP. Nanotechnology in Wheat Production and Protection. In: Dasgupta N, Ranjan S, Lichtfouse E, Eds. Environmental Nanotechnology Volume 4 Environmental Chemistry for a Sustainable World. Springer, Cham 2020; 32.
      [http://dx.doi.org/10.1007/978-3-030-26668-4_5]

[60]   Khandelwal N, Barbole RS, Banerjee SS, *et al.* Budding trends in integrated pest management using advanced micro- and nano-materials: Challenges and perspectives. J Environ Manage 2016; 184(Pt 2): 157-69.
      [http://dx.doi.org/10.1016/j.jenvman.2016.09.071] [PMID: 27697374]

[61]   Worrall EA, Hamid A, Mody KT, Mitter N, Pappu HR. Nanotechnology for plant disease management. Agronomy (Basel) 2018; 8(12): 285.
      [http://dx.doi.org/10.3390/agronomy8120285]

[62]   Barik TK, Sahu B, Swain V. Nanosilica-from medicine to pest control. Parasitol Res 2008; 103(2): 253-8.
      [http://dx.doi.org/10.1007/s00436-008-0975-7] [PMID: 18438740]

[63]   Matthews GA. Pests, pesticides and pest management. In: Mason J, Ed. Highlights in environmental research. London: Imperial College Press 2000; pp. 165-89.
      [http://dx.doi.org/10.1142/9781848160682_0006]

[64]   Ferreira MAM, Filipe JA, Coelho M, Chavaglia J. Nanotechnology applications in industry and medicine. Acta Sci Intellect 2017; 2: 31-50.

[65]   Kashyap PL, Kumar S, Srivastava AK. Nanodiagnostics for plant pathogens. Environ Chem Lett 2017; 15(1): 7-13.
      [http://dx.doi.org/10.1007/s10311-016-0580-4]

[66]   Kashyap PL, Kumar S, Jasrotia P, Singh DP, Singh GP. Nanosensors for Plant Disease Diagnosis: Current Understanding and Future Perspectives. In: Pudake R, Chauhan N, Kole C, Eds. Nanoscience for Sustainable Agriculture. Cham: Springer 2019; pp. 189-205.
      [http://dx.doi.org/10.1007/978-3-319-97852-9_9]

[67]   Boonham N, Glover R, Tomlinson J, Mumford R. Exploiting generic platform technologies for the detection and identification of plant pathogens. In: Collinge DB, Munk L, Cooke BM, Eds. Sustainable Disease Management in a European Context. Berlin: Springer 2008; pp. 355-63.
      [http://dx.doi.org/10.1007/978-1-4020-8780-6_15]

[68]   Yao KS, Li SJ, Tzeng KC, *et al.* Fluorescence silica nanoprobe as a biomarker for rapid detection of

plant pathogens. Adv Mat Res 2009; 79: 513-6.
[http://dx.doi.org/10.4028/www.scientific.net/AMR.79-82.513]

[69]   Tartaj P, del Puerto Morales M, Veintemillas-Verdaguer S. Gonz'alez-Carre ~ no T, Serna CJ. The preparation of magnetic nanoparticles for applications in biomedicine. J Phys D Appl Phys 2003; 36(13): 182.
[http://dx.doi.org/10.1088/0022-3727/36/13/202]

[70]   Li XM, Xu G, Liu Y, He T. Magnetic Fe_3O_4 nanoparticles: synthesis and application in water treatment. Nanosci Nanotechnol Asia 2011; 1: 14-24.

[71]   Ahmadov S, Ramazanov MA, Sienkiewicz A, Forro L. Uptake and intracellular trafficking of super paramagnetic iron oxide nanoparticles (SPIONs) in plants. Dig J Nanomater Biostruct 2014; 9: 1149-57.

[72]   Gorny AM, Hay FS, Wang X, Pethybridge SJ. Isolation of nematode DNA from 100 g of soil using Fe_3O_4 super paramagnetic nanoparticles. Nematology 2018; 20(3): 271-83.
[http://dx.doi.org/10.1163/15685411-00003140]

[73]   James C. Polypyrrole nanoribbon based chemiresistive immunosensors for viral plant pathogen detection. Anal Methods 2013; 5(14): 3497-502.
[http://dx.doi.org/10.1039/c3ay40371h]

[74]   Lin H-Y, Huang C-H, Lu S-H, Kuo I-T, Chau L-K. Direct detection of orchid viruses using nanorod-based fiber optic particle plasmon resonance immunosensor. Biosens Bioelectron 2014; 51: 371-8.
[http://dx.doi.org/10.1016/j.bios.2013.08.009] [PMID: 24001513]

[75]   Khaledian S, Nikkhah M, Shams-bakhsh M, Hoseinzadeh S. A sensitive biosensor based on gold nanoparticles to detect Ralstonia solanacearum in soil. J Gen Plant Pathol 2017; 83(4): 231-9.
[http://dx.doi.org/10.1007/s10327-017-0721-z]

[76]   Safarpour H, Safarnejad MR, Tabatabaei M, *et al.* Development of a quantum dots FRET-based biosensor for efficient detection of *Polymyxa betae.* Can J Plant Pathol 2012; 34(4): 507-15.
[http://dx.doi.org/10.1080/07060661.2012.709885]

[77]   Bergeson LL. (Nanosilver pesticide products: What does the future hold? Environ Qual Manage 2010; 19(4): 73-82.
[http://dx.doi.org/10.1002/tqem.20263]

[78]   Khot LR, Sankaran S, Maja JM, Ehsani R, Schuster E. Application of nanomaterials in agricultural production and crop protection: a review. Crop Prot 2012; 35: 64-70.
[http://dx.doi.org/10.1016/j.cropro.2012.01.007]

[79]   Wang L, Li X, Zhang G, Dong J, Eastoe J. Oil-in-water nanoemulsions for pesticide formulations. J Colloid Interface Sci 2007; 314(1): 230-5.
[http://dx.doi.org/10.1016/j.jcis.2007.04.079] [PMID: 17612555]

[80]   Li LS, Hu J, Yang W, Alivisatos AP. Band gap variation of size-and shape-controlled colloidal CdSe quantum rods. Nano Lett 2001; 1(7): 349-51.
[http://dx.doi.org/10.1021/nl015559r]

[81]   Barrak H, Saied T, Chevallier P, Laroche G, M'nif A, Hamzaoui AH. Synthesis, characterization, and functionalization of ZnO nanoparticles by N-(trimethoxysilylpropyl) ethylenediamine triacetic acid (TMSEDTA): Investigation of the interactions between Phloroglucinol and ZnO@ TMSEDTA. Arab J Chem 2016; 1-8.

[82]   Datnoff LE, Elmer WH, Huber DM. Mineral Nutrition and Plant Disease. St. Paul, MN: APS Press 2007.

[83]   Jo YK, Kim BH, Jung G. Antifungal activity of silver ions and nanoparticles on phytopathogenic fungi. Plant Dis 2009; 93(10): 1037-43.
[http://dx.doi.org/10.1094/PDIS-93-10-1037] [PMID: 30754381]

[84] Jung JH, Kim SW, Min JS, *et al.* The Effect of Nano-Silver Liquid against the White Rot of the Green Onion Caused by *Sclerotium cepivorum.* Mycobiology 2010; 38(1): 39-45.
[http://dx.doi.org/10.4489/MYCO.2010.38.1.039] [PMID: 23956623]

[85] Moussa SH, Tayel AA, Alsohim AS, Abdallah RR. Botryticidal activity of nanosized silver-chitosan composite and its application for the control of gray mold in strawberry. J Food Sci 2013; 78(10): M1589-94.
[http://dx.doi.org/10.1111/1750-3841.12247] [PMID: 24025030]

[86] Mishra S, Singh BR, Singh A, Keswani C, Naqvi AH, Singh HB. Biofabricated silver nanoparticles act as a strong fungicide against Bipolaris sorokiniana causing spot blotch disease in wheat. PLoS One 2014; 9(5): e97881.
[http://dx.doi.org/10.1371/journal.pone.0097881] [PMID: 24840186]

[87] Phu DV, Lang VTK, Kim Lan NT, *et al.* Synthesis and antimicrobial effects of colloidal silver nanoparticles in chitosan by γ-irradiation. J Exp Nanosci 2010; 5(2): 169-79.
[http://dx.doi.org/10.1080/17458080903383324]

[88] Jagana D, Hegde YR, Lella R. Green nanoparticles: a novel approach for the management of banana anthracnose caused by *Colletotrichum musae.* Int J Curr Microbiol Appl Sci 2017; 6(10): 1749-56.
[http://dx.doi.org/10.20546/ijcmas.2017.610.211]

[89] Ardakani AS. Toxicity of silver, titanium and silicon nanoparticles on the root-knot nematode, *Meloidogyne incognita*, and growth parameters of tomato. Nematology 2013; 15(6): 671-7.
[http://dx.doi.org/10.1163/15685411-00002710]

[90] Cromwell WA, Yang J, Starr JL, Jo YK. Nematicidal effects of silver nanoparticles on root-knot nematode in bermudagrass. J Nematol 2014; 46(3): 261-6.
[PMID: 25275999]

[91] Abdellatif KF, Abdelfattah RH, El-Ansary MSM. Green nanoparticles engineering on root-knot nematode infecting eggplants and their effect on plant DNA modification. Iran J Biotechnol 2016; 14(4): 250-9.
[http://dx.doi.org/10.15171/ijb.1309] [PMID: 28959343]

[92] Evans I, Solberg E, Huber DM. Copper and plant disease. In: Datnoff LE, Elmer WH, Huber DM, Eds. Mineral Nutrition and Plant Disease. St Paul, Minn, USA: The American Phytopathological Society 2007; pp. 177-88.

[93] Darago A. The distribution of dagger nematodes species in Hungarian wind regions and newest control options 2014.

[94] Elmer W, De La Torre-Roche R, Pagano L, *et al.* Effect of metalloid and metal oxide nanoparticles on *Fusarium* wilt of watermelon. Plant Dis 2018; 102(7): 1394-401.
[http://dx.doi.org/10.1094/PDIS-10-17-1621-RE] [PMID: 30673561]

[95] He L, Liu Y, Mustapha A, Lin M. Antifungal activity of zinc oxide nanoparticles against Botrytis cinerea and *Penicillium expansum.* Microbiol Res 2011; 166(3): 207-15.
[http://dx.doi.org/10.1016/j.micres.2010.03.003] [PMID: 20630731]

[96] Sardella D, Gatt R, Valdramidis V. Physiological effects and mode of action of ZnO nanoparticles against postharvest fungal contaminants. 109th Annual Meeting of the American Phytopathological Society. San Antonio, TX. 2017.
[http://dx.doi.org/10.1016/j.foodres.2017.08.019]

[97] Hafez EE, Hassan HS, Elkady MF, Salama E. Assessment of antibacterial activity for synthesized zinc oxide nanorods against plant pathogenic strains. Int J Sci Technol Res 2014; 3(9): 318-24.

[98] Graham JH, Johnson EG, Myers ME, *et al.* Potential of nano-formulated zinc oxide for control of citrus canker on grapefruit trees. Plant Dis 2016; 100(12): 2442-7.
[http://dx.doi.org/10.1094/PDIS-05-16-0598-RE] [PMID: 30686171]

[99]   Imada K, Sakai S, Kajihara H, Tanaka S, Ito S. Magnesium oxide nanoparticles induce systemic resistance in tomato against bacterial wilt disease. Plant Pathol 2016; 65(4): 551-60.
[http://dx.doi.org/10.1111/ppa.12443]

[100] Rico CM, Hong J, Morales MI, *et al.* Effect of cerium oxide nanoparticles on rice: a study involving the antioxidant defense system and *in vivo* fluorescence imaging. Environ Sci Technol 2013; 47(11): 5635-42.
[http://dx.doi.org/10.1021/es401032m] [PMID: 23662857]

[101] Wang Q, Ebbs SD, Chen Y, Ma X. Trans-generational impact of cerium oxide nanoparticles on tomato plants. Metallomics 2013; 5(6): 753-9.
[http://dx.doi.org/10.1039/c3mt00033h] [PMID: 23689668]

[102] Gupta K, Singh RP, Pandey A, Pandey A. Photocatalytic antibacterial performance of TiO2 and Ag-doped TiO2 against S. aureus. P. aeruginosa and E. coli. Beilstein J Nanotechnol 2013; 4: 345-51.
[http://dx.doi.org/10.3762/bjnano.4.40] [PMID: 23844339]

[103] Ingham B. X-ray scattering characterisation of nanoparticles. Crystallogr Rev 2015; 21(4): 229-303.
[http://dx.doi.org/10.1080/0889311X.2015.1024114]

[104] Lauterwasser C. Small sizes that matter: Opportunities and risks of nanotechnologies. Report in cooperation with the OECD International Futures Programme. Munchen: Allianz Center for Technology 2006; p. 45.

[105] Kah M, Hofmann T. Nanopesticide research: current trends and future priorities. Environ Int 2014; 63: 224-35.
[http://dx.doi.org/10.1016/j.envint.2013.11.015] [PMID: 24333990]

[106] Bordes P, Pollet E, Avérous L. Nano-biocomposites: Biodegradable polyester/nanoclay systems. Prog Polym Sci 2009; 34(2): 125-55.
[http://dx.doi.org/10.1016/j.progpolymsci.2008.10.002]

[107] Yan J, Huang K, Wang Y, Liu S. Study on anti-pollution nano-preparation of dimethomorph and its performance. Chin Sci Bull 2005; 50(2): 108-12.
[http://dx.doi.org/10.1007/BF02897511]

[108] Bhan S, Lalit M, Srivastava CN. Relative larvicidal potentiality of nano-encapsulated temephos and imidacloprid against Culex quinquefasciatus. J Asia Pac Entomol 2014; 17(4): 787-91.
[http://dx.doi.org/10.1016/j.aspen.2014.07.006]

[109] Campos EVR, de Oliveira JL, da Silva CMG, *et al.* Polymeric and solid lipid nanoparticles for sustained release of carbendazim and tebuconazole in agricultural applications. Sci Rep 2015; 5(1): 13809.
[http://dx.doi.org/10.1038/srep13809] [PMID: 26346969]

[110] Brunel F, El Gueddari NE, Moerschbacher BM. Complexation of copper(II) with chitosan nanogels: toward control of microbial growth. Carbohydr Polym 2013; 92(2): 1348-56.
[http://dx.doi.org/10.1016/j.carbpol.2012.10.025] [PMID: 23399164]

[111] Bhagat D, Samanta SK, Bhattacharya S. Efficient management of fruit pests by pheromone nanogels. Sci Rep 2013; 3(1): 1294.
[http://dx.doi.org/10.1038/srep01294] [PMID: 23416455]

[112] Cursino L, Li Y, Zaini PA, De La Fuente L, Hoch HC, Burr TJ. Twitching motility and biofilm formation are associated with tonB1 in *Xylella fastidiosa*. FEMS Microbiol Lett 2009; 299(2): 193-9.
[http://dx.doi.org/10.1111/j.1574-6968.2009.01747.x] [PMID: 19735464]

[113] Bernardes PC, Nélio JDA, Nilda de Fátima FS. Nanotechnology in the food industry. Biosci J 2014; 30: 1919-32.

[114] Murphy K. Nanotechnology: Agriculture's Next "Industrial" Revolution (Williston, VT: Financial Partner, Yankee Farm Credit, ACA). Spring 2008; 3-5.

[115] Zaini PA, De La Fuente L, Hoch HC, Burr TJ. Grapevine xylem sap enhances biofilm development by *Xylella fastidiosa*. FEMS Microbiol Lett 2009; 295(1): 129-34.
[http://dx.doi.org/10.1111/j.1574-6968.2009.01597.x] [PMID: 19473259]

[116] Chuan L, He P, Pampolino MF, *et al.* Establishing a scientific basis for fertilizer recommendations for wheat in China: Yield response and agronomic efficiency. Field Crops Res 2013; 140: 1-8.
[http://dx.doi.org/10.1016/j.fcr.2012.09.020]

[117] Mody VV, Cox A, Shah S, Singh A, Bevins W, Parihar H. Magnetic nanoparticle drug delivery systems for targeting tumor. Appl Nanosci 2014; 4(4): 385-92.
[http://dx.doi.org/10.1007/s13204-013-0216-y]

[118] Qian K, Shi T, Tang T, Zhang S, Liu X, Cao Y. Preparation and characterization of nano-sized calcium carbonate as controlled release pesticide carrier for validamycin against *Rhizoctonia solani*. Mikrochim Acta 2011; 173(1-2): 51-7.
[http://dx.doi.org/10.1007/s00604-010-0523-x]

[119] Xu L, Cao L-D, Li F-M, Wang X-J, Huang Q-L. Utilization of chitosan-lactide copolymer nanoparticles as controlled release pesticide carrier for pyraclostrobin against *Colletotrichum gossypii* Southw. J Dispers Sci Technol 2014; 35(4): 544-50.
[http://dx.doi.org/10.1080/01932691.2013.800455]

[120] Ilk S, Saglam N, Özgen M. Kaempferol loaded lecithin/chitosan nanoparticles: preparation, characterization, and their potential applications as a sustainable antifungal agent. Artif Cells Nanomed Biotechnol 2017; 45(5): 907-16.
[http://dx.doi.org/10.1080/21691401.2016.1192040] [PMID: 27265551]

[121] Nasseri M, Golmohammadzadeh S, Arouiee H, Jaafari MR, Neamati H. Antifungal activity of *Zataria multiflora* essential oil-loaded solid lipid nanoparticles *in-vitro* condition. Iran J Basic Med Sci 2016; 19(11): 1231-7.
[PMID: 27917280]

[122] Cao L, Zhang H, Cao C, Zhang J, Li F, Huang Q. Quaternized chitosan-capped mesoporous silica nanoparticles as nanocarriers for controlled pesticide release. Nanomaterials (Basel) 2016; 6(7): 126.
[http://dx.doi.org/10.3390/nano6070126] [PMID: 28335254]

[123] Sandhya , Kumar S, Kumar D, Dilbaghi N. Preparation, characterization, and bio-efficacy evaluation of controlled release carbendazim-loaded polymeric nanoparticles. Environ Sci Pollut Res Int 2017; 24(1): 926-37.
[http://dx.doi.org/10.1007/s11356-016-7774-y] [PMID: 27761863]

[124] Zhao P, Cao L, Ma D, Zhou Z, Huang Q, Pan C. Translocation, distribution and degradation of prochloraz-loaded mesoporous silica nanoparticles in cucumber plants. Nanoscale 2018; 10(4): 1798-806.
[http://dx.doi.org/10.1039/C7NR08107C] [PMID: 29308814]

[125] Hasheminejad N, Khodaiyan F, Safari M. Improving the antifungal activity of clove essential oil encapsulated by chitosan nanoparticles. Food Chem 2019; 275: 113-22.
[http://dx.doi.org/10.1016/j.foodchem.2018.09.085] [PMID: 30724177]

CHAPTER 9

In vitro Antifungal Efficacy of Nanomaterials against Plant Pathogenic Fungi and Oomycetes

Conor F. McGee[1,*] and **Evelyn M. Doyle**[2]

[1] *Department of Agriculture, Food and the Marine, Backweston Laboratories, Celbridge, Co. Kildare, Ireland*

[2] *School of Biology and Environmental Science, University College Dublin, Belfield, D4, Co. Dublin, Ireland*

Abstract: Nanoparticulate (NP) substances have widely documented antimicrobial properties, yet their utilisation in the biocides and pesticides industries has yet to be fully exploited. This is particularly so in the pesticides industry, where their potential has not yet been realised. This mini review identifies the emerging trends identified in research characterising the *in vitro* antimicrobial properties of NP substances against fungal and oomycete phytopathogens. Nanoparticulate substances for which there was a sufficient depth of published studies on activity against fungal and oomycete phytopathogens are covered in this review, these include chitosan, copper, magnesium, silver and zinc. All substances displayed significant activity against a range of phytopathogens, though silver and copper-based NPs appear to be the most potent at relativity low (<50 ppm) concentrations. However, as particle size and shape affect the level of exhibited toxicity, direct comparisons of activity between studies are often difficult due to the different types of NP examined. One particularly promising NP substance is the organic biodegradable substance chitosan which is considered environmentally friendly. Chitosan has also been shown to stimulate plant growth and defence in addition to possessing antifungal activity. The lack of toxicological properties marks chitosan as having particular potential for fulfilling the regulatory requirements for environmental fate and ecotoxicology necessary for gaining approval as an authorised pesticide. Another distinct problem in comparing studies is the lack of a recognised standardised growth medium/media for determining nanomaterial toxicity. A growing body of evidence suggests that the *in vitro* toxicity of certain nanoparticles is highly influenced by the properties of the growth medium, such as its pH, salinity and components. These confounding factors will be discussed and their implications for comparing nanomaterial efficacy highlighted while also providing suggestions for improving characterisation of nanomaterial efficacy. Characterisation of nanomaterial efficacy *in vitro* is a critical step in determining which nanomaterials should be progressed for further testing in higher tier tests such as simulated use trials and field trials. The aim of this chapter is to draw attention to the limitations of *in vitro* characterisation and highlight how these techniques can be improved.

* **Corresponding Author Conor F. McGee**: Department of Agriculture, Food and the Marine, Backweston Laboratories, Celbridge, Co. Kildare, Ireland; Tel: +353857277079; Email: conorfmcgee@gmail.com

Savita, Anju Srivastava, Reena Jain & Pratap Kumar Pati (Eds.)

Keywords: Chitosan, Copper, Fungi, Growth medium, *In vitro*, Magnesium, Nanoparticles, Oomycetes, Phytopathogens, Silver, Zinc.

INTRODUCTION

Fungal and oomycete plant pathogens (phytopathogens) inflict substantial economic impacts on global agriculture each year [1]. The effects of fungal and oomycete phytopathogens can be devastating to certain crops resulting in restrictions to the supply of foodstuffs. As global production of food must rise in line with the increasing human population, control of such pathogens is of the utmost importance [2]. Typically, control of these phytopathogens is undertaken using synthetic pesticides whose mechanisms of action (MOA) target molecular pathways of pathogens such as signalling pathways, sterol synthesis and cell growth [3]. However, the resistance of fungal and oomycetes to such pesticides is continually evolving, requiring the development of novel control strategies [4]. One such potential novel solution could be the utilisation of a range of different substances in nanoparticle (NP) form as pesticides [5].

There has been significant interest in the antimicrobial properties of a range of NP substances for use in applications such as biocides, medical science and disinfectants, as well as in agriculture as Plant Protection Products (PPP) [5, 6]. Nanoparticulate substances, particularly those which are metal or metalloid based, generally exhibit broad-spectrum activity inducing damage to a range of cell organelles and molecular processes [7]. This broad-spectrum activity makes NPs particularly useful for use as general pesticides for controlling the broad range of fungal and oomycete pathogens typically associated with any particular crop.

The MOA of NPs and their resulting mode of action (MoA) in microbial species has been well characterised in a number of reviews [8, 9] and will only be summarised in this mini-review. In this review, the MOA refers to the specific biochemical mediators of toxicity such as ions and reactive oxygen species (ROS), while MoA refers to observed anatomical or functional changes in target organisms. Briefly, NPs have been shown to be highly reactive, possessing high surface energies, particularly for particles with diameters <30 nm [8]. The high surface energies associated with NP substances have been shown to induce a range of processes which can result in toxicity to cells and cellular processes [8, 9]. Nanoparticulate substances have been shown to inflict physical damage to microbial cell membranes/walls that they come into direct contact with, also in addition to generating ROS from their surface [8 - 10]. Certain metal-based NPs have also been shown to exhibit increased rates of ion dissolution, particularly smaller particles with more angular shapes [11]. Ions such as those released by nanoparticles of silver (AgNP) and copper (CuNP) have been shown to be directly toxic to microorganisms [9]. Released ions have also been shown to be a source of ROS which in turn inflict high levels of cellular toxicity [9]. The MOA and MoAs of each NP substance covered in this mini review are explored in more detail in later sections.

To date, numerous studies exploring the efficacy of nanomaterials for controlling fungal and oomycete phytopathogens have been published. To the authors' knowledge the range of published studies that examined the *in vitro* effects of NPs on fungal and oomycete phytopathogens included in this review is comprehensive, though we acknowledge some studies may be overlooked. The full list of studies included in this review and the pathogens they investigated can be found in Table 1. Most often, in the *in vitro* studies, NPs are amended into a solid agar-based growth medium (AGM) which the fungi or oomycete are cultivated on. Alternatively, circular disks are saturated with an NP and are then placed on a solid growth medium to test antifungal activity *via* the disk diffusion method (DDM) or a well diffusion method (WDM) is used where aliquots of NP suspensions are deposited into "wells" bored into solid media These studies have identified a range of nanoparticulate molecules, elements and oxides that have the ability to inhibit the growth of pathogenic species *in vitro* and therefore may have potential to control diseases in the field. The most notable elements whose efficacy has shown promise in nano form are Ag, Cu and Zn, while the oxides MgO and ZnO have also shown significant promise. Nanoforms of the organic molecule chitosan have also been investigated for controlling certain pathogenic species *in vitro* and demonstrated promising efficacy. The range of fungal and oomycete plant pathogens whose growth has been inhibited by nanomaterials includes both soil-borne and foliar disease-causing species. The purpose of this chapter is to compare the studies performed investigating the efficacy of nanomaterials against plant pathogenic fungi and identify the common trends emerging from the research conducted to date.

Table 1. List of fungal and oomycete phytopathogens on which the NP substances have been tested *in vitro*.

Pathogen	Disease(s)	Host(s)	References
Alternaria alternata	Brown/Black/Leaf spot	*Citrus, Malus, Nicotiana, Pyrus, Solanum*	[27, 35, 36, 38, 43, 55, 63, 68]
Alternaria brassicicola	Black spot	*Brassica*	[32, 57, 68]
Alternaria solani	Early Blight	*Capsicum, Cucumis, Solanum*	[34, 62, 68]
Aspergillus flavus	Ear rot, Yellow mould	*Zea*	[25, 36, 51, 54, 56]
Aspergillus niger	Black mould	*Allium, Arachis, Vitis*	[25]

Pathogen	Disease(s)	Host(s)	References
Bipolarisoryzae	Brown spot disease	*Oryza*	[59]
Bipolarissorokiniana	Black point, Root rot, Spot blotch	*Hordeum, Triticum*	[16, 32, 57]
Botrytis cinerea	Grey mould	*Vitis* (notably), various hosts >200	[27, 35, 38, 40, 43, 52, 68]
Botrytis fabae	Chocolate spot	*Vicia faba*	[38]
Colletotrichum acutatum	Anthracnose	*Capsicum, Malus, Vigna*	[15, 38]
Colletotrichum dematium	Anthracnose	*Brassica, Capsicum, Vicia*	[15]
Colletotrichum gloeosporioides /Glomerella cingulate	Anthracnose	*Capsicum, Fragaria, Fabaceae,* also cereals, grasses, legumes, fruits and trees	[15, 27, 33, 48, 60, 68]
Colletotrichum higginsianum	Anthracnose	*Aloe, Brassica, Capsicum, Raphanus,*	[15]
Colletotrichum nigrum	Anthracnose	*Capsicum,*	[15]
Colletotrichum orbiculare	Anthracnose	*Capsicum, Citrullus, Cucumis,*	[15]
Corynesporacassiicola	Leaf spot	*Capsicum, Cucumis, Fabaceae, Sesamum, Solanum*	[68]
Curvularialunata	Leaf spot	*Oryza, Panicgrass, Saccharum, Zea,* many angiosperms	[24, 35]
Cylindrocarpondestructans	Root rot	*Fragaria, Panax, Paeonia*	[68]
Didymellabryoniae	Black rot	*Citrullus, Cucumis, Fabaceae, Solanum,*	[68]
Fusarium fujikuroi	Bakamae	Cereals: Including *Oryza, Zea*	[59]
Fusarium graminearum	*Fusarium* ear/head blight	*Triticum*	[47]
Fusarium oxysporum	*Fusarium*/Vascular wilt	Numerous herbaceous plants including solanaceous crops and legumes	[24, 30, 38, 49, 55, 56, 60, 62, 68]
F. oxysporumf.sp. cucumerinum	*Fusarium*/Vascular wilt	*Cucumis*	[68]
F. oxysporum f. sp. lycoperici	*Fusarium*/Vascular wilt	*Solanum*	[27, 68]
Fusarium solani	Many, including: Root rot, Soft rot, sudden death	*Solanum* mainly, but numerous other hosts	[27, 68]
Fusarium redolens	*Fusarium*/Vascular wilt	*Asparagus*	[30]

(Table 1) cont.....

Pathogen	Disease(s)	Host(s)	References
Fusarium semitectum	Brown apical necrosis (BAN), Pineapple fusariosis	*Ananas, Juglans*	[21]
Fusarium verticillioides	Ear rot	Cereals: including *Oryza, Zea*	[59]
Gibberellafujikuori	Bakanae	*Oryza,* but other cereals also susceptible	[60]
Macrophominaphaseolina	Charcoal rot	*Pinus, Zea,* notably, but over 500 other known hosts	[35, 63]
Magnaporthe grisea/Pyricularia grisea	Rice blast, Blight	*Oryza, Hordeum, Triticum,*	[16, 62]
Monilia fructicola	Brown rot	Stone fruits	[27]
Monosporascuscannonballus	Root rot, Sudden death/wilt, Mellon collapse	*Cucumis, Citrullus*	[68]
Mucor plumbeus	Mucor rot	*Citrus*	[55]
Penicillium expansum	Blue mold	*Malus*	[49, 52]
Penicillium funiculosum	Fruitlet core rot	*Ananas*	[56]
Peronospora tabacina	Blight	*Nicotiana*	[46]
Phoma glomerata	Blight, Canker, Leaf spot	*Pisum, Vitis*	[21]
Phomaherbarum	Leaf spot	Elaeis	[21]
Phragmidiummucronatum	Rose rust	*Rosaceae*	[31]
Phytophthora cactorum	Damping off, Root rot	*Fragaria, Rhododendron,* over 200 other hosts	[30]
Phytophthora capsici	Damping off, Root rot	*Solanaceae,Cucurbitaceae, Fabaceae*	[60]
Phytophthora cinnamomi	Die back, Root rot,	Broad host range including fruit crops, trees and ornamentals	[38]
Phytophthora nicotianae	Crown rot, Damping off, Root rot	Wide host range of over 255 genera	[58]
Pythium aphanidermatum	Blight, Damping off	Wide host range including *Citrus, Euphorbia, Glycine*	[29, 66, 68]
Pythium spinosum	Damping off, Root rot	*Oryza, Primula*	[68]
Raffaelea	Wilt	*Laurus, Quercus*	[22]
Rhizopus stolonifer	*Rhizopus* rot	Stone fruits	[51, 55]
Rhizoctonia solani	Collar rot, Brown patch, Root rot, Damping off, Wire stem	Wide host range including grasses, *Solanum,* cereals *Oryza*	[10, 24, 26, 30, 35, 36, 42 - 44, 53, 63, 66]

(Table 1) cont.....

Pathogen	Disease(s)	Host(s)	References
Sclerotinia minor	Soft rot, White mould	Wide host range of broadleaved flowering and vegetable crops	[26]
Sclerotium rolfsii	Southern blight, White mould	Wide host range	[25, 44]
Sclerotinia sclerotiorum	White mould	Wide host range including *Brassica, Citrus, Vicia*	[26, 35, 60]
Stemphylium lycospersici	Grey leaf spot	Many hosts including Capsicum, Carica, Solanum	[68]
Thielaviopsis basicola	Black root rot	Wide host range including *Citrus, Cucumis, Solanum*	[58]
Trichoderma harzianum	Green mould	*Agaricus bisporus*	[51]
Verticillium albo-atrum	*Verticillium* wilt	*Humulus, Malvoideae, Medicargo*	[38]
Verticillium dahliae	*Verticillium* wilt	> 300 herbaceous and woody plants	[27, 28]

Influence of Growth Media on NP Toxicity

Microbial growth media and certain growth media components have been shown to significantly influence the toxicity of antimicrobial substances *in vitro* [11 - 13]. Certain growth media may mask or accentuate the level of NP toxicity compared to what may be observed *in vivo*, but currently, the degree to which different growth media affect toxicity remains relatively uncharacterised. It is known that particular media components and the pH level affect NP stability *in vitro,* which in turn affects NP toxicity [11, 14]. However, the degree to which specific media components can affect NP substance toxicity *in vitro* has received limited attention to date, while the effect of growth media pH on NP antifungal activity has been largely ignored.

The choice of medium has been shown to strongly influence the level of *in vitro* AgNP toxicity exhibited against a selection of 6 *Colletotrichum* fungal pathogens [15]. The toxicity of AgNPs was shown to be highest when the *Colletotrichum* species were cultivated on potato dextrose agar (PDA) compared to either corn meal agar (CMA) or malt extract agar (MEA). However, the specific component in the media responsible for affecting the AgNP toxicity was not determined in the study, but the general level of AgNP toxicity was found to follow the order of PDA>CMA>MEA. This finding was supported in a subsequent much broader study investigating 17 fungal phytopathogenic species spanning 11 different genera [68]. The effects of AgNPs were once again found to be greater on PDA compared to CMA and MEA following a trend of PDA>CMA>MEA for 15 different fungal species representing 10 different genera. However, 2 *Pythium* species were found to be more resilient to AgNP toxicity when cultivated on MEA, than on either the PDA or CMA. Why the oomycete pathogens were more tolerant to AgNPs when cultivated on MEA was not determined. Although only two studies have reported these findings, the trend observed across a broad range of fungal taxa suggests that PDA facilitates AgNP toxicity to a greater extent than either CMA or MEA. How PDA compares to other growth media is currently uncharacterised, but it is known that complex growth media derived from organic sources tend to mask AgNP toxicity to microbial species to a greater degree than minimal media [12]. This is most likely due to the chelation of NPs by organic materials in the growth mediums. PDA, CMA and MEA growth mediums all contain approximately 1.5% of the gelling agent agarose, in addition to different nutrient sources while typically having acidic pHs. The nutrient sources in a typical PDA growth medium are generally the sugar dextrose and a potato infusion extract, while CMA tends to solely contain corn meal extract. MEA contains typically contains a mix of peptone, dextrose and malt extract. The typical compositions of the growth media used in studies covered in this review are given in Table **2**.

Table 2. Typical compositions of growth mediums utilised in studies covered in this review.

Growth Media	Typical Component Content					
	Agar	Carbohydrate	Protein	Extract	Salts	pH @ 25°C
IDL	-	Dextrose 0.4%	-	-	$KH_2PO_4 - 0.3\%$; $Na_2HPO_4.12H_2O - 1.7\%$; $Na_2SO_4 - 0.06\%$; $(NH_4)_2SO_4 - 0.12\%$;	pH7
LB	-	-	Tryptone – 1%	Yeast Extract – 0.5%	NaCl – 1%	pH7
CMA	1.5%	-	-	Corn Meal Extract 0.2%	-	pH 6.0 ±0.2
MEA	1.5%	-	Peptone - 0.5%	Malt extract -3%	-	pH 5.4 ± 0.2
PDA	1.5%	Dextrose – 2%	-	Potato extract 0.4%	-	pH 5.6 ± 0.2
SDA	1.5%	Dextrose – 4%	Peptone 1%	-	-	pH 5.6 ± 0.2
V8	1.5%	-	-	V8 Juice – 200 ml	$CaCO_3 - 0.2\%$	pH 7.2 ± 0.2

For certain fungal pathogens such as *Bipolarissorokiniana* and *Magnaporthe grisea*, AgNP toxicity was shown to be neutralised by the addition of simple salts, such as NaCl, into growth mediums [16]. NaCl was shown to dissociate in both V8 and oatmeal agar (OA) growth mediums, with Cl⁻ ions binding to Ag⁺ ions forming AgCl, which reduced the level of toxicity to both species *in vitro*. Certain other components such as tryptone, yeast extract and disulphide (S^{2-}) have been shown to have an even greater influence on AgNP toxicity than NaCl *in vitro* [12]. In one study, Luria-Bertani (LB) and a minimal medium (IDL) were adjusted to pH7 and then amended with specific components to determine their effect on the anti-bacterial activity of AgNPs on *Bacillus subtilis in vitro*. It was shown that AgNP toxicity decreased proportionately as the concentrations of either tryptone, yeast extract or S^{2-} were increased.

NP substances are generally screened *in vitro* to determine if they possess antifungal properties prior to deciding whether they merit conducting further higher tier *in vivo* trials. To date, studies investigating the antifungal activity against phytopathogens have generally used only one type of growth media and one pH level, as is the case for the majority of studies covered in this review. Given that the choice of growth medium has been shown to significantly affect toxicity, it may be useful to establish the influence of a selection of growth media for a broad range of NP substances in future studies to refine which growth media may be most suited to screening NP toxicity. Additionally, it has been reported that pH affects the toxicity of certain NP substances *in vivo*, particularly for AgNPs [17]. Knowing this, it is imperative to assess NP toxicity across an environmentally relevant pH range when screening for antifungal activity.

Silver based Nanomaterials

The antimicrobial properties of AgNPs are currently being exploited in a wide range of products spanning many industries [18]. As a direct consequence of the growing interest in AgNPs, there is also increasing concern regarding the potential ecotoxicological impacts of AgNPs in the environment. AgNPs are known to alter agricultural soil microbial community structures of both bacteria and fungi, in addition to reducing soil enzymatic activities [19, 20]. Additionally, key functional groups of the microbial nitrogen cycle, such as ammonia oxidisers, have been shown to be highly sensitive to AgNP contamination [17]. Regardless of these ecotoxicological concerns, the fungicidal potential of AgNPs has still attracted interest and there has been a wide range of studies exploring the possibility of controlling phytopathogens with AgNPs applications. Typically, silver particles used in most studies to date have been spherical in shape example from our laboratory shown in Figs. (**1a** and **1b**). Particle shapes are typically determined using techniques such as Transmission Electron Microscopy (TEM) which can also be used for determining particles sizes.

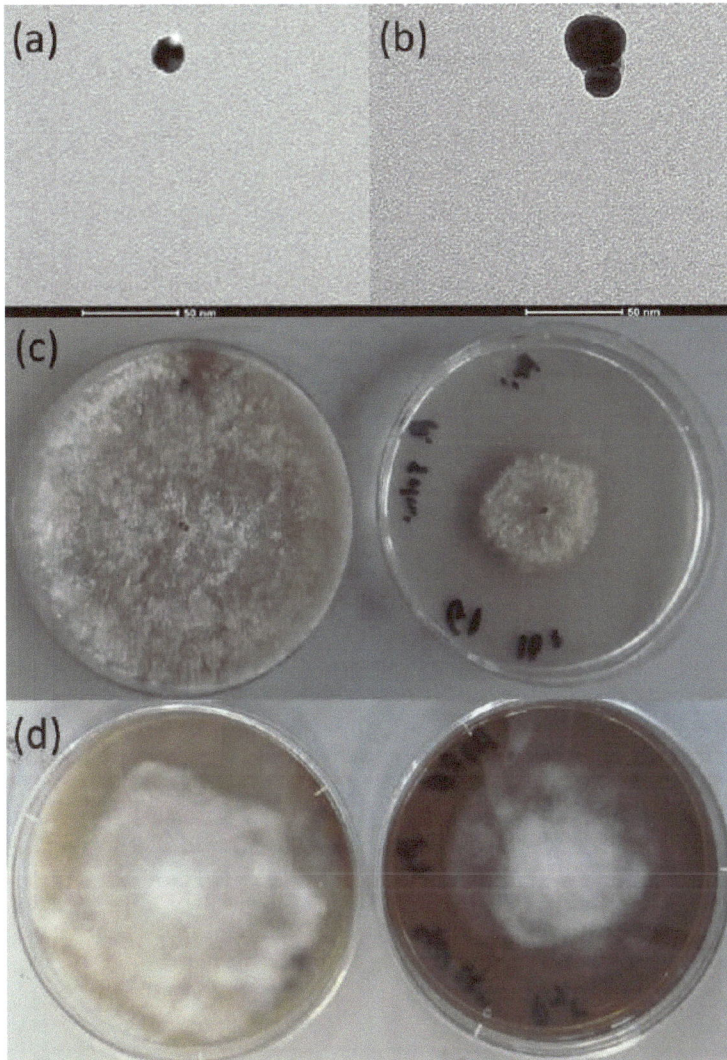

Fig. (1). AgNPs on a Formvar carbon-coated 200 copper grid imaged using a Tecnai G^2 20 TWIN electron microscope in our laboratory 1a and 1b, 50 nm scale bar depicted below. Work undertaken in our laboratory found concentrations as low as 10 ppm to reduce mycelial growth of many fungal plant pest species and wood degrading species *in vitro*.Reduced mycelial growth of *R. solani*using 10ppm AgNP amended into PDA 1c after 48 hours, control (0 ppm) on left. Reduced mycelial growth of *P. ultimum* using 10 ppm AgNP amended into PDA 1c after 48 hours, control (0 ppm) on left.

The antifungal activity of AgNPs has received much attention, though the exact mechanism of toxicity is still unclear [23]. AgNPs have been observed to physically interfere with the structure of fungal cell walls and membranes, inducing pitting in fungal hyphae [10, 21]. AgNPs are also known to disrupt cellular transport systems, such as ion efflux, which leads to an accumulation of cellular AgNPs [10, 15]. Toxicity of AgNPs is attributed to the release of Ag^+ ions which induce harmful cellular effects, while ROS are also known to be generated from the surface of AgNPs [10]. Ag^+ ions are also known to generate ROS which are in turn sources of cellular toxicity [22]. The release of Ag^+ ions has been shown to be higher from smaller and more angular AgNPs and is considered by many to be the main source of AgNP toxicity to microbial species [11]. However, some studies have shown that the toxicity of AgNPs cannot solely be attributed to the release of Ag^+ ions and there appears to be nanoscale particle effects, such as particle reactivity and increased ROS release, contributing to toxicity [23].

The antifungal activity of AgNPs has been explored against a wide range of fungal phytopathogens *in vitro* (Table **3**). AgNPs possess potent antifungal activity and have been shown to possess similar levels of efficacy *in vitro* to many conventional fungicides, such as amphotericin B and ketoconazole [24, 25]. The antifungal activity of AgNPs has been demonstrated against a wide range of fungal phytopathogens at relativity low (<50 ppm) concentrations compared to other NPs. Many of the studies covered in this review reported high levels of potency at low concentrations; these have typically investigated AgNP with small particle sizes <40 nm and often <10 nm [15, 22, 26]. The influence of AgNP size is known to have a significant effect on toxicity. The effect of AgNPs reducing mycelial growth of the fungal plant pathogens *R.solani* and *P.ultimumin vitro* in work performed in our laboratory is visually depicted in Figs. (**1c** and **1d**).

Table 3. Examples of studies examining *in vitro* antifungal activity of silver nanomaterials covered in this review.

Nanomaterial	Concentration(s)*	Pathogen(s)	Growth Medium	Refs.
Ag (32.5 nm)	DDM: 10 µl aliquot of with 20µg AgNP	*F. semitectum, P. glomerata, P. herbarum, Trichoderma sp.*	PDA	[21]
Ag (20-30 nm)	AGM: 1-100 ppm	*B. sorokiniana, M. grisea*	V8, OMA	[16]
Ag (4-8 nm)	AGM: 5-25 ppm	*Raffaelea sp.*	MEA	[22]
Ag (4-8 nm)	AGM: 3-7 ppm	*R. solani, S. sclerotiorum, S. minor*	MEA	[26]
Ag (4-8 nm)	AGM: 10-100 ppm	*C. acutatum, C. dematium, C. gloeosporioides, C. higginsianum, C. nigrum, C. orbiculare*	CMA, MEA, PDA	[15]
AgNP (10-20 nm); Chitosan (10-20 nm); Chitosan-AgNP composite (10-20 nm)	AGM–100 ppm WDM -500 µl of NP suspension (undefined conc)	*A. alternata, A. flavus, R. solani*	PDA	[36]

(Table 3) cont.....

Nanomaterial	Concentration(s)*	Pathogen(s)	Growth Medium	Refs.
Ag (7-25 nm)	AGM: 7-25 ppm	*A. Alternata, A. Brassicicola, A. solani, B. cinerea, C. cucumerinum, C. cassiicola, C. destructans, D. bryoniae, F. oxysporum f.sp. Cucumerinum, F. oxysporum f. sp. Lycoperici, F. oxysporum, F. solani, G. cingulata, M. cannonballus, P. aphanidermatum, P. spinosum, S. lycospersici*	PDA	[68]
Ag (35 nm)	WDM: 5-15 mg in 10 µl aliquots	*A. alternata, B. cinerea, C. lunata, M. phaseolina, S. sclerotiorum, R. solani*	PDA	[35]
Ag (40-50 nm)	DDM: 10 µg	*A. niger, A. flavus, S. rolfsii*	PDA	[25]
Ag (21 nm) - sodium deoxycholate capped	Spore suspension mixed with colloidal AgNP (undefined conc)	*C. gloeosporioides*	PDA	[33]
Ag (10-20 nm)	Spore suspension: 26-107 ppm AgNP	*P. mucronatum*	Water suspension	[31]
Ag (5-10 nm)	AGM: 5-50 ppm	*R. solani*	PDA	[10]
Ag (35 nm)	WDM: 10 g	*Curvularia sp., F. oxysporum, R. solani,*	PDA	[24]
Ag (12.7 nm)	AGM: 5-50 ppm	*A. solani*	PDA	[34]
Ag; Cu (size not specified)	AGM: 5-35 ppm	*F. oxysporum, F. redolens, P. cactorum, R. solani*	PDA, V8	[30]
Ag (1-30 nm)	AGM: 5-35 ppm	*P. aphanidermatum*	PDA	[29]
CuSO$_4$ (25 nm); CuO (<50nm); ZnO (<50 nm); Ag (<100nm)	AGM: 1-5000 ppm	*A. alternata, B. cinerea, M. fructicola, V. dahliae, C. gloeosporioides, F. oxysporumfspRadicisLycopersici (FORL), F. solani*	PDA	[27]
Ag (23 nm)	AGM: 20-60 ppm	*V. dahlia*	PDA	[28]
Ag (25-32 nm)	AGM: 10-100 ppm	*A. brassicicola, B. sorokiniana*	PDA	[32]

*All concentrations have been converted to ppm to allow comparison between studies

As stated previously, some studies have shown that fungal growth medium influences the level of AgNP antifungal properties [15, 68]. Many of the AgNP studies listed in Table **3** and in studies with other NP substances listed in Tables **4** - **7** have used PDA as a growth medium. However, studies have shown that the observed level of AgNP toxicity against a broad range of fungal phytopathogens is higher on PDA than on other growth media such as MEA and CMA [15, 68]. The influence of certain growth media components on *in vitro* AgNP toxicity has been well established [11].

For fungal species, the addition of common media components such as NaCl to media has been shown to neutralise the toxicity of AgNPs, with Cl⁻ ions capable of binding Ag⁺ ions and rendering them ineffective [16]. Whether PDA is the most suitable medium for screening for AgNP toxicity *in vitro* has not been conclusively established. It is likely that all *in vitro* media affect the MOA of NP substances to some extent, possibly neutralising some of the toxicity. Whether this effect is greater than what would occur *in vivo* is not yet known, however, certain studies have shown that *in vivo* AgNP concentrations required to control fungal phytopathogens are lower than those determined using synthetic media *in vitro* [28].

Interestingly, oomycete pathogens such as *Pythium aphanidermatum* and *Pythium spinosum* were shown to be far more resistant to AgNPs when grown on MEA than when cultivated on CMA or PDA [68]. The study conducted by Elshahawy*et al.* [29] also found *P. aphanidermatum* to be highly sensitive to AgNPs when cultivated on PDA. However, *Phytophthora cactorum* was found to be insensitive to up to 35 ppm of both AgNPs and CuNPs when cultivated on V8 growth medium, even though lower concentrations were found effective in controlling strain of *Fusarium redolens, Fusarium oxysporum* and *Rhizoctonia solani* which were cultivated on PDA [30]. V8 growth media is often recommended as the best choice for inducing oomycete species to produce reproductive structures such as sporangia, it is possible this medium also favours physiological tolerance to antibiosis. Oomycete species are known to be more resistant to antibiosis when there are utilisable sterols present in the growth medium [13]. The choice of medium could therefore be highly influential in determining the sensitivity of oomycete phytopathogens to NPs such as AgNPs.

AgNPs have been shown to be highly efficacious in inhibiting fungal spore germination and formation [10, 22, 31 - 33]. As is typically seen for other NP, AgNP have been shown to have significant impacts on spore germination at lower concentrations than those required to negatively affect mycelial growth [32]. Amending growth media with AgNPs has also been shown to have detrimental effects on the formation of conidia in *Raffaelea sp. in vitro* [22]. AgNPs have also been proven effective in reducing the formation of sclerotia in a wide range of pathogens, including *R. solani, Sclerotium sclerotiorum* and *Sclerotinia minor* [26]. The ability of AgNPs to inhibit spore germination indicates that they may have potential applications in inhibiting the spread of foliar pathogens on crops. AgNPs have been shown to be highly efficacious in inhibiting foliar pathogens such as *Alternaria solani* (potato early blight) [34 - 36].

However, while AgNPs may possess significant antifungal activity, like many metal or metalloid based NPs, their persistence in the environment and ecotoxicological profile is likely to be an impediment to gaining approval for use as pesticides in food production. Further research is necessary to determine the safety of many NP substances to ensure their risks can be managed sustainably prior to authorising their use in agriculture [37].

Copper Based Nanomaterials

Copper is a trace element which plays an important role in both plant and animal health [38]. However, the antimicrobial properties of copper have also been long known and exploited in agriculture [38]. The first widely used copper-based fungicide was known as the "Bordeaux mixture" a combination of copper sulphate, lime and water, which was used to control downy mildew on grapevines in the 1800s [39]. Recently, the antimicrobial properties of CuNPs have been widely explored for use in controlling human pathogens and food packaging, but agriculturally relevant plant pathogens have been largely ignored [38].

The MoA of CuNPs against fungal phytopathogens appears to be similar to other NP substances. CuNPs have been observed adhering to the surface of fungal hyphae and inducing pitting and deformations to the physical structure [30, 40]. The MOA involved in CuNP inhibition of fungal growth has been proposed to be *via* the release of Cu^+ ions and the production of ROS which disrupt cellular processes [30]. This mechanism has been observed in bacteria, for example, CuNP inhibition of *E. coli* was shown to be related to disruption of the cell membrane from direct contact with CuNPs [41]. The antibacterial toxicity of CuNPs was attributed to the release of Cu^+ ions resulting in the intercellular production of ROS [41].

To date, there have been several studies conducted on the antifungal activity of copper-based nanomaterials (Table 4). These studies have investigated a broad range of copper-based nanomaterials, including CuNP, CuONP and Cu nanocomposites and have highlighted several broad trends associated with copper-based nanomaterials, although there is still a need for further confirmatory studies. At present, copper-based nanomaterials appear to exhibit antifungal properties against a broad range of fungal phytopathogens [42-44]. Comparatively, Cu/CuO based nanomaterials appear to be generally as or more potent than other NP substances such as Ag and Zinc [27]. Copper nanomaterials appear to exhibit dose dependent toxic effects against mycelial growth of fungal phytopathogens while also demonstrating significant inhibit potential against fungal spores [27, 30].

Table 4. Examples of studies examining *in vitro* antifungal activity of copper nanomaterials covered in this review.

Nanomaterial	Concentration(s)*	Pathogen(s)	Growth Medium	Refs.
Chitosan-Cu (2-6 nm CuNP); Chitosan-Zn (1-5 nm ZnNP)	AGM: 30–100 ppm	*R. solani*	PDA	[42]
CuNP (50-100 nm); ZnNP (6-21 nm); Chitosan-Cu-Zn (16-24 nm)	AGM: 30–90 ppm	*A. alternata, R. solani, B. Cinerea*	PDA	[43]
Chitosan-Cu (2 nm CuNP)	AGM: 30-100 ppm	*R. solani, S. rolfsii,*	PDA	[44]
CuO (20-30 nm); Fe₂O₃ (40-100 nm); TiO₂ (20 nm); C60 (50 nm); Graphene sheets (500 x 1 nm)	Water suspension; 50-200 ppm	*B. cinerea*	Water	[40]
Cu (50 nm)	AGM: 100-800 ppm	*A. alternata, B. cinereal, B. fabae, C. acutatum, F. oxysporum, P. cinnamomi, V. albo-atrum*	PDA	[38]
Cu; Ag (size not specified)	AGM: 5-35 ppm	*F. oxysporum, F. redolens, P. cactorum, R. Solani*	PDA, V8	[30]
Cu (25 nm); CuO (<50nm); ZnO (<50 nm); Ag (<100nm)	AGM: 1-5000 ppm	*A. alternata, B. cinerea, M. fructicola, V. dahliae, C. gloeosporioides, F. oxysporumfspRadicisLycopersici (FORL), F. solani*	PDA	[27]

*All concentrations have been converted to ppm to comparability between studies

CuNPs may be more potent than copper oxidised form (CuONPs) in inhibiting mycelial growth of fungal pathogens *in vitro* [27]. In a study investigating the antifungal properties of range of NPs including ZnO, Ag, Cu, CuO against 8 fungal phytopathogens *in vitro*, CuNPs were found to be the most potent inhibitors of mycelial growth when added to PDA [27]. The general order of antifungal activity was found to be CuNPs>ZnONPs>AgNPs>CuONPs. However, characterisation of the NPs indicated that the particle size range associated with the CuNPs (25 nm) was smaller than that of the CuONPs (<50 nm), which may have contributed to the greater toxicity of the CuNPs. Regardless, the greater level of toxicity displayed by CuNPs compared to CuO-NPs was broadly in line with previous findings on a range of different organisms [27].

Copper based nanomaterials have generally exhibited greater antifungal activity than other NP substances in *in vitro* tests against phytopathogens. As already mentioned, certain CuNPs have been observed to be more inhibitive of mycelial growth than ZnONP and AgNP [27]. Copper based nanomaterials have also displayed greater toxicity when combined with chitosan nanocomposite materials [42]. A chitosan-CuNP material was found to be more potent at a similar concentration *in vitro* than a chitosan-ZnONP in inhibiting mycelial growth of *Rhizoctonia solani*. Al-Dhabaan *et al.* [43] found that ZnONPs and CuNPs could work synergistically in a chitosan nanocomposite material for inhibiting mycelial growth of *A. alternata, R. solani* and *Botrytis cinerea*. The antifungal activity of a chitosan-CuNPs nanocomposite material against the sclerotia forming fungi *R. solani* and *Sclerotium rolfsii* was also observed by Rubina *et al.* [44] where mycelial growth and sclerotia were found to be inhibited in a dose dependent manner by chitosan-CuNP.

While CuONP have been shown to be less potent than CuNP, ZnONP and AgNP, they have been observed to be more potent than certain other NP substances. CuONP were found to be more potent than Fe_2O_3, TiO_2, C60 (50 nm) and graphene sheets in inhibiting mycelial growth of *B. cinerea* [40]. The study of Aleksandrowicz-Trzcinska *et al.* [30] reported that CuNPswere be less effective than AgNPs in controlling mycelial growth of the pathogens *R. solani*(strain), *F. redolens and M. giganteus* across a concentration range of 5-35 ppm on a PDA growth medium. However, particle sizes were not characterised in the study so the influence of this property cannot be determined. Interestingly, a second strain of *R. solani* tested in this study was found to be insensitive to both CuNP and AgNP, indicating the presence of a possibly resistance mechanism. Additionally, the oomycete *P. cactorum* was not found to be affected by either CuNP or AgNP in this study, though this species was cultivated on a different growth medium (V8) to all the other species, and this may have influenced the results. Mycelial growth of the oomycete *Phytophthora cinnamomi* was found to be inhibited by 100 ppm of CuNPs when cultivated on a PDA growth medium [38].

Copper-based nanomaterials generally exhibit a dose-dependent effect against the growth of fungal phytopathogens *in vitro*. Dose-dependent effects of copper-based nanomaterials on *R. solani* have typically been observed across a concentration range of 30-100 ppm [42 - 44]. Dose-dependent effects have also been observed against a range of other fungal phytopathogens [27, 43, 44], though certain concentrations of CuNPs (100 ppm) have been observed to stimulate mycelial growth of particular fungal phytopathogens such as *Alternariaalternata, Botrytis fabae, F. oxysporumf.sp. ciceris* and *F. oxysporumf.sp. melonis* [38]. However, at higher (200-800 ppm) concentrations, antifungal effects of CuNPs were observed against the selected fungal phytopathogens [38].

Generally, copper-based nanomaterials have exhibited a greater level of inhibitory activity against fungal spores than mycelial growth. Banik and Perez-de-Luque [38] found fungal spores to be absent from cultures of certain fungi when cultivated on a growth medium amended with CuNPs, when mycelial growth was still observed at higher concentrations of CuNPs. The study of Malandrakis *et al.* [27] found CuNPs to be 10-100 times more potent in inhibiting spore germination of 7 fungal phytopathogens than inhibiting mycelial growth.

Copper based nanomaterials appear to be quite potent and exhibit toxicity against a broad range of fungal pathogens. However, copper-based pesticides have a long history of use in agriculture and subsequently, copper based resistance mechanisms have developed in many fungal and oomycete phytopathogens [45]. The build-up of copper resistance mechanisms in fungal and oomycete pathogens may limit the potential for copper-based NPs that mediate their effects *via* Cu^+ release. Similar to silver, persistence in the environment and resulting ecotoxicological profile may become an issue for copper-based NP pesticides.

Zinc Based Nanomaterials

Zinc is an environmentally ubiquitous element composing 0.004% of the earth's crust [46]. It is an essential plant and animal nutrient which is considered to be potentially safe for use in the food industry as a pesticide [46 - 48]. The relative toxicological safety of zinc compounds such as ZnO has led to their approval for use in food packaging and fortifying cereal-based foods by the US FDA [46]. Several studies have also found ZnONPs to be less toxic to plant life and beneficial soil bacteria than other substances in NP form, such as AgNPs [47].

To date, the majority of studies investigating the antifungal properties of Zn have focused on ZnO (Table **5**). ZnONPs generally form aggregates during their production process, reducing the overall surface areas of preparations [49]. Preparations of NPs often require commonly used dispersants and/or ultrasonication to break up aggregates into stable preparations of NPs for practical use [49, 50]. Dispersed preparations of ZnONPs possess greater surface area which results in increased activity of particles and the expression of their novel nanoscale properties [47]. This has often been observed for NPs where smaller sized particles displayed a higher degree of antifungal activity [51].

Table 5. Examples of studies examining *in vitro* antifungal activity of zinc nanomaterials covered in this review.

Nanomaterial	Concentration(s)*	Pathogen(s)	Growth Medium	Refs.
ZnO (70 nm)	AGM: 224-976 ppm	*B. cinerea; P. expansum*	PDA	[52]
ZnO (25 and 40 nm)	DDM: 325-975 ppm OD: 325-1302 ppm	*A. flavus, R. stolonifer, T. harzianum*	SDA	[51]
ZnO (30 nm); MgO (50 nm)	AGM: 0.1-0.5 ml of particle suspension (undefined conc)	*A. alternata, F. oxysporum, M. plumbeus R. stolonifer*	PDA	[55]
ZnO (70 nm)	AGM: 2-12 ppm	*F. oxysporum, P. expansum*	PDA	[49]
ZnO (<100 nm); ZnO (>1000 nm)	AGM: 100-500 ppm	*F. graminearum*	Mung Bean Agar	[47]
ZnO (20-25 nm)	OD: 20-100 ppm	*A. flavus*	PDA	[54]
ZnO (12 nm)	WDM: 25-100 ppm	*A. flavus, F. oxysporum, P. funiculosum*	PDA	[56]
ZnO (50 nm); Zn (264 nm); ZnO (1886 nm)	Suspension: 5-70 ppm	*P. tabacina*	Water	[46]
Chitosan-Cu (2-6 nm CuNP); Chitosan-Zn (1-5 nm ZnNP)	AGM: 30–100 ppm	*R. solani*	PDA	[42]
CuNP (50-100 nm); ZnNP (6-21 nm); Chitosan-Cu-Zn (16-24 nm)	AGM: 30–90 ppm	*A. alternata, B. cinerea, R. solani,*	PDA	[43]
ZnO (<50 nm), Ag (<100nm), Cu (25 nm), CuO (<50nm)	AGM: 1-5000 ppm	*A. alternata, B. cinerea, C. gloeosporioides, M. fructicola, F oxysporumfspRadicisLycopersici (FORL), F. solani, V. dahliae,*	PDA	[27]
ZnO (30 nm)	AGM: 0.15-15 ppm	*R. solani*	PDA	[53]
Zn (29-37 nm)	AGM: 10-100 ppm	*A. brassicicola, B. sorokiniana*	PDA	[57]
ZnO (51, 53, 63 & 77 nm); MgO (52 & 96 nm); MgO-ZnO (139, 161, 219); ZnO-Mg(OH) (54, 71, 88 & 98)	AGM: 2-5000 ppm	*C. gloeosporioides*	PDA	[48]

*All concentrations have been converted to ppm to comparability between studies

The MOA and MoA of ZnO are currently not completely characterised, although several observations on toxicity have been made. The direct contact of ZnONPs with the surface of bacterial cells have been attributed to disruption of cell membrane permeability resulting in leakage of cytoplasmic material [50, 52]. The mechanism by which ZnONP causes toxicity in bacteria has been proposed to be related to the formation of free radicals such as ROS on the surface of ZnONPs directly damaging to the cell membrane [52 - 54]. Protein leakage from fungal hyphae exposed to ZnONPs has also been observed for certain phytopathogens [49, 51]. The release of Zn^+ ions from ZnO has also been suggested as another mechanism against fungal phytopathogens [47]. However, a study conducted by Dimkpa *et al.* [47] demonstrated that Zn^+ release was similar between micron and nano sized ZnO particles *in vitro* and attributed the greater level of antifungal activity exhibited by ZnONPs to nanoscale properties. The study of Navale *et al.* [54] revealed that the oxidation capacity of ZnONPs toward the antioxidant glutathione (GSH) was possibly the underlying mechanism of action for antimicrobial activity. ROS producing activity of ZnONP has also been shown to be increased by the presence of UV light [54].

To date, the studies investigating the antifungal activity of ZnNPs and ZnONPs against phytopathogens have identified several broad trends. In comparative tests, the antifungal properties of ZnONPs tend to be less than similar sized particles of other metallic substances such as Cu and MgO [27, 42, 55]. However, differences in particles sizes between substances often make direct comparisons of antifungal properties difficult, given the influence of particle size on toxicity [27]. For ZnO, size has been shown to influence certain properties, with smaller sized particles displaying greater antifungal activity than larger particles [46, 47, 51]. Broadly ZnONP antifungal activity has been shown to be dose-dependent against a broad range of phytopathogens [27, 49, 53, 56]. *In vitro* assessment of antifungal efficacy has shown fungal spores to be more suspectable to ZnO than mycelial growth for many phytopathogens [27, 55, 57].

Comparative assessments of antifungal activity among NP substances have generally revealed lower levels of efficacy associated with ZnO NP than other metallic/metalloid NP. Wani and Shah [55] demonstrated that MgONPs displayed a higher level of inhibition against fungal spore germination than ZnONPs. The MgONPs, which were larger (50 nm) than the ZnO NPs (30 nm), were also more potent when tested against a range of fungal pathogens grown on PDA. However, la Rosa-Garcia *et al.* [48] found ZnONP to be more effective at inhibiting spore germination of *Colletotrichum gloeosporioides* than MgONP. These results may not be necessarily contradictory but may indicate possible species/strain-specific effects for certain NP substances. The la Rosa-Garcia study [48] did indicate that certain *C. gloeosporioides* strains were more resistant than others to both nanomaterials. Composite materials consisting of chitosan combined with NP of either Cu or Zn have also been shown to be potent in controlling the mycelial growth of *R. solani* when cultivated on a PDA growth medium; however, the chitosan-Cu composite material was shown to be more potent the chitosan-Zn [42]. CuNPs were also shown to be more potent than ZnONPs in controlling the growth of a range of fungi on PDA, however, ZnONP were more potent than AgNP and CuO [27]. The fact that ZnONPs were more potent than the AgNPs was surprising although possibly a result of the larger particle AgNP size (<100 nm) than ZnO (<50 nm). Overall ZnONPs appear to possess broad antifungal properties, albeit, with generally lower degree of efficacy than other NP substances.

In a study investigating the influence of ZnO particle size on antifungal activity, ZnONPs were shown to be more efficacious than zinc micron particles (ZnOMPs)against *Fusarium graminearum* [47]. The mechanism involved in the inhibition of *F. graminearum* by ZnONP could not be explained by the level of Zn^+ dissolution, which was similar to that of ZnOMP. It was suggested that it was more likely to be a result of nanoscale properties influencing ROS production. ZnOMP (1886 nm) were also found to be less effective than ZnONP (50 nm) in suppressing spore germination of the blue mould inducing pythiaceae phytopathogen *Peronospora tabacina*, interestingly ZnONPs were also found to be more effective than ZnNP [46]. The difference in antifungal activity could possibly be attributed to the differences in particle size as the ZnNPs (264 nm) were far larger than the ZnONP (50 nm). The particle size effect has also been demonstrated for particles within the nano-scale range, ZnONPs with APS of 25 nm were shown to be far more potent in controlling mycelial growth than ZnONP APS 40nm against a range of fungal pathogens.

ZnONPs have been shown to exhibit dose-dependent inhibition of fungal mycelial growth and spore germination in many studies. ZnONPs were shown to exhibit dose-dependent inhibition of mycelial growth against 3 fungal phytopathogens (*Aspergillus flavus, Rhizopus stolonife r*and *Trichodermaharzianum)* on Sabourand dextrose agar (SDA) and in a SD liquid medium [51], this property was observed for two different nanoscale particle sizes. Dose-dependent effects of ZnONPs have generally been observed at concentrations lower than 100 ppm. Lahuf *et al.* [53] reported dose-dependent inhibition of mycelial growth of *R. solani* in a concentration range of 0.15-15 ppm. A concentration range from 2-12 ppm was shown to inhibit mycelial growth of the pathogens *B. cinerea* and *Penicillium expansum* on PDA [49]. Dose-dependent inhibition of mycelial growth was also observed on PDA for the fungi *A. flavus, F. oxysporum* and *Penicillium funiculosum* across a concentration range of 20-100 ppm [54, 56]. Malandrakis *et al.* [27] confirmed that the dose-dependent antifungal activity of ZnO was a broadly observed property as they demonstrated this characteristic against 7 fungal phytopathogens *in vitro*.

ZnONPs have been shown to be very potent in preventing germination and development of certain fungal spores. Cultivation of *P. expansum* on a PDA medium amended with ZnONP was shown to prevent the development of conidiophores [52]. The level of spore germination was also shown to be reduced in a dose-dependent manner by ZnONP for *A. alternata, F. oxysporum, R. stolonifer* and *Mucorplumbeus* [55]. The level of *P. tabacina* spore germination inhibition was shown to be broadly similar for both ZnNP and ZnONP across a 5-70 ppm concentration range [46]. Spores have been shown to be more susceptible to ZnO than fungal mycelial growth for a range of species. Fungal spores were found to be a 10-100 times more susceptible to ZnONP *in vitro* for 7 fungal phytopathogens when cultivated on a PDA growth medium [27]. Kriti *et al.* [57] found ZnNP particles to completely inhibit spore germination of *Alternaria brassicicola* and *B. sorokiniana*at concentrations of 10 and 20 ppm, respectively. However, substantial inhibition of mycelial growth was generally observed at higher concentrations, typically $\geq$100 ppm.

Zinc based nanomaterials have been shown to exhibit significant antifungal activity at low concentrations against a range of phytopathogens. As a ubiquitous substance with relatively low levels of associated ecotoxicity or mammalian toxicity, zinc appears to have potential for use in commercial pesticides. The relative safety of ZnONPs has led to proposed applications in many industries in low does [27, 46, 55], although further testing is necessary to determine its full lifecycle and environmental fate. The relative safety and eco-friendly nature of ZnO make it potentially suitable candidate as an active substance in fungicides [27, 48].

Magnesium based Nanomaterials

Magnesium is an essential nutrient required for normal plant and animal physiological development. It is considered safe in low doses, and its NP form has attracted interest as an antimicrobial agent given its safety and stability [48, 55]. MgONPs are considered relatively safe compared to other heavy metal-based NPs given the body's ability to metabolise and efficiently eliminate the MgO degradation products (Mg^+ and OH^-) [58].

MgONPs have been shown to possess significant antimicrobial activity against a range of bacterial and fungi [59]. The MOA of MgONPs has been attributed to the production of ROS from the surface of NPs interacting with microbial cell wall/membranes [58]. The increased level of cellular ROS leads to disruption of cellular processes and damage to the cell organelles [58]. The antifungal effects of MgONPs on fungal phytopathogens have been relativity underexplored compared to many other types of substances in NP forms [58].

Although there have been few studies investigating the antifungal properties of MgONPs, these compounds have been shown to possess substantial antifungal properties against a range of fungal phytopathogens (Table **6**). MgONPs have been shown to be more effective in inhibiting mycelial growth of *Phytophthora nicotianae* and *Thielaviopsis basicola* than larger micron-sized particles [58]. The reduction in mycelial growth was shown to be dose-dependent over a 125-500 ppm concentration range [58]. Mycelial growth of the phytopathogens *Botrytisoryzae, Fusarium fujikuroi* and *Fusarium verticillioides* has also been shown to be inhibited in a dose-dependent manner by MgONPs and the nanocomposite sepiolite, a magnesium-silicate clay [59].

Table 6. Examples of studies examining *in vitro* antifungal activity of magnesium nanomaterials covered in this review.

Nanomaterial	Concentration(s)*	Pathogen(s)	Growth Medium	Refs.
MgO (50 nm); ZnO (30 nm)	AGM: 0.1-0.5 ml of particle suspension (undefined conc)	*A. alternata, F. oxysporum, M. plumbeus, R. stolonifer*	PDA	[55]
ZnO (51, 53, 63 & 77 nm); MgO (52 & 96 nm); MgO-ZnO (139, 161, 219); ZnO-Mg(OH) (54, 71, 88 & 98)	AGM: 2-5000 ppm	*C. gloeosporioides*	PDA	[48]
MgO (100 nm)	AGM: 125-500 ppm	*P. nicotianae, T. basicola*	PDA, OA	[58]
MgO (10-20 nm); MgO-sepoilite (10-1000 nm)	AGM: 180-350 ppm	*B. oryzae, F. fujikuroi, F. verticillioides*	PDA	[59]

*All concentrations have been converted to ppm to comparability between studies.

This particle sized based toxicity has been attributed to the higher level of ROS production from the surface of MgONPs than MgOMPs [58]. The ROS production was proposed to induce toxicity *via* increased oxidative stress [58]. Microscope examination of fungal hyphae revealed the interaction of ZnONPs induced morphological deformations and melanisation in hyphae [48, 58, 59]. This hyphal swelling and vascular expansion appeared to be characteristic of MgONP exposure [48].

MgONPs have been shown to be particularly effective in suppressing spore germination. Spore germination of *A. alternata, F. oxysporum, M. plumbeus* and *R. stolonifer* was shown to be inhibited by MgONPs to a greater extent than ZnONPs *in vitro* [55]. However, for certain species, such as *C. gloeosporioides*, MgONPs were less effective than ZnONPs in inhibiting spore germination [48]. Interestingly, nanocomposites of ZnONP-MgONP were found to be less effective in inhibiting spore germination than either MgONPs and ZnONPs possibly due to an antagonistic effect of Mg^+ on ZnONPs *in vitro* [48].

Magnesium based NPs could be considered to potentially possess less environmentally concerning characteristics than AgNPs or copper-based nanomaterials and thus they may be more suited for utilisation in agricultural pesticides. The research to date would indicate that magnesium-based NPs may have suitable antifungal properties but the data is limited. Further evidence is required to determine the range of fungal and oomycetes pathogens that may be susceptible to MgNPs or MgONPs. Like many other NP substances, the influence of the components that make up the growth medium on *in vitro* toxicity has not been characterised, nor has their efficacy across a broad pH range. These limitations need to be addressed in future research.

Chitosan based Nanomaterials

Chitosan is a naturally occurring linear polysaccharide polymer composed of randomly distributed D-glucosamine and N-acetyl glucosamine residues [60]. Production of chitosan is achieved chemically through the partial deacetylation of chitin, a naturally occurring substance, the main sources of which are the cuticles of various crustaceans, particularly shellfish such as crabs and shrimps [44, 60, 61]. The production of chitosan *via* deacetylation typically generates preparations of particles that vary in size and stability [62]. The heterogenous nature of chitosan particle preparations generally results in variability in both chemical and biological properties [62]. Currently, larger bulk particles of chitosan have limited use in agricultural applications due to their low levels of solubility in water [63]. This has led to investigations focusing on improving chitosan production methods with the aim of producing preparations of consistent smaller particle sizes with less heterogeneity [62].

As chitosan is a non-toxic biodegradable polymer produced from natural sources, it possesses many advantages such as the availability of replenishable substrates, biocompatibility with many systems, biodegradability, and ecological safety [44, 60, 63]. The biocompatibility of chitosan with plant systems has been demonstrated in several studies. Chitosan has also been shown to induce systemic disease resistance in various plants enhancing their natural disease suppressive capabilities [60]. Additionally, combinations of chitosan NPs with traditional fungicides such as hexaconazole and dazomet, and other NPs such as copper, have been demonstrated to induce positive growth effects on plants as well as enhanced disease suppression of fungal pathogens [64 - 66]. There appears to be great potential for using chitosan NPs either as a stand-alone fungicide or as composite material embedded with other NPs [42, 43]. Chitosan can also encapsulate material for packaging traditional fungicides [65].

The antifungal properties of chitosan NP preparations against a variety of fungal plant pathogens have been demonstrated in several studies to date (Table 7). There have been several *in-vitro* studies investigating the antifungal properties of chitosan-based NP against both oomycete and fungal plant pathogens. Chitosan and composite materials composed of porous chitosan particles embedded with metallic/metalloid NP have shown considerable antifungal potential [36, 42, 44, 63]. This antifungal activity has been tested against both above and below ground phytopathogens. The antifungal activity of chitosan has been shown to be dependent on physicochemical characteristics of particles such as their weight, deacetylation degree and pH of their solution [36]. Currently, the antifungal MOAs of chitosan particles have been attributed to the disruptive effects these particles have on the structural integrity of the fungal cell wall and cell membrane [36, 66].

Table 7. Examples of studies examining *in vitro* antifungal activity of chitosan nanomaterials covered in this review.

Nanomaterial	Concentration(s)*	Pathogen(s)	Growth Medium	Refs.
AgNP (10-20 nm); Chitosan (10-20 nm); Chitosan-AgNP composite (10-20 nm)	AGM – 100 ppm; WDM - 500 µl NP suspension (undefined conc)	*A. alternata, A. flavus, R. solani*	PDA	[36]
Chitosan (192 nm); Chitosan-Cu (196 nm); Chitosan-saponin (374 nm)	AGM: 10 – 1000 ppm	*A. alternata, M. phaseolina, R. solani*	PDA	[63]
Chitosan (1—20 nm)	DDM: 100µg	*A. solani, F. oxysporum, P. grisea*	Unspecified	[62]
Chitosan-Cu (2 nm CuNP)	AGM: 30 - 100 ppm	*R. solani, S. rolfsii,*	PDA	[44]
Chitosan-Cu (2-6 nm CuNP); Chitosan-Zn (1-5 nm ZnNP)	AGM: 30 – 100 ppm	*R. solani*	PDA	[42]
CuNP (50-100 nm); ZnNP (6-21 nm); Chitosan-Cu-Zn (16-24 nm)	AGM: 30 – 90 ppm	*A. alternata, B. cinerea, R. solani,*	PDA	[43]
Chitosan (100-1000 nm)	AGM: 100-5000 ppm	*C. gelosporidies, F. oxysporum, G. fujikuori P. capsici, S. sclerotiorum*	PDA	[60]

(Table 7) cont.....

Nanomaterial	Concentration(s)*	Pathogen(s)	Growth Medium	Refs.
Chitosan-Cu (164 nm)	AGM: 500-1000 ppm	*P. aphanidermatum, R. solani*	PDA	[66]
*All concentrations have been converted to ppm to comparability between studies				

The effect of chitosan particle size on antifungal activity has been explored in two notable studies [60, 62]. The Oh *et al.* [60] study demonstrated that a chitosan preparation consisting of nano-scale chitosan possessed a greater level of antifungal activity than a chitosan preparation that consisted mostly of micron-sized particles. This is a typical property observed for many types of nano-scale substances, where smaller-sized particles generally display a greater level of toxicity than larger particles. The influence of chitosan particle size on antifungal activity was also demonstrated by Sathiyabama and Partasarathy [62], where inhibition of mycelial growth was more pronounced on media amended with chitosan NP than media amended with bulkier polymers of chitosan. This effect of chitosan particle size on antifungal activity has been relatively unexplored to date, though the toxicity of chitosan particles against cell lines has been shown to be size dependent with chitosan NP inducing higher levels of ROS than larger polymers of chitosan [67].

Several studies have also shown chitosan NPs inhibiting the growth of phytopathogens synergistically with a range of metallic and metalloid NPs. Chitosan sheets embedded with CuNPs have received the most attention to date, while combinations with Ag, Zn and saponin have also been tested. Combinations of chitosan and Cu have been shown to be particularly effective in controlling the mycelial growth of phytopathogens such as *A. alternata* and *R. saloni*, with dose-dependent reductions in growth observed [42 - 44, 63]. For both *R. solani* and *S. rolfsii*, sclerotia generation was also observed to be suppressed in a dose dependent manner *in vitro* by chitosan-Cu nanocomposites [44]. Chitosan-Cu nanocomposites have also been shown to possess greater antifungal activity than similar sized preparations of chitosan-Zn when tested against *A. alternata, B. cinerea* and *R. solani* on a PDA growth medium [42, 43]. A nanocomposite consisting of both chitosan-Cu-Zn was shown to have greater antifungal activity than preparations of chitosan-Cu or chitosan-Zn alone, though the majority of the additional antifungal activity conferred by the metalloid elements appeared to originate from the Cu [43]. Chitosan-Ag composites have also been shown to possess greater antifungal activity than either Ag or chitosan NP on their own [36]. Chitosan-saponin has been shown to be relatively ineffective for controlling pathogens such as *A. alternata, Macrophominaphaseolina* and *R. solani* compared to chitosan-Cu composites [63].

In addition to true fungi, oomycetes such as the pythiaceae have also been shown to be susceptible to inhibition by chitosan NPs *in vitro*. Chitosan-Cu combinations have been shown to be effective at controlling pythiaceae phytopathogens on PDA, with mycelial growth of *P. aphanidermatum* and *Phytophthora capsici* observed to be suppressed by a Chitosan-Cu nanocomposite and a chitosan NP, respectively [60, 66].

The antifungal activity of chitosan NPs appears to occur through the physical interaction of chitosan particles with fungal hyphae. Microscopic examination of fungal hyphae exposed to chitosan NP has revealed irregular growth patterns and hyphal deformation [44, 66]. The structural deformations have been demonstrated to compromise cell membrane/wall integrity resulting in leakage of cytoplasmic content [44, 66]. This effect has been observed in both fungi and pythiaceae and has been shown to increase over the duration of the exposure [66].

To date, the effects of chitosan on fungal pathogens have only been tested on PDA growth medium and no comparison of the effects of different growth media has been undertaken. Further investigation into these factors will help reveal how to best employ these substances in field conditions.

CONCLUSION

Several NP substances have been shown to possess potent inhibitory activity against a range of fungal and oomycete pathogens. Potency varies between substances, with metallic based nanomaterials, such as Ag and Cu, demonstrating significantly greater efficacy than other elements/oxides such as Zn/ZnO and Mg/MgO. However, the emerging toxicological/ecotoxicological profiles of NP such as Ag would indicate potential problems with registering such substances as Plant Protection Products. NPs of substances with more benign toxicity, such as the organic substance chitosan and metallic oxides MgO and ZnO may prove to be more likely to gain regulatory approval as PPP. This review of the literature would indicate that significant *in vitro* efficacy has been demonstrated by low concentrations of these substances against a range of phytopathogens. The *in vitro* efficacy of these substances would indicate that they possess broad-spectrum activity given the range of pathogens they have been shown to supress. The biodegradability of chitosan would also suggest that this substance may have lesser environmental impacts than metallic NP. Additionally, chitosan's stimulating effects on plant growth and defence highlight additional benefits in using this substance. Further characterisation of chitosan's *in vitro* efficacy against a wider range of pathogens is, however, necessary to determine its full potential. Further to this, more detailed characterisation of the influence of growth mediums in *in vitro* testing would facilitate a more reliable comparison of NP substance antifungal activity. Ultimately, consistent control of diseases on crops in field conditions will reveal the true efficacy of these NPs pesticidal properties.

CONSENT FOR PUBLICATION

Not applicable.

CONFLICT OF INTEREST

The authors declare no conflict of interest, financial or otherwise.

ACKNOWLEDGEMENT

Declared none.

REFERENCES

[1]   Sikes BA, Bufford JL, Hulme PE, Cooper JA, Johnston PR, Duncan RP. Import volumes and biosecurity interventions shape the arrival rate of fungal pathogens. PLoS Biol 2018; 16(5): e2006025.
[http://dx.doi.org/10.1371/journal.pbio.2006025] [PMID: 29851948]

[2]   Dietzel K, Valle D, Fierer N, U'Ren JM, Barberán A. U'Ren JM, Barberán A. Geographical distribution of fungal plant pathogens in dust across the United States. Front Ecol Evol 2019; 7: 304.
[http://dx.doi.org/10.3389/fevo.2019.00304]

[3]   FRAC Fungicide Resistance Management 2020. https://www.frac.info/fungicide-resistance-manage-ment/by-frac-mode-of-action-group

[4]   Hawkins NJ, Bass C, Dixon A, Neve P. The evolutionary origins of pesticide resistance. Biol Rev Camb Philos Soc 2018; 94(1): 135-55.
[http://dx.doi.org/10.1111/brv.12440] [PMID: 29971903]

[5]   Camara MC, Campos EVR, Monteiro RA, do Espirito Santo Pereira A, de Freitas Proença PL, Fraceto LF. Development of stimuli-responsive nano-based pesticides: emerging opportunities for agriculture. J Nanobiotechnology 2019; 17(1): 100.
[http://dx.doi.org/10.1186/s12951-019-0533-8] [PMID: 31542052]

[6]   Varier KM, Gudeppu M, Chinnasamy A, *et al.* Nanoparticles: antimicrobial applications and its prospects InAdvanced nanostructured materials for environmental remediation. Cham: Springer 2019; pp. 321-55.

[7]   Yah CS, Simate GS. Nanoparticles as potential new generation broad spectrum antimicrobial agents. Daru 2015; 23(1): 43.
[http://dx.doi.org/10.1186/s40199-015-0125-6] [PMID: 26329777]

[8]   Auffan M, Rose J, Bottero JY, Lowry GV, Jolivet JP, Wiesner MR. Towards a definition of inorganic nanoparticles from an environmental, health and safety perspective. Nature nanotechnology. 2009(10);634-41.[9] Tripathi DK, Tripathi A, Singh S, Singh Y, Vishwakarma K, Yadav G, Sharma S, Singh VK, Mishra RK, Upadhyay RG, Dubey NK. Uptake, accumulation and toxicity of silver nanoparticle in autotrophic plants, and heterotrophic microbes: a concentric review. Front Microbiol 2017; 8: 7.

[10]   Nejad MS, Bonjar GH, Khatami M, Amini A, Aghighi S. *In vitro* and *in vivo* antifungal properties of silver nanoparticles against Rhizoctonia solani, a common agent of rice sheath blight disease. IET nanobiotechnology 2016; 11(3): 236-40.

[11]   De Leersnyder I, De Gelder L, Van Driessche I, Vermeir P. Revealing the importance of aging,

environment, size and stabilization mechanisms on the stability of metal nanoparticles: a case study for silver nanoparticles in a minimally defined and complex undefined bacterial growth medium. Nanomaterials (Basel) 2019; 9(12): 1684.
[http://dx.doi.org/10.3390/nano9121684] [PMID: 31775314]

[12] De Leersnyder I, De Gelder L, Van Driessche I, Vermeir P. Influence of growth media components on the antibacterial effect of silver ions on Bacillus subtilis in a liquid growth medium. Sci Rep 2018; 8(1): 9325.
[http://dx.doi.org/10.1038/s41598-018-27540-9] [PMID: 29921908]

[13] McGee CF, Gaffney MT, Doyle O. Influence of exogenous cholesterol in Pythiaceae resistance to inhibition by Trichoderma antibiosis. Eur J Plant Pathol 2016; 145(4): 1013-8.
[http://dx.doi.org/10.1007/s10658-016-0870-5]

[14] Liu J, Dai C, Hu Y. Aqueous aggregation behavior of citric acid coated magnetite nanoparticles: Effects of pH, cations, anions, and humic acid. Environ Res 2018; 161: 49-60.
[http://dx.doi.org/10.1016/j.envres.2017.10.045] [PMID: 29101829]

[15] Lamsal K, Kim SW, Jung JH, Kim YS, Kim KS, Lee YS. Application of silver nanoparticles for the control of colletotrichum species in vitro and pepper anthracnose disease in field. Mycobiology 2011; 39(3): 194-9.
[http://dx.doi.org/10.5941/MYCO.2011.39.3.194] [PMID: 22783103]

[16] Jo YK, Kim BH, Jung G. Antifungal activity of silver ions and nanoparticles on phytopathogenic fungi. Plant Dis 2009; 93(10): 1037-43.
[http://dx.doi.org/10.1094/PDIS-93-10-1037] [PMID: 30754381]

[17] McGee CF, Clipson N, Doyle E. Exploring the Influence of Raising Soil pH on the Ecotoxicological Effects of Silver Nanoparticles and Micron Particles on Soil Microbial Communities. Water Air Soil Pollut 2020; 231(4): 1-8.
[http://dx.doi.org/10.1007/s11270-020-04540-y]

[18] Pachapur VL, Dalila Larios A, Cledón M, Brar SK, Verma M, Surampalli RY. Behavior and characterization of titanium dioxide and silver nanoparticles in soils. Sci Total Environ 2016; 563-564: 933-43.
[http://dx.doi.org/10.1016/j.scitotenv.2015.11.090] [PMID: 26725442]

[19] McGee CF, Storey S, Clipson N, Doyle E. Soil microbial community responses to contamination with silver, aluminium oxide and silicon dioxide nanoparticles. Ecotoxicology 2017; 26(3): 449-58.
[http://dx.doi.org/10.1007/s10646-017-1776-5] [PMID: 28197855]

[20] McGee CF, Storey S, Clipson N, Doyle E. Concentration-dependent responses of soil bacterial, fungal and nitrifying communities to silver nano and micron particles. Environ Sci Pollut Res Int 2018; 25(19): 18693-704.
[http://dx.doi.org/10.1007/s11356-018-2087-y] [PMID: 29705905]

[21] Gajbhiye M, Kesharwani J, Ingle A, Gade A, Rai M. Fungus-mediated synthesis of silver nanoparticles and their activity against pathogenic fungi in combination with fluconazole. Nanomedicine 2009; 5(4): 382-6.
[http://dx.doi.org/10.1016/j.nano.2009.06.005] [PMID: 19616127]

[22] Kim SW, Kim KS, Lamsal K, et al. An in vitro study of the antifungal effect of silver nanoparticles on oak wilt pathogen Raffaelea sp. J Microbiol Biotechnol 2009; 19(8): 760-4.
[PMID: 19734712]

[23] Abramenko NB, Demidova TB, Abkhalimov EV, Ershov BG, Krysanov EY, Kustov LM. Ecotoxicity of different-shaped silver nanoparticles: Case of zebrafish embryos. J Hazard Mater 2018; 347: 89-94.
[http://dx.doi.org/10.1016/j.jhazmat.2017.12.060] [PMID: 29291521]

[24] Balashanmugam P, Balakumaran MD, Murugan R, Dhanapal K, Kalaichelvan PT. Phytogenic synthesis of silver nanoparticles, optimization and evaluation of in vitro antifungal activity against human and plant pathogens. Microbiol Res 2016; 192: 52-64.

[http://dx.doi.org/10.1016/j.micres.2016.06.004] [PMID: 27664723]

[25]   Elumalai EK, Vinothkumar P. Role of silver nanoparticle against plant pathogens. Nano Biomed Eng 2013; 5(2).
[http://dx.doi.org/10.5101/nbe.v5i2.p90-93]

[26]   Min JS, Kim KS, Kim SW, *et al.* Effects of colloidal silver nanoparticles on sclerotium-forming phytopathogenic fungi. Plant Pathol J 2009; 25(4): 376-80.
[http://dx.doi.org/10.5423/PPJ.2009.25.4.376]

[27]   Malandrakis AA, Kavroulakis N, Chrysikopoulos CV. Use of copper, silver and zinc nanoparticles against foliar and soil-borne plant pathogens. Sci Total Environ 2019; 670: 292-9.
[http://dx.doi.org/10.1016/j.scitotenv.2019.03.210] [PMID: 30903901]

[28]   Jebril S, Jenana RK, Dridi C. Green synthesis of silver nanoparticles using Melia azedarach leaf extract and their antifungal activities: In vitro and in vivo. Mater Chem Phys 2020; 248: 122898.
[http://dx.doi.org/10.1016/j.matchemphys.2020.122898]

[29]   Elshahawy I, Abouelnasr HM, Lashin SM, Darwesh OM. First report of Pythium aphanidermatum infecting tomato in Egypt and its control using biogenic silver nanoparticles. J Plant Prot Res 2018; 58(2)

[30]   Aleksandrowicz-Trzcińska M, Szaniawski A, Olchowik J, Drozdowski S. Effects of copper and silver nanoparticles on growth of selected species of pathogenic and wood-decay fungi in vitro. For Chron 2018; 94(2): 109-16.
[http://dx.doi.org/10.5558/tfc2018-017]

[31]   Gado EA, El-Deeb B, Ali EF, Mostafa NY, Bazaid SA. Evaluation of silver nanoparticles for the control of phragmidium species *in vitro* and taif rose rust disease in field. Res J Pharm Biol Chem Sci 2016; 7(3): 886-96.

[32]   Kriti A, Ghatak A, Mandal N. Inhibitory potential assessment of silver nanoparticle on phytopathogenic spores and mycelial growth of bipolarissorokiniana and alternaria brassicicola. Int J Curr Microbiol Appl Sci 2020; 9(3): 692-9.
[http://dx.doi.org/10.20546/ijcmas.2020.903.083]

[33]   Shanmugam C, Gunasekaran D, Duraisamy N, Nagappan R, Krishnan K. Bioactive bile salt-capped silver nanoparticles activity against destructive plant pathogenic fungi through in vitro system. RSC Advances 2015; 5(87): 71174-82.
[http://dx.doi.org/10.1039/C5RA13306H]

[34]   Ismail AW, Sidkey NM, Arafa RA, Fathy RM, El-Batal AI. Evaluation of in vitro antifungal activity of silver and selenium nanoparticles against Alternaria solani caused early blight disease on potato. Biotechnology Journal International. 2016; pp. 1-1.

[35]   Krishnaraj C, Ramachandran R, Mohan K, Kalaichelvan PT. Optimization for rapid synthesis of silver nanoparticles and its effect on phytopathogenic fungi. Spectrochim Acta A Mol Biomol Spectrosc 2012; 93: 95-9.
[http://dx.doi.org/10.1016/j.saa.2012.03.002] [PMID: 22465774]

[36]   Kaur P, Thakur R, Choudhary A. An in vitro study of the antifungal activity of silver/chitosan nanoformulations against important seed borne pathogens. Int J Sci Technol Res 2012; 1(6): 83-6.

[37]   Iavicoli I, Leso V, Beezhold DH, Shvedova AA. Nanotechnology in agriculture: Opportunities, toxicological implications, and occupational risks. Toxicol Appl Pharmacol 2017; 329: 96-111.
[http://dx.doi.org/10.1016/j.taap.2017.05.025] [PMID: 28554660]

[38]   Banik S, Luque AP. In vitro effects of copper nanoparticles on plant pathogens, beneficial microbes and crop plants. Span J Agric Res 2017; 15(2): 23.
[http://dx.doi.org/10.5424/sjar/2017152-10305]

[39]   Millardet PM. Traitement du Mildou et du Rot par le mélange de chaux et sulfat de cuivre. 1885.

[40]   Hao Y, Cao X, Ma C, *et al.* Potential applications and antifungal activities of engineered nanomaterials against graymold disease agent Botrytis cinerea on rose petals. Front Plant Sci 2017; 8: 1332.
[http://dx.doi.org/10.3389/fpls.2017.01332] [PMID: 28824670]

[41]   Chatterjee AK, Chakraborty R, Basu T. Mechanism of antibacterial activity of copper nanoparticles. Nanotechnology 2014; 25(13): 135101.
[http://dx.doi.org/10.1088/0957-4484/25/13/135101] [PMID: 24584282]

[42]   Abd-Elsalam KA, Vasil'kov AY, Said-Galiev EE, *et al.* Bimetallic blends and chitosan nanocomposites: novel antifungal agents against cotton seedling damping-off. Eur J Plant Pathol 2018; 151(1): 57-72.

[43]   Al-Dhabaan FA, Shoala T, Ali AA, Alaa M, Abd-Elsalam K. Chemically-produced copper, zinc nanoparticles and chitosan–bimetallic nanocomposites and their antifungal activity against three phytopathogenic fungi. Agric Technol Thail 2017; 13(5): 753-69.

[44]   Rubina MS, Vasil'kov AY, Naumkin AV, *et al.* Synthesis and characterization of chitosan–copper nanocomposites and their fungicidal activity against two sclerotia-forming plant pathogenic fungi. J Nanostructure Chem 2017; 7(3): 249-58.
[http://dx.doi.org/10.1007/s40097-017-0235-4]

[45]   Lamichhane JR, Osdaghi E, Behlau F, Köhl J, Jones JB, Aubertot JN. Thirteen decades of antimicrobial copper compounds applied in agriculture. A review. Agron Sustain Dev 2018; 38(3): 28.
[http://dx.doi.org/10.1007/s13593-018-0503-9]

[46]   Wagner G, Korenkov V, Judy JD, Bertsch PM. Nanoparticles composed of Zn and ZnO inhibit Peronospora tabacina spore germination in vitro and P. tabacina infectivity on tobacco leaves. Nanomaterials (Basel) 2016; 6(3): 50.
[http://dx.doi.org/10.3390/nano6030050] [PMID: 28344307]

[47]   Dimkpa CO, McLean JE, Britt DW, Anderson AJ. Antifungal activity of ZnO nanoparticles and their interactive effect with a biocontrol bacterium on growth antagonism of the plant pathogen Fusarium graminearum. Biometals 2013; 26(6): 913-24.
[http://dx.doi.org/10.1007/s10534-013-9667-6] [PMID: 23933719]

[48]   la Rosa-García D, Susana C, Martínez-Torres P, *et al.* Antifungal activity of ZnO and MgO nanomaterials and their mixtures against Colletotrichum gloeosporioides strains from tropical fruit. Journal of Nanomaterials 2018.

[49]   Yehia RS, Ahmed OF. In vitro study of the antifungal efficacy of zinc oxide nanoparticles against Fusarium oxysporum and Peniciliumexpansum. Afr J Microbiol Res 2013; 7(19): 1917-23.
[http://dx.doi.org/10.5897/AJMR2013.5668]

[50]   Brayner R, Dahoumane SA, Yéprémian C, *et al.* ZnO nanoparticles: synthesis, characterization, and ecotoxicological studies. Langmuir 2010; 26(9): 6522-8.
[http://dx.doi.org/10.1021/la100293s] [PMID: 20196582]

[51]   Gunalan S, Sivaraj R, Rajendran V. Green synthesized ZnO nanoparticles against bacterial and fungal pathogens. Prog Nat Sci 2012; 22(6): 693-700.
[http://dx.doi.org/10.1016/j.pnsc.2012.11.015]

[52]   He L, Liu Y, Mustapha A, Lin M. Antifungal activity of zinc oxide nanoparticles against Botrytis cinerea and Penicillium expansum. Microbiol Res 2011; 166(3): 207-15.
[http://dx.doi.org/10.1016/j.micres.2010.03.003] [PMID: 20630731]

[53]   Lahuf AA, Alfarttoosi HA, Al-Sweedi TM, Middlefell-Williams JE. Evaluation of an integration between the nanosized zinc oxide and two cultivars for the control of damping-off disease in sunflower crop. Res Crops 2019; 20(1): 174-9.

[54]   Navale GR, Thripuranthaka M, Late DJ, Shinde SS. Antimicrobial activity of ZnO nanoparticles against pathogenic bacteria and fungi. Sci Med Central 2015; 3: 1033.

[55]　Wani AH, Shah MA. A unique and profound effect of MgO and ZnO nanoparticles on some plant pathogenic fungi. J Appl Pharm Sci 2012; 2(3): 4.

[56]　Narendhran S, Sivaraj R. Biogenic ZnO nanoparticles synthesized using L. aculeata leaf extract and their antifungal activity against plant fungal pathogens. Bull Mater Sci 2016; 39(1): 1-5.
[http://dx.doi.org/10.1007/s12034-015-1136-0]

[57]　Kriti A, Ghatak A, Mandal N. Antimycotic efficacy of Zinc nanoparticle on dark-spore forming Phytopathogenic fungi. J Pharmacogn Phytochem 2020; 9(2): 750-4.

[58]　Chen J, Wu L, Lu M, Lu S, Li Z, Ding W. Comparative study on the fungicidal activity of metallic MgO nanoparticles and macroscale MgO against soilborne fungal phytopathogens. Front Microbiol 2020; 11: 365.
[http://dx.doi.org/10.3389/fmicb.2020.00365] [PMID: 32226420]

[59]　Sidhu A, Bala A, Singh H, Ahuja R, Kumar A. Development of MgO-sepoilite nanocomposites against phytopathogenic fungi of rice (*Oryzae sativa*): A green approach. ACS Omega 2020; 5(23): 13557-65.
[http://dx.doi.org/10.1021/acsomega.0c00008] [PMID: 32566820]

[60]　Oh JW, Chun SC, Chandrasekaran M. Preparation and *in vitro* characterization of chitosan nanoparticles and their broad-spectrum antifungal action compared to antibacterial activities against phytopathogens of tomato. Agronomy (Basel) 2019; 9(1): 21.
[http://dx.doi.org/10.3390/agronomy9010021]

[61]　Younes I, Rinaudo M. Chitin and chitosan preparation from marine sources. Structure, properties and applications. Mar Drugs 2015; 13(3): 1133-74.
[http://dx.doi.org/10.3390/md13031133] [PMID: 25738328]

[62]　Sathiyabama M, Parthasarathy R. Biological preparation of chitosan nanoparticles and its *in vitro* antifungal efficacy against some phytopathogenic fungi. Carbohydr Polym 2016; 151: 321-5.
[http://dx.doi.org/10.1016/j.carbpol.2016.05.033] [PMID: 27474573]

[63]　Saharan V, Mehrotra A, Khatik R, Rawal P, Sharma SS, Pal A. Synthesis of chitosan based nanoparticles and their *in vitro* evaluation against phytopathogenic fungi. Int J Biol Macromol 2013; 62: 677-83.
[http://dx.doi.org/10.1016/j.ijbiomac.2013.10.012] [PMID: 24141067]

[64]　Choudhary RC, Kumaraswamy RV, Kumari S, *et al.* Cu-chitosan nanoparticle boost defense responses and plant growth in maize (Zea mays L.). Sci Rep 2017; 7(1): 9754.
[http://dx.doi.org/10.1038/s41598-017-08571-0] [PMID: 28851884]

[65]　Maluin FN, Hussein MZ, Yusof NA, *et al.* Phytotoxicity of chitosan-based agronanofungicides in the vegetative growth of oil palm seedling. PLoS One 2020; 15(4): e0231315.
[http://dx.doi.org/10.1371/journal.pone.0231315] [PMID: 32315346]

[66]　Vanti GL, Masaphy S, Kurjogi M, Chakrasali S, Nargund VB. Synthesis and application of chitosan-copper nanoparticles on damping off causing plant pathogenic fungi. Int J Biol Macromol 2020; 156: 1387-95.
[http://dx.doi.org/10.1016/j.ijbiomac.2019.11.179] [PMID: 31760011]

[67]　Jesus S, Marques AP, Duarte A, *et al.* Chitosan nanoparticles: Shedding light on immunotoxicity and hemocompatibility. Front Bioeng Biotechnol 2020; 8: 100.
[http://dx.doi.org/10.3389/fbioe.2020.00100] [PMID: 32154232]

[68]　Kim SW, Jung JH, Lamsal K, Kim YS, Min JS, Lee YS. Antifungal effects of silver nanoparticles (AgNPs) against various plant pathogenic fungi. Mycobiology 2012; 40(1): 53-8.
[http://dx.doi.org/10.5941/MYCO.2012.40.1.053] [PMID: 22783135]

CHAPTER 10

Cosmetic and Medical Applications of Fungal Nanotechnology

Babita Singh[1,*], **Sonali Singhal**[1] and **Tanzeel Ahmed**[2]

[1] *ICAR-National Research Centre for Integrated Pest Management, New Delhi-110012, India*

[2] *School of Biotechnology, IFTM University, Lodhipur-Rajput, NH-24, Delhi Road, Moradabad, Uttar Pradesh-244001, India*

Abstract: Nanotechnology is the science of manipulating atoms and molecules in the nanoscale - 80,000 times smaller than the width of a human hair. Nanotechnology is a revolutionary technology that is being used in many fields all over the world as it finds applications in automobiles, electronics, material science, *etc*. Fungal nanotechnology has great prospects for developing new products with industrial, agricultural, medicinal, and consumer applications in a wide range of areas. Nanotechnology has applications in the field of cosmetics, which are known as nanocosmetics. Various types of nanomaterials are employed in cosmetic and medical applications *i.e.* inorganic nanoparticles, Silica (SiO_2), Carbon Black, Nano-Organic materials, Nano-Hydroxyapatite, Gold, and Silver Nanoparticles, Nanoliposomes, *etc*. NPs have been explored and identified as carriers for drug delivery. New drug delivery systems based on nanotechnology have been applied in the treatment of human diseases, such as cancer, diabetes, microbial infections, and gene therapy. The benefits of these treatments are that the drug is targeted to diseased cells, and its safety profile is enhanced by the reduced toxic side effects to normal cells. In general, NPs can be conjugated with different types of drugs to deliver bioactive compounds to the target site by various methods, such as the use of nanotubes, liposomes, quantum dots, nanopores, and dendrimers. It is employed in fuel cell applications that involve polymers in the proton exchange membrane, binder for the electrodes, and matrix for bipolar plates.

Keywords: Fungal Nanotechnology, Nanotechnology, Nanocosmetics, Nanoparticles, Nanosensors, Nanocosmaceuticals.

INTRODUCTION

Nanotechnology is the study of controlling particles and atoms inside the nanoscale - multiple times less than the width of an individual's hair. The world

* **Corresponding Author Babita Singh**: ICAR-National Research Centre for Integrated Pest Management-New Delhi-110012; Tel: +91 9870816511; Email: romikasingh1311@gmail.com

Savita, Anju Srivastava, Reena Jain & Pratap Kumar Pati (Eds.)

commercial center for items that contain nanomaterials is anticipated to prevail in $2.6 trillion by 2015.

The utilization of nanotechnology has extended its boundaries across different surges of science, from gadgets to medication [1]. Nanotechnology holds colossal potential inside the analysis, treatment, and avoidance of COVID-19. Nanotechnology could help the battle against COVID-19 through various methodologies, such as staying away from viral tainting and shower by (a) plan of contamination safe individual defensive gear (PPE) to reinforce the security the wellbeing of medical services laborers and improvement of viable antiviral sanitizers and surface coatings, which are prepared to inactivate the infection and forestall its spread; (b) plan of exceptionally explicit and delicate nano-based sensors to rapidly recognize the disease or immunological reaction; (c) advance-ment of most recent medications, with upgraded movement, diminished poisonousness and supported delivery, additionally as tissue-focus, for example, to the lungs; and (d) advancement of a nano-based inoculation to zest up humoral and cell safe reactions [2].

Nanotechnology is a progressive innovation that is being utilized in many fields all around the world as it discovers applications in autos, gadgets, material science and so on. Contagious nanotechnology has extraordinary possibilities for growing new items with modern, agrarian, restorative, and customer applications in a wide scope of regions. The natural and clinical examination local area of the world is focussing on how they can make tranquilizes less expensive, more compelling, how could decrease results of medications how might deal with infections like a malignancy in the best manner, how could foster new devices and gadgets which can diminish agony and draining which is normal during medical procedures and tasks. The fields of synthetic designing, agri-food, organic chemistry, drugs, diagnostics, and clinical gadget advancement all utilize contagious items, with parasitic nanomaterials as of now utilized in applications going from drug improvement to the food business and agrarian biotechnology. Contagious specialists are climate agreeable, clean, non-toxic specialists, for the metal nanoparticles combination and utilize both intracellular and extracellular techniques [3]. A few nanoparticles, for example, gold nanoparticles, silver nanoparticles, different nanoparticles integrated by numerous universal parasitic species, for example, *Trichoderma, Fusarium, Penicillium, Rhizoctonia, Pleurotus* and *Aspergillus, Beauveriabassiana*, and so on the effortlessness of increasing and downstream handling and the presence of contagious mycelia which manage the cost of an expanded surface region give key benefits [4]. Understanding the diversity of parasites in arranged environments, as well as their interactions with various microorganisms, creatures, and plants, enables genuine and imaginative mechanical turns of events and the application of metal nanoparticles in a variety of fields, including farming, catalysis, biomedical biosensors, and beauty care products.

NANOTECHNOLOGY IN COSMETICS

Nanotechnology has applications in the field of beauty care products, which are known as nanocosmetics. This far and wide impact of nanotechnology in the restorative enterprises is because of the expanded properties accomplished by the particles at the nano level including straightforwardness, shading, solvency, and so on the various sorts of nanomaterials worked in beauty care products incorporate nanosomes, liposomes, fullerenes, strong lipid nanoparticles and so on through the utilization of nanotechnology applications in beauty care products we can make corrective items which helps in eliminating wrinkles and dim spots and has brilliance impact with sunscreen, moisturizes the skin and firms, lights up the sensitive skin around the eyes, supports collagen union, skin fix.

As of late, issues over the wellbeing of such nanocosmetics were raised and have constrained the corrective enterprises to restrict the utilization of nanotechnology in beauty care products and for installing laws to go through an undeniable security appraisal before they go into the market. The makeup business was among quick to carry out nanotechnological standards in item improvement. More than 1,000 enrolled nanotechnology-put together items concerning the worldwide market in 2009, over 13% were named items for corrective use [5].

The uses of nanotechnology and nanomaterials are found in numerous corrective items, including creams, hair care items, makeup, and sunscreen. In cosmetics and skin health management, there are two primary uses for nanocosmetics:

1. Nanoparticles in beauty care products as UV channels.
2. Nanoparticles in beauty care products as medication conveyance specialists.

In the broader aspect of nanotechnology, NPs can be divided into two groups:-

Chitin and its deacetylated subsidiary chitosan are of extraordinary interest to the cosmeceutical business because of their exceptional biological and mechanical properties. A chitin nanofibril is an illustration of a nanocrystal got from the scavenger exoskeleton, disposing of carbonate and protein segments while as yet being viewed as protected to utilize. Chitin nanofibrils in emulsions can produce the development of a hygroscopic sub-atomic film that dials backwater, vanishing and adds to skin hydration .

Types of Nanoparticles in Cosmetics

Inorganic Nanoparticles

Inorganic nanoparticles are incorporated from inorganic components (Ag, Au, Ti, and so on), while the natural ones are combined from polymers. These nanoparticles are non-harmful, hydrophilic, biocompatible, and profoundly stable, contrasted with natural nanoparticles. Quite possibly, the most generally utilized inorganic nanoparticles for sunscreens is TiO_2, and in nanoscale it has a higher sun assurance factor (SPF) which makes it more effective, and has a superior corrective outcome because of its straightforwardness, contrasted with TiO_2 color. Generally, on the lookout, organizations use words, for example, "sheer" or "undetectable" when nanoscaleTiO_2 or ZnO are utilized. It is accounted for that nanoscale TiO_2 and ZnO show incredible benefits over numerous items at bigger than nano-measurements. Miniature TiO_2 and ZnO are utilized as fixings in sunscreens because of their UVA and UVB assimilation capacities. Nanoparticles of ZnO and TiO_2 are additionally generally utilized in sunscreens as UV channels beginning at the size of 20 nm. They show better scattering and leave a superior restorative outcome. As to security, inward breath of high groupings of ZnO nanoparticles has been accounted for to cause wellbeing harm. Nonetheless, an alternate course of openness (*i.e.*, the dermal course), for ZnO focuses on ordinary sunscreen recipes, is viewed as protected since entrance into the practical epidermis is not accounted for .

Silica (SiO₂)

Silica nanoparticles have drawn interest from the corrective business since they show hydrophilic surfaces preferring extended flow and low creation cost. Nano silica is utilized to improve the adequacy, surface, and timeframe of realistic usability of restorative items. It adds sponginess and goes about as an enemy of building up specialists. It has been demonstrated that silica nanoparticles may assist with working on the appearance and circulation of shades in lipsticks and keep colors from relocating into the almost negligible difference of lips. Silica nanoparticles are settled nanodispersions with a reach size of 5 to 100 nm, which can convey lipophilic and hydrophilic substances to their site of activity by exemplification. Silica nanoparticles can be found in leave-on and wash-off surface-level items for hair, skin, lips, face, and nails, and an increment of silica nanoparticles presence in superficial items is expected. Results are disputable concerning the wellbeing of silica-based nanoparticles and factors, for example, size and surface adjustments ought to be considered while evaluating harmfulness. Along these lines, suppositions in regards to the utilization and openness of silica nanoparticles in beauty care products are as yet uncertain, and further long-openness tests are required.

Nanocarbon

Carbon Black, CI 77266, is a known restorative fixing that is regularly utilized as a colorant for eye enlivening corrective items, skin items, and mascaras. It is now permitted in the EU in its nanoform, and it is used as a colourant at a maximum concentration of 10%. In comparison to micron-sized nanoparticles, carbon black nanoparticles had a larger proclivity for beginning cytotoxicity, aggravation, and adjusted phagocytosis in human monocytes. When there is no danger of inward breath, it is viewed as protected to be utilized in superficial items in the EU.

Buckyballs

Buckminsterfullerene, C60, is maybe the most notorious nanomaterial and is around 1 nm in distance across. It has discovered its direction into some extravagant face creams. The inspiration is to gain by its ability to act as a strong forager of free revolutionaries [7].

Nano-Organic Materials

Tris-Biphenyl Triazine is extremely proficient and it has an entirely photostable channel. For the plan of sunscreens photostable channel settling on it an alternate decision fixing. Tris-Biphenyl Triazine (nano) functions as a wide-going UV channel, appropriate for sunscreen items and against maturing face care items. It presents significant photostability and is an approved UV channel in Europe. Methylene bis-benzotriazolyltetramethylbutylphenol (nano), is an approved UV channel in the EU market and can be utilized at groupings of up to 10% w/w for dermally put on surface-level items. In light of SCCS, methylene bis-benzotriazolyltetramethylbutylphenol does not make a danger to people whenever applied on sound, flawless skin. Notwithstanding, SCCS uncovered concerns in regards to possible aggravation impacts and likely bioaccumulation in chosen tissues.

Nano-Hydroxyapatite

Nano-hydroxyapatite is utilized in oral consideration and restorative items, being joined in different items for the treatment of excessive dental touchiness and finish remineralization, and is viewed as a promising and safe alternative for oral consideration items. Nano-hydroxyapatite particles have been incorporated into oral consideration items, like dentifrices and mouthwashes, and because of its remineralization and desensitization properties, nano-hydroxyapatite could be an option in contrast to fluoride toothpaste.

Gold and Silver Nanoparticles

Gold and silver nanoparticles, aside from the various applications that they have, likewise show antibacterial and antifungal properties. Gold and silver nanoparticles are utilized in superficial items like antiperspirants and are hostile to maturing creams. Silver nanoparticles utilized as an additive in beauty care products and silver nanoparticles stayed stable, without displaying sedimentation, for more than 1 year [8]. Silver nanoparticles also demonstrated adequate protection against tiny organisms and parasites and did not penetrate human skin. The use of gold and silver nanoparticles in surface-level plans is integrating differences between silver and gold nanoparticles into the creation of cream, according to a study. Silver nanoparticles were acquainted with the cream combination agglomerate, yet gold nanoparticles did not agglomerate after prologue to cream blends. They ascribed this wonder to the more prominent worth of the electrodynamic likely situated on the outside of gold nanoparticles. In light of a model dermal film study, they have announced worries over the infiltration of nanoparticles into the skin for tests with nanoparticles centralization of 110–200 mg/kg. Because of the unpredictable creation of corrective creams, it is not difficult to describe the essential gold nanomaterials *in situ* [9].

Nanoliposomes

Liposomes at the nanoscale are called nanoliposomes. They are concentric bilayered vesicles in which the fluid volume is encased by a lipid bilayer of phospholipids. Nanoliposomes are biodegradable and biocompatible, addressing a profoundly versatile fixing class for the corrective field. They are utilized as defensive transporters of dynamic fixings (*e.g.*, nutrients), for expanding skin penetrability and for saturating purposes. They can be utilized for scent conveyance in antiperspirants, body-shower antiperspirants, and lipsticks. Notwithstanding their promising provisions, low medication stacking, low reproducibility, and physicochemical shakiness issues have restricted their business applications in beauty care products [10].

Lipid Nanoparticles exist in two Varieties: Strong Lipid Nanoparticles (SLN) and Nanostructured Lipid Carriers (NLC)

The original lipid nanoparticles were presented as strong lipid nanoparticles (SLN), the second, further developed age as nanostructured lipid transporters (NLC). Indistinguishable from the liposomes, the lipid nanoparticles (NLC) showed up as items first on the corrective market. The section depicts an outline of the corrective advantages of lipid nanoparticles that implies improvement of compound strength of actives, film arrangement, controlled impediment, skin hydration, upgraded skin bioavailability, and actual security of the lipid nanoparticles as effective details. NLC are available as concentrates to be utilized as superficial excipients, exceptional plan difficulties for these items are examined. NLC showed up likewise in various completed corrective items around the world. An outline of these items is given, including their embellishments because of the lipid nanoparticles, lipids utilized for their creation, and consolidated corrective actives [11 - 13].

Nanoemulsions

Nanotechnology involves mechanical improvements on the nanometer scale, generally 0.1-100nm. Nanoemulsions have as of late, become progressively significant as expected vehicles for the controlled conveyance of beauty care products and the streamlined scattering of dynamic fixings, specifically skin layers. Because of their lipophilic inside, Nanoemulsions are more appropriate for the vehicle of lipophilic mixtures than liposomes. Like liposomes, they support the skin infiltration of dynamic fixings and consequently increment their focus in the skin. Another benefit is the little estimated bead with its high surface region permitting viable vehicle of the dynamic to the skin. Moreover, nanoemulsions acquire expanding interest because of their bioactive impacts. This might decrease the transepidermal water misfortune (TEWL), showing that the boundary capacity of the skin is strengthened. Nanoemulsions are worthy in beauty care products because there is no innate creaming, sedimentation, flocculation, or mixture that is seen with macroemulsions. Oil-in-water emulsions (O/W emulsions) assume a significant part in beauty care products: they are basic in the definition of such items as body salves, skin creams, and sunscreens. A somewhat later, however quickly developing field of utilization is NanoGel frameworks and Emulsion-based moist disposable clothes, nanoemulsions that are liberated from emulsifiers dependent on polyethylene glycol (PEG). Such mixes are exceptionally appealing in the developing business sector for impregnating emulsions for saturated tissue. It is a somewhat later yet quickly developing field of use: emulsion-based moist disposable clothes for such applications as child care and make-up removal [14].

Nanocapsules

Nanocapsules are polymeric NM cases that are encircled by a sleek or water stage. Nanocapsules are utilized in beauty care products for the security of fixings, for diminishing synthetic scents, and for settling incongruence issues between plan parts. Polymeric nanocapsule suspensions can be straightforwardly applied on the skin as a result or consolidated into semisolid plans as a fixing. The level of skin entrance of fixing can be adjusted by the polymer and the surfactant utilized as crude materials. Balanced out poly-l-lactic corrosive nanocapsules with a breadth of around 115 nm was ready through nanoprecipitation, and a supported arrival of scent, by entangling aroma atoms in a polymeric nano-transporter was accomplished. This sort of exemplification of atoms in biocompatible nanocapsules can assume a huge part in the eventual fate of antiperspirant items [15].

Dendrimers

Dendrimers are another class of nano-sized, unimolecular, and profoundly stretched nanostructures with globular, standard fanning and underlying balance. The absolute number of series of branches decides the age of the dendrimer. The original dendrimer has one series of branches, while there are two series on account of second-age dendrimers. Dendrimers are tiny, with an estimated breadth of 2–20 nm. A wide assortment of dendrimers exists, and natural properties like polyvalence, monodispersity, solvency, low cytotoxicity, synthetic solidness, self-collecting, and electrostatic communications render dendrimers a promising transporter for drug conveyance with high selectivity and exactness. These nanostructures are being utilized in different cosmeceutical items like sunscreen, shampoos, against skin inflammation cream, and hair-styling gels, antiperspirants. The dendrimers shave shown potential in the compelling conveyance of dermal arrangements through the skin boundary. An *in vitro* study showed an expanded volume and infiltration of chlorhexidinedigluconate (CHG) into the skin after pretreatment of skin with PAMAM dendrimers. These discoveries were helpful as far as further developing treatment viability against bacterial contamination of the skin. The dendrimer-based pretreatment likewise upgrades the skin pervasion of chlorhexidinedigluconate. Biodegradable polymers, for example, poly α-esters, polysaccharides, poly amidoamine, and poly alkyl cyanoacrylates dendrimers can fill in as an epitomizing specialist for individual consideration and beauty care products details. Natural thickness is one of the significant dendrimer attributes that make them helpful for corrective definitions. Numerous restorative organizations, like Dow Chemical Company, L'Oréal, Revlon, and Unilever, have a large number of dendrimer-based corrective plans for applications in skin, nail, and hair care items. L'Oreal has been utilizing terminal hydroxyl functionalized polyester dendrimers joined by film-framing polymers in superficial definitions for skin applications. A further developed power by the utilization of dendrimers has been seen in different L'Oreal beauty care products items [16].

Cubosomes

Cubosomes are discrete, sub-micron, nanostructured particles of the bi-constant cubic fluid glasslike stage. It is framed by the self gathering of fluid translucent particles of specific surfactants when blended in with water and a microstructure at a specific proportion. Cubosomes offer a huge surface region low thickness and can exist at practically any weakening level. They have high warmth solidness and are equipped for conveying hydrophilic and hydrophobic particles. Joined with the minimal expense of the crude materials and the potential for controlled delivery through functionalization, they are an appealing decision for restorative applications just as for drug conveyance [18].

Hydrogels

They are 3D hydrophilic polymer networks that swell in water or organic liquids without dissolving because of synthetic or actual cross-joins. They can anticipate future changes and change their property appropriately to forestall the harm [19].

Mineral-Based Cosmetic Ingredients with Nano-Sized Dimensions

Some surface-level items, like sunscreens, utilize mineral-based materials and their exhibition relies upon their molecule size. In sunscreen items, titanium dioxide, and zinc oxide, in the size scope of 20 nm, are utilized as effective UV channels. Their primary benefit is that they give wide UV assurance and do not cause cutaneous unfavorable wellbeing impacts.

Applications of Nanoparticles in Cosmetic

The United States Federal Food, Drug, and Cosmetic Act define beauty care products as "materials intended to be scoured, poured, sprinkled, or showered on the human body for purifying, embellishing, increasing engaging qualities, or changing the lookThe term 'cosmeceutical' is used to describe a remedial item that has been proven to be beneficial. On the global market, cosmetics have experienced steady growth. Nanotechnology has recently emerged as the best methodology in the cosmeceutical industry, presenting more modest particles (<100 nm) that may enter the skin and be easily retained, getting at the specified tissue without difficulty. Along these lines, nanoparticles are regularly utilized in the makeup business, being utilized in various items and details. Nanotechnology is currently generally utilized in beauty care products and dermatological items, like cleansers, against wrinkle creams, aromas, toothpaste, lipsticks, lotions, sunscreens, hair care items, skin cleaning agents, and nail care bosom cream. As per, NPs are for the most part grouped into eight-item classes as far as their size and usefulness. These are liposomes, nanocapsules, strong lipid nanoparticles, nanocrystals, dendrimers, cubosomes, niosomes, and nanogold and nanosilver. As of late, significant consideration has zeroed in on eco-accommodating innovations for the creation of metal nanoparticles like gold, silver, and platinum. The innovation is called eco-friendly because the specialists utilized, like microbes, organisms, yeasts, and plants, are the biofactories for the NPs [21].

Preservatives in Cosmetics using Silver Nanoparticles

Additives are important components in the design and manufacture of beauty care products to prevent essential microbiological contamination, as well as to prevent optional microbial tainting after manufacture when the purchaser opens and closes the compartments during day-to-day use. Phenoxyethanol and parabens are regularly utilized in beauty care products. Be that as it may, these antibacterial mixtures briefly aggravate the skin as well as increment skin affectability to UV light. In this manner, for a long time specialists have been hoping to supplant these synthetic compounds with other safe options. Ag-NPs are currently normally utilized as additives because of their antimicrobial properties. These NPs are widely utilized in beauty care products like antiperspirants, face packs, and against maturing creams. Ag-NPs are currently employed as additives in toothpaste and shampoos as a result of their antibacterial properties. The endophytic parasite class *Penicillium* is used to combine Ag-NPs. Tannins, saponins, terpenoids, and flavonoids are among the phytochemicals found in Penicillium. In the transformation of silver particles into NPs, these chemicals can act as reducing and covering agents. The covering experts, for example, the amide and carbonyl groupings seen in Ag-NPs of 10–60 nm biosynthesized from Fusariumsemitectum, be stable for 6 months. Covering specialists is imperative to stay away from the agglomeration of NPs and they can likewise offer soundness to the item. These stable features can be used to enhance restorative products as well as tangible items. Qualities of the object, because they maintain the item's uniform look and prevent sedimentation for more than a year nano zinc oxide and titanium dioxide can improve the feel and spread capability of restorative definitions in superficial things. Aside from that, they can also show that they have better sun insurance than their non-nano counterparts. In addition, when compared to the use of silver in the first state, nanosilver can enhance the particle's antibacterial capabilities.

Cosmetics with Antimicrobials

Silver NPs (PAg-NPs) from *Penicillium* spp. show strong antibacterial capabilities, with a high ability to inhibit *E. coli* and *P. aeruginosa* growth at 100 l culture filtrate/1 ml microbe stock.

Various studies of Ag-NPs have revealed that these NPs may have antibacterial properties against *E. coli, B. subtilis, V. cholera, P. aeruginosa*, and *S. aureus*.

When administered to infectious cells, Ag-NPs can disrupt the parasite envelope design, causing serious harm to contagious cells such as *Candida albicans, Candida tropicalis, Candida glabrata, Candida parapsilosis*, and *Candida* spp. Antimicrobial movement was also discovered by Kokura *et al.*, who found that Ag-NPs with a low convergence of 1.0 ppm had antimicrobial properties in bacterial (*E. coli, P. aeruginosa, S. aureus*) and parasitic (*C. albicans, A. niger, P. citrium, A. pullulans*) removal from sifted kitchen and seepage squander water. Ttitaniumoxide (TiO$_2$) NPs have been used in sunscreens, brightening creams, morning and night creams, and skin milks, among other applications. When TiO$_2$ NPs delivered by Aspergillusflavus were administered at a dosage of 40 μg ml−1, *E. coli* development was inhibited [22].

It is hard to decide the number of nanomaterials utilized in superficial items because the meaning of what comprises a nanomaterial is as of now developing. The utilization of the prefix "nano" in superficial promoting and naming may not agree with how the term is utilized by administrative specialists [5, 23].

- Appearance upgrade
- Dental care
- Hair-care
- Skincare

Cosmetic Antioxidants and Anti-Inflammatory Agents

Because of their advantages, nanoparticles have recently been remembered for a huge variety of remedial goods. They work on tangible properties and security of items, give better inclination, and improve better sun insurance. PAg-NPs showed solid cell reinforcement movement, demonstrated by DDPH (1,1-diphenyl-2-picrylhydrazyl) and FRAP (Ferric Reducing Ability of Plasma) tests. They additionally showed mitigating exercises, and hence, they have been considered as added substances in surface-level items to profit from their normal capacities. Moreover, the Ag-NPs demonstrated to essentially build wound recuperating properties of some surface-level items. The mitigating impact of PAg-NPs is displayed by their job in expanding layer adjustment. There are just a couple of reports accessible in regards to the calming exercises of endophytic parasitic Ag-NPs. Notwithstanding, the primary concern was the danger of NPs poisonousness inferable from their nanosize and anticipated infiltration through the skin. It has recently been discovered that roughly 0.002-0.02 ppm of AgNPs can reach the epidermis and are not harmful at these concentrations. AgNPs are said to be flushed out of the circulation system at these concentrations, indicating that they are not toxic.

Nanomaterials in Cosmetics and Methods of Engineering

Nanotechnology is presently utilized in the improvement of basically three spaces of the beauty care products creation measure:

1. Formulation
2. Packaging
3. Manufacturing gear

Instances of current or planned employment of nanotechnology in bundling incorporate antibacterial coatings (nanosilver, zinc oxide, or magnesium oxide), air/dampness hindrances (nanoclays, for example, montmorillonite) to expand the period of usability of the item, and different radio-recurrence distinguishing proof labels which permit observing of temperature and dampness profiles during capacity and harm during transportation. The employments of nanotechnology in assembling hardware may not be clear to the purchaser; however can offer enormous advantages for the creation cycle. Non-tacky materials for hardware, simple to-clean surfaces, low grating subtleties, and further developed energy productivity are a couple of instances of advantages that can be acquired by utilizing nanotechnology for assembling restorative items. The biggest part of progressing nano-related innovative work concerns the definition parts of corrective items.

The most widely recognized materials utilized in nanotechnology are silver (299 items), carbon (82), titanium/titania (50), silicon/silica (35), zinc/zinc oxide (30), and gold (27). This load of materials is available in presently showcased corrective items, including sunscreens. With regards to the current section, the term NP is utilized from a conventional perspective, including every one of its spaces of utilization in a corrective item, for example

  i. As the dynamic fixing
 ii. As the conveyance vehicle/nanocarrier
iii. As a plan help

NPs have distinctive morphology and can be a natural, inorganic, and blended beginning. As dynamic fixings, NPs can, for instance, offer actual UV safeguarding, skin brightening, hair, and dental consideration impacts. Utilized as a plan helps, NPs can work as insoluble thickening specialists because of their novel rheology profile, which can't be obtained for the material in the mass structure. They can additionally upgrade the item's optical properties, like straightforwardness, reflectivity, and shading, in a way that isn't workable for similar material in the mass structure. Restorative nanocarriers, which don't have any organic action of their own, are utilized to convey nutrients, cancer prevention agents, synthetic UV channels, hostile to skin inflammation or against maturing substances in the objective layer of skin where they are then broken up or dissolved, after which the impact of the dynamic fixings is the same as that of the atomic structure. Nanocarriers offer a few benefits which are shown in the further developed viability, low poisonousness, improved penetration, substance dependability, and controlled pace of conveyance of the dynamic fixings. Different kinds of nanocarrier frameworks utilized in surface-level plans have been surveyed in a study [20], which shows instances of nanocarriers utilized as conveyance vehicles. Grid-type nanocarriers, as nanoclays or mesoporous silica NPs, are non-empty nanospheres (or particles of other morphology), which gradually convey their heap either employing dissemination from the lattice or through framework disintegration. Depo-type nanocarriers, likewise alluded to as nanocapsules, comprise a center, wherein the dynamic fixings are suspended, and a shell encases the center. A layer-by-layer (LbL) NP is like a depo-type NP, then again, actually is shell comprising of different, exchanging layers of oppositely charged polyelectrolytes. LbL NPs can be empty or contain another NP. A dendrimer NP, or fractal polymeric tree, comprises an expanded engineered polymer, wherein the dynamic fixing is entangled in the interstices between the branches. There is additionally a sufficient assortment of NPs from greasy substances including both grid-type nanocarriers, like strong lipid NPs (SLNPs) and nanostructured lipid transporters (NLCs), and depo-type nanocarriers, for example, liposomes. A liposome comprises a bilayer phospholipid vesicle with a watery center. It ought to be noticed that monolayered phospholipid vesicles settled by a co-surfactant highlighting a hydrophobic center are known as nanotopes different kinds of nanocarrier frameworks will be examined in more detail underneath. It ought to be referenced that climate responsive NPs, which can off-load their payload because of a particular upgrade, like temperature, pH, or a particular atom, have been in the focal point of examination. One ongoing model incorporates a particle engraved, a reversible polymeric organization comprising of nanovacuoles, equipped for atomic acknowledgment. The presence of and restricting with the particular atom makes reversible inward burdens that burst the polymeric organization at the nanovacuoles. The atom engraved reversible polymeric organization, which comprises a glycosidic type sugar spine, can be utilized for the controlled conveyance of restorative dynamic fixings [5].

Applications of Nanoparticles in Medical Fields

Drug Delivery

Drug Distribution NPs have been studied and identified as transporters for drug delivery over the last decade. New medication conveyance frameworks dependent on nanotechnology have been applied in the therapy of human illnesses, like malignancy, diabetes, microbial contaminations, and quality treatment. The advantages of these medicines are that the medication is focused on infected cells, and the diminished poisonous incidental effects improve its security profile on ordinary cells. As a general rule, NPs can be formed with various sorts of medications to convey bioactive mixtures to the objective site by different techniques, like the utilization of nanotubes, liposomes, nanopores, and dendrimers. Au-NPs, for example, is appropriate for the planning of medicine conveyance frameworks due to their safety in terms of toxicity and immunocompatibility. Nanomaterials combined with an organic technique could be used to treat diabetes mellitus as an optional treatment. Au-NPs showed transaminase and basic phosphatase and diminishing uric corrosive in a diabetic mouse model. At a modest fixation of 8 μg/mL, Au-NPs coupled with vancomycin by *Trichoderma viride* were linked to the microbial surface by ionic connection and viably smothered the development of vancomycin-safe *Staphylococcus aureus*. TEM analysis revealed that the vancomycin-bound Au-NP had reached the bacterial layer, demonstrating cell crossing of *S. aureus* (Fayaz *et al.*). In another study, Sun *et al.* used a covalent connection to stack doxorubicin into bacterial magnetosomes, and these magnetosomes suppressed cancer development by 86.8%. Brown *et al.* discovered that Au-NPs functionalized with a thiolated polyethylene glycol monolayer covered by a carboxylate bunch improved the delivery of the anticancer drug oxaliplatin [24].

Healing of Wounds

Robert Burrell was the first to develop a commercial nanosilver molecule product for clinical use. It was used to treat a variety of injuries, including burns, ulcers, and epidermal necrolysis. Ag-NPs made from *Aspergillus niger* were found to be promising wound healers, acting against pathogenic microscopic organisms and balancing the cytokines involved in wound healing this biogenic silver plan, in combination with enoxaparin, improved twisted recuperation in an *in-vivo* evaluation of Ag-NPs given by *Fusarium oxysporum*. The benefits gained included a shorter time for the separation of fibroblasts into hyperactive cells (myofibroblasts) as compared to regular injury dressings and a shorter period for the flaming cycle when compared to standard injury dressings.

Antibacterial Activity

With pandemics and the increased resistance of microbes to a variety of commonly used antimicrobials, NPs have recently been considered true-to-form replacements to commonly utilized portion structures. Ag-NPs consolidated from the development *Aspergillus niger* showed extraordinary inhibitory development against Gram-positive organisms like *S. aureus* and Gram-negative microorganisms like *E. coli*. The edible mushroom *Agaricus bisporus* was used as a bioreductant to transport Ag-NPs. The delivered nanometal particle has antibacterial action against human microorganisms such as *E. coli*, *Proteus vulgaris*, and *Klebsiella* spp. According to Duran, extracellular Ag-NPs generated by *Fusarium* sp. killed *S. aureus* and were thus useful in the treatment of material surfaces. With an insignificant inhibitory center (MIC) of 30 g/mL, Ag-NPs organized by *T. viride* were able to suppress *E. coli* growth. . *Aspergillus clavatus* was employed as a biofactory for the manufacture of Ag-NPs, having antibacterial activity against *Candida albicans*, *Penicillium fluorescens*, and *Escherichia coli*. The made Ag-NPs went probably as antibacterial subject matter experts, with the ability to suffocate the growth of living organisms such as *E. coli*, *S. aureus*, and *P. aeruginosa*. Another study by Govindappa *et al*. found that the Ag-NPs produced by *Penicillium* effectively inhibit the growth of *E. coli* and *P. aeruginosa*, as evidenced by the use of SEM. Furthermore, when administered at high core interests, the Ag-NPs demonstrated robust cell support, quieting, and anti-lipoxygenase development, as well as tyrokinase inhibitory activities. In a study, endophytic parasite supernatant from *Alternaria* sp., as well as *Raphanus sativus*, was used to create an unpretentious, quick, and one-adventure approach for Ag-NP combination.

Antifungal Activity

In the field of medicine, researchers have made numerous attempts to develop antimicrobial specialists through the discovery of novel bioactive specialists and innovative product designs that may be used in the clinical treatment of illnesses caused by harmful bacteria and parasites. As a result, Ag-NPs have been shown to have exceptional potential as antimicrobial specialists, and they have been used in the design of clinical items for the prevention of auxiliary medical clinic disorders. When applied at groups of 0.11–1.75 g/mL, *Aspergillus tubingensis* and *Bionectria ochroleuca* mixed Ag-NPs that exhibited antifungal motility, and these Ag-NPs may be used in emergency clinic disorders caused by *Candida* sp. To obtain NPs produced by *Fusarium oxysporum*, a green science strategy (which included microbial and nanotechnology) was used. The supplied particles exhibited strong antifungal activity and inhibited the growth of human infectious pathogens such as *Candida* spp. and *Cryptococcus neoformans*. These NPs were found to be capable of harming parasite cells' cell dividers and cytoplasmic layers. In addition, Ag-NPs produced by Schizophyllum collectively demonstrated remarkable antifungal activity against dermatophytic parasitic microbes such as *Trichophytonsimii, Trichophyton mentagrophytes*, and *Trichophytonrubrum.* *Arthroderma fulvum* developed the antifungal movement of Ag- NPs against eleven parasitic pathogens, including *Candida* spp., *Aspergillus* spp., and *Fusarium*s pp When administered at concentrations ranging from 0.125 to 4.00 g/mL, the Ag- NPs demonstrated significant antifungal activity against all pathogenic growths tested.

Antiviral Activity

Antifungal, antibacterial, and antiviral properties are all found in nano metals [26]. Ag-NPs combined with *Aspergillus fumigatus* were found to have antiviral properties against HIV-1. In different researches, the antiviral properties of Ag-NPs combined from *Aspergillus* sp., noting that at concentrations of 30 to 180 ppm, the Ag-NPs reduced the number of viral plaques, whereas at higher fixations, 210 to 240 ppm, the Ag-NPs showed absolute hindrance of viral particles in the bacterial host, resulting in total viral replication inactivation. Ag-NPs of a size of 1–10 nm presented a challenge to HIV-1, blocking viral attachment to the host cell surface, according to one study. In a study, Ag-NPs with a size of 1–10 nm presented a challenge to HIV-1 and prevented viral attachment to the host cell surface.

Disease Nanotechnology

Formal meanings of nanotechnological gadgets commonly highlight the prerequisites that the actual gadget or its fundamental parts be man-made, and in the 1–1,00 nm range in no less than one measurement. Malignancy-related instances of nano-advancements incorporate injectable medication conveyance nanovectors, for example, liposomes for the treatment of bosom disease naturally focused on, nanosized attractive reverberation imaging (MRI) contrast specialists for intraoperative imaging with regards to neuro-oncological intercessions and novel nanoparticle-based techniques for the high-particularity discovery of DNA and protein. In the reasonable universe of nanotechnology, George Whitesides puts less tough restrictions on the specific measurements and characterizes the 'right' size in bionanotechnology in a functional design. Robert Langer and partners contend comparatively with regards to tranquilize conveyance applications. In concordance with these methodologies, this current audit's essential methodology is that the characterizing provisions of disease nanotechnology are implanted in their advancement potential for patient consideration [27].

Malignancy Therapy

Malignant growth is the main source of death around the world. Chemotherapy has prompted great outcomes, however, as a rule, cells created protection from the chemotherapy specialists. As a result, scientists have worked hard to develop solutions that are both biocompatible and practicable for the treatment of disease patients while also reducing the side effects of the synthetic substances used. Through the age of free oxygen radicals, studies have revealed that biogenic Ag-NPs can trigger apoptotic pathways *in vitro*. Likewise, Ag-NPs have sparked renewed interest in their use in the diagnosis and treatment of human cancer, and as a result, these particles are being viewed as potential antitumor and antiproliferative agents. They are also thought to be antiangiogenic in the breast cancer cell lines MCF7 and T47D, biosynthesized Ag-NPs produced by the yeast *Cryptococcus laurentii* (BNM 0525) showed a significant antitumor effect. They discovered that following 24-hours of exposure to Ag-NP arrangements at concentrations ranging from 1 to 10 g/mL, bosom disease cell development was inhibited, film spillage was induced, and Ag-NP anticancer movement was delivered in shaken stock societies. At Ag-NP centralizations of 10 to 100 g/mL, MTT cytotoxicity analysis revealed cell passings of 27.2 percent to 64 percent in human laryngeal cancer cells (HEP-2).

OTHER APPLICATIONS

Fuel Cell Applications

Polymers are used in the proton exchange membrane, electrode binder, and bipolar plate matrix in fuel cell applications. Carbon black particles (0.5–1.0 m), Pt catalyst particles (2–5 nm), and a polymeric binder are commonly used in electrodes (usually Nafion). The performance of platinum nanoparticles placed upon single-walled carbon nanotubes with nafion as a binder was shown to be superior to carbon black-based electrodes. Numerous studies have noticed the use of nanoparticles in proton exchange membranes to increase mechanical qualities as well as proton conductivity. Nanoparticles are also used in direct methanol fuel cells.

Electrical/Electronics, Optoelectronics, and Sensors

The creation of current electrical and optoelectronic devices is deeply rooted in nanotechnology. On a dimensional scale, electronic devices have now reached the nanoscale. Polymer-based nanocomposites' applicability in these domains is rather broad, spanning a large range of potential applications and nanocomposites types. One type of nanocomposite that is gaining a lot of attention is conjugated polymers with carbon nanotubes. According to a recent study, photovoltaic (PV) cells and photodiodes, supercapacitors, sensors, printed conductors, light-emitting diodes (LEDs), and field-effect transistors are all viable applications [29].

Fungi-produced metal nanoparticles have a lot of potential as sensors for optical and electrical systems. *Trichoderma viride* developed Ag-NPs, which were successfully used in biosensor and bioimaging applications. These Ag-NPs were used for blue-orange light emission at wavelengths of 320–520 nm, and they were completely characterized using EDX (Energy Dispersive X-ray) and XRD examinations [30]. The use of yeast to synthesize Au-Ag alloy nanoparticles found that these Ag-NPs were five times more sensitive than previous vanillin sensors. This chapter demonstrated the tremendous potential of Ag-NPs as sensors in the quantitative assessment of vanillin synthesis from the vanilla bean and vanilla tea [31]. The activity of the enzyme glucose oxidase (GOx), which is used to evaluate the glucose content of commercial glucose injections, was boosted by the Au-NPs. The Au-NP-GOx-based biosensor's action is based on Au-NPs' extraordinarily sensitive sensing [32].

CONCLUSION

The chapter is concluded that the application of fungal nanotechnology in cosmetics, this emerging technology is about to turn into a vital tool for knowledge-based investigations of recent cosmetic and medical products. Nowadays, nanotechnology is being established for both generating new packing materials and manufacturing equipment for cosmetics and medical fields. In cosmetics and medical, nanomaterials are already beings utilized or being proposed for coming application as vehicles of active substances. The cosmeceutical nanomaterials are rapidly and widely developing, thus we can expect a range of novel nanotechnology-based cosmetic products on the shelves of our nearest drug stores in the near future.

CONSENT FOR PUBLICATION

Not applicable.

CONFLICT OF INTEREST

The authors declare no conflict of interest, financial or otherwise.

ACKNOWLEDGEMENT

Declared none.

REFERENCES

[1]   Raj S, Jose S, Sumod US, Sabitha M. Nanotechnology in cosmetics: Opportunities and challenges. J

Pharm Bioallied Sci 2012; 4(3): 186-93.
[http://dx.doi.org/10.4103/0975-7406.99016] [PMID: 22923959]

[2]   Estefania VR. Campos, anderson E. S. Pereira, JhonesLuiz de Oliveira, Lucas BragançaCarvalho, Mariana Guilger-Casagrande, Renata de Lima, Leonardo FernandesFraceto., How can nanotechnology help to combat COVID-19? opportunities and urgent need. J Nanobiotechnology 2020; 18(1): 125.
[http://dx.doi.org/10.1186/s12951-020-00685-4]

[3]   Prasad R. Fungal Nanotechnology: Applications in Agriculture, Industry, and Medicine. DOI 2017.
[http://dx.doi.org/10.1007/978-3-319-68424-6]

[4]   Mariana   GuilgerCasagrande.   TaisGermano   Costa,   TatianePasquoto-Stigliani,   Leonardo FernandesFraceto and Renata de Lima., Biosynthesis of silver nanoparticles employing Trichodermaharzianum with enzymatic stimulation for the control of Sclerotiniasclerotiorum, 9:14351 | 2019.
[http://dx.doi.org/10.1038/s41598-019-50871-0]

[5]   Mihranyan Albert, Ferraz Natalia, Strømme Maria. Current status and future prospects of nanotechnology in cosmetics. 0079-6425/$ - see front matter 2011 Elsevier Ltd. All rights reserved. 2011.
[http://dx.doi.org/10.1016/j.pmatsci.2011.10.001]

[6]   Mauricio M D, Guerra Ojeda S, Marchio P. S. L. Vallles, M. Aldasoro , I. Escribano Lopez, J. R. Herance, M. Rocha, J. M. Vila , V. M. Victor, Nanoparticles in medicine: a focus on vascular oxidative stress, oxidative medicine and cellular longevity Volume 2018, Article ID 6231482, 20.

[7]   Bakry Rania. Rania Bakry, Rainer M Vallant Muhammad Najam-ul-Haq, Matthias Rainer, ZoltanSzabo, Christian W Huck, and Günther K Bonn.Medicinal applications of fullerenes,Int J Nanomedicine. 2007; 2(4): 639-49.
[PMID: 18203430]

[8]   Kokura S, Handa O, Takagi T, Ishikawa T, Naito Y, Yoshikawa T. Silver nanoparticles as a safe preservative for use in cosmetics. Nanomedicine 2010; 6(4): 570-4.
[http://dx.doi.org/10.1016/j.nano.2009.12.002] [PMID: 20060498]

[9]   JolantaPulit-Prociak. MarcinBanach. Silver nanoparticles – a material of the future…?, 2016.
[http://dx.doi.org/10.1515/chem-2016-0005]

[10]   GeorgiosFytianos, Abbas Rahdar,and George Z. Kyzas., Nanomaterials in Cosmetics: Recent Updates. 2020.
[http://dx.doi.org/10.3390/nano10050979]

[11]   Müller RH, Petersen RD, Hommoss A, Pardeike J. Nanostructured lipid carriers (NLC) in cosmetic dermal products 2007.

[12]   Jensen LB, Magnussson E, Gunnarsson L, Vermehren C, Nielsen HM, Petersson K. Corticosteroid solubility and lipid polarity control release from solid lipid nanoparticles. Int J Pharm 2010; 390(1): 53-60.
[http://dx.doi.org/10.1016/j.ijpharm.2009.10.022] [PMID: 19836439]

[13]   Müller RH, Shegokar R, Keck CM. 20 years of lipid nanoparticles (SLN and NLC): present state of development and industrial applications. Curr Drug Discov Technol 2011; 8(3): 207-27.
[http://dx.doi.org/10.2174/157016311796799062] [PMID: 21291409]

[14]   Surbhi Sharma & Kumkum Sarangdevot, Nanoemulsions For cosmetics, research scientist. (R&D), Roha Dyechem PVT. LTD 2012.

[15]   Poletto Fernanda. Ruy Carlos R Beck, Adriana R Pohlmann, Polymeric nanocapsules: Concepts and applications, 2011.
[http://dx.doi.org/10.1007/978-3-642-19792-5_3]

[16]   Muhammad Bilal, Hafiz M. N. Iqbal, New insights on unique features and role of nanostructured materials in cosmetics, 2020.

[http://dx.doi.org/10.3390/cosmetics7020024]

[17]   ShreyaKaul. NehaGulati, DeepaliVerma, Siddhartha Mukherjee, and UpendraNagaich., Role of nanotechnology in cosmeceuticals: A review of recent advances., Journal of Pharmaceutics Volume 2018, Article ID 3420204 2018; 19.

[18]   Santos AC, Morais F, Simões A, *et al.* Ana Claudia Santos, Francisca Morais, Ana Simoes, Irina Pereira, Joana A. D. Sequeira, Miguel Pereira-Silva, Francisco Veiga&AntónioRibeiro., Nanotechnology for the development of new cosmetic formulations. Publisher: Taylor & Francis Journal. Expert Opin Drug Deliv 2019; 16(4): 313-30.
       [http://dx.doi.org/10.1080/17425247.2019.1585426]

[19]   Morales ME, Gallardo V, Clarés B, García MB, Ruiz MA. Study and description of hydrogels and organogels as vehicles for cosmetic active ingredients. J Cosmet Sci 2009; 60(6): 627-36. [PubMed]. [PMID: 20038351]

[20]   Wu X, Richard H. Guy, Applications of nanoparticles in topical drug delivery and in cosmetics., University of Kentucky, College of Pharmacy, Lexington, KY 40536-0082, U.S.A, 2009.

[21]   Patel A. Parixit Prajapati1, Rikishaoghra, Overview on application of nanoparticles in cosmetics., Asian Journal of Pharmaceutical Sciences and Clinical Research, 40-55, ISSN: 2249-2135 , 2011.

[22]   Rajakumar G, Rahuman AA, Roopan SM, *et al.* Fungus-mediated biosynthesis and characterization of TiO$_2$ nanoparticles and their activity against pathogenic bacteria. Spectrochim Acta A Mol Biomol Spectrosc 2012; 91: 23-9.
       [http://dx.doi.org/10.1016/j.saa.2012.01.011] [PMID: 22349888]

[23]   Linda M. Katz, KapalDewan, Robert L. Bronaugh., Nanotechnology in cosmetics 2015.
       [http://dx.doi.org/10.1016/j.fct.2015.06.020]

[24]   Hesham A. El Enshasy, Nagib A. El Marzugi, Elsayed A. Elsayed, Ong Mei Ling, Roslinda Abd Malek, Afif Najiha Kepli, Nor Zalina Othman, and Solleh Ramli Medical and Cosmetic Applications of Fungal Nanotechnology: Production, Characterization, and Bioactivity. 2018; 21-59.
       [http://dx.doi.org/10.1007/978-981-10-8666-3_2]

[25]   Sarabjeet Singh Suri. HichamFenniri, Baljit Singh. Nanotechnology-based drug delivery systems, Journal of Occupational Medicine and Toxicology http://www.occup-med.com/content/2/1/162007; 2:16.
       [http://dx.doi.org/10.1186/1745-6673-2-16]

[26]   Yuan Y-G, Zhang S. Ji-Yoon Hwang ,Keun Kong. Silver Nanoparticles Potentiates Cytotoxicity and Apoptotic Potential of Camptothecin in Human Cervical Cancer Cells., Oxidative Medicine and Cellular Longevity Volume 2018, Article ID 6121328 2018; p. 21.
       [http://dx.doi.org/10.1155/2018/6121328]

[27]   Ferrari M. Cancer nanotechnology: opportunities and challenges. Nat Rev Cancer 2005; 5(3): 161-71.
       [http://dx.doi.org/10.1038/nrc1566] [PMID: 15738981]

[28]   Aleksandra Zielonka, Magdalena KlimekOchab, Fungal synthesis of size-defined nanoparticles. 29 August 2017.

[29]   Paul DR, Robeson LM. Polymer nanotechnology: Nanocomposite. Received 19 February 2008, Revised 2 April 2008, Accepted 4 April 2008, Available online 13 April 2008.

[30]   Fayaz AM, Balaji K, Girilal M, *et al.* Biogenic synthesis of silver nanoparticles and their synergistic effect with antibiotics: a study against gram-positive and gram-negative bacteria. Nanomedicine 2010; 6(1): 103-9.
       [http://dx.doi.org/10.1016/j.nano.2009.04.006]

[31]   Zheng D, Hu C, Gan T, Dang X, Hu S. Preparation and application of a novel vanillin sensor based on biosynthesis of Au-Ag alloy nanoparticles. Sens Actuators B Chem 2010; 148(1): 247-52.
       [http://dx.doi.org/10.1016/j.snb.2010.04.031]

[32]   Thibault S, Aubriet H, Arnoult C, Ruch D. Gold nanoparticles and a glucose oxidase based biosensor: an attempt to follow-up aging by XPS. Mikrochim Acta 2008; 163(3-4): 211-7.
[http://dx.doi.org/10.1007/s00604-008-0028-z]

CHAPTER 11

Nanobiotechnological Strategies for Detection of Fungi and Mycotoxins in Food and Feed

Sofia Agriopoulou[1,*], Eygenia Stamatelopoulou[1], Vasiliki Skiada[1] and **Theodoros Varzakas[1]**

[1] *Department of Food Science and Technology, University of the Peloponnese, Antikalamos, 24100 Kalamata, Greece*

Abstract: Mycotoxins are poisonous compounds that are produced by toxigenic fungi as secondary metabolites. Their production can occur at any stage of food and feed supply, including harvesting, storage, processing and distribution, contaminating a plethora of foodstuffs. As mycotoxins exert their toxic properties at a very low level, usually at µg/kg level, their early and fast detection in food materials is necessary. The early detection of mycotoxins and fungi contamination could pose the elimination or reduction of possible threats associated with the consumption of mycotoxin-contaminated food. Contamination of food with mycotoxins can be prevented by monitoring and control at various critical stages of the food chain at the pre-and post-harvest stage. Given the widespread use and rapid development of nanotechnology in a variety of fields, it is considered that the application of many nanomaterials in the detection of mycotoxins will be a pioneering strategy. Except for conventional methods, such as gas chromatography and high-performance liquid chromatography coupled with mass spectrometry or other detectors, various nanotechnological approaches are used for the detection of fungi and mycotoxins. Nanobiosensors, nanoparticles, nanowires, nanorods, and nanodiagnostic kits are used for the rapid detection of mycotoxin in food analysis. In addition, the electronic nose and electronic tongue are highly helpful in detecting a variety of mycotoxins and mycotoxin-producing fungi in food and feed. The main purpose of this chapter is to describe the role of nanobiotechnology and nanomaterials in the detection of fungi and mycotoxins in food and animal feed.

Keywords: Detection, Electronic nose, Electronic tongue, Food quality, Fungi, Mycotoxins, Nanobiotechnology, Nanobiosensors, Nanodiagnostic kits, Nanoparticles, Nanorods, Nanowires.

* **Corresponding Author Sofia Agriopoulou:** Department of Food Science and Technology, University of the Peloponnese, Antikalamos, 24100 Kalamata, Greece; E-mail: sagriopoulou@gmail.com

Savita, Anju Srivastava, Reena Jain & Pratap Kumar Pati (Eds.)

INTRODUCTION

Mycotoxins are a group of toxic fungal metabolites found in a wide range of food and feed products worldwide, with carcinogenic, genotoxic, teratogenic, nephrotoxic, and hepatotoxic effects [1, 2]. Although some fungi are beneficial to man, producing useful metabolites that contain pharmacologically active drugs for treating man and animals, some are responsible for human infections of the eyes, nails, hair, and skin, some fungi are pathogenic and cause extensive damage and agricultural losses to many crops, and some species produce mycotoxins [3]. Fungal infection is a huge threat to the economy due to reduced input and exports of foods and feedstuff affected by fungal growth [4]. Contamination of agricultural products with mycotoxins is a major problem mainly in tropical and subtropical regions, where climate and poor storage conditions favour the growth of fungi and the production of mycotoxins [5]. Many agricultural products such as nuts, fresh and dried fruits and vegetables, cereals such as maize, rice, and wheat, flours, liquids such as wine, grape juice and beer, milk and dairy products, spices and herbs, coffee and cocoa, and feed can be contaminated with mycotoxins at all stages of the food and feed chain, both pre-and post-harvest [2, 6, 7]. The main producers of mycotoxins are the fungi of the genera of *Aspergillus*, *Fusarium*, *Penicillium*, *Claviceps* and *Alternaria* [8].

The major classes of mycotoxins that are of the greatest agroeconomic importance are aflatoxins, ochratoxins, fumonisins, trichothecenes, emerging Fusarium mycotoxins, enniatins, ergot alkaloids, Alternaria toxins, and patulin, causing diseases both in humans and animals, known as mycotoxicosis [2 - 9]. To date, about 500 mycotoxins have been identified, while another 1000 are expected to be identified. Concerning so-called masked mycotoxins, the risk is extremely high, as there is no established method for their determination [10] and as their reconvert to their native form in the digestive tract [11].

Among all mycotoxins, aflatoxins are extremely dangerous to the health of humans and animals, as showing carcinogenic, mutagenic, teratogenic and immunosuppressive actions [12]. Consumption of mycotoxins contaminated food and feed by humans and animals respectively cause chronic or acute toxicity, a disease commonly attributed to the term mycotoxicosis [13] and even death [14]. Cardiotoxicity, central nervous system disorders, gastrointestinal tract damage, nephrotoxicity, and hepatotoxicity are some negative health impacts that are caused by mycotoxins [15]. The US Centers for Disease Control (CDC) estimates that more than 4.5 billion people are chronically exposed to high levels of contaminants by mycotoxins food [16].

The European Union (EU), the US Food and Drug Administration (FDA) and many authorities from other countries have established regulatory guidelines and maximum levels for many mycotoxins of major concern including aflatoxins (AFs) such as aflatoxin AFB1 (AFB1), aflatoxin B2 (AFB2), aflatoxin G1 (AFG1), aflatoxin G2 (AFG2) and aflatoxin M1 (AFM1), ochratoxin A (OTA), deoxynivalenol (DON), fumonisins (FBs), T-2 and HT-2 toxins (T-2, HT-2), zearalenone (ZEN) and patulin (PAT), for both imports and exports of contaminated food and feed [17, 18].

Numerous efforts have been made by many researchers to develop methods for detecting mycotoxins in food and feed, with many of these methods focusing primarily on chromatography and immunology [19]. Conventional techniques such as thin-layer chromatography (TLC), high-performance liquid chromatography (HPLC) in combination with different detectors like ultraviolet (UV), fluorescence detector (FLD) or diode array (DAD), liquid chromatography coupled with mass spectrometry (LC–MS), liquid chromatography-tandem mass spectrometry (LC–MS/MS) and gas chromatography-tandem mass spectrometry (GC–MS/MS) are the main chromatographic techniques, which are used for detection of mycotoxins. As it concerns immunochemical techniques, enzyme-linked immunosorbent assay (ELISA) and lateral-flow devices (LFDs) are the main representatives [8]. Prerequisites are required for the application of all these techniques, such as sampling, extraction, and cleanup methods [20]. Moreover, all the above-mentioned techniques are very accurate, but they have some limitations that are associated with these techniques, such as the high cost, the extensive and lengthy sample preparation, which takes a long time, and the need for bulky instruments and skilled personnel [21], limiting their use in the feed and food industries [22]. The development of new, simple, robust, reliable, inexpensive and fast techniques by researchers is the need of the hour [23, 24].

As it concerns, the conventional method for analyzing the load of toxicogenic fungi is based on the plate counting method which is done by plating dilutions on a single culture plate and then recording colonies [25]. In addition, regular fungal identification is mainly based on phenotypic methods, which often require expertise in morphological analysis and the result is always expressed in the order of magnitude such as 1×10^n CFU/g [26]. Moreover, DNA-based detection, molecular assays, like, polymerase chain reaction (PCR) or quantitative PCR (qPCR), enzyme-linked immunosorbent assay, is also used in early detection for fungal spoilage organisms [27].

Nowadays more sophisticated techniques such as spectroscopy techniques namely Raman spectroscopy, Fourier transform infrared (FTIR) spectroscopy and hyperspectral imaging (HSI) have been used by researchers for detecting mycotoxins and toxigenic fungi in foodstuffs as they have possessed much potential application in rapid analysis for food quality and safety [28, 29].

Research on mycotoxins and mycotoxigenic fungi has been carried out by various sciences such as food science, toxicology, applied and analytical chemistry, veterinary sciences, mycology, phytology and agricultureand much less from nanotechnology, although nanotechnology is a rapidly growing field [30] with a plethora of nanomaterials for potential applications in the agricultural sector [31]. Nanotechnology is the use and manipulation of structures and materials in the nanoscale range, usually 100 nm or less in at least one dimension, where 1 nm is 1 × 10^{-9}m, for various purposes [32]. Structures, materials and systems can be developed through nanotechnology and can be applied in various fields, including food agriculture and medicine [33]. Different names such as nanosensors, nanoparticles, nanodispersions, nanolaminates, nanotubes, nanowires, nanoarrays, buckyballs, quantum dots, and other terms are used for these materials. The main feature of nanoscale materials is their extremely small size with a very large ratio of surface area to volume. The result of this property is the appearance of improved optical, electrical, mechanical, and functional characteristics which differ from their normal attributes at the macro scale, making them ideal for applications of this new interdisciplinary technology [34, 35]. Biotechnology using techniques and knowledge from the field of biology aims to produce beneficial products by manipulating genetic molecular, and cellular functions. Nanotechnology and biotechnology have been combined to produce a dynamically evolving discipline of "nanobiotechnology" that allows the study of biological systems at the molecular level [36].

The detection of mycotoxin in food and feed materials is essential to confirm the quality and safety of food. Except for conventional methods, various nanotechnological approaches are used for the detection of fungi and mycotoxins. In the context of this chapter, the discussion will elucidate the role and relationship of nanobiotechnology with some aspects of the nanostructured materials and their use in the detection of fungi and mycotoxins in food and feed.

Nanomaterials Classification

Nanomaterials are classified in four categories according to the number of dimensions, which at least one of them must be on the nanoscale. There are zero-dimension nanomaterials, (nanoparticles, quantum dots, carbon dots), one-dimension nanomaterials, (nanotubes, nanowires, nanoribbons, nanorods, nanobelts, and hierarchical nanostructures), two-dimension nanomaterials (branched structures, nanoplates, nanoprisms, nanowalls, nanosheets, and nanodiscs) and three-dimension nanomaterials (nanocoils, nanoballs (dendritic structures), nanopillars, nanocones, and nanoflowers) [37]. The main physicochemical properties that differentiate various nanomaterials are size, shape materials, nature, and surface functionalization, among others, as shown in Fig. (**1**).

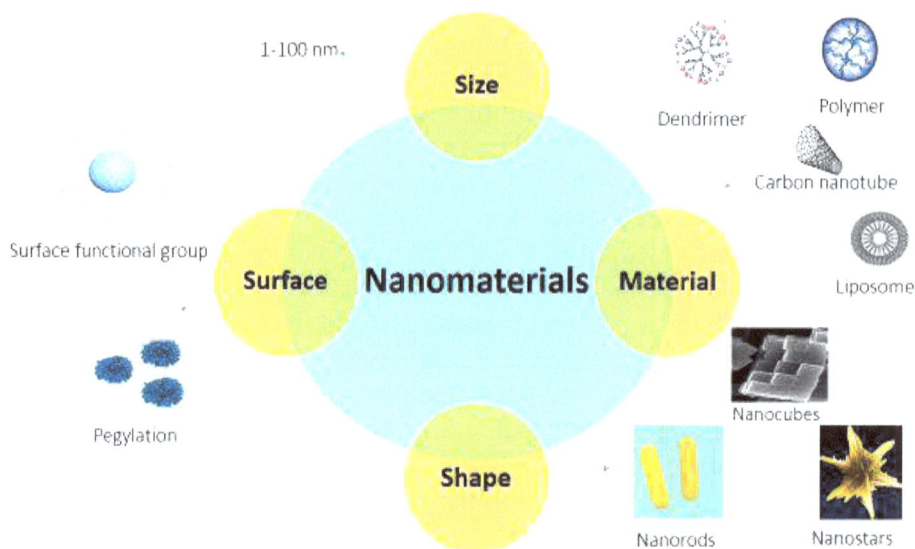

Fig. (1). A summary of physicochemical properties of nanomaterials.

The first three categories have three dimensions, the two and one dimension in nanometers, respectively. The nanomaterials that belong to the fourth category, are made of building units that have dimensions in the nanoscale but the final material has no size in the nanoscale. One more classification of nanomaterials can be done according to their chemical synthesis. There are carbon-based nanomaterials, metals based nanomaterials and metal oxides nanomaterials, dendrimers and composites, and biomaterials [38]. Different forms of nanomaterials used in the detection of fungi and mycotoxins in food and feed are shown in Fig. (**2**).

Nanobiosensors and Mycotoxins Detection

To overcome all the problems, that arise from the use of conventional methods for detecting mycotoxins, much attention has been paid to the toxin sensing technology and the instrument development by many other means [23]. A biosensor is an analytical device that comprises a biological sensing element (*a probe*), also called the recognition element, to measure the different biological processes (a bio-component or a bio-receptor such as isolated enzymes, DNA, whole-cell, or cell surface, tissues, aptamers, molecularly imprinted polymers, antibodies, oligonucleotides, lectins, organelles, *etc.*,) that will interact specifically to the analyte, a suitable transducing system, (*a transducer*) (optical, electrochemical, mechanical, piezoelectric elements, *etc.*) for converting the biochemical response into a recognizable electrochemical/electric signal, that are proportional to the target concentration, and an output system (*a detector*) [30, 39 - 43]. Based on signal transduction methods, biosensors may be classified into six groups: electrochemical (conductimetric and impediments, potentiometric, amperometric), thermal (isothermal titration calorimetry, heat-sensitive change in polymer film colour), optical (surface plasmon response, UV-visible absorbance, light-based potentiometric, luminescence and fluorescence, total internal reflection), mass (piezoelectric and magnetoelastic), magnetic, and micromechanical. A biosensor has remarkable advantages in comparison to standard analytical methods as it is simple, small in size, a portable instrument for a quick on-spot analysis, easy to operate, detects quickly, not expensive, offers high-performance analysis, and has high specificity and sensitivity [21, 44 - 46]. The "nanobiosensor" is the result of the integration of a nanomaterial with a receptor or transducer or receptor–transducer interface, which has significant improvements in sensitivity, selectivity, speed and the ability to multiplex complex sample analysis [47, 48].

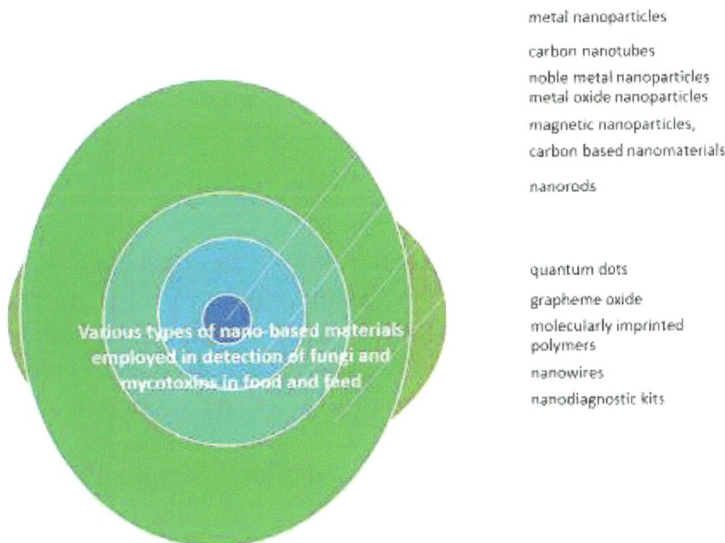

Fig. (2). Various types of nano-based materials are employed in the detection of fungi and mycotoxins in food and feed.

For a nanosensor to be characterized as an ideal, it must have characteristics, and in particular, it must be highly specific for the analysis. In addition, it must be stable under normal storage conditions and the responses received must be accurate, precise, reproducible and linear throughout the useful analytical range and also be free of electrical noise. Moreover, it must be small in size, biocompatible, non-toxic and non-antigenic and must be inexpensive, portable and capable of being used by semi-skilled personnel [30]. Nanosensors have either their size, or its sensitivity, or the sensor distance from the object to the nanoscale and can be used to detect contaminants, mycotoxins, gases, pesticides and microorganisms in foods [49]. Fig. (**3**) represents a nanobiosensor.

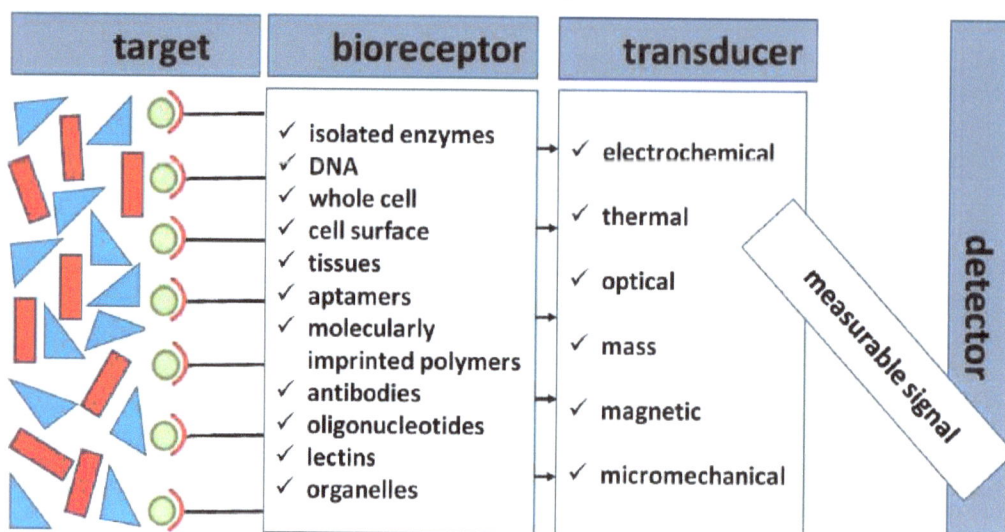

target	bioreceptor	transducer	
	✓ isolated enzymes ✓ DNA ✓ whole cell ✓ cell surface ✓ tissues ✓ aptamers ✓ molecularly imprinted polymers ✓ antibodies ✓ oligonucleotides ✓ lectins ✓ organelles	✓ electrochemical ✓ thermal ✓ optical ✓ mass ✓ magnetic ✓ micromechanical	measurable signal / detector

Fig. (3). Schematic representation of nanobiosensor. Adapted from [82].

With the development of nanotechnology, nanomaterials are playing an increasingly important role in the development of biosensors. The use of specialized biological receptors with nanoparticles that have both unique optical, electrical and electrochemical properties and a large surface to volume ratio due to their small size, contribute significantly to the development of an interesting perspective in the field of biosensors. Nanobiosensors undoubtedly are characterized as very promising detection devices. Metal nanoparticles (MNPs), carbon nanotubes (CNTs), nanorods, quantum dots (QDs), graphene oxide (GO), molecularly imprinted polymers (MIPs) and nanowires are some nanomaterials that have been successfully used in nanobiosensors [50, 51]. Nanomaterials function primarily in sensors, (i) as optical signal generation detectors (ii) as fluorescence quenches, (iii) as alternatives to enzyme labels, (iv) as signal amplifiers, (v) as mediators and (vi) as immobilizer supporters.

The improvement that nanoparticles can bring to nanosensors to improve their activity is mainly at the level of the bioreceptor and at the level of the transducer. Specifically, at the bioreceptor level, nanoparticles can act as loads for biomolecules and signalling molecules to increase detectable analytes. In addition, the modification of the electrode with nanoparticles makes it easier the transport of electrons between the solid surface and the attached biomolecule, which improves the sensitivity of the sensor [52, 53].

Nanobiosensors, except for direct detection of mycotoxins, can also detect any gases released due to microbial growth within the packaged food [54]. Electrochemical immunosensor based on poly (3,4-ethylene dioxythiophene) modified with gold nanoparticle was used by Sharma *et al.* to detect AFB1. The immunosensor displayed a LOD of 0.0045 ng/mL and LOQ of 0.0156 ng/mL [55].

Nanobiosensors and Mycotoxigenic Fungi Detection

Nanobiosensor-based technology provides a very useful tool in mycotoxigenic fungi detection. Diagnostic systems based on nanobiosensors are characterized as non-destructive, minimally invasive, cheap, and easy-to-use systems. Moreover, they have improved detection limit, sensitivity, specificity and achieved on-the-spot detection of pathogens. Nanoparticles, when used as a sensing element either alone or in combination can detect pathogens directly or indirectly by detecting growth-promoting precursors [27].

Detection of *Aspergillus niger* in food was achieved by synthesis of CuO nanoparticles and fabrication of nanostructural layer biosensors [56]. Tin oxide (SnO_2) nanobiosensor based on thin-film and nanopowder was also developed, for *Aspergillus niger* by Azhir *et al.* [57]. *Fusarium oxysporum* causing infection in *Arabidopsis thaliana* and tomato was detected by the use of a sensitive nanosensor [58].

Aptamer-Based Biosensors for Mycotoxin Detection

Aptamers offer special opportunities for the development of highly sensitive and selective sensors for mycotoxin determination [59]. Among all bio-receptors, aptamers seem to be very promising elements, as they are cheaper than antibodies, they can be easily regenerated and synthesized *in-vitro*. They can be easily modified and labelled with other receptors or chemical groups such as fluorescent dyes, enzymes, biotin, *etc.* They can be stored easily and they have a high affinity towards the target. Aptamers are composed of single-stranded oligonucleotides (DNA or RNA), they are produced by a process called the systematic evolution of ligands by exponential enrichment process (SELEX) [24] and interact with analytes with antibody-like ability. These aptamers are folded in a well-defined three-dimensional structure and have significant advantages over antibodies. The combination of nanomaterials with aptamers provides their use in nanoapta-sensors [21].

Various nanomaterials such as noble metal NPs, metal oxide NPs, magnetic NPs, and carbon-based nanomaterials (CNTs and graphene) have been incorporated into aptasensors for the development of new sensors based on nanomaterials [60]. These nanomaterials used in nano-aptasensors to detect mycotoxins play an important role in immobilizing, generating and enhancing the signal in the fluorescence quenching and act as an alternative to the enzyme labels [24, 61 - 66].

Metallic Nanoparticles (MNPs)

Metal-based nanoparticles (Ag, Au, Cu, Zn, Ni, Fe, TiO_2) have been widely used to immobilize biomolecules in sensor construction because their nanoscale surfaces accelerate the antibody-antigen reaction and provide biocompatible environments for biomolecule stabilization, resulting in an enhanced signal for immunoassay [52]. Moreover, functional groups such as CH_3, –OH, –COOH, –NH_2, or –$CONH_2$ are easily attached to nanoparticles facilitating various modifications [67, 68].

Among all metallic nanoparticles, gold nanoparticles (AuNPs)are currently used in an abundance of biomedical applications due to their size-dependent chemical, electronic and optical properties [69]. They exhibit properties such as high biocompatibility with bioelectric devices, reactivity, selectivity, and specificity, convenience in surface functionalization/modification and increased signal due to surface plasmon resonance (SPR)and have been proven to be powerful tools in nano-analysis applications [4, 52, 70].

AuNPs have been used to detect mycotoxins such as ochratoxin A [71], and aflatoxins [72], with the utility of nano-aptasensors. A small gold nanoparticle immunochromatographic strip was recently developed for AFB1 detection in peanuts, corn, rice, and bread. The application of AuNPs had the potential to detect AFB1 by monitoring a visible colour change from red to purple-blue, with a detection limit of 2 ng/mL in a 96-well plate [73]. AuNPs have excellent optical properties, displaying different colours depending on their size, shape and state of aggregation and are also used in colourimetric immunosensors [69]. In a very recent study, engineered gold nanoparticles were synthesized as multicolour labels for fumonisin B1 (FB1), ZEN, OTA, and AFB1detection on the immunochromatographic test strip nanosensors with a detection limit of 3.27, 0.70, 0.10, and 0.06 ng/mL, respectively [74].

Magnetic Nanoparticles (MNPs)

The use of magnetic nanoparticles (MNPs) is extremely effective in detecting mycotoxins in combination with traditional analysis techniques such as HPLC and LC-MS/MS. Magnetic solid-phase extraction (MSPE) is a new type of solid-phase extraction (SPE), which is one of the universal sample preparation techniques used for mycotoxins extraction [75]. Magnetite (Fe_3O_4) is the most common magnetic nanoparticle of MSPE. Moreover, some other components such as alumina, silica, manganese oxide, carbon NMs and MIPs, chitosan, and surfactants are combined with magnetite for the improvement of the absorption properties of MSPE [37].

MSPE have a high surface-area-to-volume ratio and a shorter diffusion distance for analyzers. For this reason, the sample processing time is reduced as rapid separation of magnetic materials and solutions through the application of a magnetic field is achieved [76, 77].

Magnetic nanoparticles have been used for extraction and determination of Ochratoxin A(OTA) in cereals [78] for extraction and analysis of AFB1, AFB2 in cereal products [79], for determination of AFM1 in liquid milk samples [80] and for the extraction of AFB1, AFB2, AFG1, and AFG2 in cereal and nut samples [81].

Carbon-Based Nanoparticles

Fullerenes, carbon nanotubes (CNTs) which divided into two general categories, in single-walled carbon nanotubes (SWCNTs-one rolled graphene sheet), and multi-walled carbon nanotubes (MWCNTs-two or more rolled graphene sheets), carbon dots (CDs), and graphene (Gr), are the main carbon-based nanomaterials [21, 52, 82]. Through the study of carbonaceous materials, the researchers discovered that carbon atoms were able to form particles of zero or almost zero dimension with limited size (1-10 nm) [83].

Fullerenes are defined as any molecule made of only carbon atoms. Fullerenes having spherical geometry are called buckyballs. Fullerenes are the molecular form of carbon and especially the third allotropic form of carbon, while the other two are diamond and graphite. Cylindrical fullerenes are given the name nanotubes [84].

Carbon nanotubes, well-characterized and synthesized as long thin cylinders, were developed in 1991 by Sumio Iijima [85]. CNTs have significant advantages over other materials as they are characterized by their unique prompt properties of high surface area, hollow shape, more catalytic in nature, biocompatibility, high mechanical strength and chemical stability and high electrical conductivity, indicating their possible application for biosensor development [24 - 44].

A label-free sensing system was recently developed by Zhang *et al.* with the use of MWCNTs for AFB1 detection. A bio-compatible film based on CNT-GNPs nanocomposite modified the gold electrode. The nanocomposite film showed high electron transfer capacity, achieving high sensitivity ranging from 0.05 to 25 ng/mL with a LOD of 0.03 ng/mL [86]. An amperometric aptasensor for ochratoxin A detection has been established by employing SWCNTs as signal-enhancing material in a recent study. The aptasensor was tested at OTA concentrations between 0-45 nM and used to detect OTA in serum and juice samples, recording a LOD of 134 and 58 pM, respectively [87].

Carbon dots (C-dots) belong to the category of new fluorescent nanomaterials, having gained attention today due to their excellent properties, such as satisfactory light stability, favourable biocompatibility, low toxicity and good water solubility [88]. In order to produce high yielding beans, C-dots methods known as top-down methods and bottom-up methods are used. In the first category, C-dots are cutting or exfoliating from large carbon materials such as graphite, carbon nanotubes, activate carbon, or carbon fibre and in the second category, C-dots arise from the treatment of organic precursors such as citric acid, glucose, polymer, or some natural materials chemically or physically [89]. The nitrogen-doped C-dots (N,C-dots) possess N-containing organic functional groups on the surface of the C-dots and they are characterized by rich photophysical properties [90]. N,C-dots were used to construct a novel aptasensor by Wang *et al.* to detect AFB1. The aptasensor exhibited high sensitivity, achieving a LOD of 5 pg/mL [89].

Individual layers of carbon atoms form graphene, which is the allotropic two-dimensional (2D) form of carbon [67]. Due to the two-dimensional structure of graphene, its application becomes problematic and the exploitation of graphene is more difficult in some of its applications, so for this reason, it has been processed in various forms, such as quantum dots [91]. In addition, graphene oxide (GO), is used in fluorescence-based nanobiosensors, to increase analytical sensitivity [92]. In a quenching aptamer-assay using exfoliated functional GO, ZEN was detected with 0.5 ng/mL LOD value in beer and wine [93]. An electrochemical immunosensor to quantify patulin in apple juice samples using a glassy carbon electrode modified with GO was developed, with LOD 9.8 pg mL^{-1} [94].

Quantum Dots (QDs)

Quantum dots are composed of atoms from groups II-VI or III-V of the periodic table (mainly are composed of compounds formed by Zn and Cd with Te and Se) and they have particle sizes usually below 10nm. They are semiconductor nanocrystals, which are used as fluorescent labels and fluorescent donor/acceptor [52, 95]. They have a diameter that ranges between 1-20 nm. QDs are significantly resistant to photo-bleaching and chemical degradation, they have a high quantum yield, high molar extinction coefficients, broad absorption spectra, as well as narrow and symmetric emission bands (30-50 nm) [95]. QDs have been used in the biosensing field for OTA detection as fluorescence labels [96].

Quantum dots nanobeads (QDNBs) have higher fluorescence intensity and stronger tolerance than QDs, and they are prepared by self-assembly of high-quality CdSe/ZnS quantum dots (QDs) [94]. QDNBs were used to develop an immunochromatographic test strip for simultaneous quantitative detection of fumonisin B1, DON, and ZEA in grains [97] and for the simultaneous qualitative detection of ZEN, OTA and fumonisin B1 (FB1) in corn samples [94].

Nanowires

Nanowires result from the hybridization of semiconductor nanoparticles and biomolecules to be used in a nanobiosensor [23]. Nanowires belong to the growing family of nanomaterials, which also include nanotubes, nanoparticles and nanorods [98]. Nanowires can be used as immobilization platforms in aptasensors [99], as signal molecules and as aptamer nanocarrier [100]. Tellurium nanowires were used as sacrificial templates to synthesize *in-situ* porous platinum nanotubes (p-PtNTs). After that, an electrochemical ZEN aptasensor was contacted as composite prepared from p-PtNTs/AuNPs and thionine-labelled GO, achieving a LOD of 0.17 pg·mL^{-1} [101]

Nanorods

Gold nanorods (GNRs) have received much attention due to their geometric anisotropy and unique optical properties. In particular, GNRs exhibit tunable longitudinal surface plasmon resonance (SPR) depending on the GNR aspect ratio. A new optical aptasensor based on GNR side-by-side assembly for one-step determination of ochratoxin A was used by Hu *et al.* achieving a limit of detection of 0.22 ng/mL. Landry and their co-workers synthesized zinc oxide (ZnO) nanorod sensors to detect the fungus *Aspergillus fumigatus,* among many other fungi and bacteria. The design of this sensor was based on changing the conductivity of ZnO nanorods in contact with fungi and bacteria [102].

Nanodiagnostic Kits with Nanomaterials-based Immunoassays for Mycotoxin Detection in the Field

Real-time diagnostic techniques can contribute to the early diagnosis of phytopathogens in the field. In an environment with active changes in plant pathogens and many climate change issues (*e.g.*, high temperature and modified rainfall amount and distribution), the early detection of phytopathogens with the help of sensors plays a prominent role [67, 103]. A diagnostic kit can be used in the field to detect potential pathogens that may contaminate the crop. Rapid detection can inhibit the growth of microorganisms, as detection takes place in real-time [13].

Various emerging nanomaterials (*e.g.*, metal nanomaterials, carbon nanomaterials, magnetic particles, silica NPs, and quantum dots) have caused great interest and have been widely used to improve immunoassay efficiencies, especially sensitivity [60]. In addition, nanomaterials-based immunoassays have achieved to become largely applicable in the detection of mycotoxins [23, 52, 104 - 108].

For instance, a multiplex dipstick immunoassay was developed by Lattanzio *et al.* for real-time detection of major *Fusarium* toxins, namely zearalenone, T-2 and HT-2 toxins, DON and FBs in wheat, oats and maize at or below their respective EU maximum residue limits, proving to be a very fast, cheap and easy-to-use technique for detecting mycotoxin contamination in cereals [109]. In a very recent study, Huang *et al.* developed an Au nanosphere (AuNSs)-based immuno-chromatographic test strip (ICSs) for the simultaneous detection of DON and FB1 in grain, which exhibited a detection limit of 20.0 ng/mL for both mycotoxins [110]. Table 1 presents recent studies of the use of nanobiotechnology on the detection of mycotoxins in food.

Table 1. Recent studies of the use of nanobiotechnology on detection mycotoxins on food.

Mycotoxin	Food	Detection Method	Detection Strategy	Results	Limit of Detection	Recovery Rate	References
Citrinin (CIT) Zearalenone (ZEN)	Spiked corn	Dual fluorescent immunochromatographic assay (DF-ICA) based on europium nanoparticles (EuNPs)	Formation of the DF-ICA probe using two specific monoclonal antibodies	Effective, short reaction time (25 min), flexible, hypersensitivity	0.06 ng/mL 0.11 ng/mL	86.6–114.4%	[111]
Aflatoxin M1 (AFM1)	Milk samples	Palladium nanoparticles (PdNPs) - Based fluorescence resonance energy transfer aptasensor	Excellent fluorescence quenching ability of PdNPs towards 5-carboxyfluorescein (FAM) with negligible non-specific fluorescence quenching	Highly sensitive, simple configuration	4.6 pM	NA**	[112]
Ochratoxin A (OTA)	Spiked wine	Gap-electrical homogeneous aptasensor based on Au@Fe$_3$O$_4$ nanocomposites coupled with a bi-enzyme-aided system	Degradation of DNA on the surface of Au@Fe$_3$O$_4$ nanocomposites and formation of a conductive network with high conductivity	Highly sensitive, and selective	0.85 ng/mL	91.8–115.6%	[113]

(Table 1) cont.....

Mycotoxin	Food	Detection Method	Detection Strategy	Results	Limit of Detection	Recovery Rate	References
Patulin (PAT)	Spiked apple juice/apple wine	Nb.BbvCI* powered DNA walking machine-based Zr-MOFs (Metal-organic frameworks)-labelled electro-chemical aptasensor using Pt@AuNRs/Fe-MOFs/PEI-rGO (Polyethyleneimine-reduced graphene oxide) as electrode modification material	PAT takes the aptamer away from the DNA walking chain (wDNA), and wDNA can repeatedly dissociate the MB@Zr-MOFs-cDNA under the Nb.BbvCI power to amplify the detection signal	Satisfactory detection range, detection limit, specificity, long-term stability	4.14×10^{-5} ng/mL	87–101%	[114]
Citrinin (CIT) Aflatoxin B1 (AFB1) Ochratoxin A (OTA)	Aqueous solutions/ methanolic extract of real foodstuff	Ultrasensitive visual detection using yellow-light emitting Carbon dot (CD) and Congo red dye	The higher binding affinity of citrinin with diammonium citrate and urea derived Carbon dot in presence of Mg^{2+}	Simple, inexpensive, robust and highly reproducible no need for any equipment or trained personnel	6.6-10 pmol/mL	75%	[115]
Ochratoxin A	Spiked malt	Electrochemical immunosensor modified with the self-assembled monolayer (SEM)	Ultra-sensitive detection of OTA based on indirect competitive principle using the Differential Pulse Voltammetry (DPV) method	Highly sensitive	0.08 ng/mL	71.77–97.18%	[116]
Fumonisin B1 (FB1) Zearalenone (ZEN) Ochratoxin A (OTA) Aflatoxin B1 (AFB1)	Spiked corn	Multicolour immunochromatographic test strip (ICTS) nanosensor	Four-color AuNPs synthesis through an AuNP morphology	Size-engineering strategy, high sensitivity and excellent specificity of the developed polychrome nanosensors ICTS for individual and simultaneous detection, rapid diagnostic, sensitive detection of multiple target analytes	3.27 ng/mL 0.70 ng/mL 0.10 ng/mL 0.06 ng/mL	88.36–112.49% 88.28–104.68% 82.36–116.23% 85.32–112.19%	[74]
Fumonisins (FBs)	Spiked maize	Nanomagnetic beads (NMBs) and antibody-biotin-streptavidin-horseradish peroxidase (HRP) sensor	Amplified signal based on NMB solid-phase detector and an antibody-biotin-streptavidin-HRP sensor	Fast analysis (22 minutes) without preprocessing and handling process, rapid and sensitive detection	<0.21 ng/mL	100.6–107.3%	[117]
Aflatoxin M1 (AFM1)	Spiked milk	Flexible dispense-Printed electrochemical immunosensor	Single-walled carbon nanotube (SWCNTs) electrodes coated with specific antibodies were used to make flexible biosensors	Inexpensive, fast and easy to use, suitable for application at milk collection points or milk processing lines	0.02 μg/L	NA**	[118]
Patulin (PAT)	Complex media	Mimetic Ag nanoparticle/Zn-based MOF nanocomposite (AgNPs@ZnMOF) capped with a molecularly imprinted polymer	The combination between molecularly imprinted polymer and outstanding peroxidase-like activity of novel AgNPs@ZnMOF nanocomposite were used for fluorescent sensing	Good sensitivity	0.06 μM	NA**	[119]

Nb.BbvCI*-nicking endonuclease
NA**-Not available

Electronic Nose (E-nose)

Electronic nose technology has developed rapidly in recent years and can offer fast detection, repeatability, high performance and evaluation range. An electronic nose system consists of a gas sensor panel which, after testing, collects the taste information of the sample and the data is sent to a pattern recognition system. The pattern recognition system, after processing the data, extracts the scan results for specific properties of the samples [120]. A unique fingerprint is created for each food, which is characteristic of its aroma and taste [121, 122]. The electronic nose mimics the sensation of mammalian odour-producing a complex response unique to each odour [123], can detect volatile compounds at low concentrations, sometimes not noticeable by the human nose [124].

Moreover, electronic nose technology is a reliable tool and is a non-invasive detection technique [28], that is used to detect microbial infections by toxic fungi and mycotoxins in the food industry [125 - 127]. Nose technology consists of many sensors (electrochemical, mass sensors, piezoelectric sensors), converting audio signals into electrical signals used for the analysis of volatile organic compounds (VOCs) [125]. VOCs are metabolic products of fungi and are formed in both primary and secondary metabolism as side-products [128]. It is also known that e-nose sensors are nonspecific and non-selective sensors but can be highly sensitive, as they are responding to a range of different compounds [129], interacting with volatile compounds present in the headspace of a sample [130].

A new, fast, non-invasive and easy-to-use method was developed by Lippolis *et al.* using an e-nose method based on oxide metal sensors to detect DON in unprocessed durum wheat. In this study, 525 naturally contaminated wheat were analyzed. The e-nose method was able to distinguish infection levels with a high detection rate (R=82.1%) making it a useful tool for high screening of DON contamination in durum wheat [131]. In addition, a metal oxide semiconductor (MOS)-based electronic nose has been developed to distinguish different levels of DON contaminated wheat bran samples [132] and to achieve prediction of ochratoxin A-producing strains of *Penicillium* on dry-cured meat [133]. Moreover, Ottoboni *et al.* used the e-nose method in combination with lateral flow immunoassays to rapidly detect AFs and FBs in 161 samples of single infection and co-infected maize. The assessment was performed using discriminant function analysis (DFA). The overall cross-validation rate of a sample correctly classified by the DFA model for AFs was 81% while for FBs 85% made the electronic nose a rapid tool for the detection of AFs and FBs or both in maize kernel contamination [129].

In addition, electronic noses can distinguish between mouldy and non moldy samples. The electronic nose has been used to classify oat samples as mouldy, weakly musty, and strongly musty, and also to predict ergosterol levels and fungal CFU in oats, rye, barley, and wheat [134]. OTA, Citrinin (CIT), and ergosterol production in wheat have also been detected by the electronic nose [135]. Electronic noses have been used for the early detection of different types of indigenous mould contamination in green coffee [136], and for early discrimination and growth tracking of *Aspergillus spp.* contamination in rice kernels [7]. E-nose techniques have been used for the detection of fungal infection and they are based on identifying specific volatile organic compounds (VOCs) associated with the growth of fungi on cereal grains. As chemical changes in the composition of VOCs are caused, a different biochemical pattern is created by the growth of fungi which is directly related to the concentration of mycotoxins [130].

In fresh produce, the application of e-nose techniques has been also successfully implemented for the detection of fungi that produce mycotoxins. A conductive polymer-based electronic nose is constructed for early detection of *Penicillium digitatum* in post-harvest oranges [124]. In a very recent study, Jia *et al.* detected and recognized mouldy apples inoculated with *Penicillium expansum* and *Aspergillus niger*, with the use of an electronic nose. In addition, the same electronic nose system was able to distinguish apples inoculated with different moulds, proving that this device is suitable for fast, cheap and non-destructive fungal detections [120]. The combination of the two sensing techniques, hyperspectral imaging and electronic nose was implemented for the evaluation of fungal contamination in strawberries during decay [137]. Fungi contamination from common spoilage fungi, such as *Botrytis cinerea*, *Monilinia fructicola* and *Rhizopus stolonifer* was evaluated by e-nose with the analysis of volatile compounds generated in the fungi-inoculated peaches [138]. Table **2**. presents some applications of electronic nose and electronic tongue on mycotoxins and mycotoxigenic fungi detection on food and feed.

Table 2. Application of electronic nose and electronic tongue on mycotoxins and mycotoxigenic fungi detection on food and feed.

Mycotoxin/fungi	Food	Detection method	Result/Limit of detection	Computational analysis	References
Penicillium expansum and *Aspergillus niger*	Fresh and mouldy Golden Delicious apples	Gas sensors of the PEN3 electronic nose system	Prediction performance 72.0-96.3%	Backpropagation neural network (BPNN)	[120]
Deoxynivalenol (DON)	Ground wheat	E-nose based on metal oxide (MOS) sensors	Recognition percentage 82.1%	Discriminant function analysis (DFA)	[131]

(Table 2) cont.....

Mycotoxin/fungi	Food	Detection method	Result/Limit of detection	Computational analysis	References
Aflatoxin and Fumonisin	Maize	E-nose in combination with lateral flow immunoassays (LFIAs)	Recognition percentage Co-contaminated maize 67% Regular contaminated maize 65% Single contaminated maize 61%	Discriminant function analysis (DFA)	[129]
Aspergillus sp.	Spiked brown rice grain	E-nose coupled with chemometrics	Coefficient of determination, R^2 = 0.969 Root mean square error, RMSE = 0.31 Log CFU/g	Principal component analysis (PCA) linear discriminant analysis (LDA) and support vector machine (SVM)	[139]
Ochratoxin A (OTA)	*Aspergillus carbonarius* cultured grape-based medium	E-nose technology and GC–MS	Excellent prediction performance on the accumulation of OTA	Partial least squares-discriminant analysis (PLS-DA) using GC–MS volatile data and e-nose signal data	[125]
Deoxynivalenol (DON)	Spiked durum wheat whole-grain	E-nose equipped with metal-oxide-semiconductor (MOS) sensors	Prediction error rate 0-3.28% at: non-contaminated wheat samples contaminated wheat samples below the EC limit (DON < 1,750 µg kg⁻¹) contaminated above the EC limit (DON > 1,750 µg kg⁻¹)	Principal component analysis (PCA)	[140]
Fusarium poae	Contaminated wheat	SPME-GC-MS and e-nose	Very good discrimination capability	Principal component analysis (PCA)	[141]
Ochratoxin A (OTA) Deoxynivalenol (DON)	Barley grains	E-nose and gas chromatography combined with mass spectrometry (GC-MS)	5 mg kg⁻¹	Principal component analysis (PCA) and partial least squares (PLS)	[135]
Aflatoxins (AFs)	Contaminatedmaize	E-nose equipped with 10 metal-oxide-semiconductor (MOS)	100% classification of maize contaminated and non-contaminated with aflatoxins	Principal component analysis (PCA) and linear discriminant analysis (LDA)	[142]

(Table 2) cont.....

Mycotoxin/fungi	Food	Detection method	Result/Limit of detection	Computational analysis	References
Aflatoxin B1(AFB1) Ochratoxin A (OTA)	Extract of non-infected corn seeds	E- tongue based on nanocrystalline silicon oxide (nc-SiO$_2$) immunosensors with anti-AFB1 and anti-OTA antibodies	0.1 fg mL^{-1}	Principal component analysis (PCA) and partial least squares discriminate analysis (PLS-DA)	[143]
Aflatoxin B1 (AFB1) Aflatoxin B2(AFB2) Ochratoxin A (OTA)	Extract of non-infected corn seeds	E- tongue based on nanocrystalline silicon oxide-based impedance immunosensors	0.1 fg mL^{-1}	Incremental fuzzy algorithm	[144]

Electronic Tongue (E- tongue)

Electronic tongue is a set of multiple sensors based on non-specific or low-selectivity detection units. Using electroanalytical methods, it exploits their transverse sensitivity to identify a liquid medium. Multifactorial statistical techniques are required for qualitative and/or quantitative analysis as well as an increased amount of data. The term was coined in 2005 by the International Association of Pure and Applied Chemistry (IUPAC), with different detection methods typically based on voltammetric, potentiometric, detection and biosensor detection [145]. The electronic tongue is considered a model analogous to e-nose. Their difference is mainly in the analysis of liquids rather than in the analysis of gaseous samples [146].

The use of nanocrystalline silicon oxide (nc-SiO$_2$) immunosensor was suggested to detect a single food toxin from a mixture of other toxins. The limit of detection was 0.1 fg/mL by impedance spectroscopy method but there was no possibility of detecting multiple toxins. Recently, Ghosh *et al.* developed the silicon oxide nanocrystalline (nc-SiO$_2$) based on an e-tongue array for the simultaneous detection of multiple toxins with hypomolecular sensitivity. AFB1 and OTA were spiked into corn seeds and the optimized e-tongue system was able to quantify 0.1 fg/mL from AFB1 and OTA [143]. The same research group, in another study, improved the quantification accuracy of multiple food toxins by more than 90% using an upgraded e-tongue system capable of quantifying 0.1 fg/mL AFB1 and OTA with an error of only 10% and 20%, respectively, which is a remarkable achievement in the field of detection of food toxins [143] making the e-tongue method, sensitive, low cost, with minimal dependence on the operator and with possibilities for further development [143 - 146].

Recently, Xu *et al.* developed the generation of a dual fluorescent immuno-chromatic assay (DF-ICA) using a conjugate of europium nanoparticles (EuNPs) with two monoclonal antibodies (mAb) was examined. EuNPs offer high fluorescence, long fluorescence life and non-toxic effect on the sample taken.

A competitive immunochromatographic assay based on EuNPs-mAb allowed the detection and quantification under optimal conditions of the mycotoxins CIT and ZEN in spiked maize samples. The limits of detection (LODs) of CIT and ZEN were 0.06 and 0.11 ng/mL and the recovery rates were from 86.3% to 111.6% and from 86.6% to 114.4%, respectively. In addition, DF-ICA was validated by HPLC assays to obtain good consistency [111].

Jin *et al.* developed a new dual near-infrared fluorescence-based lateral flow immunosensor for the determination of ZEN and DON in maize using anti-zearalenone and anti-deoxynivalenol antibodies as detection reagents. The low detection limits (0.55 μg/kg and 3.8 μg/kg) as well as the high recovery rate (81.7% to 107.3%) of ZEN and DON after a spike in maize, indicating that the developed immunosensor provides effective accuracy in detecting mycotoxins in agricultural products [147].

CONCLUSION

Naturally occurring mycotoxins in foodstuffs cause, in addition to safety issues, a great economic concern. Much effort is devoted to the development of novel, rapid, inexpensive, simple and sensitive multiple mycotoxin screening methods. Mycotoxins are the most important contaminants in foods and feeds, so there is a strong need for continuous and onsite monitoring and it is necessary to detect them to ensure the safety and quality of the foods, to protect humans' health and to improve food trade. The requirement to implement well-developed techniques for mycotoxins detection is an urgent challenge, especially in foods that are complex samples, which may be contaminated with more than one mycotoxin. The emerging nanotechnology has led to the discovery of novel nanomaterials for the detection of mycotoxins and mycotoxigenic fungi, providing fast, low cost and promising approaches to improve the sensitivity and simplify the detection. Several nanotechnology-enabled detection sensors have been developed for mycotoxins and mycotoxigenic fungi detection. Emerging nanomaterials are incorporated with novel recognition elements such as antibodies, peptides, or aptamers and are used in various sensitive detection sensors or lateral flow formats in diagnostic approaches. The simultaneous detection and identification of several mycotoxins in one single test, reducing time and costs per analysis, is one of the most attractive options. In addition, for rapid and accurate diagnosis, e-nose and e- tongue seem to be promising screening methods to detect mycotoxins and mycotoxigenic fungi as early detection is an essential control measure for ensuring storage longevity and food safety. As the Hazard Analysis and Critical

Control Point (HACCP) concept requires mycotoxins detection at each level of food production, the use of nanobiotechnological devices for their early detection could significantly contribute to safeguarding consumers' health, ensuring safe and healthy food and feed.

CONSENT FOR PUBLICATION

Not applicable.

CONFLICT OF INTEREST

The authors declare no conflict of interest, financial or otherwise.

ACKNOWLEDGEMENT

Declared none.

REFERENCES

[1]   Afsah-Hejri L, Hajeb P, Ehsani RJ. Application of ozone for degradation of mycotoxins in food: A review. Compr Rev Food Sci Food Saf 2020; 19(4): 1777-808.
[http://dx.doi.org/10.1111/1541-4337.12594] [PMID: 33337096]

[2]   Agriopoulou S, Stamatelopoulou E, Varzakas T. Advances in occurrence, importance, and mycotoxin control strategies : Prevention and detoxification in foods. Foods 2020; 9(2): 137.
[http://dx.doi.org/10.3390/foods9020137]

[3]   Ayofemi Olalekan Adeyeye S. Aflatoxigenic fungi and mycotoxins in food: a review. Crit Rev Food Sci Nutr 2020; 60(5): 709-21.
[http://dx.doi.org/10.1080/10408398.2018.1548429] [PMID: 30689400]

[4]   Thipe VC, Keyster M, Katti KV. Sustainable nanotechnology: Mycotoxin detection and protection. In: Abd-Elsalam KA, Prasad R, Eds. Nanobiotechnology Applications in Plant Protection, Nanotechnology in the Life Sciences. Springer 2018; pp. 323-49.
[http://dx.doi.org/10.1007/978-3-319-91161-8_12]

[5]   Agriopoulou S. Physicochemical study using liquid chromatography technique on the effect of ozone on aflatoxins' degradation and elimination, in pure substrates and foods 2015.

[6]   Varzakas T. Quality and safety aspects of cereals (Wheat) and their products. Crit Rev Food Sci Nutr 2016; 56(15): 2495-510.
[http://dx.doi.org/10.1080/10408398.2013.866070] [PMID: 25830822]

[7]   Gu S, Wang J, Wang Y. Early discrimination and growth tracking of *Aspergillus* spp. contamination in rice kernels using electronic nose. Food Chem 2019; 292: 325-35.
[http://dx.doi.org/10.1016/j.foodchem.2019.04.054] [PMID: 31054682]

[8]   Agriopoulou S, Stamatelopoulou E, Varzakas T. Advances in analysis and detection of major mycotoxins in foods. Foods 2020; 9(4): 1-23.
[http://dx.doi.org/10.3390/foods9040518] [PMID: 32326063]

[9]   Agriopoulou S. Enniatins : An Emerging Food Safety Issue. EC Nutr 2016; 3: 1142-6.

[10]   Berthiller F, Maragos C, Dall'Asta C. Introduction to masked mycotoxins. In: DallAsta C, Berthiller F, Eds. Masked mycotoxins in food: Formation, occurrence and toxicological relevance. Royal Society of Chemistry 2016; pp. 1-13.

[11]   Freire L, Sant'Ana AS. Modified mycotoxins: An updated review on their formation, detection, occurrence, and toxic effects. Food Chem Toxicol 2018; 111: 189-205.
[http://dx.doi.org/10.1016/j.fct.2017.11.021] [PMID: 29158197]

[12]   Agriopoulou S, Koliadima A, Karaiskakis G, Kapolos J. Kinetic study of aflatoxins' degradation in the presence of ozone. Food Control 2016; 61: 221-6.
[http://dx.doi.org/10.1016/j.foodcont.2015.09.013]

[13]   Ingle AP, Gupta I, Jogee P, Rai M. Role of nanotechnology in the detection of mycotoxins: A smart approach. In: Rai M, Abd-Elsalam K, Eds. Nanomycotoxicology Treating nycotoxins in the nano way. Elsevier 2019; pp. 11-33.

[14]   Mousavi Khaneghah A, Fakhri Y, Gahruie HH, Niakousari M, Sant'Ana AS. Mycotoxins in cereal-based products during 24 years (1983–2017): A global systematic review. Trends Food Sci Technol 2019; 91: 95-105.
[http://dx.doi.org/10.1016/j.tifs.2019.06.007]

[15]   Sobral MMC, Faria MA, Cunha SC, Ferreira IMPLVO. Toxicological interactions between mycotoxins from ubiquitous fungi: Impact on hepatic and intestinal human epithelial cells. Chemosphere 2018; 202: 538-48.
[http://dx.doi.org/10.1016/j.chemosphere.2018.03.122] [PMID: 29587235]

[16]   Abd-Elsalam KA, Hashim AF, Alghuthaymi MA, Said-Galiev E. Nanobiotechnological strategies for toxigenic fungi and mycotoxin control. Food Preserv 2017; pp. 337-64.
[http://dx.doi.org/10.1016/B978-0-12-804303-5.00010-9]

[17]   Haque MA, Wang Y, Shen Z, Li X, Saleemi MK, He C. Mycotoxin contamination and control strategy in human, domestic animal and poultry: A review. Microb Pathog 2020; 142: 104095.
[http://dx.doi.org/10.1016/j.micpath.2020.104095] [PMID: 32097745]

[18]   Kunz BM, Wanko F, Kemmlein S, Bahlmann A, Rohn S, Maul R. Development of a rapid multi-mycotoxin LC-MS/MS stable isotope dilution analysis for grain legumes and its application on 66 market samples. Food Control 2020; 109: 106949.
[http://dx.doi.org/10.1016/j.foodcont.2019.106949]

[19]   Tittlemier S, Cramer B, Dall'Asta C, *et al.* Developments in mycotoxin analysis: An update for 2017–2018. World Mycotoxin J 2019; 12(1): 3-29.
[http://dx.doi.org/10.3920/WMJ2018.2398]

[20]   Zhang L, Dou XW, Zhang C, Logrieco AF, Yang MH. A review of current methods for analysis of mycotoxins in herbal medicines. Toxins (Basel) 2018; 10(2): 10.
[http://dx.doi.org/10.3390/toxins10020065] [PMID: 29393905]

[21]   Rhouati A, Bulbul G, Latif U, Hayat A, Li ZH, Marty JL. Nano-aptasensing in mycotoxin analysis: Recent updates and progress. Toxins (Basel) 2017; 9(11): 1-23.
[http://dx.doi.org/10.3390/toxins9110349] [PMID: 29143760]

[22]   Zhang W, Tang S, Jin Y, *et al.* Multiplex SERS-based lateral flow immunosensor for the detection of major mycotoxins in maize utilizing dual Raman labels and triple test lines. J Hazard Mater 2020; 393: 122348.
[http://dx.doi.org/10.1016/j.jhazmat.2020.122348] [PMID: 32143157]

[23]   Rai M, Jogee PS, Ingle AP. Emerging nanotechnology for detection of mycotoxins in food and feed. Int J Food Sci Nutr 2015; 66(4): 363-70.
[http://dx.doi.org/10.3109/09637486.2015.1034251] [PMID: 26001087]

[24]   Goud KY, Reddy KK, Satyanarayana M, Kummari S, Gobi KV. A review on recent developments in optical and electrochemical aptamer-based assays for mycotoxins using advanced nanomaterials. Microchim Acta 2020; p. 187.

[25]   ISO. Microbiology of food and animal feeding stuffs — Horizontal method for the enumeration of microorganisms — Colony-count technique at 30 °C. 2003; 3: 1-9.

[26] Marinach-Patrice C, Lethuillier A, Marly A, *et al.* Use of mass spectrometry to identify clinical *Fusarium* isolates. Clin Microbiol Infect 2009; 15(7): 634-42.
[http://dx.doi.org/10.1111/j.1469-0691.2009.02758.x] [PMID: 19456834]

[27] Kashyap PL, Kumar S, Jasrotia P, Singh DP, Singh GP. Nanosensors for plant disease diagnosis: currentunderstanding and future perspectives. In: Pudake R, Chauhan N, Kole C, Eds. Nanoscience for Sustainable Agriculture. Springer 2019; pp. 189-205.
[http://dx.doi.org/10.1007/978-3-319-97852-9_9]

[28] Xing F, Yao H, Liu Y, Dai X, Brown RL, Bhatnagar D. Recent developments and applications of hyperspectral imaging for rapid detection of mycotoxins and mycotoxigenic fungi in food products. Crit Rev Food Sci Nutr 2019; 59(1): 173-80.
[http://dx.doi.org/10.1080/10408398.2017.1363709] [PMID: 28846441]

[29] Xing F, Yao H, Hruska Z. Detecting peanuts inoculated with toxigenic and atoxienic *Aspergillus flavus* strains with fluorescence hyperspectral imagery. Sens Agric Food Qual SafIX 2017; 10217: 1021701.

[30] Singh A, Prasad SM. Nanotechnology and its role in agro-ecosystem: a strategic perspective. Int J Environ Sci Technol 2017; 14(10): 2277-300.
[http://dx.doi.org/10.1007/s13762-016-1062-8]

[31] Rajwade JM, Chikte RG, Paknikar KM. Nanomaterials: new weapons in a crusade against phytopathogens. Appl Microbiol Biotechnol 2020; 104(4): 1437-61.
[http://dx.doi.org/10.1007/s00253-019-10334-y] [PMID: 31900560]

[32] Agriopoulou S. Nanotechnology in Food Packaging. EC Nutr 2016; 42: 118-42.

[33] Rauta PR, Mohanta YK, Nayak D, Mohanta YK, Mohanta TK, Nayak D. Recent developments on nanotechnology in agriculture. Nanotechnol Biol Med 2019; pp. 79-86.

[34] Naseer B, Srivastava G, Qadri OS, Faridi SA, Islam RK, Younis K. Importance and health hazards of nanoparticles used in the food industry. Nanotechnol Rev 2018; 7(6): 623-41.
[http://dx.doi.org/10.1515/ntrev-2018-0076]

[35] McNamee SE, Bravin F, Rosar G, Elliott CT, Campbell K. Development of a nanoarray capable of the rapid and simultaneous detection of zearalenone, T2-toxin and fumonisin. Talanta 2017; 164: 368-76.
[http://dx.doi.org/10.1016/j.talanta.2016.11.032] [PMID: 28107943]

[36] Qamar SA, Asgher M, Khalid N, Sadaf M. Nanobiotechnology in health sciences: Current applications and future perspectives. Biocatal Agric Biotechnol 2019; 22: 101388.
[http://dx.doi.org/10.1016/j.bcab.2019.101388]

[37] Azzouz A, Kailasa SK, Lee SS, *et al.* Review of nanomaterials as sorbents in solid-phase extraction for environmental samples. TrAC -. Trends Analyt Chem 2018; 108: 347-69.
[http://dx.doi.org/10.1016/j.trac.2018.08.009]

[38] Feregrino-Perez AA, Magaña-López E, Guzmán C, Esquivel K. A general overview of the benefits and possible negative effects of the nanotechnology in horticulture. Sci Hortic (Amsterdam) 2018; 238: 126-37.
[http://dx.doi.org/10.1016/j.scienta.2018.03.060]

[39] Santana Oliveira I, da Silva AG Junior, de Andrade CAS, Lima Oliveira MD. Biosensors for early detection of fungi spoilage and toxigenic and mycotoxins in food. Curr Opin Food Sci 2019; 29: 64-79.
[http://dx.doi.org/10.1016/j.cofs.2019.08.004]

[40] Santos AO, Vaz A, Rodrigues P, Veloso ACA, Venâncio A, Peres AM. Thin films sensor devices for mycotoxins detection in foods: Applications and challenges. Chemosensors (Basel) 2019; 7(1): 3.
[http://dx.doi.org/10.3390/chemosensors7010003]

[41] Patel A, Patra F, Shah N, Khedkar C. Application of nanotechnology in the food industry: Present

status and future prospects. Imp of Nanoscience Food Ind 1-27.2018;

[42] Alhamoud Y, Yang D, Fiati Kenston SS, *et al.* Advances in biosensors for the detection of ochratoxin A: Bio-receptors, nanomaterials, and their applications. Biosens Bioelectron 2019; 141: 111418.
[http://dx.doi.org/10.1016/j.bios.2019.111418] [PMID: 31228729]

[43] Economou A, Karapetis SK, Nikoleli GP, Nikolelis DP, Bratakou S, Varzakas TH. Enzyme-based sensors. In: Toldrá F, Leo ML, Eds. Advances in food diagnostics. John Wiley & Sons Ltd 2017; pp. 231-50.
[http://dx.doi.org/10.1002/9781119105916.ch9]

[44] Younis MR, Wang C, Younis MA, Xia X. Use of biosensors for mycotoxins analysis in food stuff. Nanobiosensors 2020; pp. 171-201.
[http://dx.doi.org/10.1002/9783527345137.ch8]

[45] Chauhan N, Jain U, Soni S. Sensors for Food Quality Monitoring. In: Pudake R, Chauhan N, Kole C, Eds. Nanoscience for Sustainable Agriculture. Cham: Springer 2019; pp. 601-26.
[http://dx.doi.org/10.1007/978-3-319-97852-9_23]

[46] Nikoleli GP, Nikolelis DP, Siontorou CG, Karapetis S, Varzakas T. Novel biosensors for the rapid detection of toxicants in foods. Adv Food Nutr Res 2018; 84: 57-102.
[http://dx.doi.org/10.1016/bs.afnr.2018.01.003] [PMID: 29555073]

[47] Khiyami MA, Almoammar H, Awad YM, Alghuthaymi MA, Abd-Elsalam KA. Plant pathogen nanodiagnostic techniques: forthcoming changes? Biotechnol Biotechnol Equip 2014; 28(5): 775-85.
[http://dx.doi.org/10.1080/13102818.2014.960739] [PMID: 26740775]

[48] Omanović-Mikličanin E, Maksimović M. Nanosensors applications in agriculture and food industry. Bull Chem Technol Bosnia Herzegovina 2016; 47: 59-70.

[49] Adeyeye SAO, Fayemi OE. Nanotechnology and food processing: between innovations and consumer safety. J Culin Sci Technol 2019; 17(5): 435-52.
[http://dx.doi.org/10.1080/15428052.2018.1476276]

[50] Devi Lamabam S, Thangjam R. Emerging trends in the application of nanobiosensors in the food industry. Novel Approaches of nanotechnology in Food. Elsevier 2016; pp. 663-96.
[http://dx.doi.org/10.1016/B978-0-12-804308-0.00019-4]

[51] Munawar V, Garcia-Cruz A, Majewska M, Karim K, Kutner W, Piletsky SA. Sensors and Actuators B : Chemical Electrochemical determination of fumonisin B1 using a chemosensor with a recognition unit comprising molecularly imprinted polymer nanoparticles. Sens Actuators B Chem 2020; 321: 128552.
[http://dx.doi.org/10.1016/j.snb.2020.128552]

[52] Xue Z, Zhang Y, Yu W, *et al.* Recent advances in aflatoxin B1 detection based on nanotechnology and nanomaterials-A review. Anal Chim Acta 2019; 1069: 1-27.
[http://dx.doi.org/10.1016/j.aca.2019.04.032] [PMID: 31084735]

[53] Ragavan KV, Rastogi NK. Graphene-copper oxide nanocomposite with intrinsic peroxidase activity for enhancement of chemiluminescence signals and its application for detection of Bisphenol-A. Sens Actuators B Chem 2016; 229: 570-80.
[http://dx.doi.org/10.1016/j.snb.2016.02.017]

[54] Rashidi L, Khosravi-Darani K. The applications of nanotechnology in food industry. Crit Rev Food Sci Nutr 2011; 51(8): 723-30.
[http://dx.doi.org/10.1080/10408391003785417] [PMID: 21838555]

[55] Sharma A, Kumar A, Khan R. Electrochemical immunosensor based on poly (3,4-ethylenedioxythiophene) modified with gold nanoparticle to detect aflatoxin B_1. Mater Sci Eng C 2017; 76: 802-9.
[http://dx.doi.org/10.1016/j.msec.2017.03.146] [PMID: 28482593]

[56] Etefagh R, Azhir E, Shahtahmasebi N. Synthesis of CuO nanoparticles and fabrication of

nanostructural layer biosensors for detecting *Aspergillus niger* fungi. Sci Iran 2013; 20: 1055-8.

[57]   Azhir E, Etefagh R, Shahtahmasebi N, Mohammadi M, Amiri D, Sarhaddi R. *Aspergillus niger* biosensor based on tin oxide (SnO$_2$) nanostructures: nanopowder and thin film. Indian J Sci Technol 2012; 5(7): 3010-2.
[http://dx.doi.org/10.17485/ijst/2012/v5i7.14]

[58]   Lau HY, Wang Y, Wee EJH, Botella JR, Trau M. Field demonstration of a multiplexed point-of-care diagnostic platform for plant pathogens. Anal Chem 2016; 88(16): 8074-81.
[http://dx.doi.org/10.1021/acs.analchem.6b01551] [PMID: 27403651]

[59]   Evtugyn G, Hianik T. Aptamer-based biosensors for mycotoxin detection. Nanomycotoxicology. Elsevier 2019; pp. 35-70.

[60]   Zhang X, Li G, Wu D, Liu J, Wu Y. Recent advances on emerging nanomaterials for controlling the mycotoxin contamination : From detection to elimination. Food Frontiers 2020; pp. 1-22.

[61]   Tian F, Zhou J, Jiao B, He Y. A nanozyme-based cascade colorimetric aptasensor for amplified detection of ochratoxin A. Nanoscale 2019; 11(19): 9547-55.
[http://dx.doi.org/10.1039/C9NR02872B] [PMID: 31049533]

[62]   Sun S, Zhao R, Feng S, Xie Y. Colorimetric zearalenone assay based on the use of an aptamer and of gold nanoparticles with peroxidase-like activity. Mikrochim Acta 2018; 185(12): 535.
[http://dx.doi.org/10.1007/s00604-018-3078-x] [PMID: 30406298]

[63]   Taghdisi SM, Danesh NM, Beheshti HR, Ramezani M, Abnous K. A novel fluorescent aptasensor based on gold and silica nanoparticles for the ultrasensitive detection of ochratoxin A. Nanoscale 2016; 8(6): 3439-46.
[http://dx.doi.org/10.1039/C5NR08234J] [PMID: 26791437]

[64]   Goud KY, Kailasa SK, Kumar V, *et al.* Progress on nanostructured electrochemical sensors and their recognition elements for detection of mycotoxins: A review. Biosens Bioelectron 2018; 121: 205-22.
[http://dx.doi.org/10.1016/j.bios.2018.08.029] [PMID: 30219721]

[65]   Yang X, Qian J, Jiang L, *et al.* Ultrasensitive electrochemical aptasensor for ochratoxin A based on two-level cascaded signal amplification strategy. Bioelectrochemistry 2014; 96: 7-13.
[http://dx.doi.org/10.1016/j.bioelechem.2013.11.006] [PMID: 24355136]

[66]   Rahi S, Choudhari P, Chormade V. Aflatoxin and ochratoxin A detection: Traditional and currentmethods. In: Satyanarayana T, Deshmukh SK, Deshpande MV, Eds. Advancing Frontiers in Mycology and Mycotechnology. Springer 2019; pp. 377-404.

[67]   Shoala T. Carbon nanostructures: detection, controlling plant diseases and mycotoxins. Carbon nanomaterials for agri food and environmental applications. Elsevier 2020; pp. 261-77.
[http://dx.doi.org/10.1016/B978-0-12-819786-8.00013-X]

[68]   Dos Santos CA, Ingle AP, Rai M. The emerging role of metallic nanoparticles in food. Appl Microbiol Biotechnol 2020; 104(6): 2373-83.
[http://dx.doi.org/10.1007/s00253-020-10372-x] [PMID: 31989225]

[69]   Aldewachi H, Chalati T, Woodroofe MN, Bricklebank N, Sharrack B, Gardiner P. Gold nanoparticle-based colorimetric biosensors. Nanoscale 2017; 10(1): 18-33.
[http://dx.doi.org/10.1039/C7NR06367A] [PMID: 29211091]

[70]   Hossain MZ, Maragos CM. Gold nanoparticle-enhanced multiplexed imaging surface plasmon resonance (iSPR) detection of Fusarium mycotoxins in wheat. Biosens Bioelectron 2018; 101: 245-52.
[http://dx.doi.org/10.1016/j.bios.2017.10.033] [PMID: 29096362]

[71]   Liu Y, Yu J, Wang Y, Liu Z, Lu Z. An ultrasensitive aptasensor for detection of Ochratoxin A based on shielding effect-induced inhibition of fluorescence resonance energy transfer. Sens Actuators B Chem 2016; 222: 797-803.
[http://dx.doi.org/10.1016/j.snb.2015.09.007]

[72] Sabet FS, Hosseini M, Khabbaz H, Dadmehr M, Ganjali MR. FRET-based aptamer biosensor for selective and sensitive detection of aflatoxin B1 in peanut and rice. Food Chem 2017; 220: 527-32.
[http://dx.doi.org/10.1016/j.foodchem.2016.10.004] [PMID: 27855935]

[73] Sojinrin T, Liu K, Wang K, *et al.* Developing gold nanoparticles-conjugated aflatoxin B1 antifungal strips. Int J Mol Sci 2019; 20(24): 1-16.
[http://dx.doi.org/10.3390/ijms20246260] [PMID: 31842251]

[74] Wu Y, Zhou Y, Huang H, *et al.* Engineered gold nanoparticles as multicolor labels for simultaneous multi-mycotoxin detection on the immunochromatographic test strip nanosensor. Sens Actuators B Chem 2020; 316: 128107.
[http://dx.doi.org/10.1016/j.snb.2020.128107]

[75] Yang Y, Li G, Wu D, *et al.* Recent advances on toxicity and determination methods of mycotoxins in foodstuffs. Trends Food Sci Technol 2020; 96: 233-52.
[http://dx.doi.org/10.1016/j.tifs.2019.12.021]

[76] Li N, Du J, Wu D, *et al.* Recent advances in facile synthesis and applications of covalent organic framework materials as superior adsorbents in sample pretreatment. TrAC -. Trends Analyt Chem 2018; 108: 154-66.
[http://dx.doi.org/10.1016/j.trac.2018.08.025]

[77] Hashemi B, Zohrabi P, Shamsipur M. Recent developments and applications of different sorbents for SPE and SPME from biological samples. Talanta 2018; 187: 337-47.
[http://dx.doi.org/10.1016/j.talanta.2018.05.053] [PMID: 29853056]

[78] Mashhadizadeh MH, Amoli-Diva M, Pourghazi K. Magnetic nanoparticles solid phase extraction for determination of ochratoxin A in cereals using high-performance liquid chromatography with fluorescence detection. J Chromatogr A 2013; 1320: 17-26.
[http://dx.doi.org/10.1016/j.chroma.2013.10.062] [PMID: 24210301]

[79] Hashemi M, Taherimaslak Z, Rashidi S. Application of magnetic solid phase extraction for separation and determination of aflatoxins B1 and B2 in cereal products by high performance liquid chromatography-fluorescence detection. J Chromatogr B Anal Technol. J Chromatogr B Analyt Technol Biomed Life Sci 2014; 960: 200-8.
[http://dx.doi.org/10.1016/j.jchromb.2014.03.035] [PMID: 24814005]

[80] Taherimaslak Z, Amoli-Diva M, Allahyary M, Pourghazi K. Magnetically assisted solid phase extraction using Fe_3O_4 nanoparticles combined with enhanced spectrofluorimetric detection for aflatoxin M1 determination in milk samples. Anal Chim Acta 2014; 842: 63-9.
[http://dx.doi.org/10.1016/j.aca.2014.05.007] [PMID: 25127653]

[81] Karami-Osboo R, Mirabolfathi M. A Novel dispersive nanomagnetic particle solid-phase extraction method to determine aflatoxins in nut and cereal samples. Food Anal Methods 2017; 10(12): 4086-93.
[http://dx.doi.org/10.1007/s12161-017-0975-2]

[82] Sertova N, Ignatova M. Detection of mycotoxins through nanobiosensors based on nanostructured materials : A review. Int J Biosci 2019; 6655: 106-15.

[83] Wang Y, Xia Y. Optical, electrochemical and catalytic methods for in-vitro diagnosis using carbonaceous nanoparticles: a review. Mikrochim Acta 2019; 186(1): 50.
[http://dx.doi.org/10.1007/s00604-018-3110-1] [PMID: 30612201]

[84] Yadav J. Fullerene: Properties, synthesis and application. Res Rev J Phys 2017; 6: 1-6.

[85] Iijima S. Helical microtubules of graphitic carbon. Nature 1991; 354(6348): 56-8.
[http://dx.doi.org/10.1038/354056a0]

[86] Zhang S, Shen Y, Shen G, Wang S, Shen G, Yu R. Electrochemical immunosensor based on Pd-Au nanoparticles supported on functionalized PDDA-MWCNT nanocomposites for aflatoxin B1 detection. Anal Biochem 2016; 494: 10-5.
[http://dx.doi.org/10.1016/j.ab.2015.10.008] [PMID: 26521980]

[87] Abnous K, Danesh NM, Alibolandi M, Ramezani M, Taghdisi SM. Amperometric aptasensor for ochratoxin A based on the use of a gold electrode modified with aptamer, complementary DNA, SWCNTs and the redox marker Methylene Blue. Mikrochim Acta 2017; 184(4): 1151-9.
[http://dx.doi.org/10.1007/s00604-017-2113-7]

[88] Lim SY, Shen W, Gao Z. Carbon quantum dots and their applications. Chem Soc Rev 2015; 44(1): 362-81.
[http://dx.doi.org/10.1039/C4CS00269E] [PMID: 25316556]

[89] Wang B, Chen Y, Wu Y, *et al.* Aptamer induced assembly of fluorescent nitrogen-doped carbon dots on gold nanoparticles for sensitive detection of AFB1. Biosens Bioelectron 2016; 78: 23-30.
[http://dx.doi.org/10.1016/j.bios.2015.11.015] [PMID: 26584079]

[90] Deng J, Lu Q, Hou Y, *et al.* Nanosensor composed of nitrogen-doped carbon dots and gold nanoparticles for highly selective detection of cysteine with multiple signals. Anal Chem 2015; 87(4): 2195-203.
[http://dx.doi.org/10.1021/ac503595y] [PMID: 25594515]

[91] Papageorgiou DG, Kinloch IA, Young RJ. Mechanical properties of graphene and graphene-based nanocomposites. Prog Mater Sci 2017; 90: 75-127.
[http://dx.doi.org/10.1016/j.pmatsci.2017.07.004]

[92] Yugender Goud K, Hayat A, Satyanarayana M, *et al.* Aptamer-based zearalenone assay based on the use of a fluorescein label and a functional graphene oxide as a quencher. Mikrochim Acta 2017; 184(11): 4401-8.
[http://dx.doi.org/10.1007/s00604-017-2487-6]

[93] Caglayan MO, Şahin S, Üstündağ Z. Detection strategies of zearalenone for food safety : A review. Crit Rev Anal Chem 2020; 1-20.
[http://dx.doi.org/10.1080/10408347.2020.1797468] [PMID: 32715728]

[94] Duan H, Li Y, Shao Y, Huang X, Xiong Y. Multicolor quantum dot nanobeads for simultaneous multiplex immunochromatographic detection of mycotoxins in maize. Sens Actuators B Chem 2019; 291: 411-7.
[http://dx.doi.org/10.1016/j.snb.2019.04.101]

[95] Xu X, Xu C, Ying Y. Aptasensor for the simple detection of ochratoxin A based on side-by-side assembly of gold nanorods. RSC Advances 2016; 6(56): 50437-43.
[http://dx.doi.org/10.1039/C6RA04439E]

[96] Yao J, Xing G, Han J, *et al.* Novel fluoroimmunoassays for detecting ochratoxin A using CdTe quantum dots. J Biophotonics 2017; 10(5): 657-63.
[http://dx.doi.org/10.1002/jbio.201600005] [PMID: 27243787]

[97] Hou S, Ma J, Cheng Y, Wang H, Sun J, Yan Y. Quantum dot nanobead-based fl uorescent immunochromatographic assay for simultaneous quantitative detection of fumonisin B1, dexyonivalenol, and zearalenone in grains. Food Control 2020; 117: 107331.
[http://dx.doi.org/10.1016/j.foodcont.2020.107331]

[98] Grieshaber D. MacKenzie1 R, Vörös J, Reimhult E. Electrochemical biosensors - Sensor principles and architectures. Sensors (Basel) 2008; 81400-58.

[99] Hayat A, Catanante G, Marty JL. Current trends in nanomaterial-based amperometric biosensors. Sensors (Basel) 2014; 14(12): 23439-61.
[http://dx.doi.org/10.3390/s141223439] [PMID: 25494347]

[100] He B, Yan XA. 'signal-on' voltammetric aptasensor fabricated by hcPt@AuNFs/PEI-rGO and Fe₃O₄NRs/rGO for the detection of zearalenone. Sens Actuators B Chem 2019; 290: 477-83.
[http://dx.doi.org/10.1016/j.snb.2019.04.005]

[101] He B, Yan X. An amperometric zearalenone aptasensor based on signal amplification by using a composite prepared from porous platinum nanotubes, gold nanoparticles and thionine-labelled

graphene oxide. Mikrochim Acta 2019; 186(6): 383.
[http://dx.doi.org/10.1007/s00604-019-3500-z] [PMID: 31140009]

[102] Landry CJ, Burns FP, Baerlocher F, Khashayar G. Novel Solid-State Microbial Sensors Based on ZnO Nanorod Arrays. Adv Funct Mater 2018; 28(19): 1706309.
[http://dx.doi.org/10.1002/adfm.201706309]

[103] Conte G, Fontanelli M, Galli F, Cotrozzi L, Pagni L, Pellegrini E. Mycotoxins in feed and food and the role of ozone in their detoxification and degradation : An Update. Toxins (Basel) 2020; 12(8): 486.
[http://dx.doi.org/10.3390/toxins12080486] [PMID: 32751684]

[104] Chen C, Yu X, Han D, *et al.* Non-CTAB synthesized gold nanorods-based immunochromatographic assay for dual color and on-site detection of aflatoxins and zearalenones in maize. Food Control 2020; 118: 107418.
[http://dx.doi.org/10.1016/j.foodcont.2020.107418]

[105] Liu Z, Hua Q, Wang J, *et al.* A smartphone-based dual detection mode device integrated with two lateral flow immunoassays for multiplex mycotoxins in cereals. Biosens Bioelectron 2020; 158: 112178.
[http://dx.doi.org/10.1016/j.bios.2020.112178] [PMID: 32275211]

[106] Wang X, Niessner R, Tang D, Knopp D. Nanoparticle-based immunosensors and immunoassays for aflatoxins. Anal Chim Acta 2016; 912: 10-23.
[http://dx.doi.org/10.1016/j.aca.2016.01.048] [PMID: 26920768]

[107] Xu S, Zhang G, Fang B, Xiong Q, Duan H, Lai W. Lateral flow immunoassay based on polydopamine-Coated gold nanoparticles for the sensitive detection of zearalenone in maize. ACS Appl Mater Interfaces 2019; 11(34): 31283-90.
[http://dx.doi.org/10.1021/acsami.9b08789] [PMID: 31389683]

[108] Yu Q, Li H, Li C, Zhang S, Shen J, Wang Z. Gold nanoparticles-based lateral flow immunoassay with silver staining for simultaneous detection of fumonisin B1 and deoxynivalenol. Food Control 2015; 54: 347-52.
[http://dx.doi.org/10.1016/j.foodcont.2015.02.019]

[109] Lattanzio VMT, Nivarlet N, Lippolis V, *et al.* Multiplex dipstick immunoassay for semi-quantitative determination of *Fusarium* mycotoxins in cereals. Anal Chim Acta 2012; 718: 99-108.
[http://dx.doi.org/10.1016/j.aca.2011.12.060] [PMID: 22305904]

[110] Huang X, Huang X, Xie J, Li X, Huang Z. Rapid simultaneous detection of fumonisin B_1 and deoxynivalenol in grain by immunochromatographic test strip. Anal Biochem 2020; 606: 113878.
[http://dx.doi.org/10.1016/j.ab.2020.113878] [PMID: 32755601]

[111] Xu Y, Ma B, Chen E, *et al.* Dual fluorescent immunochromatographic assay for simultaneous quantitative detection of citrinin and zearalenone in corn samples. Food Chem 2021; 336: 127713.
[http://dx.doi.org/10.1016/j.foodchem.2020.127713] [PMID: 32768909]

[112] Li H, Yang D, Li P, *et al.* Palladium nanoparticles-Based fluorescence resonance energy transfer aptasensor for highly sensitive detection of Aflatoxin M1 in milk. Toxins (Basel) 2017; 9(10): 318.
[http://dx.doi.org/10.3390/toxins9100318]

[113] Wang Y, Zhang X, Zhan Y, Li J, Nie H, Yang Z. Au@Fe$_3$O$_4$ nanocomposites as conductive bridges coupled with a bi-enzyme-aided system to mediate gap-electrical signal transduction for homogeneous aptasensor mycotoxins detection. Sens Actuators B Chem 2020; 321: 128553.
[http://dx.doi.org/10.1016/j.snb.2020.128553]

[114] He B, Dong X. Nb.BbvCI powered DNA walking machine-based Zr-MOFs-labeled electrochemical aptasensor using Pt@AuNRs/Fe-MOFs/PEI-rGO as electrode modification material for patulin detection. Chem Eng J 2021; 405: 126642.
[http://dx.doi.org/10.1016/j.cej.2020.126642]

[115] Mandal S, Das P. Ultrasensitive visual detection of mycotoxin citrinin with yellow-light emitting

carbon dot and Congo red. Food Chem 2020; 312: 126076.
[http://dx.doi.org/10.1016/j.foodchem.2019.126076] [PMID: 31896461]

[116] Sun C, Liao X, Huang P, *et al.* A self-assembled electrochemical immunosensor for ultra-sensitive detection of ochratoxin A in medicinal and edible malt. Food Chem 2020; 315: 126289.
[http://dx.doi.org/10.1016/j.foodchem.2020.126289] [PMID: 32014670]

[117] Yang H, Zhang Q, Liu X, *et al.* Antibody-biotin-streptavidin-horseradish peroxidase (HRP) sensor for rapid and ultra-sensitive detection of fumonisins. Food Chem 2020; 316: 126356.
[http://dx.doi.org/10.1016/j.foodchem.2020.126356] [PMID: 32045810]

[118] Abera BD, Falco A, Ibba P, Cantarella G, Petti L, Lugli P. Development of flexible dispense-printed electrochemical immunosensor for aflatoxin m1 detection in milk. Sensors (Basel) 2019; 19(18): 1-11.
[http://dx.doi.org/10.3390/s19183912] [PMID: 31514303]

[119] Bagheri N, Khataee A, Habibi B, Hassanzadeh J. Mimetic Ag nanoparticle/Zn-based MOF nanocomposite (AgNPs@ZnMOF) capped with molecularly imprinted polymer for the selective detection of patulin. Talanta 2018; 179: 710-8.
[http://dx.doi.org/10.1016/j.talanta.2017.12.009] [PMID: 29310298]

[120] Jia W, Liang G, Tian H, Sun J, Wan C. Electronic nose-based technique for rapid detection and recognition of moldy apples. Sensors (Basel) 2019; 19(7): 1-11.
[http://dx.doi.org/10.3390/s19071526] [PMID: 30934812]

[121] Beghi R, Buratti S, Giovenzana V, Benedetti S, Guidetti R. Electronic nose and visible-near infrared spectroscopy in fruit and vegetable monitoring. Rev Anal Chem 2017; 36(4): 1-24.
[http://dx.doi.org/10.1515/revac-2016-0016]

[122] Sanaeifar A, Zaki Dizaji H, Jafari A, Guardia M. Early detection of contamination and defect in foodstuffs by electronic nose: A review. Trends Analyt Chem 2017; 97: 257-71.
[http://dx.doi.org/10.1016/j.trac.2017.09.014]

[123] Casalinuovo IA, Di Pierro D, Coletta M, Di Francesco P. Application of electronic noses for disease diagnosis and food spoilage detection. Sensors (Basel) 2006; 6(11): 1428-39.
[http://dx.doi.org/10.3390/s6111428]

[124] Gruber J, Nascimento HM, Yamauchi EY, *et al.* A conductive polymer based electronic nose for early detection of Penicillium digitatum in post-harvest oranges. Mater Sci Eng C 2013; 33(5): 2766-9.
[http://dx.doi.org/10.1016/j.msec.2013.02.043] [PMID: 23623094]

[125] Zhang X, Li M, Cheng Z, Ma L, Zhao L, Li J. A comparison of electronic nose and gas chromatography-mass spectrometry on discrimination and prediction of ochratoxin A content in *Aspergillus carbonarius* cultured grape-based medium. Food Chem 2019; 297: 124850.
[http://dx.doi.org/10.1016/j.foodchem.2019.05.124] [PMID: 31253256]

[126] Wang L, Hu Q, Pei F, Mugambi MA, Yang W. Detection and identification of fungal growth on freeze-dried Agaricus bisporus using spectra and olfactory sensors. J Sci Food Agric 2020; 100(7): 3136-46.
[http://dx.doi.org/10.1002/jsfa.10348] [PMID: 32096232]

[127] Eifler J, Martinelli E, Santonico M, Capuano R, Schild D, Di Natale C. Differential detection of potentially hazardous *Fusarium* species in wheat grains by an electronic nose. PLoS One 2011; 6(6): e21026.
[http://dx.doi.org/10.1371/journal.pone.0021026] [PMID: 21695232]

[128] Capuano R, Paba E, Mansi A, *et al. Aspergillus* species discrimination using a gas sensor array. Sensors (Basel) 2020; 20(14): 1-12.
[http://dx.doi.org/10.3390/s20144004] [PMID: 32708481]

[129] Ottoboni M, Pinotti L, Tretola M, *et al.* Combining E-nose and lateral flow immunoassays (LFIAs) for rapid occurrence/co-occurrence aflatoxin and fumonisin detection in maize. Toxins (Basel) 2018; 10(10): 416.

[http://dx.doi.org/10.3390/toxins10100416] [PMID: 30332757]

[130]   Orina I, Manley M, Williams PJ. Non-destructive techniques for the detection of fungal infection in cereal grains. Food Res Int 2017; 100(Pt 1): 74-86.
[http://dx.doi.org/10.1016/j.foodres.2017.07.069] [PMID: 28873744]

[131]   Lippolis V, Pascale M, Cervellieri S, Damascelli A, Visconti A. Screening of deoxynivalenol contamination in durum wheat by MOS-based electronic nose and identification of the relevant pattern of volatile compounds. Food Control 2014; 37: 263-71.
[http://dx.doi.org/10.1016/j.foodcont.2013.09.048]

[132]   Lippolis V, Cervellieri S, Damascelli A, *et al.* Rapid prediction of deoxynivalenol contamination in wheat bran by MOS-based electronic nose and characterization of the relevant pattern of volatile compounds. J Sci Food Agric 2018; 98(13): 4955-62.
[http://dx.doi.org/10.1002/jsfa.9028] [PMID: 29577312]

[133]   Lippolis V, Ferrara M, Cervellieri S, *et al.* Rapid prediction of ochratoxin A-producing strains of Penicillium on dry-cured meat by MOS-based electronic nose. Int J Food Microbiol 2016; 218: 71-7.
[http://dx.doi.org/10.1016/j.ijfoodmicro.2015.11.011] [PMID: 26619315]

[134]   Jonsson A, Winquist F, Schnürer J, Sundgren H, Lundström I. Electronic nose for microbial quality classification of grains. Int J Food Microbiol 1997; 35(2): 187-93.
[http://dx.doi.org/10.1016/S0168-1605(96)01218-4] [PMID: 9105927]

[135]   Olsson J, Börjesson T, Lundstedt T, Schnürer J. Detection and quantification of ochratoxin A and deoxynivalenol in barley grains by GC-MS and electronic nose. Int J Food Microbiol 2002; 72(3): 203-14.
[http://dx.doi.org/10.1016/S0168-1605(01)00685-7] [PMID: 11845819]

[136]   Sberveglieri V, Comini E, Emilia R, Amendola V, Emilia R. Electronic nose for the early detection of different types of indigenous mold contamination in green coffee. Proc Int Conf SensTechnol. 461-5.
[http://dx.doi.org/10.1109/ICSensT.2013.6727696]

[137]   Liu Q, Sunk K, Zhao N, *et al.* Information fusion of hyperspectral imaging and electronic nose for evaluation of fungal contamination in strawberries during decay. Postharvest Biol Technol 2019; 153: 152-60.
[http://dx.doi.org/10.1016/j.postharvbio.2019.03.017]

[138]   Liu Q, Zhao N, Zhou D, *et al.* Discrimination and growth tracking of fungi contamination in peaches using electronic nose. Food Chem 2018; 262: 226-34.
[http://dx.doi.org/10.1016/j.foodchem.2018.04.100] [PMID: 29751914]

[139]   Jiarpinijnun A, Osako K, Siripatrawan U. Visualization of volatomic profiles for early detection of fungal infection on storage Jasmine brown rice using electronic nose coupled with chemometrics. Measurement 2020; 157: 107561.
[http://dx.doi.org/10.1016/j.measurement.2020.107561]

[140]   Campagnoli A, Cheli F, Polidori C, *et al.* Use of the electronic nose as a screening tool for the recognition of durum wheat naturally contaminated by deoxynivalenol: a preliminary approach. Sensors (Basel) 2011; 11(5): 4899-916.
[http://dx.doi.org/10.3390/s110504899] [PMID: 22163882]

[141]   Presicce DS, Forleo A, Taurino AM, *et al.* Response evaluation of an E-nose towards contaminated wheat by Fusarium poae fungi. Sens Actuators B Chem 2006; 206(118): 433-8.
[http://dx.doi.org/10.1016/j.snb.2006.04.045]

[142]   Cheli F, Campagnoli A, Pinotti L, Savoini G, Orto VD. Electronic nose for determination of aflatoxins in maize. Biotechnol Agron Soc Environ 2009; 13: 39-43.

[143]   Ghosh H, Das R, Roychaudhuri C. Optimized nanocrystalline silicon oxide impedance immunosensor electronic tongue for subfemtomolar estimation of multiple food toxins. IEEE Trans Instrum Meas 2017; 66(5): 964-73.

[http://dx.doi.org/10.1109/TIM.2016.2625978]

[144] Samanta N, Member CRS. Nanocrystalline silicon oxide impedance immunosensors for sub-femtomolar mycotoxin estimation in corn samples by incremental fuzzy approach. IEEE Sens J 2016; 16(4): 1069-78.
[http://dx.doi.org/10.1109/JSEN.2015.2496998]

[145] Shimizu FM, Braunger ML, Riul A. Heavy metal/toxins detection using electronic tongues. Chemosensors (Basel) 2019; 7(3): 1-19.
[http://dx.doi.org/10.3390/chemosensors7030036]

[146] Logrieco A, Arrigan DWM, Brengel-Pesce K, Siciliano P, Tothill I. DNA arrays, electronic noses and tongues, biosensors and receptors for rapid detection of toxigenic fungi and mycotoxins: a review. Food Addit Contam 2005; 22(4): 335-44.
[http://dx.doi.org/10.1080/02652030500070176] [PMID: 16019803]

[147] Jin Y, Chen Q, Luo S, *et al.* Dual near-infrared fluorescence-based lateral flow immunosensor for the detection of zearalenone and deoxynivalenol in maize. Food Chem 2021; 336: 127718.
[http://dx.doi.org/10.1016/j.foodchem.2020.127718] [PMID: 32763741]

CHAPTER 12

Early Detection of Crop Fungal Pathogens for Disease Management using DNA and Nanotechnology Based Diagnostics

Rizwana Rehsawla[1], Apurva Mishra[2], Rajinder S. Beniwal[3], Neelam R. Yadav[1,*] and R.C. Yadav[1]

[1] *Department of Molecular Biology, Biotechnology and Bioinformatics, CCS Haryana Agricultural University, Hisar-125004, India*

[2] *Topological Molecular Biology Laboratory, Department of Microbiology and Molecular Genetics, University of California, Davis, California 95616, USA*

[3] *Department of Plant, Pathology, CCS Haryana Agricultural University, Hisar-125004, India*

Abstract: Virulent fungal plant pathogens are a serious threat to crop productivity and are considered a major limitation to food security worldwide. To meet these challenges, pathogen detection is crucial for taking appropriate measures to curb yield losses. Disease diagnosis at an early stage is one of the best strategies for crop protection. Earlier, traditional methods were used to diagnose and manage fungal diseases, which included visual scouting of the disease symptoms and spray of fungicides. The utility of immunoassays for early detection and precise identity has been appreciably stepped forward following the improvement of enzyme-connected immunosorbent assay (ELISA) and monoclonal antibodies. Nucleic acid-based diagnostic techniques have turnout to be the preferred type because of their greater speed, specificity, sensitivity, reliability, and reproducibility. The biosensor eliminates the need of sample preparation and can be used for on-site detection of fungal pathogens at latent infection stages so that preventive measures can be taken. Currently, multiple human and animal diseases have been detected with the help of biosensors. However, reports on plant pathogen detection using biosensors are still in infancy. Despite many applications of antibodies, there are also multiple drawbacks, including high cost, low physical and chemical stability, and the ethical issues associated with their use. Now, DNA based biosensors are gaining popularity because of their sensitive and precise detection of DNA target sequences. Immunological and DNA-based techniques combined with nanotechnology offer highly sensitive and selective gel-free detection methods, and the lab-on-chip (LOC) feature of biosensors makes them a very reliable tool in crop protection.

Keywords: Biosensors, Diagnostic techniques, DNA based biosensors, Fungal Pathogen, Immunosensor, Nanotechnology.

* **Corresponding Author Neelam R. Yadav**: Department of Molecular Biology, Biotechnology and Bioinformatics, CCS Haryana Agricultural University, Hisar-125004, India; Tel: +918053013070; Email: nryadav58@gmail.com

Savita, Anju Srivastava, Reena Jain & Pratap Kumar Pati (Eds.)

INTRODUCTION

Global food security is the major concern of the world due to the growing population in recent years. Agricultural efficiency and yields are limited worldwide by both abiotic (non-biological) and biotic (biological) factors such as water shortages, salinity, drought, heat stress, and environmental degradation, vulnerability to weather-related disasters and plant pathogen, insect pest attack. Plant diseases caused by pathogen attack may lead to devastating effect on social life, economy and ecology globally, harming biodiversity as well. Approximately 20% to 40% of losses in major agricultural crops are due to biotic stress *i.e.* by plant pathogen attack during sowing, germination, growth and development, which pose a wide spectrum of challenges to agricultural scientists in crop production. These are considered to be the major obstacle to global food security [1]. The term "pathogen" is defined as the infectious biological entity that could be bacteria, viruses, fungi, nematodes or any other microorganism that can infect human beings, animals and plants and lead to the chain of infections, diseases and symptoms and make them sick [2]. Plant pathogens that cause disease in plants are also known as phytopathogens. Most of the plant pathogens cause harm to plants but there are some trans-kingdom pathogens that can make immunocompromised people sick [3]. New, old and emerging plant infectious diseases are the main reason for low agricultural productivity. Table **1** shows the fungal pathogens which adversely affect crop productivity. The higher degree of transmission, occurrence and extremity of the pathogens makes them harmful to the world food supply [4]. Worldwide, pathogens and pests are causing losses to the tune of 10% to 28% in wheat, 25% to 41% in rice, 20% to 41% in maize, 8% to 21% in potato, and 11% to 32% in soybean [5]. Usually, a number of plant pathogenic organisms show their symptoms after some days or months. Thus, this helps in routine laboratory diagnosis and pathogen detection. But nowadays, early as well as fast detection for pathogen become necessary for several reasons, like climate change, globalization, increased human mobility, international trade, pesticides resistant pathogens. Due to these reasons, the timely spray for eradication of pathogen becomes no longer endurable for farmers. The development of the agriculture sector is only possible by using modern technologies with the minimum damage to agroecology [6]. There are enormous challenges for plant disease detection, research and practical applications [7]. Moreover, there is an urgent need to take preventive measures to protect crops that are economically important and for this, on-site diagnostic methods are required. The on-site diagnosis will help in developing proper field operable strategies for early plant pathogen control that will prevent the spread of disease and control by assisting in better understanding of disease epidemiology and also improving the design cultivar of choice [8].

Table 1. List of the plant fungal pathogens that cause serious yield losses in crops.

Crops	Fungal Pathogen	Common Name
Grasses and Cereals like Barley	*Blumeria graminis*	Powdery mildew
Cereals	*Fusarium graminearum*	Blight, Root and Crown rot and Head blight on small grain Cereals
Wheat	*Mycosphaerella graminicola*	Septoria tritici blotch
	Pythium ultimum	Damping off
	Puccinia spp.	Rust
	Melamp soralini	Rust
	Septoria tritici	Blotch
Barley	*Puccinia spp.*	Rust
Rice	*Magnaporthe oryzae*	Blast
Maize, Teosinte	*Ustilago maydis*	Smut
Corn	*Pythium ultimum*	Damping off
Tomato	*Fusarium oxysporum*	Wilt
	Phytophthora infestans	Late blight
	Phytophthora parasitica	Rot disease of root and stem
Potato	*Phytophthora infestans*	Late blight
	Pythium ultimum	Damping off
Soybean	*Pythium ultimum*	Damping off
Cabbage, Kale, Lettuce	*Albugo candida*	White rust
Linseed	*Melampsoralini*	Rust
Peppers	*Phytophthora capsici*	Blight, root rot and stem rot
Eggplant	*Phytophthora parasitica*	Rot disease of root and stem
Tobacco	*Phytophthora parasitica*	Rot disease of root and stem
	Fusarium oxysporum	Wilt
Legumes	*Fusarium oxysporum*	Wilt
Cucurbits	*Fusarium oxysporum*	Wilt
Sugar beet	*Albugo candida*	White rust
Woody and herbaceous plants	*Colletotrichum spp.*	Anthracnose
Grapes	*Botrytis cinerea*	Grey mould
	Plasmopara viticola	Downy mildew
Pineapple, Papaya	*Phytophthora parasitica*	Rot disease of root and stem

(Table 1) cont.....

Crops	Fungal Pathogen	Common Name
Pines, Azaleas, Camelia, Boxwood	*Phytophthora cinnamomi*	Root rot; dieback
Flax	*Melamps oralini*	Rust
Oak	*Phytophthora ramorum*	Sudden oak death
Arabidopsis	*Hyaloperonospora arabidopsidis*	Downy mildew
Fir	*Pythium ultimum*	Damping off

Checking plant health and detecting plant pathogen at early stages are crucial measures to lower the disease escalation and facilitate effective prevention practices. Earlier visual disease assessment was achieved by using traditional disease diagnosis methods. The traditional approach involves the use of pesticides to control plant disease. These pesticides cost the environment and cause a detrimental effect on human health and are not always able to control the spread and severity of the disease. Therefore, even with the application of pesticides in agriculture, the crop losses continuously persist, causing hike in the production costs, poor quality products [9]. This is a very time-consuming, labor intensive, costly, and error prone technique. These methods are too subjective for disease diagnosis [10]. To control disease and pathogens, new effective and innovative techniques are required to address upcoming challenges and trends in agricultural production that require more precision than ever before [7].

WHAT HAPPENS WHEN PATHOGENS ATTACK A PLANT?

The plants are always under the threat of being attacked by different types of infectious plant pathogens and other harsh environmental conditions because they are present in open land in harsh environmental conditions without moving. For their survival and protection from the disease, plants have developed different interactive strategies with these pathogens and communicate in different ways sometimes these interactions lead to the development of disease. During the course of evolution, plants have evolved a variety of mechanisms that provides them immunity to fight biotic and abiotic stress. However, like animals plant immune system lack circulating cells such as antibodies and macrophages and the signal responses are very limited [11]. When a disease-causing organism attacks a plant, a complex immune response is generated by plants, which results in the production of disease-specific proteins and other molecules. These proteins and other molecules involved in plant defense and help plant to withstand the infection and help in limiting the spread of infection. In this process, pathogens also produce proteins and toxins to facilitate and manifestation of their infection in the plants before symptoms of disease appear.

These molecules play an important role in diagnostic assays for the early diagnosis of plant disease [9]. Fig. (**1**) describes the various methods available for plant pathogen detection.

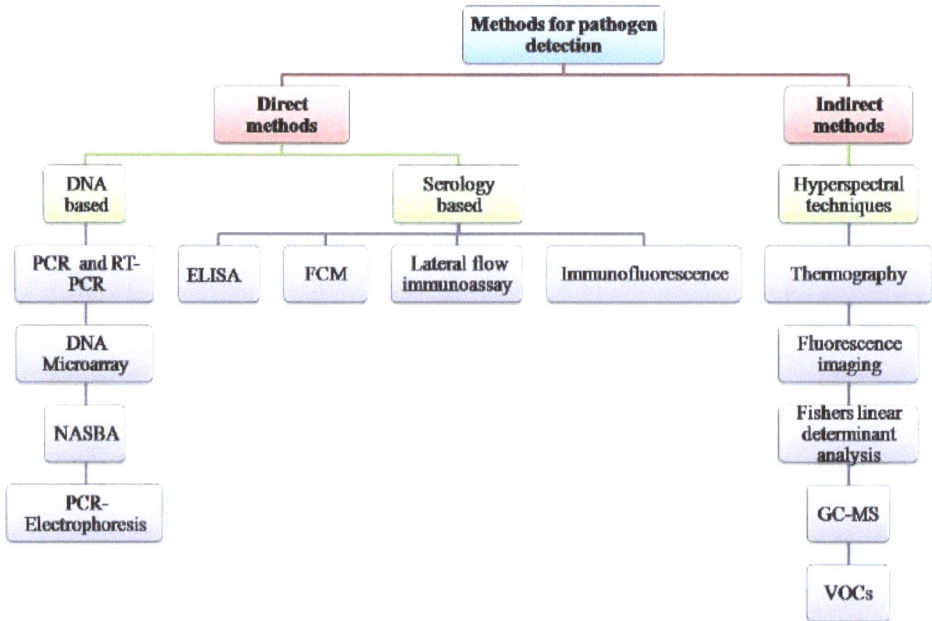

Fig. (1).   Schematic representation of various direct and indirect methods used in pathogen detection.

DIRECT METHODS FOR PATHOGEN DETECTION

DNA-based and serological based methods come under the direct method of pathogen detection. In spite of traditional visual observation of symptoms, these methods are more error free and detailed [12]. Currently the most frequent and readily used molecular techniques for plant pathogen detection are "enzyme-linked immunosorbent assay (ELISA)" based on antigen-antibody interactions and polymerase chain reaction (PCR) based on hybridization (Table **2**). In ELISA-based systems, an antibody is immobilized on the surface, which is conjugated with an enzyme or fluorescent dye and exposed to the sample containing target analyte, which binds to the specific antibody and antigen-antibody interaction takes place that confirms the existence of a particular plant pathogen. Similarly, in the PCR based methods, the first step is DNA isolation from the pathogen or plant sample infected with the pathogen and after that PCR is done with the help of specific primers to amplify the DNA to obtain millions of copies. There are many limitations associated with PCR based detection as it requires sample preparation, DNA extraction, primer designing, false results, amplification process, thermocycler and sample evaporation, is inefficient in multiple pathogen detection and is not field operable or does off-site detection. There are many attempts made to overcome these limitations. For example, to address the temperature control system or to eradicate the need for thermocycler, isothermal amplification assays have been developed, such as loop mediated isothermal amplification (LAMP) assay [13] and nucleic acid sequence-based amplification (NASBA). The first report of using LAMP protocol was by Notomi *et al.* [14] for the hepatitis B virus that causes acute and chronic liver infection [15]. In plants, the Loop-LAMP (Loop-mediated isothermal amplification) procedures were initially designed to detect *Fusarium graminearum* which caused rot disease in cereals. Similarly, to detect the multiple pathogen, multiplex PCR platforms have been developed to identify different pathogens in a sample by using different primer sets and each pathogen is detected in a separate channel. Although DNA-based methods such as PCR, DNA microarray, FISH, NASBA, DNA fingerprinting and serological methods such as IF, ELISA, GC-MS and FCM have revolutionized diagnosis, despite this, these are relatively sophisticated to operate, costly, require experts, time taking and are not very reliable at asymptomatic or during early infection, especially in case of pathogen with wide-range [15, 16]. Presently, scientists are devoting their knowledge, time and energy in developing affordable and portable PCR-based sensors to combat plant pathogens.

Table 2. Plant fungal pathogen detection by direct methods.

Pathogen	Type of the Pathogen	Method	References
Alternaria solani	Fungus	LAMP	[17]
Botrytis cinerea	Fungus	ELISA, IF	[16]
Fusarium graminearum	Fungus	PCR	[18]
Lycopersicon esculentum	Fungus	RT-PCR	[19]
Phhakopsorapachayrhizi	Fungus	IF	[20]
Phytophthora ramorum	oomycete	RT-PCR	[21]
Puccinia striiformis, Puccinia triticina	Fungus	LAMP	[13, 22]

Primer Based Plant Pathogen Detection

Detection of pathogen and disease diagnosis is the prime objective for disease management. It is based upon the cause and effect of the disease. There are several methods of diagnosis like identification and taxonomical study of the pathogen. But these methods are not sufficient. The advancement in the bio-technological approaches put a high impact on plant disease diagnosis, which increases the effectiveness and precise diagnosis. For effective management of disease accurate detection of pathogen is indispensable not only up to genus level but up to species and strain level. Many methods are available, but nowadays, molecular methods are more reliable for plant pathogens. Polymerase chain reaction (PCR) invented by Kary B. Mullis [23] provides a simple method for the amplification of small fragments of DNA. This method is used to make a large number of copies of a particular DNA segment precisely and quickly using primers that are considered as starting point of DNA synthesis. Primer based PCR technology is time saving and cheap and purity of DNA is not a hurdle as poor quality DNA can be amplified; it is rapid and most important it provides high quality data for accurate diagnosis. Primer based technique has been applied in plant disease diagnosis that helps in understanding plant and pathogen relation and its correlation with the environment. It is also useful in distinguishing between different pathogens that are closely related. Primer based technology also helps in expression analysis and direct detection of pathogen. Table **3** shows a number of plant pathogens detected by primer-based PCR technology.

Table 3. Genes/ Genomic sequences used for fungal pathogen detection.

Crop	Fungi	Disease	Gene	References
Cereals	*R. solani AG2-1/AG2-8 R. oryzae genotype 1*	Sheath blight of rice, black scurf of potato, wirestem disease, Rhizoctonia root rot	Rs2.1/8 RoGr1 β- tubulin	[24, 25]
Rice, Maize and Wheat	*Bipolaris oryzae*	Helminthosporium	ITS1 ITS2	[26]
Wheat	*Puccinia striiformis f.sp. tritici*	Stripe rust	Pst gene for BAC clone	[27]
Barley	*Blumeriagraminis f. sp. Hordei*	Fungal infection	GAPDH	[28]
Chickpea	*Ascochyta rabiei*	Ascochyta blight	UBC-702, 708, 726	[29]
Soybean	*Cercospora kikuchii*	leaf blight and purple seed stain diseases	CTB6 gene	[30]
Safflower	*Fusarium oxysporumf.sp. melonis*	Safflower wilt	Fef1 Fef2	[31]
Cotton, Tomato, Potato, Soybean, Jack bean, Eggplant and Chickpea	*Sclerotium rolfsii*	Seedling damping-off, basal rot, dry collar rot, foot rot, stem blight, boll rot and root rot	RADP primer C3 OPA02 OPC20	[32]
Potato, Oak	*Phytophthora spp.*	Potato leaf blight, sudden oak disease	ITS4 ITS 6	[33]
Grapes, Berries, Vegetables and Ornamental crops	*Botrytis cinerea*	Grey mould	C729 BC108, BC563	[34]

(Table 3) cont.....

Crop	Fungi	Disease	Gene	References
Mango	*Colletotrichum gloeosporioides*	Anthracnose disease	MKCg	[35]
Apple	*Venturia inequalis*	Apple scab	Beta tubulin I Beta tubulin II	[36]

Besides so many advantages of Primer-based detection methods, a number of limitations are associated with this technique. For example, in the PCR method, sometimes primers fail to amplify, they give false positive results and sometimes it is difficult to distinguish between vital and non-vital inoculum. Hence, for effective and accurate diagnosis, we need to follow the criteria for selection, designing and precise handling of the primers and their protocol [37].

INDIRECT METHODS OF PATHOGEN DETECTION

Indirect techniques of pathogen detection involve imaging techniques such as thermography, hyperspectral techniques, a non-invasive imaging technique known as fluorescence imaging, electronic nose, and volatile organic compound for plant pathogen detection. These methods are used to find out the effect of pathogen attacks on the function of the plant. They cannot diagnose the pathogen; they only measure the changes in the physiology of plants after infection [8]. These techniques have been used to determine morphological temperature, water evaporation rate change from plants and volatile organic compounds released by infected plants [38] as given in Table **4**. For example, the volatile compound detection technique is used to differentiate between healthy and diseased plants. The VOCs produced and the amount released by healthy and unhealthy plants are indicative of plant conditions like health, germination, presence of stress conditions that may affect crop yields. Many other factors that affect the release of VOCs are humidity, available moisture, light, temperature and soil condition. Thus, administering the released VOCs from plants provides necessary information about plant health. The volatile compound profile is affected in the diseased plant and is sensed by an electronic nose connected with a gas sensor. This technique is used in identifying the soft rot in potato tubers which causes post-harvest loss during storage [39]. But these techniques are very responsive to minute environmental changes and also shortfall for each type of disease.

Table 4. List of Plant fungal pathogen detected by indirect methods.

Plant	Pathogen	Common Name	Method	References
Potato	*Erwinia*	Soft Rot	VOCs, Electronic nose	[39]
Wheat	*Fusarium graminearum*	Scab (Fusarium head blight)	Hyperspectral imaging	[40]

Plant	Pathogen	Common Name	Method	References
Wheat	*Blumeriagraminis f. sp. Tritici*	Powdery mildew	Fishers linear determinant analysis (FLDA	[41]
Wheat	*Fusarium graminearum*	Head blight	IR-Thermography	[42]
Tomato	*Phytophthora infestans*	Late blight disease	Hyperspectral remote sensing	[43]
Carrot	*Botrytis cinerea*	Soft rot, grey mold	GC-MS	[44]
Sweet orange	*Phytophthora spp.*	Brown rot	Fluorescence and reflectance Hyperspectral imaging	[45]
Citrus	*Guignardiacitricarpa*	Citrus black spot	Hyperspectral imaging	[46]
Grapevine	*Guignardabidwellii*	Black rot	Fluorescence imaging	[47]
Grapevine	*Plasmopara viticola*	Downy mildew	Thermography	[48]
Sugar beet	*Cercosporabeticola*	Cerospora leaf spot	Spectral angle mapper (SAM)	[49]
Mango	*Lasiodiplodiatheobromae and*	Stem end rot	GC-MS	[50]
	Colletotrichum gloeosporioides	Anthracnose		
Apple	*Venturia inaequalis*	Apple scab	Hyperspectral Reflectance and Fluorescence Imaging	[51]

In addition, most of the methods described above are not able to provide instant detection in real time that makes them inappropriate and untimely for on-field detection.

Innovative Approaches for Plant Pathogen Detection

Molecular detection methods mentioned above are not sufficient to diagnose pathogen at an early stage as they have some limitations in pathogen detection such as low concentration in materials like seeds and insect vectors, cross contamination, blocking of target DNA amplification, cross-amplification of non-target DNA; requirement of specialized laboratories, and sophisticated equipment's. Another limitation is related to on-site detection as they are not field operable [4]. Additionally, diagnostic methods to detect *Phytophthora ramorum* mainly infecting oak tree and *P. kernoviae* infecting European beech and *Rhododendron ponticum*, have been designed using immunoassay in lateral-flow devices. In this method, target DNA is extracted directly on the lateral flow device and a piece from the membrane directly labeled with primers is subjected to isothermal amplification by LAMP assay and after that, amplified products are detected in a lateral-flow-device format by a sandwich type immunoassay containing a reagent that specifically binds to the amplified product and a positive line on the test strip indicate the presence of the pathogen [52]. In this method, several limitations exist, including the difficulty of selecting species-specific robust genomic sequences, complicating the nature of LAMP primer design. These kinds of analytical techniques need a series of steps, time, price and sophisticated instrumentation. Many lateral flow devices are developed for plant pathogenic microorganisms. Mainly, these devices are developed for the well characterized pathogen and for specialized antibodies and their commercial use is based on this property. In addition, this method is not sufficient or efficient for multiple pathogen detection. Their high cost and single pathogen detection are the main problems with LFDs devices. Regardless, commercialized devices are present for single pathogen detection; no device is present to detect multiple pathogen on-sites. To overcome these hurdles, there is an urgent need of device which can directly and quickly detect pathogens on the field. The main problem with portable devices is that they require alternative amplification technique instead of conventional one. For this reason, scientists are focusing on the development of innovative biosensors and presently, such portable biosensors have emerged and are being widely used in agriculture, clinical, environmental studies and food industries. The most fascinating property of biosensors is that they are rapid, simple, cost-effective and can be used in remote areas where expensive and sophisticated instrumentation facilities are not present [8].

BIOSENSOR: THE NEED OF THE HOUR

Agricultural productivity is always at the risk of damage by a natural threat like pathogenic attack and other abiotic stress conditions [53]. Hence, loss in crop production should be minimized to increase profits for farmers and consumers alike. Also, during the import and export of agricultural produce to national or international borders, the need for biosecurity is essential. Technologies that can precisely and timely identify both identified and unidentified pathogens and new strains of already known pathogens are needed to be developed. Low cost methods are much needed for rapid and accurate disease diagnosis. To overcome this hurdle in pathogen detection biosensors were developed. A sensing device that is able to monitor and convert any biological response into an electrical, physical and chemical signal is known as a biosensor. It is an integrated receptor–transducer device, as described in Fig. (**2**). Biosensors consist of biorecognition components which are highly selective and specific. They bind with the target analytes and after their specific binding a transducer component converts the binding event to electrical or optical signals. The biological material could be nucleic acid, antibody, enzyme, hormone, microorganism, organelle or whole cell, synthetic bio receptors and nanoparticle. Due to the eye-catching feature of the biosensors in pathogen detection, this particular discipline has become the most studied say in research or development [54].

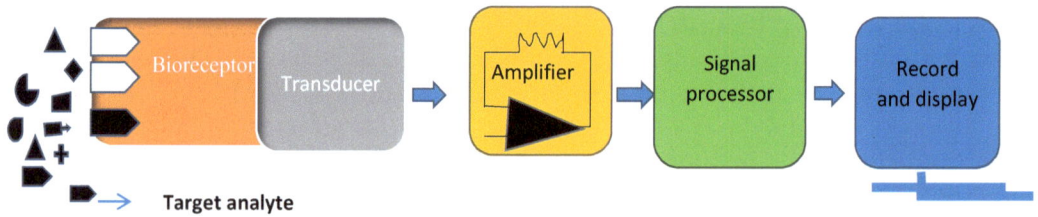

Target analyte

Fig. (2). Working Principle of biosensor.

Pathogen biosensors are based on biological recognition receptors that enhance the specificity of the detection and help in the accurate detection of the phytopathogen. A wide range of various biosensors have been developed for the diagnosis.

- Antibody based
- DNA based
- Enzyme based
- Phage based
- Nano based

Biosensors play a major part in the diagnosis because of their specificity and selective nature of detection compared to the older techniques. Novel sensors are based on the host response analysis for *e.g.* differential mobility spectrometer and LFDs they detect the infection directly on the field and at an early stage delivering results very fast [10]. The lab on chip feature of biosensors makes it very reliable tool in agriculture, food industry, biotechnology and medicine. The field of biosensor nanotechnology has emerged as a powerful diagnostic approach for pathogen detection as it revolutionizes the sector of diagnosis, agriculture and food industries [55]. Also it helps in environmental diagnostics and post-harvest applications including processing, storage and packaging. There are lots of examples on use of nano-biosensors on animal/human pathogen detection, but in plant-pathogen disease diagnostics, the application of nano-biosensor use is still in its infancy. In brief, nanoparticles are different from their bulk counterparts, which, when reduced to nanosize (1-100 nm) become suitable for diagnostic probes as they acquired certain properties on the size reduction that are not present with their bulk counterpart [56]. There are many nanoparticles synthesized for different studies say gold, silver, lipid based, polymer based, carbon nanoparticles. Out of them gold nanomaterial is most frequently used as labels in diagnosis due to a unique physiochemical properties that allow detection at very low concentrations. The advantages of using nanoparticles as probe in DNA analysis is due to their detection sensitivity and shortened detection time and lower cost clearly make this technology enable in preventing global threat to crop production and food security, as well as to integrate the development and deployment of genome-enabled technologies (Fig. **3**). The nanomaterials have high surface area to volume, high photoelectronic conductivity, optic properties and plasmonic properties that make them promising for sensor development. Metals and their oxides, quantum dots, carbon nanotubes, graphene, nanowire, nanofilms as well as polymers are used as nanoparticles

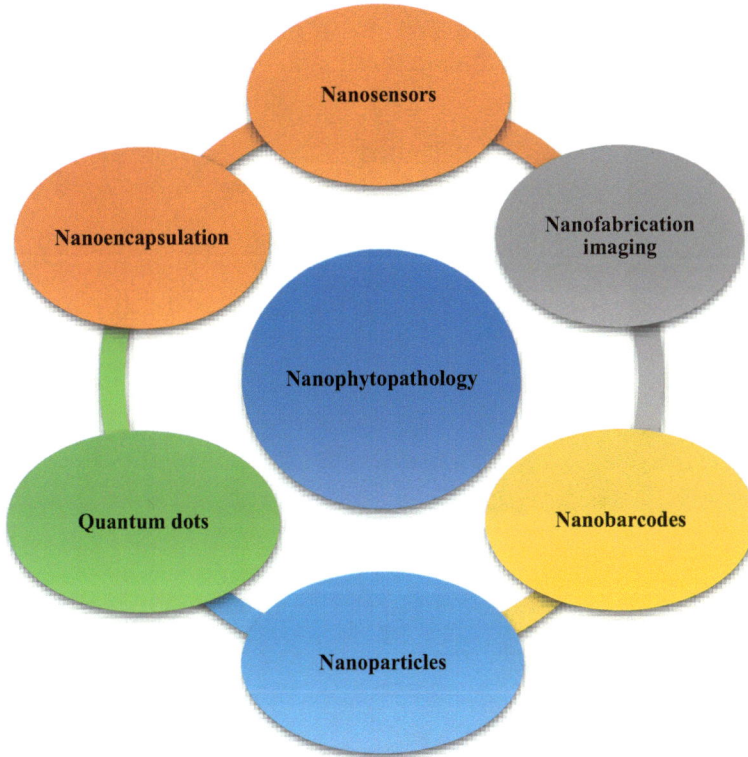

Fig. (3). Nanotechnology in plant pathogen detection.

Biological materials such as DNA, RNA, antibodies and many other molecules have been coupled with nanoparticles and used for diagnosis [16]. There are many examples of nanosensors using biological material for recognition for example *Xanthomonas axonopodis*a bacterium causes, bacterial leaf spot disease is detected by using antibody based nano biosensor. They combined the silica nanoparticles with the antibody to detect the bacteria that infect plants of solanaceae family [57]. Similarly, Singh *et al.* [58] developed an optical nano-immunosensor for detection of fungal teliospores of karnal bunt disease caused by *Tilletia indica* in wheat using surface plasmon resonance (SPR).

ANTIBODY BASED BIOSENSORS

The basic principle behind the antibody-based biosensor is the antigen-antibody interactions [16]. The antibodies can be generated for various target analytes. The inherent selectivity and the sensitivity of the antibodies make them very popular recognition elements in many sensing devices. Earlier traditional immunoassay formats such as competitive, displacement, sandwich, and enzyme-linked immunosorbent assays (ELISA) were used. The use mainly depends upon the application [59]. Antibody-based biosensors or immunosensors can allow fast and label free detection of pathogens qualitatively and quantitatively (Fig. **4**).

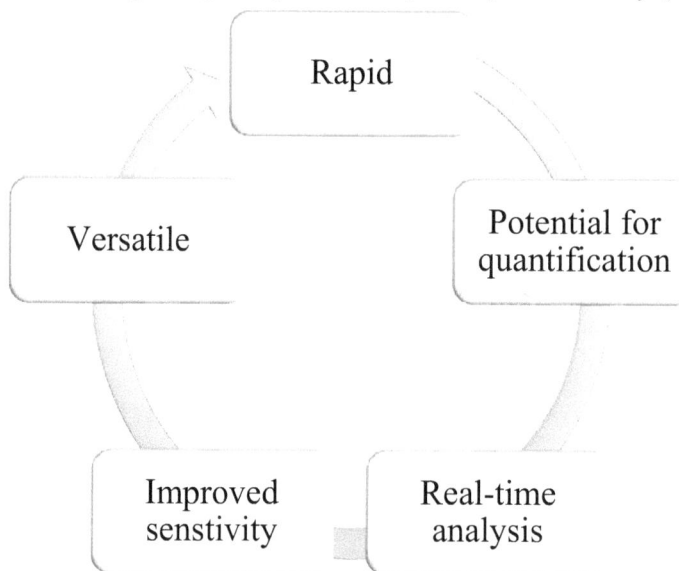

Rapid

Versatile

Potential for quantification

Improved senstivity

Real-time analysis

Fig. (4). Advantages of Biosensors in Pathogen detection.

There are lots of examples biosensors based on antibody in plant pathogen diagnosis. In the antibody-based biosensors different types of electrochemical transducers are used, such as amperometric, potentiometric, impedimetric and conductometric. Non-electrochemical transducers such as surface plasmon resonance (SPR) cantilever-based sensors and quartz crystal microbalance (QCM) for affinity biosensing have been also developed. Among them, SPR based immunosensors are most popular tool and frequently used in the detection. The research on SPR began more than 80 years ago by Wood in 1902 [60]. Later its principle for sensing was introduced [61, 62]. In SPR-based sensors, analyte is attached to the metal surface such as gold and a change in refractive index can be measured [63]. It is a surface-sensitive optical technique to examine interactions of molecules present very close to the surface of transducer. It offers label free, rapid, real time analysis, identifying specific analytes without sample preparation that make SPR an ideal tool for analysis [64 - 66]. The first two plant pathogenic viruses which were detected by biosensor based detection are Tobacco mosaic virus (TMV) and Cowpea mosaic virus (CPMV). The biosensor deals with the epitope mapping of monoclonal antibodies(mAbs) of these two viruses [67]. Similarly, Candresse *et al.* [68] developed a SPR based immunosensor to study immunological variability and epitope characterization of Lettuce mosaic virus (LMV). They have used 19 monoclonal antibodies (mAbs) on the basis of their recognition specificity against 15 LMV isolates. Surface plasmon resonance analysis showed that LMV had a significant immunological difference in the structure of LMV epitope.

The first report on fungal plant pathogen detection was SPR-based given by Skottrup *et al.* [69]. A surface plasmon resonance (SPR) immunosensor for detection of *P. infestans* sporangia was developed using monoclonal antibody (phyt/G1470 mAb) against *P. infestans* G1470 mAb by SPR. The assay provided a detection limit of 2.2 × 106 sporangia/ml and analysis time of 75 min. For *Puccinia striiformis f.sp tritici* (Pst), the causal agent of devastating yellow rust disease in wheat, Label free detection was done using SPR technology. They used urediniospore (pathogenic fungal spore) and, by hybridoma technology prepared Pst mAb8. The detection limit of 3.1×10^5 urediniospores/ml was achieved [70]. Similarly, SPR based biosensor was developed that detected the *Phakopsora pachyrhizi*, a pathogenic fungus causing rust disease in Asian Soybean [71]. They developed an antibody against this fungus and used as biological recognition element with detection limit of 3.5-28 mg/ml and detection limit of 800 ng/ml.

QCM-based sensor consists of a quartz crystal resonator. The basic principal of the sensor is that it measures the change in frequency of the crystal and determines the variation in mass per unit area. The sensor's reactivity is based on the width of the crystal. The thinner is the crystal more is the sensitivity. QCM has attained unusual importance in the fields of material science, environmental monitoring, biosensors and electrochemistry. The QCM crystal is typically modified with antibodies which act as a recognition element [72, 73]). Eun *et al.* [72] for the first time used the antibody as recognition element in QCM sensor to detect plant viruses. QCM based immuno-sensors are more favorable to detect pathogen than protein.

However, cantilever-based sensors are based upon the principal of resonance frequency change. When target analyte and the sensor surface combine, there occurs some change in resonance frequency and this change is measure by the sensor. For sensing, the cantilever is coated with a specific sensory layer which must be able to recognize target molecules in compatible manner [74]. These sensors have been used for pathogen detection due to their ability of detecting small molecules like nucleic acid, proteins and other molecules [75]. These are small, portable sensors and are cost effective. This property of cantilever based biosensors enable multi target analysis. However, there is no report showing multiplexed pathogen detection [67]. Nugaeva *et al.* [76] developed a cantilever-based biosensor for detecting *Aspergillus niger,* a fungal plant pathogen with a response range of 10^3 cfu/ml. In the study, they have used several cantilevers and each cantilever was individually coated with polyclonal antibodies produced against the fungus. The fungal spores were immobilized on the surface of cantilever. The detection is based on resonance frequency shift. The best part of the sensor was that it could distinguish between dead and viable spores. An immuno-affinity fluorimetric biosensor was developed by Garcia *et al.* [77] for the identification and quantification of aflatoxins. The aflatoxins are formed by common fungi *Aspergillus flavus, Aspergillus parasiticus* and *Aspergillus nomius.* The fungus mainly infects maize, cottonseed, peanuts, onion, apricots, nuts, grains and spices and cause contamination of food. The aflatoxins are not good for the human health; their ingestion or inhalation may lead to severe medical problems such as immunosuppression. Earlier studies have shown that traditional molecular detection methods have some limitations in pathogen detection, such as detection problems at low concentration in materials like seeds [4]. Infected seeds act as vector for long-distance dissemination of the pathogen. Seed borne diseases cause reduction in yield, quality of grain and also deteriorate the market value. The failure of conventional diagnosis methods led to the development of advance molecular approaches [78]. Due to invisible characteristic symptoms it is very difficult to diagnose the disease in the field and not the biochemical assay is sensitive toward the detection of the seed disease and the traditional methods used are very time consuming [79]. To overcome this problem, a portable cell biosensor with antibodies specific to the target on the sensor membrane was developed by Perdikaris *et al.* [80]. The sensor used a high throughput bioelectric recognition assay. This sensor was used to detect potato virus Y, rattle virus in tobacco and mosaic virus of cucumber. Also the sensor is used for detecting mix infection. Basically when the infectious virus particles attached to the specific antibodies which were designed for a particular pathogen, their attachment caused a virus-specific change on the cellular membrane. This change is detected by the sensor and it takes around 70 minutes for 96 samples to be detected.

In wheat, fungus *Tilletia indica* causes Karnal bunt disease in the kernels and leads to the 0.2-0.5% loss to the crop and also affect the economy as many countries around the globe are not accepting the wheat from Indian subcontinent due to this disease in wheat produce. There are many conventional and molecular techniques to detect the infectious teliospore of Karnal bunt. The major drawback of these techniques is that the teliospores of *Tilletia indica* closely resemble with teliospores of other fungi that causes bunt disease in other crops like rye and rice.

So, there comes the need of specific diagnostic technique in order to determine the teliospores of *Tilletia indica* correctly and specifically identifying only contaminating fungus [81]. Singh *et al.* [82] developed a label free real time SPR based biosensor for detecting karnal bunt disease. This technique help in specific identification of fungal teliospore of karnal bunt and could be used by the seed certification lab and plant quarantine department. They developed an anti-teliosporic antibody and the antibody was immobilized on a SAM. The sample was loaded onto the antibody leading to antigen-antibody interaction. The chip was highly sensitive and detected 2.5 teliospores equal to 625 pg. Nanotechnology based immunosensor known as a fiber optic particle plasmon resonance (FOPPR) for label-free detection of Cymbidium mosaic potyvirus virus (CymMV) or Odontoglossum ringspot tobamovirus virus (ORSV) of orchids has been reported [83]. They used gold nanorods for direct sensing of the analyte, which overcame the limitation of signal interpretation in the visible spectrum. The detection limit of this nanosensor for both the viruses in leaf saps were 48 and 42 pg/mL, respectively compared to the ELISA with detection limit of 1200 pg/mL for both viruses. The total time taken by the sensor to differentiate between uninfected and infected orchid was 10 min. An immunochromatographic test strip for the detection of *Pantoea stewartii subsp. Stewartii* (Pss) causing wilt disease in corn seed was developed [84]. The disease was first detected in New York, USA, by Stewart in 1895 [85]. Monoclonal antibody was developed by mouse immunization process for the lateral flow immunoassay. The developed Mabs highly sensitive for the pathogen and cross reactivity with other pathogens. The detection limit of the strip was 1×105 cfu/mL for Pss.

Despite many applications of antibodies as a sensory device, there are lots of drawbacks associated with the use of antibodies for portable devices. The antibodies secreted *in vivo* have higher cost of production as these are extracted from animals and many ethical issues are associated with the production of antibodies *in vivo*. *In vitro* synthesized antibodies show low specificity, low physical and chemical stability. To overcome these limitations, synthesis of artificial or chemically synthesized affinity binders are used that results in the development of aptamers. Aptamers are chemically synthesized antibodies and have the potential for modification for better stability and regeneration. Other alternatives to antibodies are molecularly enzyme–substrate reactions, molecularly imprinted polymers, aptamers, small molecule probes, protein-based probes, DNA probes and antimicrobial peptides. The major limitation of this approach depends upon the quality, size and storage conditions of the antibody that could affect the stability of antibodies [4].

DNA BASED BIOSENSORS

DNA sensors are based upon the principle of complementary base pairing of the sequences and thus provide more sensitive, specific, and fast detection of target when compared to antibody-based assays [86]. Nucleic acids have a wide range of biological, chemical and physical activities and are used extensively in bioanalytical assays and biosensors. In DNA based biosensors, oligonucleotides are used as a sensing element with known fragments or bases of DNA or RNA [87]. Many DNA-based probes are coupled with detection techniques like Surface plasmon resonance (SPR), conductance impedance and fluorescence resonance energy transfer (FRET). The genetic information analysis is of great significance in many areas. The study of gene sequence can help in understanding the gene polymorphism and rapid detection of genetic mutation, screening for targets known to play a role in disease and help in early diagnosis of disease in the asymptomatic stage. Due to their faster, simpler and cheaper sequence-specific information these sensors are more reliable. The recent advances in diagnostics open new opportunities for DNA sensors. In DNA based sensors, single stranded DNA work as a probe and immobilized on the surface of electrodes using different techniques of immobilization like physical adsorption and covalent attachment. After immobilization the target is hybridized on the surface and an electro indicator is used for measuring the hybridization. DNA probes used in the sensors are modified either chemically or enzymatically with materials like radioactive material, chemiluminophore or ligands, biotin, by attaching functional group at one end of the probe, HRP *etc*. to detect the signal as nucleic acid itself does not able to emit any signal on hybridization. The use of biochip technologies put an end to the role of PCR [87, 88]. The recognition properties of nucleic acids are extremely beneficial and for this reason, they are of great interest analytical applications [89, 90] and powerful biosensing tool [91, 92]. Different techniques such as Label-free, voltammetry, optical, electrochemiluminescence and fluorescent quartz crystal microbalance have been developed for the detection of DNA hybridization as DNA biosensor [4].

Eun *et al.* [93] developed a piezoelectric DNA-biosensor for two orchid viruses; Cymbidium mosaic potyvirus virus (CymMV) and Odontoglossum ringspot tobamovirus virus (ORSV), which causes floral necrosis and ring spots on leaves, respectively. This is the first DNA biosensor report for the detection of plant viruses. This sensor was based on the principal of hybridization (DNA-RNA hybridization). Specific oligonucleotide probes with their 5′- phosphate end modified with a mercaptohexyl group were designed. The probes then immobilized onto quartz crystal microbalance (QCM) and test samples of viruses were drop cast. The hybridization between oligonucleotide probes and test solutions containing viral RNA took place giving response range of the sensor. In purified RNA sample, the sensor detected as low as1 ng, while for crude sample the detection limit was10 ng. Zezza *et al.* [94] first time developed a SPR-based DNA sensor for the detection of a common wheat pathogen *F. culmorum* which causes fusarium head blight, premature ear senescence, yield loss (70%) and reduction in grain quality. PCR amplified products of the fungus used for detection gave a detection limit of 0.06 pg. Papadakis *et al.* [95], for the first time, developed an acoustic-based biosensor known as the Quartz Crystal Microbalance in combination with multiplex PCR reaction without using any fluorescent label to sense three plant pathogens, *i.e. Ralstonia solanacearum, Pseudomonas syringaepv tomato* and *Xanthomonas campestris pv. vesicatoria.* The sensor showed ability to detect three pathogens simultaneously. The NiO/ITO electrode was used by Kaur *et al.* [96] for the detection of free cholesterol and low densitylipoproteins. They concluded that NiO is a potential candidate for reagent less biosensors. Tak *et al.* [97] detect meningitis by electrochemical DNA biosensor using Ni-ZnO thin film. Electrochemical studies (CV and DPV) showed low detection limit of 5 ng/µl with a wide liner range of detection (5 ng/µl- 200 ng/µl). Cai *et al.* [98] developed a DNA walker and DNA nanoflowers based electrochemical biosensor for the detection of Staphylococcus aureus with great sensitivity. DNA nanoflowers provide binding sites for electroactive redox mediator *i.e.* methylene blue (MB) and so provide a strong signal.

APTASENSORS

The aptamer technology is a newly developed diagnosis and analysis method in food safety, agricultural monitoring, pesticides and nutrient control strategies. Aptamers are 15-40 bases long synthetic single stranded DNA/RNA oligonucleotides discriminated by systematic evolution of exponential enrichment technology (SELEX) process. These selectively bind with organic and inorganic targets of low molecular weight, proteins, or even entire cells [15]. The aptamers are synthetic so they can be easily manipulated or modify after the SELEX process to increase the selectivity and affinity with its target molecule for better recognition [88].

As we know that antibodies have high affinity and specificity which is why they are frequently used in biosensors as the bio-recognition elements. However, they also have some shortcomings, such as they are unmanageable, have a complex selection process are costly and require high environmental prerequisite for detection. Therefore, to overcome these limitations for detection of pathogenic entities, new bio-recognition elements with high specificity known as aptamers are needed to explore [99]. Once aptamers are selected, they can be synthesized with high reproducibility and purity. They can be classified as:

- DNA aptamers
- RNA aptamers
- Peptide aptamers [87]

Aptasensor is developed by immobilizing aptamers on a surface using encapsulation or entrapment by adding chemical linkers and bioconjugate terminus like thiol group to one end of the structure. Aptamers are then exposed to a sample with different targets and selectively bind the target, and signal transduction is measured. The most common detection strategies coupled with the aptasensors are SPR and quartz crystal microbalance (QCM)). Different types of aptasensors have been developed based on different approaches like electrochemical, optical, mechanical, and acoustic. Aptamers are mainly used in food security [100] and for human disease diagnosis. The aptasensors are also used for the detection of mycotoxins, a toxic chemical produced by fungi in food. Gold nanoparticle based aptasensor was developed by Hosseini *et al*. [101] to detect aflatoxin B1. It is based on colour change from yellow to purple when aflatoxin is present in the sample and causes destabilization of AuNP-aptasensor. Similarly, Luan *et al*. [102] developed colorimetric AuNP-aptasensor for aflatoxin B2. Also, there are antiviral and antimicrobial aptamers used for human disease diagnosis like HIV, influenza, hepatitis. Aptamers that are selected against virus surface proteins are used for inhibition of the virus and also used to develop diagnostic tool which directly inhibits the key enzyme necessary for viral life cycle [103]. There are a large number of aptamers that have been isolated that specifically bind to infectious microbes such as *Staphylococcus, Escherichia coli, Salmonella, Bacillus thuringiensis, Campylobacter jejuni, Mycobacterium tuberculosis* and *Campylobacter coli* [104]. Further, they are used in vaccine development to overcome the major limitation of vaccine that are cost and time. The detection, identification, and quantification of disease-causing microbes and viruses are essential for their early diagnosis. In plant diagnostics, there are very few examples of aptamer-based plant pathogen detection. For example, Balogh *et al*. [105] develop a diagnostic aptamer-based sensor for apple stem pitting virus (ASPV). In this detection method, the antibody with apple stem pitting virus specific aptamers were used. They developed a protocol for measuring protein concentrations and virus identification in complex sample matrix by double oligonucleotide sandwich-enzyme-linked oligonucleotide assay (DOS-ELONA). The identification of virus infected plant protein extracts was done by selected aptamers with SPR and dot blot and coat proteins were detected with Western blot analysis. The aptamers are expected to rival antibodies as recognition molecules of virus diagnostics due to their small size, flexible selection, sensitivity and stability. Krivitsky *et al*. [106] developed an aptamer-based electrochemical biosensor for the detection of soybean rust. They developed a specialized aptamer unit that is effective against *P. pachyrhizi* rust spores. The amplification of the analytical detection of spores of interest is achieved by combining an aptamer with an enzymatic process.

PLANT PATHOGEN DIAGNOSIS USING DIRECT DNA SEQUENCING

The entire genomic information of an organism is stored in DNA or RNA. The genetic makeup is unique for each organism. The early detection of a pathogen is the most critical weapon in crop protection. DNA/RNA sequencing can be applied for pathogen detection. With the help of genomic analysis, the genomic information gathered can be used for knowing the process and gene involved in infection, thereby helping in identification of unknown plant pathogen. New technology has been developed to overcome this limitation *i.e.* next generation sequencing (NGS). The sequencing can help to understand the genetic sequence of an organism and it can be used for creating enough data information for plant pathogens to detect known and unknown plant pathogens and diseases caused by them without the need for prior knowledge of its molecular sequence. Many of the existing detection assays are available but have some limitations like they are limiting in identifying single microbe at a time.This latest technology can help in identifying culturable as well as unculturable organisms like obligate parasite (*Puccinia striiformis*), completely new unexpected pathogens, or new variants of existing pathogens and helps in detection of the pathogen within a few days [107, 108]. Genome of tomato, turnip, rapeseed, papaya, cucumber, soybean, pecan, cacao, apple, chickpea, Arabidopsis, date palm and many others has been sequenced by using NGS. Applications of sequencing for large genome and high-copy repeat interspersion in plants are still in their early stages, but have been used for different plants such as bread wheat, sorghum, pine, black cottonwood and particularly for crops of maize, conifers, *etc*. Through NGS technology, a variety of identified and unidentified plant pathogens have been detected [8] as detailed in Table 5.

Table 5. List of Plant fungal pathogen detected by NGS technology.

Pathogen	Type	Disease	Crop	References
Pyrenophora teres f.sp. teres	Fungus	Net blotch	Barley	[109]
Phytophthora infestans	Fungus	Late blight	Potato and tomato	[110]
C. pseudonaviculata	Fungus	Buxus blight	Buxus	[111]

The sequencing can help discover new genetic markers that can be useful in breeding [112]. Currently, new efforts are made to develop single molecule sequencing (SMS) for example, nanopore based sequencing. In SAM, DNA strands are immobilized in a flow cell and DNA polymerase and fluorescently labelled nucleotides are added one at a time in cyclic manner for the detection.

Nanopore technology can be used to identify single nucleic acid polymer. It has several advantages such as quick, long read lengths, fast run times, cheap and portable, so it can be used in every laboratory for routine diagnosis. Many companies have commercialized the sequencing systems like helicos biosciences developed the first true single molecule sequencing [113]. Similarly, pacific biosciences developed real time sequencing system. Recently, for third generation sequencing, oxford nanopore technology along with IBM planned a silicon based nanopore. Attempts at data analysis for plant sample sequencing are currently in progress. Although DNA sequencing encourages the use of future portable devices for plant pathogen detection, it still has several limitations that hinder the on-site adoption of sequencing technologies globally. The main hurdle is data analysis preventing the use of sequencing technologies in detection due to non-availability of network connection for sending data to a central database, mainly in developing countries [114]. Another limitation is the time needed for the on-site pathogen detection. There are technologies that help in shorter sequence reads but at the same time they deteriorate accuracy of genome analysis. To address this problem, iPlant Collaborative Company presents infrastructure and tools to resolve the problems of data analysis of NGS technology for plant samples [115]. There is an urgent need to develop novel field operable sequencing approach to diagnose pathogen. MinION, the smallest not larger than a thumb drive and most portable NGS sequencer was designed which detects both RNA viruses, bacteria in plants and insect vectors. It is able to detect Plum pox virus from both peaches and aphid vectors. In addition, the MinION was able to detect *Candidatus Liberibacter asiaticus* from psyllids [116]. This lays the groundwork for a field usable NGS system that could detect both pathogens and vectors from a variety of environmental samples.

MOLECULARLY IMPRINTED POLYMERS (MIPS)

Molecularly imprinted polymers (MIPs), synthetic receptors, are recognised as a target as around them, synthetic polymers are synthesised [117]. Maintaining the shape of the target allows binding the target in a sample through bonding [118]. Other materials like metals, silica, quantum dots and gels can be used to fuse with MIPs to improve sensitivity and time for detection. Synthesised polymers and cheaper materials lower the costs of development of MIPs. Variability may reduce the reliability and precision using MIPs as sensors [119] but developing epitopes rather than whole molecules can make them a good choice for on-site detection. However, further associated issues should be worked out to make MIPs more cost-effective for site testing so that they provide uniform, accurate results with low variation [120].

CELL PHONE-BASED BIOSENSORS

Cell phones are one of the most widely used electronic devices worldwide. Modern cell phones are smart enough and have the capability to be a prime component of biosensors. They have certain features which make them capable of doing sensing [121]. They have high-resolution imaging, sophisticated user interfaces and advanced communication and data-processing capability. At present, auxiliary reusable devices (ARDs) and auxiliary disposable devices (ADDs) are two strategies for combining cell phones with biosensors. ARDs are specifically designed for certain cell phone brands and models such as for Samsung Galaxy SII whereas ADDs have generic designs that are compatible with a number of phone brands and models in which both the biosensing part and coupling system are disposable. ARDs assist commercial immunochromatographic assays and are applied to detect tuberculosis, HIV, and malaria. On the other hand, ADD system coupled with a SPR based biosensor is used for studying biomolecular interactions. This system was tested on various phones like Nokia, Apple and android for the detection of biomarker for cancer, kidney and inflammatory disorders. The role of cell phone-based biosensor in clinical research for human disease diagnosis has revolutionized the idea of using the cell phone-based biosensor in agriculture for monitoring and diagnosis of disease causing pathogens. Chern *et al.* [122] developed a small molecule sensing, cell phone enabled fluorescent bead assay using transcription factor. The developed biosensor is very sensitive, antibody free, less expensive and fast. The change in fluorescent colour produced by the biosensor can be detected by eye or using a less expensive pocket friendly camera. The future biosensing devices can take advantage of the smart features of the phones. The importance of the early diagnosis and the devices which are field operable, portable, handy and easy to use calls for the use of such technologies. The farmers cannot perform the sophisticated laboratory process for diagnosis. So, for on-site detection of the pathogen, cell phone-based biosensors are the most promising sensors as they can be easily used by farmers and also pocket friendly. However, there are obstacles to integrate biosensors with cell phones, and a future positive scenario would require active collaboration between phone manufacturers and ARD/ADD developers to develop hardware and software for POCT applications [123].

EPILOGUE

The fungal pathogens are like hackers as they disrupt plant cells and act as specialized infiltrators. They cause quality deficiency and upto 25% of crop losses accounting the major threat for food security. The eradication of pathogens and enhancing for the productivity enhancement of effective pathogen control policy is the major world concern nowadays. Preventing yield losses require timely and precise identification and detection of pathogens. Previously, conventional methods were used to diagnose the disease that included visual counting of the disease symptoms and pesticide spray. Calendar based schedule of pesticide spray becomes no more effective due to the emergence of new pathogens as in some plants, the symptoms appear very late and till the time disease become epidemic. There are lots of examples in the history which indicate how devastating a plant disease could be. The early detection of the plant diseases, even in symptomless condition, is crucial for taking up appropriate control measures [15]. This calls for an urgent need for rapid, reliable, sensitive, cost effective and easy to use diagnostic methods for fungal pathogen detection. The world has made tremendous progress in the field of science and technology and with this advancement, scientists are developing new approaches for pathogen detection and understanding the relation between the plant and pathogen [124]. Most of the serological and nucleic acid based approaches meet these requirements and overcome the limitation of conventional methods. The advance methods of disease diagnosis involve PCR, ELISA, GC-MS, Flow cytometry, thermography and many more. Many diagnostic kits have been developed for the detection of pathogens which promise to result in 24 hours. However, these approaches help in the detection of the pathogen but majority of these are confined to specialized laboratories, require sophisticated instruments as well as labour intensive and time consuming. Moreover, these are costly, highly trained and specialized staff is required for the detection and on-site detection is not possible with these approaches [125]. There is high demand for approaches which are more users friendly, on-site, robust, most important low cost and perform real time monitoring. The diagnostics market is increasing rapidly and covers a wide range of disciplines like biotechnology, nanotechnology, molecular biology, bioinformatics and physics altogether. Scientists have spent an immense amount of time in research on diagnostics to develop small, portable instruments, having an on-site application and relatively inexpensive. To overcome the entire hurdles in the pathogen detection, biosensors were developed. They are mostly used in the medicine and are highly successful; however their potential success in the food, environment and agriculture is still to be established. The future of pathogen detection lies in the successful competency of biosensors with the developed DNA and immunoassay techniques. Biosensors would become a promising and attractive alternative, but they still have to be subjected to some modifications, improvements and proper validation for on-field use. The sensor needs to be portable and simple so that it can directly be used by farmers to check the presence of pathogen and to detect the degree of infection [8]. It should have the ability to diagnose a large number of pathogens with several known pathogen sources, be accurate and cost-effective with laboratory-scale technologies, and have a minimum number of manipulation steps. Both immunobiosensor and DNA based biosensor are promising for phytopathology but nucleic acid biosensors are of major interest due to their simple, fast and cost-effective nature to obtain sequence specific information [126]. They are more sensitive, effective and specific in comparison to antibody-based biosensors as these detect single target pathogen within a complex mixture of different analyte, detect multiple targets and also unculturable pathogens. The recognition layer for nucleic acid can be used for multiple applications and synthesized readily [4]. DNA based biosensors are advantageous over antibody based due to the use of nucleic acid amplification techniques, which allows to detect plant pathogen in symptomless condition, before the appearance of disease symptoms [10]. Present techniques for rapid diagnosis of new or formerly detected diseases show that progress in this area, particularly for unidentified samples, is slow compared to what is needed. The increasing numbers of analytes require monitoring, control and high sensitivity and effort in the development of different sensors as diagnostic tool. However, plant pathogen diagnostics are expected to improve by involving emerging techniques in the coming years. Multiplexing assays to detect several pathogens simultaneously is also a major advantage of emerging lab-based methods. In comparison, point-of-care techniques such as isothermal amplification, biosensors, nano-biosensors, cell phone (CP) based devices, paper-based devices (PBD), lab-on-chip (LOC), and robotics are easy to use by a non-skilled person, require less handling and provide simple readouts [127].

CONCLUSION

Monitoring plant health and detecting pathogens areessential to reduce disease spread and facilitate effective management practices. When plant diseases are correctly diagnosed and identified early, the need-based treatments also translate to economic and environmental gains. Crop losses can be minimized, and specific treatments can be tailored to combat specific pathogens. Portable diagnostic equipment, bio-barcoded DNA sensors based on nanoparticles and QDs have been used in the multiple detection of plant pathogens and toxigenic fungi. To date, mobile diagnostic tests have been developed to quickly detect plant diseases and can be used to prevent epidemics. These nano-based diagnostic kits not only increase the rate of pathogen detection but also increase the accuracy of the diagnosis. In addition, a combination of nanotechnology and microfluidic systems has been effectively applied to molecular plant pathology and can be adapted to detect specific pathogens and toxins. Nano-scale devices with novel properties could be used to make smart agricultural systems in the near future. The nanosmart devices are both an early warning system and a protective system. Nano devices will be made available during the next decade, which can quickly and very cheaply produce thousands of measurements. Future prospects for diagnosing plant diseases will continue with the miniaturization of nano scale biochip technology. Nano phytopathology can be used as a tool for understanding plant-pathogenic interactions, providing new methods for crop protection.

CONSENT FOR PUBLICATION

Not applicable.

CONFLICT OF INTEREST

The authors declare no conflict of interest, financial or otherwise.

ACKNOWLEDGEMENT

Declared none.

REFERENCES

[1]   Velásquez AC, Castroverde CDM, He SY. Vela ́squez, AC, Castroverde CDM, He SY. Plant–Pathogen Warfare under Changing Climate Conditions. Curr Biol 2018; 28(10): 619-34.
[http://dx.doi.org/10.1016/j.cub.2018.03.054]

[2]   Kivirand K, Rinken T. Introductory Chapter: Why Do We Need Rapid Detection of pathogens IntechOpen. Tartu, Estonia: University of Tartu 2018.

[3]   University of Florida (US). Plant pathogens Emerging plant pathogen institute 2006.

[4]   Khater M, de la Escosura-Muñiz A, Merkoçi A. Biosensors for plant pathogen detection. J Biosens Bioelectron 2017; 93: 72-86.
[http://dx.doi.org/10.1016/j.bios.2016.09.091] [PMID: 27818053]

[5]   Savary S, Willocquet L, Pethybridge SJ, Esker P, McRoberts N, Nelson A. The global burden of pathogens and pests on major food crops. Nat Ecol Evol 2019; 3(3): 430-9.
[http://dx.doi.org/10.1038/s41559-018-0793-y] [PMID: 30718852]

[6]   Manjunatha SB, Biradar DP, Aladakatti YR. Nanotechnology and its applications in agriculture: A review. J Farm Sci 2016; 29(1): 1-13.

[7]   Mahlein AK, Kuska MT, Behmann J, Polder G, Walter A. Hyperspectral sensors and imaging technologies in phytopathology: state of the art. Annu Rev Phytopathol 2018; 56(1): 535-58.
[http://dx.doi.org/10.1146/annurev-phyto-080417-050100] [PMID: 30149790]

[8]   Nezhad AS. Future of portable devices for plant pathogen diagnosis. Lab Chip 2014; 14(16): 2887-904.
[http://dx.doi.org/10.1039/C4LC00487F] [PMID: 24920461]

[9]   Mendez MJ, Aimar SB, Aparicio VC, *et al.* Glyphosate and Aminomethylphosphonic acid (AMPA) contents in the respirable dust emitted by an agricultural soil of the central semiarid region of Argentina. Aeolian Res 2017; 29: 23-9.
[http://dx.doi.org/10.1016/j.aeolia.2017.09.004]

[10]   Martinelli F, Scalenghe R, Davino S, *et al.* Advanced methods of plant disease detection. A review. Agron Sustain Dev 2015; 35(1): 1-25.
[http://dx.doi.org/10.1007/s13593-014-0246-1]

[11]   Doughari JH. An Overview of Plant Immunity. J Plant Pathol Microbiol 2015; 6(11): 1-11.

[12]   Alahi MEE, Mukhopadhyay SC. Detection Methodologies for Pathogen and Toxins: A Review. Sensors (Basel) 2017; 17(8): 1885.

[http://dx.doi.org/10.3390/s17081885] [PMID: 28813028]

[13]   Aggarwal R, Sharma S, Gupta S, Manjunatha C, Singh VK, Kulshreshtha D. Gene-based analysis of Puccinia species and development of PCR-based marker to detect Puccinia striiformis f. sp. tritici causing yellow rust of wheat. J Gen Plant Pathol 2017; 83(4): 205-15.
[http://dx.doi.org/10.1007/s10327-017-0723-x]

[14]   Notomi T, Okayama H, Masubuchi H, *et al.* Loop-mediated isothermal amplification of DNA. Nucleic Acids Res 2000; 28(12): E63.
[http://dx.doi.org/10.1093/nar/28.12.e63] [PMID: 10871386]

[15]   Donoso A, Valenzuela S. In-field molecular diagnosis of plant pathogens: recent trends and future perspectives. Plant Pathol 2018; 67(7): 1451-61.
[http://dx.doi.org/10.1111/ppa.12859]

[16]   Fang Y, Ramasamy RP. Current and Prospective Methods for Plant Disease Detection. Biosensors (Basel) 2015; 5(3): 537-61.
[http://dx.doi.org/10.3390/bios5030537] [PMID: 26287253]

[17]   Khan M, Wang R, Li B, Liu P, Weng Q, Chen Q. Comparative Evaluation of the LAMP Assay and PCR-Based Assays for the Rapid Detection of *Alternaria solani*. Front Microbiol 2018; 9: 2089.
[http://dx.doi.org/10.3389/fmicb.2018.02089] [PMID: 30233554]

[18]   Schilling AG, Moller EM, Geiger HH. Polymerase chain reaction-based assays for species specific detection of Fusarium culmorum, F. graminearum and F. avenaceum. Mol Plant Pathol 1996; 86: 515-22.

[19]   Lievens B, Brouwer M, Vanachter AC, Cammue BP, Thomma BP. Real-time pcr for detection and quantification of fungal and oomycete tomato pathogens in plant and soil samples. Plant Sci 2006; 171(1): 155-65.
[http://dx.doi.org/10.1016/j.plantsci.2006.03.009]

[20]   Baysal-Gurel F, Ivey MLL, Dorrance A, *et al.* An Immunofluorescence Assay to Detect Urediniospores of Phakopsora pachyrhizi. Plant Dis 2008; 92(10): 1387-93.
[http://dx.doi.org/10.1094/PDIS-92-10-1387] [PMID: 30769566]

[21]   Tomlinson JA, Boonham N, Hughes KJD, Griffin RL, Barker I. On-site DNA extraction and real-time PCR for detection of Phytophthora ramorum in the field. Appl Environ Microbiol 2005; 71(11): 6702-10.
[http://dx.doi.org/10.1128/AEM.71.11.6702-6710.2005] [PMID: 16269700]

[22]   Manjunatha C, Sharma S, Kulshreshtha D, *et al.* Rapid detection of Puccinia triticina causing leaf rust of wheat by PCR and loop mediated isothermal amplification. PLoS One 2018; 13(4): e0196409.
[http://dx.doi.org/10.1371/journal.pone.0196409] [PMID: 29698484]

[23]   Mullis KB. Ferre' F, Gibbs RA (Eds.). Polymerase chain reaction. Birkhäuser 1994; p. 458.

[24]   Okubara PA, Schroeder KL, Paulitz TC. Identification and quantification of *Rhizoctonia solani* and R. oryzae using real-time polymerase chain reaction. Phytopathology 2008; 98(7): 837-47.
[http://dx.doi.org/10.1094/PHYTO-98-7-0837] [PMID: 18943261]

[25]   Budge GE, Shaw MW, Colyer A, Pietravalle S, Boonham N. Molecular tools to investigate Rhizoctonia solani distribution in soil. Plant Pathol 2009; 58(6): 1071-80.
[http://dx.doi.org/10.1111/j.1365-3059.2009.02139.x]

[26]   Weikert-Oliveira RCB, Resende MA, Valerio HM, Caligiorne RB, Paiva E. Genetic variation among pathogens causing Helminthosporium diseases of rice, maize and wheat. Fitopatol Bras 2002; 27(6): 238-46.
[http://dx.doi.org/10.1590/S0100-41582002000600015]

[27]   Ma J, Chen X, Wang M, Kang Z. Constructing Physical and Genomic Maps for Puccinia striiformisf sptritici, the Wheat Stripe Rust Pathogen, by Comparing Its EST Sequences to the Genomic Sequence of P graminis f sp tritici, the Wheat Stem Rust Pathogen. Comp FunctGenom 2009; pp. 1-13.

[28]   Komínková E, Dreiseitl A, Malečková E, Doležel J, Valárik M. Genetic Diversity of Blumeria graminis f. sp. hordei in Central Europe and Its Comparison with Australian Population. PLoS One 2016; 11(11): e0167099.
[http://dx.doi.org/10.1371/journal.pone.0167099] [PMID: 27875588]

[29]   Chongo G, Gossen BD, Buchwaldt L, Adhikari T, Rimmer SR. Genetic diversity of Ascochyta rabiei in Canada. Plant Dis 2004; 88(1): 4-10.
[http://dx.doi.org/10.1094/PDIS.2004.88.1.4] [PMID: 30812454]

[30]   Chanda AK, Ward NA, Robertson CL, Chen ZY, Schneider RW. Development of a quantitative polymerase chain reaction detection protocol for Cercosporakikuchii in soybean leaves and its use for documenting latent infection as affected by fungicide applications. Phytopathology 2014; 104(10): 1118-24.
[http://dx.doi.org/10.1094/PHYTO-07-13-0200-R] [PMID: 24805074]

[31]   Haegi A, Catalano V, Luongo L, *et al.* A newly developed real-time PCR assay for detection and quantification of Fusarium oxysporum and its use in compatible and incompatible interactions with grafted melon genotypes. Phytopathology 2013; 103(8): 802-10.
[http://dx.doi.org/10.1094/PHYTO-11-12-0293-R] [PMID: 23464901]

[32]   Rasu T, Sevugapperumal N, Thiruvengadam R, Ramasamy S. Morphological and genomic variability among Sclerotium rolfsii populations. Bioscan •••; 8(4): 1425-30. [Supplement on Genetics and Plant Breeding].

[33]   Grünwald NJ, Martin FN, Larsen MM, *et al.* Phytophthora-ID.org: a sequence based Phytophthora identification tool. Plant Dis 2011; 95(3): 337-42.
[http://dx.doi.org/10.1094/PDIS-08-10-0609] [PMID: 30743500]

[34]   Rigotti S, Viret O, Gindro K. Two new primers highly specific for the detection of botrytis cinerea pers. Fr Phytopathol Mediterr 2006; 45(3): 253-60.

[35]   Kamle M, Pandey BK, Kumar P, Kumar M, Kamle M. A species-specific PCR based assay for rapid detection of mango anthracnose pathogen colletotrichum gloeosporioidespenz. and Sacc. J Plant Pathol Microbiol 2013; 4(184): 10-4172.

[36]   Koenraadt H, Somerville SC, Jones AL. Characterization of mutations in the beta-tubulin gene of benomyl-resistant field strains of Venturia inaequalis and other plant pathogenic fungi. Phytopathology 1992; 82(11): 1348-54.
[http://dx.doi.org/10.1094/Phyto-82-1348]

[37]   Kumar S, Archak S, Tyagi RK, *et al.* Evaluation of 19460 wheat accessions conserved in the Indian national Genebank to identify new sources of resistance to rust and spot blotch diseases. PLoS One 2016; 11(12): e0167702.
[http://dx.doi.org/10.1371/journal.pone.0167702] [PMID: 27942031]

[38]   Fang Y, Umasankar Y, Ramasamy RP. Electrochemical detection of p-ethylguaiacol, a fungi infected fruit volatile using metal oxide nanoparticles. Analyst (Lond) 2014; 139(15): 3804-10.
[http://dx.doi.org/10.1039/C4AN00384E] [PMID: 24895939]

[39]   Chang Z, Lv J, Qi H, *et al.* Bacterial Infection Potato Tuber Soft Rot Disease Detection Based on Electronic Nose. Open Life Sci 2017; 12(1): 379-85.
[http://dx.doi.org/10.1515/biol-2017-0044]

[40]   Delwiche SR, Kim MS. Biological Quality and Precision Agriculture II. Proc SPIE 2000; 4203: 13-20.
[http://dx.doi.org/10.1117/12.411752]

[41]   Yuan L, Zhang J, Zhao J, Du S, Huang W, Wang J. Discrimination of yellow rust and powdery mildew in wheat at leaf level using spectral signatures.
[http://dx.doi.org/10.1109/Agro-Geoinformatics.2012.6311599]

[42]   Al Masri A, Hau B, Dehne HW, Mahlein AK, Oerke EC. Impact of primary infection site of Fusarium species on head blight development in wheat ears evaluated by IR-thermography. Eur J Plant Pathol

2016; 147(4): 855-68.
[http://dx.doi.org/10.1007/s10658-016-1051-2]

[43]   Zhang M, Qin Z, Liu X, Ustin SL. Detection of stress in tomatoes induced by late blight disease in California, USA, using hyperspectral remote sensing. ITC J 2003; 4(4): 295-310.
[http://dx.doi.org/10.1016/S0303-2434(03)00008-4]

[44]   Vikram A, Lui LH, Hossain A, Kushalappa AC. Metabolic fingerprinting to discriminate diseases of stored carrots. Ann Appl Biol 2006; 148(1): 17-26.
[http://dx.doi.org/10.1111/j.1744-7348.2005.00036.x]

[45]   Sighicelli M, Colao F, Lai A, Patsaeva S. Monitoring post-harvest orange fruit disease by fluorescence and reflectance hyperspectral imaging. Acta Hortic 2009; 817(817): 277-84.
[http://dx.doi.org/10.17660/ActaHortic.2009.817.29]

[46]   Bulanon DM, Burks TF, Kim DG, Ritenour MA. Citrus black spot detection using hyperspectral image analysis. Agric Eng Int 2013; 15(3): 171.

[47]   Konanz S, Kocsányi L, Buschmann C. Advanced Multi-Color Fluorescence Imaging System for Detection of Biotic and Abiotic Stresses in Leaves. Agriculture 2014; 4(2): 79-95.
[http://dx.doi.org/10.3390/agriculture4020079]

[48]   Stoll M, Schultz HR, Baecker G, Berkelmann-Loehnertz B. Berkelmann-LoehnertzB. Early pathogen detection under different water status and the assessment of spray application in vineyards through the use of thermal imagery. Precis Agric 2008; 9(6): 407-17.
[http://dx.doi.org/10.1007/s11119-008-9084-y]

[49]   Mahlein AK, Steiner U, Hillnhütter C, Dehne HW, Oerke EC. Hyperspectral imaging for small-scale analysis of symptoms caused by different sugar beet diseases. Plant Methods 2012; 8(1): 3.
[http://dx.doi.org/10.1186/1746-4811-8-3] [PMID: 22273513]

[50]   Moalemiyan M, Vikram A, Kushalappa AC, Yaylayan V. Volatile metabolite profiling to detect and discriminate stem-end rot and anthracnose diseases of mango fruits. Plant Pathol 2006; 55(6): 792-802.
[http://dx.doi.org/10.1111/j.1365-3059.2006.01443.x]

[51]   Delalieux S, Auwerkerken A, Verstraeten WW, *et al.* Hyperspectral Reflectance and Fluorescence Imaging to Detect Scab Induced Stress in Apple Leaves. Remote Sens (Basel) 2009; 1(4): 858-74.
[http://dx.doi.org/10.3390/rs1040858]

[52]   Tomlinson JA, Dickinson MJ, Boonham N. Rapid detection of Phytophthora ramorum and P. kernoviae by two-minute DNA extraction followed by isothermal amplification and amplicon detection by generic lateral flow device. Phytopathology 2010; 100(2): 143-9.
[http://dx.doi.org/10.1094/PHYTO-100-2-0143] [PMID: 20055648]

[53]   Fletcher RA, Gill A, Davis TD, Sankhla N. Triazoles as plant growth regulators and stress protectants. Hortic Rev (Am Soc Hortic Sci) 2000; 24: 55-138.

[54]   Kawamura A, Miyata T. Biosensors.Biomaterials Nanoarchitectonics. Elsevier Inc. 2016; pp. 157-76.
[http://dx.doi.org/10.1016/B978-0-323-37127-8.00010-8]

[55]   Khiyami MA, Almoammar H, Awad YM, Alghuthaymi MA, Abd-Elsalam KA. Plant pathogen nanodiagnostic techniques: forthcoming changes? Biotechnol Biotechnol Equip 2014; 28(5): 775-85.
[http://dx.doi.org/10.1080/13102818.2014.960739] [PMID: 26740775]

[56]   Sharon M, Choudhary AK, Kumar R. Nanotechnology in agricultural diseases and food safety. J Phytol 2010; 2(4): 83-92.

[57]   Yao KS, Li SJ, Tzeng KC, *et al.* Fluorescence Silica Nanoprobe as a Biomarker for Rapid Detection of Plant Pathogens. Adv Mat Res 2009; 79–82: 513-6.
[http://dx.doi.org/10.4028/www.scientific.net/AMR.79-82.513]

[58]   Singh S, Singh M, Agrawal VV, Kumar A. An attempt to develop surface plasmon resonance based

immunosensor for Karnal bunt (*Tilletia indica*) diagnosis based on the experience of nano-gold based lateral flow immuno-dipstick test. Thin Solid Films 2010; 519(3): 1156-9.
[http://dx.doi.org/10.1016/j.tsf.2010.08.061]

[59]   Sapsforda KE, Bradburneb C, Delehanty JB, Medintz IL. Sensors for detecting biological agents. Mater Today 2008; 11(3): 38-49.
[http://dx.doi.org/10.1016/S1369-7021(08)70018-X]

[60]   Wood RW. On a remarkable case of uneven distribution of light in a diffraction grating spectrum. Lond Edinb Dublin Philos Mag J Sci 1902; 64(21): 396-402.
[http://dx.doi.org/10.1080/14786440209462857]

[61]   Liedberg B, Nylander C, Lunström I. Surface plasmon resonance for gas detection and biosensing. Sens Actuators 1983; 4: 299-304.
[http://dx.doi.org/10.1016/0250-6874(83)85036-7]

[62]   Prabowo BA, Purwidyantri A, Liu KC. Surface Plasmon Resonance Optical Sensor: A Review on Light Source Technology. Biosensors (Basel) 2018; 8(3): 80.
[http://dx.doi.org/10.3390/bios8030080] [PMID: 30149679]

[63]   Zeng C, Huang X, Xu J, *et al.* Rapid and sensitive detection of maize chlorotic mottle virus using surface plasmon resonance-based biosensor. Chin J Anal Chem 2013; 440(1): 18-22.
[http://dx.doi.org/10.1016/j.ab.2013.04.026] [PMID: 23660014]

[64]   Vaisocherová H, Mrkvová K, Piliarik M, Jinoch P, Steinbachová M, Homola J. Surface plasmon resonance biosensor for direct detection of antibody against Epstein-Barr virus. Biosens Bioelectron 2007; 22(6): 1020-6.
[http://dx.doi.org/10.1016/j.bios.2006.04.021] [PMID: 16797175]

[65]   Shankaran DR, Gobi KV, Miura N. Recent advancements in surface plasmon resonance immunosensors for detection of small molecules of biomedical, food, and environmental interest. Sens. Sens Actuators B Chem 2007; 121(1): 158-77.
[http://dx.doi.org/10.1016/j.snb.2006.09.014]

[66]   Carlsson J, Gullstrand C, Westermark GT, Ludvigsson J, Enander K, Liedberg B. An indirect competitive immunoassay for insulin autoantibodies based on surface plasmon resonance. Biosens Bioelectron 2008; 24(4): 882-7.
[http://dx.doi.org/10.1016/j.bios.2008.07.018] [PMID: 18722764]

[67]   Skottrup PD, Nicolaisen M, Justesen AF. Towards on-site pathogen detection using antibody-based sensors. Biosens Bioelectron 2008; 24(3): 339-48.
[http://dx.doi.org/10.1016/j.bios.2008.06.045] [PMID: 18675543]

[68]   Candresse T, Lot H, German-Retana S, *et al.* Analysis of the serological variability of Lettuce mosaic virus using monoclonal antibodies and surface plasmon resonance technology. J Gen Virol 2007; 88(Pt 9): 2605-10.
[http://dx.doi.org/10.1099/vir.0.82980-0] [PMID: 17698673]

[69]   Skottrup P, Nicolaisen M, Justesen AF. Rapid determination of Phytophthora infestans sporangia using a surface plasmon resonance immunosensor. J Microbiol Methods 2007; 68(3): 507-15.
[http://dx.doi.org/10.1016/j.mimet.2006.10.011] [PMID: 17157943]

[70]   Chen XM. Epidemiology and control of stripe rust (Puccinia striiformisf. sp. tritici) on wheat. Can J Plant Pathol 2005; 27(3): 314-37.
[http://dx.doi.org/10.1080/07060660509507230]

[71]   Mendes RK, Carvalhal RF, Stach-Machado DR, Kubota LT. Surface plasmon resonance immunosensor for early diagnosis of Asian rust on soybean leaves. Biosens Bioelectron 2009; 24(8): 2483-7.
[http://dx.doi.org/10.1016/j.bios.2008.12.033] [PMID: 19200709]

[72]   Eun AJC, Huang L, Chew FT, Li SFY, Wong SM. Detection of two orchid viruses using quartz crystal

microbalance (QCM) immunosensors. J Virol Methods 2002; 99(1-2): 71-9.
[http://dx.doi.org/10.1016/S0166-0934(01)00382-2] [PMID: 11684305]

[73]   Vashist SK, Vashist P. Recent advances in quartz crystal microbalance-based sensors. Sensors (Basel) 2011; 1-13.

[74]   Chaudhary M, Gupta A. Microcantilever-based sensors. Def Sci J 2009; 59(6): 634-41.
[http://dx.doi.org/10.14429/dsj.59.1569]

[75]   Campbell GA, Mutharasan R. Detection of Bacillus anthracis spores and a model protein using PEMC sensors in a flow cell at 1 mL/min. Biosens Bioelectron 2006; 22(1): 78-85.
[http://dx.doi.org/10.1016/j.bios.2005.12.002] [PMID: 16423521]

[76]   Nugaeva N, Gfeller KY, Backmann N, *et al.* An antibody-sensitized microfabricated cantilever for the growth detection of *Aspergillus niger* spores. Microsc Microanal 2007; 13(1): 13-7.
[http://dx.doi.org/10.1017/S1431927607070067] [PMID: 17234032]

[77]   Garcia MNV, Mottram T. Biosensor technology addressing agricultural problems. Biosyst Eng 2003; 84(1): 1-12.
[http://dx.doi.org/10.1016/S1537-5110(02)00236-2]

[78]   Majumder D, Rajesh T, Suting EG, Debbarma A. Detection of seed borne pathogens in wheat: recent trends. AJCS 2013; 7(4): 500-7.

[79]   Blakemore EJA, Law JR, Reeves JC. PCR identification of Erwinia stewartii and its comparison with two other methods. J. Seed Sci Technol 1999; 27(1): 385-96.

[80]   Perdikaris A, Vassilakos N, Yiakoumettis I, Kektsidou O, Kintzios S. Development of a portable, high throughput biosensor system for rapid plant virus detection. J Virol Methods 2011; 177(1): 94-9.
[http://dx.doi.org/10.1016/j.jviromet.2011.06.024] [PMID: 21781989]

[81]   Behl S, Jarrar M, Gaur S. Need for rapid and novel diagnostic techniques for quarantined fungal pathogens of wheat such as karnal bunt (tilletia indica). Int J Recent Sci Res 2015; 6(5): 4252-6.

[82]   Singh S, Singh M, Taj G, Gupta S, Kumar A. Development of surface plasmon resonance (spr) based immuno-sensing system for detection of fungal teliospores of karnal bunt (*Tilletia indica*), a quarantined disease of wheat. Biosens Bioelectron 2012; 3(4): 1-8.

[83]   Lin HY, Huang CH, Lu SH, Kuo IT, Chau LK. Direct detection of orchid viruses using nanorod-based fiber optic particle plasmon resonance immunosensor. Biosens Bioelectron 2014; 51: 371-8.
[http://dx.doi.org/10.1016/j.bios.2013.08.009] [PMID: 24001513]

[84]   Feng M, Kong D, Wang W, Liu L, Song S, Xu C. Development of an immunochromatographic strip for rapid detection of *Pantoea stewartii subsp. stewartii*. Sensors (Basel) 2015; 15(2): 4291-301.
[http://dx.doi.org/10.3390/s150204291] [PMID: 25686315]

[85]   Stewart FC. A Bacterial Disease of Sweet Com. NY Agric Exp Stn Bull 1897; 130: 422-39.

[86]   Srinivasan B, Tung S. Development and Applications of Portable Biosensors. J Lab Autom 2015; 20(4): 365-89.
[http://dx.doi.org/10.1177/2211068215581349] [PMID: 25878051]

[87]   Karunakaran C, Rajkumar R, Bhargava K. Introduction to Biosensors. 2015.
[http://dx.doi.org/10.1016/B978-0-12-803100-1.00001-3]

[88]   Kavita V. DNA Biosensors-A Review. J Biomed Eng 2017; 7(2): 1-5.

[89]   Paleček E, Bartošík M. Electrochemistry of nucleic acids. Chem Rev 2012; 112(6): 3427-81.
[http://dx.doi.org/10.1021/cr200303p] [PMID: 22372839]

[90]   Paleček E. Past, present and future of nucleic acids electrochemistry. Talanta 2002; 56(5): 809-19.
[http://dx.doi.org/10.1016/S0039-9140(01)00649-X] [PMID: 18968559]

[91]   Vercoutere W, Akeson M. Biosensors for DNA sequence detection. Curr Opin Chem Biol 2002; 6(6): 816-22.

[http://dx.doi.org/10.1016/S1367-5931(02)00395-2] [PMID: 12470736]

[92]   Berney H, West J, Haefele E, Alderman J, Lane W, Collins JK. A DNA diagnostic biosensor: Development, characterisation and performance. Sens Actuators B Chem 2000; 68(1-3): 100-8.
[http://dx.doi.org/10.1016/S0925-4005(00)00468-8]

[93]   Eun AJC, Huang L, Chew FT, Fong-Yau Li S, Wong SM. Detection of two orchid viruses using quartz crystal microbalance-based DNA biosensors. Phytopathology 2002; 92(6): 654-8.
[http://dx.doi.org/10.1094/PHYTO.2002.92.6.654] [PMID: 18944263]

[94]   Zezza F, Pascale M, Mulè G, Visconti A. Detection of Fusarium culmorum in wheat by a surface plasmon resonance-based DNA sensor. J Microbiol Methods 2006; 66(3): 529-37.
[http://dx.doi.org/10.1016/j.mimet.2006.02.003] [PMID: 16563535]

[95]   Papadakis G, Skandalis N, Dimopoulou A, Glynos P, Gizeli E. Bacteria murmur: Application of an acoustic biosensor for plant pathogen detection. PLoS One 2015; 10(7): e0132773.
[http://dx.doi.org/10.1371/journal.pone.0132773] [PMID: 26177507]

[96]   Kaur G, Tomar M, Gupta V. Nanostructured NiO-based reagentless biosensor for total cholesterol and low density lipoprotein detection. Anal Bioanal Chem 2017; 409(8): 1995-2005.
[http://dx.doi.org/10.1007/s00216-016-0147-z] [PMID: 28078419]

[97]   Tak M, Gupta V, Tomar M. An electrochemical DNA biosensor based on Ni doped ZnO thin film for meningitis detection. J Electroanal Chem (Lausanne) 2017; 792: 8-14.
[http://dx.doi.org/10.1016/j.jelechem.2017.03.032]

[98]   Cai R, Zhang S, Chen L, Li M, Zhang Y, Zhou N. Self-assembled DNA nanoflowers triggered by a DNA walker for highly sensitive electrochemical detection of *Staphylococcus aureus*. ACS Appl Mater Interfaces 2021; 13(4): 4905-14.
[http://dx.doi.org/10.1021/acsami.0c22062] [PMID: 33470807]

[99]   Xian WY, Zhong YZ, Yan SC. B YY. Application of Aptamer Based Biosensors for Detection of Pathogenic Microorganisms. Chin J Anal Chem 2012; 40(4): 634-42.
[http://dx.doi.org/10.1016/S1872-2040(11)60542-2]

[100]   McKeague M, Bradley CR, De Girolamo A, Visconti A, Miller JD, Derosa MC. Screening and initial binding assessment of fumonisin b(1) aptamers. Int J Mol Sci 2010; 11(12): 4864-81.
[http://dx.doi.org/10.3390/ijms11124864] [PMID: 21614178]

[101]   Hosseini M, Khabbaz H, Dadmehr M, Ganjali MR, Mohamadnejad J. Aptamer-based colorimetric and chemiluminescence detection of aflatoxin B1 in foods samples. Acta Chim Slov 2015; 62(3): 721-8.
[http://dx.doi.org/10.17344/acsi.2015.1358] [PMID: 26466094]

[102]   Luan Y, Chen J, Xie G, *et al.* Visual and microplate detection of aflatoxin B2 based on NaCl-induced aggregation of aptamer-modified gold nanoparticles. Mikrochim Acta 2015; 182(5-6): 995-1001.
[http://dx.doi.org/10.1007/s00604-014-1420-5]

[103]   Oluwaseun AC, Phazang P, Sarin NB. Biosensors: a fast-growing technology for pathogen detection in agriculture and food sector, biosensing technologies for the detection of pathogens - a prospective way for rapid analysis. J Intechopen 2018; pp. 1-16.

[104]   Davydova A, Vorobjeva M, Pyshnyi D, Altman S, Vlassov V, Venyaminova A. Aptamers against pathogenic microorganisms. Crit Rev Microbiol 2016; 42(6): 847-65.
[http://dx.doi.org/10.3109/1040841X.2015.1070115] [PMID: 26258445]

[105]   Balogh Z, Lautner G, Bardóczy V, Komorowska B, Gyurcsányi RE, Mészáros T. Selection and versatile application of virus-specific aptamers. FASEB J 2010; 24(11): 4187-95.
[http://dx.doi.org/10.1096/fj.09-144246] [PMID: 20624933]

[106]   Krivitsky V, Granot E, Avidor Y, Borberg E, Voegele RT, Patolsky F. Rapid Collection and Aptamer-Based Sensitive Electrochemical Detection of Soybean Rust Fungi Airborne Urediniospores. ACS Sens 2021; 6(3): 1187-98.
[http://dx.doi.org/10.1021/acssensors.0c02452] [PMID: 33507747]

[107] Berger P. wight MM, Rivera Y, Nicholson J. Next Generation Sequencing Applications for Plant Protection and quarantine. National plant board annual meeting USDA.

[108] Bronzato Badial A, Sherman D, Stone A, *et al.* Transcriptome amplification coupled with nanopore sequencing as a surveillance tool for plant pathogens in plant and insect tissues. Plant Dis 2018; 102(8): 1648-52.
[http://dx.doi.org/10.1094/PDIS-04-17-0488-RE] [PMID: 30673417]

[109] Hisano H, Sakamoto K, Takagi H, Terauchi R, Sato K. Exome QTL-seq maps monogenic locus and QTLs in barley. BMC Genomics 2017; 18(1): 125.
[http://dx.doi.org/10.1186/s12864-017-3511-2] [PMID: 28148242]

[110] Gao L, Tu ZJ, Millett BP, Bradeen JM. Insights into organ-specific pathogen defense responses in plants: RNA-seq analysis of potato tuber-Phytophthora infestans interactions. BMC Genomics 2013; 14(1): 340.
[http://dx.doi.org/10.1186/1471-2164-14-340] [PMID: 23702331]

[111] Malapi-Wight M, Salgado-Salazar C, Demers JE, Clement DL, Rane KK, Crouch JA. Sarcococca Blight: Use of Whole-Genome Sequencing for Fungal Plant Disease Diagnosis. Plant Dis 2016; 100(6): 1093-100.
[http://dx.doi.org/10.1094/PDIS-10-15-1159-RE] [PMID: 30682271]

[112] Pérez-de-Castro AM, Vilanova S, Cañizares J, *et al.* Application of genomic tools in plant breeding. Curr Genomics 2012; 13(3): 179-95.
[http://dx.doi.org/10.2174/138920212800543084] [PMID: 23115520]

[113] Gupta PK. Single-molecule DNA sequencing technologies for future genomics research. Trends Biotechnol 2008; 26(11): 602-11.
[http://dx.doi.org/10.1016/j.tibtech.2008.07.003] [PMID: 18722683]

[114] Loit K, Adamson K, Bahram M, *et al.* Relative Performance of MinION (Oxford Nanopore Technologies) *versus* Sequel (Pacific Biosciences) Third-Generation Sequencing Instruments in Identification of Agricultural and Forest Fungal Pathogens. Appl Environ Microbiol 2019; 85(21): 1-38.
[http://dx.doi.org/10.1128/AEM.01368-19] [PMID: 31444199]

[115] Goff SA, Vaughn M, McKay S, *et al.* The iPlant collaborative: Cyberinfrastructure for plant biology. Front Plant Sci 2011; 2(1): 34.
[http://dx.doi.org/10.3389/fpls.2011.00034] [PMID: 22645531]

[116] Bronzato Badial A, Sherman D, Stone A, *et al.* Nanopore Sequencing as a Surveillance Tool for Plant Pathogens in Plant and Insect Tissues. Plant Dis 2018; 102(8): 1648-52.
[http://dx.doi.org/10.1094/PDIS-04-17-0488-RE] [PMID: 30673417]

[117] Cassedy A, Mullins E, O'Kennedy R. Sowing seeds for the future: The need for on-site plant diagnostics. Biotechnol Adv 2020; 39: 107358.
[http://dx.doi.org/10.1016/j.biotechadv.2019.02.014] [PMID: 30802484]

[118] Wackerlig J, Lieberzeit PA. Molecularly imprinted polymer nanoparticles in chemical sensing – synthesis, characterisation and application. Sens Actuators B Chem 2015; 207: 144-57.
[http://dx.doi.org/10.1016/j.snb.2014.09.094]

[119] Uzun L, Turner APF. Molecularly-imprinted polymer sensors: realising their potential. Biosens Bioelectron 2016; 76: 131-44.
[http://dx.doi.org/10.1016/j.bios.2015.07.013] [PMID: 26189406]

[120] Li S, Cao S, Whitcombe MJ, Piletsky SA. Progress in polymer science size matters: challenges in imprinting macromolecules. Prog Polym Sci 2016; 39(1): 145-63.
[http://dx.doi.org/10.1016/j.progpolymsci.2013.10.002]

[121] Neethirajan S. Recent advances in wearable sensors for animal health management. Sens Biosensing Res 2017; 12: 15-29.

[http://dx.doi.org/10.1016/j.sbsr.2016.11.004]

[122] Chern M, Garden PM. Baer RC, Galagan JE, Dennis AM. Transcription factor based smallâ molecule sensing with a rapid cell phone enabled fluorescent bead assay. Angewandte Chemie International Edition 2020.
[http://dx.doi.org/10.1002/anie.202007575]

[123] Preechaburana P, Suska A, Filippini D. Biosensing with cell phones. Trends Biotechnol 2014; 32(7): 351-5.
[http://dx.doi.org/10.1016/j.tibtech.2014.03.007] [PMID: 24702730]

[124] López MM, Bertolini E, Olmos A, *et al.* Innovative tools for detection of plant pathogenic viruses and bacteria. Int Microbiol 2003; 6(4): 233-43.
[http://dx.doi.org/10.1007/s10123-003-0143-y] [PMID: 13680391]

[125] Capote N, Bertolini E, Olmos A, Vidal E, Martínez MC, Cambra M. Direct sample preparation methods for the detection of Plum pox virus by real-time RT-PCR. Int J Microbiol 2009; 12(1): 1-6.
[PMID: 19440977]

[126] Cesewski E, Johnson BN. Electrochemical biosensors for pathogen detection. Biosens Bioelectron 2020; 159: 112214.
[http://dx.doi.org/10.1016/j.bios.2020.112214] [PMID: 32364936]

[127] Rani A, Donovan N, Mantri N. Review: The future of plant pathogen diagnostics in a nursery production system. Biosens Bioelectron 2019; 145: 111631.
[http://dx.doi.org/10.1016/j.bios.2019.111631] [PMID: 31574353]

CHAPTER 13

Potential Role of Nanotechnology in the Wood Industry to Develop Resistance against Fungi

Saloni Bahri[1] and **Somdutta Sinha Roy**[*, 1]

[1] *Department of Botany, Miranda House, University of Delhi, Delhi, India*

Abstract: Wood properties can be changed using nanomaterials that can penetrate deeply into the wood substrate. This capacity of nano-based materials can be utilized in changing wood properties in a way that is very effective in their long-term use. Nanotechnology can certainly change the future of the wood industry by increasing the functional life of wood products as well as usability under various conditions. But its full potential to make wood resistant against fungi has still not been explored. Research is underway but there is still a long way to go. Studies carried out on the use of nanoparticles have clearly shown the negative impact of nanoparticles on human health as well as on the environment. This issue needs to be addressed.

Keywords: Biocides, Nanotechnology, Self-cleaning surface, Scratch resistance, Termite resistance, Wood industry.

INTRODUCTION

Wood is an important plant product required for the very existence of mankind. Wood is a source of fuel and myriad of the products obtained/manufactured, which is derived from the forests partly or wholly. Durability of wood refers to its natural ability to resist the attack by fungi, bacteria and insects. Wood as such and its different forms such as seasoned lumber, logs, *etc.* are susceptible to the attack by different types of pathogens causing immense loss to the wood industry. In order to protect the wood against the damaging effects of pathogens, it is either seasoned or treated with various preservatives. The important chemical preservatives are copper naphthenate, zinc chloride, creosote, Wolman salts, pentachlorophenol or compounds of chromium, mercury and arsenic. To increase the durability of wood, generally two methods are used (i) protective coating with weather resistant paints, varnishes or lacquers, and (ii) immersing woods in open tanks, or spraying or injecting them with chemicals [1].

---

* **Corresponding author Somdutta Sinha Roy:** Department of Botany, Miranda House, University of Delhi, Delhi, India; Tel: +919958855454; E-mail: somduttasinha.roy@mirandahouse.ac.in

Savita, Anju Srivastava, Reena Jain & Pratap Kumar Pati (Eds.)

In nature, the decay of wood by fungi is an important ecological process. The Xylophagous fungi are able to decompose wood and release nutrients locked up in wood back in the soil. The wood-rotting fungi are able to degrade various proportions of cellulose and lignin of the wood leading to their classification into categories such as Brown rot, White rot or Soft rot (Fig. **1**). The brown rot fungi are known to break down cellulose and hemicellulose that are the basic cell wall components [2]. The white rot fungi are known to attack and break down lignin present in the secondary wall which results in a white coloured decaying wood. These fungi are known to produce the laccase enzyme, which is involved in lignin degradation. Soft rot fungi, on the other hand, produce cellulase enzyme, which breaks down wood and makes it hollow and spongy from inside [3]. Understanding the biology and ecology of these fungi becomes important to develop strategies against them.

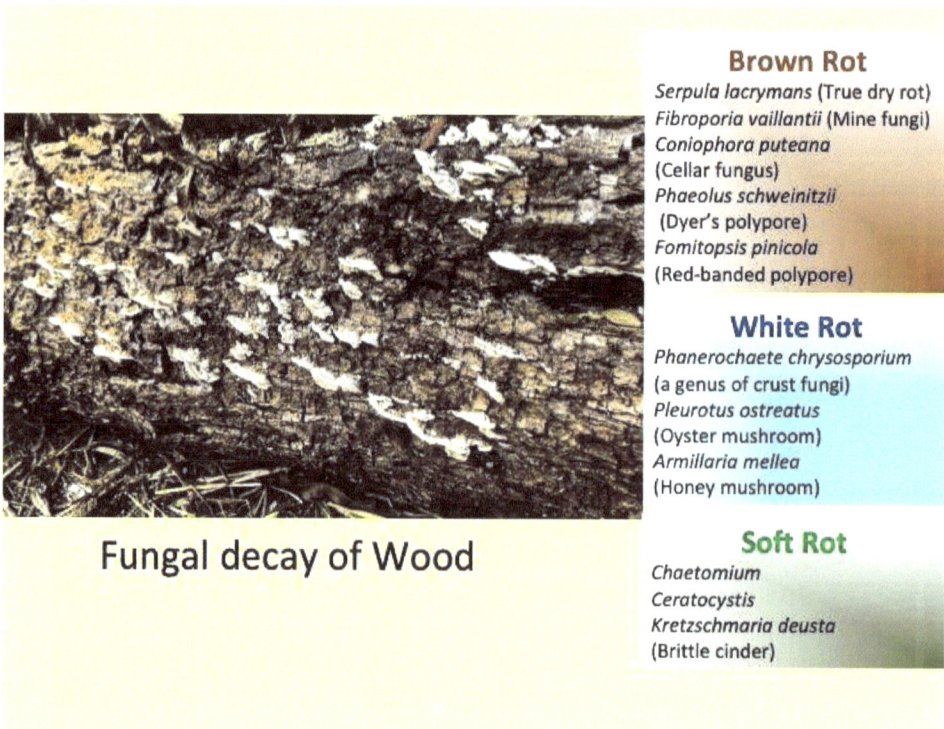

Brown Rot
Serpula lacrymans (True dry rot)
Fibroporia vaillantii (Mine fungi)
Coniophora puteana
(Cellar fungus)
Phaeolus schweinitzii
 (Dyer's polypore)
Fomitopsis pinicola
(Red-banded polypore)

White Rot
Phanerochaete chrysosporium
(a genus of crust fungi)
Pleurotus ostreatus
(Oyster mushroom)
Armillaria mellea
(Honey mushroom)

Soft Rot
Chaetomium
Ceratocystis
Kretzschmaria deusta
(Brittle cinder)

Fungal decay of Wood

Fig. (1).　Different genera of fungi responsible for the decay of wood. (Photo: by author).

Nanotechnology can play a significant role in wood preservation (Fig. **2**) [4, 5]. It is an emerging field that utilizes nanoparticles ranging in size from 1 to 100 nm. Through nanotechnology, nanosized wood preservatives e.g. copper, zinc oxide and silver can be prepared. These can then be applied to the wood directly *via* vacuum pressure treatment. Such treatment increases the time for which preservative is retained in the wood and thus increases its durability [6, 7].

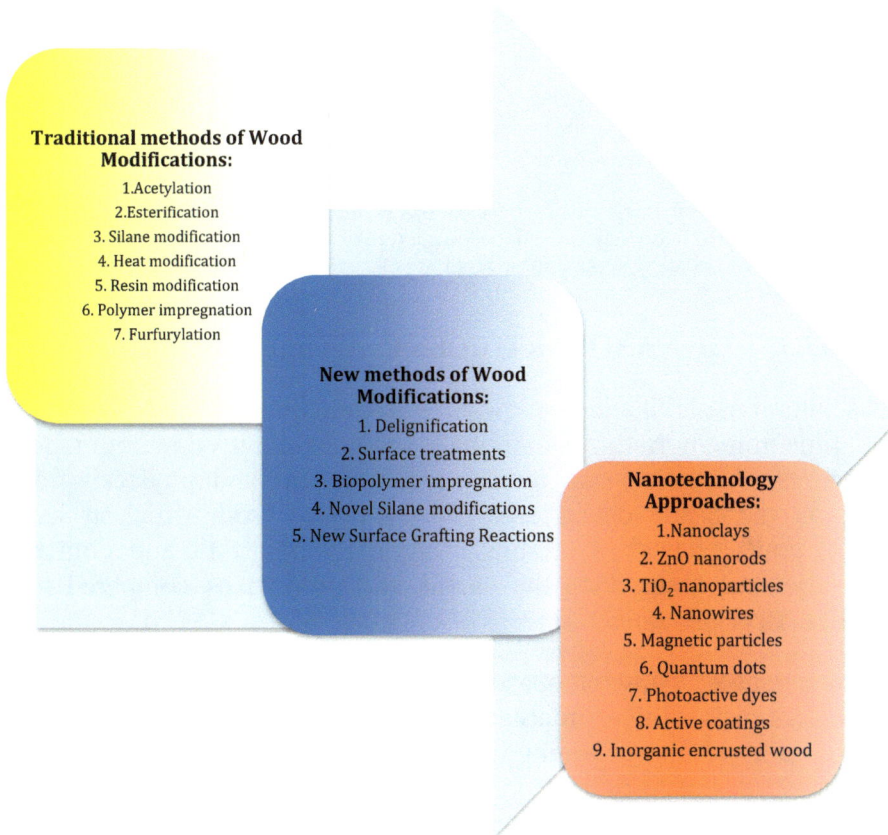

Traditional methods of Wood Modifications:
1. Acetylation
2. Esterification
3. Silane modification
4. Heat modification
5. Resin modification
6. Polymer impregnation
7. Furfurylation

New methods of Wood Modifications:
1. Delignification
2. Surface treatments
3. Biopolymer impregnation
4. Novel Silane modifications
5. New Surface Grafting Reactions

Nanotechnology Approaches:
1. Nanoclays
2. ZnO nanorods
3. TiO_2 nanoparticles
4. Nanowires
5. Magnetic particles
6. Quantum dots
7. Photoactive dyes
8. Active coatings
9. Inorganic encrusted wood

Fig. (2). Different methods of wood modification. (After Spear *et al.* 2021) [4].

The other methods of protecting the wood against fungi include the use of nanomaterials to modify the surface of the wood to increase its resistance against weathering [8, 9], and different nanocarriers can be used to encapsulate hydrophobic biocides to improve water dispensability [10]. Types of polymeric nanocarriers are shown in Fig. (**3**) [11]. In spite of these proposed applications of nanotechnology in the wood industry, the potential risks associated with nanoparticles must be assessed critically [12].

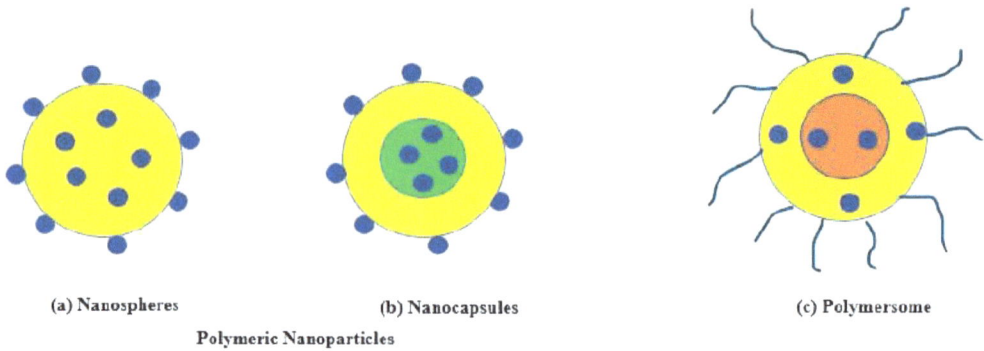

(a) Nanospheres   (b) Nanocapsules   (c) Polymersome

Polymeric Nanoparticles

Fig. (3). Different types of polymeric nanocarriers: **(a, b)** Polymeric nanoparticles have active ingredients which are either encapsulated in or conjugated to polymers. **(c)** Polymersome: It has a hydrophilic inner core and is made up of hydrophilic-hydrophobic block copolymers that are arranged in a lipophilic bilayer vesicular system. (Based on Papadopoulos *et al.* 2019) [11].

APPLICATIONS OF NANOPARTICLES IN WOOD TECHNOLOGY

Wood is a natural material and a renewable resource that has found a multitude of uses from time immemorial. It is extremely versatile and used in both indoor and outdoor spaces. It has a wide range of mechanical and physical properties depending on the tissue characteristics of the tree from which it has been extracted. Overall, it has a remarkable strength to weight ratio and a much lower energy requirement to produce the usable end product as compared to other materials like steel or concrete [13].

A major advantage of using nanoparticle-based technology remains that the size of nanoparticles makes them suitable to penetrate the pores present naturally on the wood surface and thus changing the surface properties of the material. Using nanomaterials for the preservation of wood has its own advantages and disadvantages when compared to conventional methods (Fig. **4**) [11].

Primary structure of wood consists of tracheid and fiber cells, which are long, tubular and close-ended cells. Like any other plant cell, their primary wall is composed of cellulose with the impregnation of natural water-resistant materials like lignin and suberin. But, in spite of several structural advantages, the porous nature of the wood remains one of the major factors of its disadvantage. We can partially overcome this by application of varnishes and paints, but even with their application, wood could not be used for outdoor applications [14].

ADVANTAGES OF NANOMATERIALS IN WOOD PRESERVATION

Hydrophobicity

- Rough hydrophobic surface can be created using Nanocomposite coatings without any effect on softness and abrasion resistance of the wood. The pore size and space available within the cell wall gets reduced with the impregnation of nanomaterials, which can be used for the absorption of water molecules.

UV-Protection

- There are many applications of the surface modification with nano-sized inorganic fillers due to their nontoxicity and stability under exposure to UV. Their large surface to volume ratio is effective towards improving the poor photoresistance property of the wood.

Fire Performance

- Wood is chemically made up of cellulose, hemicellulose, and lignin. This chemical composition makes it undergo thermal degradation when it is in contact with an ignition source. Nanoparticles alone or in combination with other fire-retardant chemicals can not only lower the ignitability of the wood but can also limit the leaching of fire-retardant chemicals.

Antimicrobial

- The decay resistance of wood can be enhanced by utilization of nanoparticles as they reduce the moisture availability in the wood either by preventing the moisture absorption or by blocking the flow path of liquid water.

Mechanical Properties

- The mechanical properties of wood are dependent on environmental agents, its isotropic properties, and sometimes on the nature of the treatment given to make wood attributes better. The wood impregnation with nanoparticles improves its hardness by filling the cavities of the wood.

Disadvantages of Nanomaterials in Wood Preservation

- Nanotechnology is now widely used in treatment of wood. But there are increasing discussions and concerns regarding the potential health and environmental risks of nanomaterials. The study on potential effects of nanoparticles may be delayed as not much information is available about nanomaterials.

Fig. (4). Advantages of nanomaterials in preservationof wood.(Based on Papadopoulos *et al.* 2019) [11].

SELF-CLEANING SURFACES

Self-cleaning surfaces can be described as surfaces that can get rid of microbes, dust, *etc.*, without the application of any external cleaners. This property can be associated with extreme hydrophobicity of the surfaces and their ability to get rid of dirt and grime. Inspiration for self-cleaning surfaces comes from nature and is based on the surface structure of Lotus leaves. Lotus leaves show remarkable properties in which the water droplet falling on the surface rolls off, collecting the dirt along with it. This is possible because of nanoparticle sized wax structures on the surface of the lotus leaf which makes the angle between the surface of the leaf and water droplet more than 150 degrees, resulting in extreme hydrophobicity. Literature has mentions of nano-TiO_2 and nano-SiO_2 coated surfaces exhibiting this property [14].

SCRATCH AND WEATHERING RESISTANCE

Most of the wood products are coated with products like polishes and varnishes to increase their functional life and usability. But, unfortunately, even the paints applied are affected by the environment and lose the shine and the luster due to scratches and weathering of the paints. To mitigate these problems, paint companies came up with a unique solution to add nanoparticles to their paints, which made the surfaces more chip resistant. Silicon nanoparticles were used initially, but though this improved the scratch resistance but the paint would chip easily. Later, this problem could be solved with aluminum silicate nanoparticles, which had excellent chip and scratch resistance and also gave a glossy finish to the final product [15].

BIOCIDES

As wood is a natural product and, naturally occurring organisms like fungi and bacteria very easily degrade it. These organisms have evolved over millennia and produce enzymes, which are able to degrade cellulose, lignin and suberin present in the wood. To protect the wood products, therefore, we have to coat them with material that protects the wood fibers from the attack of such biological agents. In addition to these microscopic agents, termites are a major problem related to wood as these organisms are able to eat into the wood. Thus, the treatment of wood with biocides that are able to prevent these biological agents are essential for producing durable and long lasting wood products. Presently, a lot of work is being going on nanomaterial-based biocide treatment of the wood. There can be several polymer-based carrier systems for the delivery of the active ingredient or the biocide.

The delivery of the active biocide can be done by several well-established techniques like: nanoprecipitation, emulsion-diffusion, double emulsification, emulsion-coacervation, layer-by-layer method, *etc* [12]. Additionally, Liu *et al.* (2001) [16] had shown that fungicides like 4,5-dichloro-2-n-octyl-4-isothiazolone (RH-287) and chlorpyrifos could easily be delivered using nanomaterial matrices of common plastics like PVC (polyvinyl chloride). Other potential nanocarriers are nanotubules like carbon nanotubules (CNT's) and Halloysite, which is a naturally occurring aluminosilicate clay nanotubular material. Some experiments with nanotubes loaded with biocides have shown promising results by reducing water absorption by capillary action and making wood surface resistant to bacterial growth. Wei *et al.* (2018) [17] demonstrated that silicon dioxide nanoparticles modified with poly-(dimethylsiloxane) and (heptadecafluoro-1,1,2,2-tetradecyl) trimethoxysilane and sprayed on cellulose-based surface formed a super hydrophobic surface that resist water, dust, and thoroughly prevent fungal attachment to treated wood (Yao *et al.* 2017) [18]. Many such procedures have been discussed in literature where nanoparticle treatment is combined with hydrophobization treatments (Fig. **5**) [19] to generate woods that are resistant to fungal degradation.

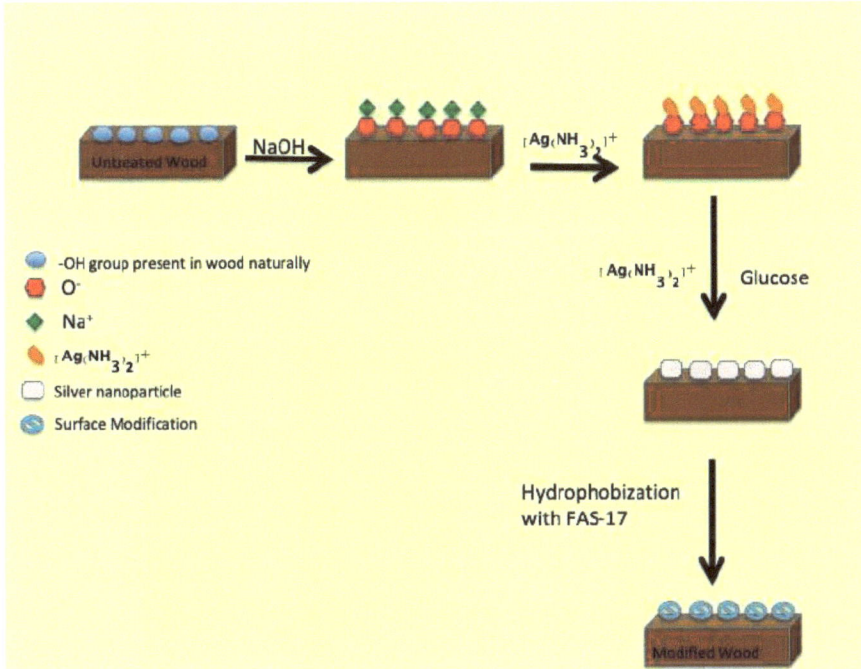

Fig. (5). Schematic diagram of the preparation of a super-hydrophobic wood surface obtained by treating them with silver nanoparticles (Ag NPs) followed by (heptade cafluoro-1,1,2,2-tetradecyl) trimethoxysilane(CF3(CF2)7CH2CH2Si(OCH3)3, shortly FAS-17 (Based on Gao *et al.* 2018) [19].

NANOTECHNOLOGY IN THE WOOD INDUSTRY TO DEVELOP RESISTANCE AGAINST FUNGI

It is evident from all the studies in recent times that nanotechnology has a huge potential in changing the surface properties of wood owing to its small size, which can penetrate deep into the structure of the wood. Wood rotting fungi, like *Gloeophyllum trabeum* and *Trametes versicolor* causes huge problems in the longevity of the wood products. Several studies have tried to address this by using biocides specific for these fungi [16]. Also, nanoparticles are known to resist the decay of the wood in two ways, one by blocking the absorption of water or by blocking the channel of water flow inside the wood [20]. It has been shown that

Zinc oxide nanoparticles applied with Melamine-Urea formaldehyde (MUF) glue for the manufacturing of particle boards have a good antimicrobial property and are resistant to molds such as *Aspergillus niger* and *Penicillium brevicompactum* as well as the brown-rot fungus *Coniophora puteana* [20].

A study showed nano-micronized copper particles can penetrate the microstructure of wood and enhance the biocidal properties against fungal decomposers [21]. Incorporation of TiO_2 nanoparticles in the varnish was studied for nine tropical wood species and it showed that this decreased the water absorption and overall longevity of the wood [22].Some of the recent reports have indicated that the use of plant extracts along with copper nanoparticles show even better results than Cu nanoparticles only [23] and this may open up new avenues of research. Filpo *et al.* [24] in 2013, used titanium dioxide nanoparticle solution for treating different types of wood. These treated wood samples were kept in contact with two wood decay causing fungal species namely, *Mucor circinelloides* (brown-rot) and *Hypocrealixii* (white-rot).The results showed that the treatment of different wood samples with titanium dioxide nanoparticles prevented fungal attack over long period of time (Fig. **6**).

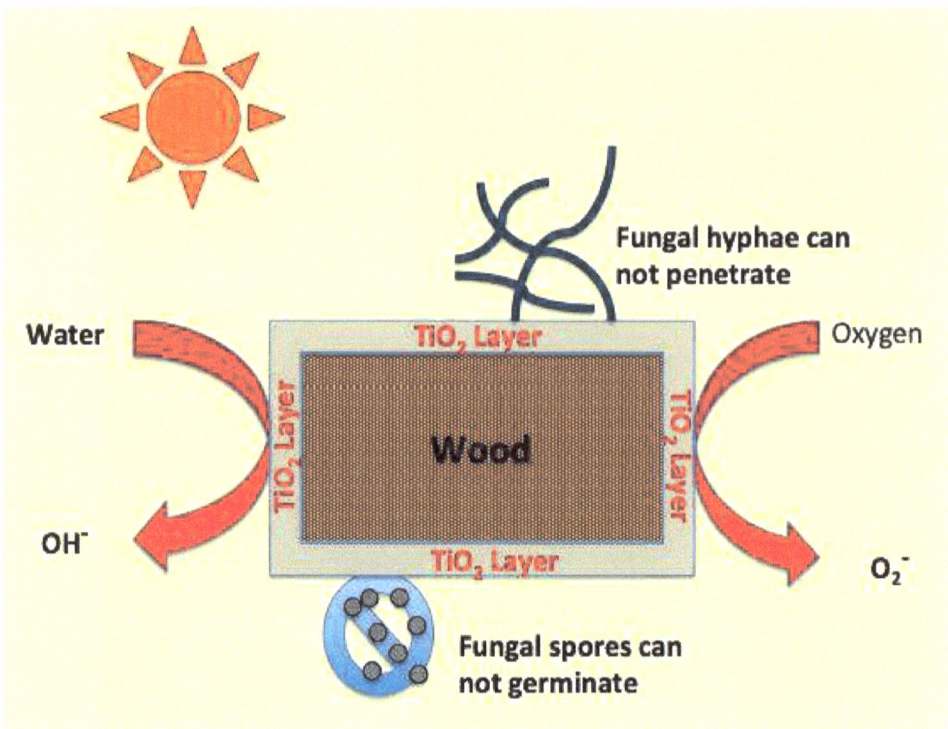

Fig. (6). Titanium dioxide nanoparticles provide complete protection to the wood from various environmental factors including attack by fungi. (Based onFilpo *et al.* 2013) [24].

CONCLUSION

Nanotechnology offers a great potential for wood protection. It can have a major impact on the wood industry. But there are several reports available now that indicate negative effects of nanomaterials on living organisms as well as on their surroundings. More research work needs to be carried out in this direction so that full potential of nanotechnology in wood preservation can be exploited taking into consideration the possible effects of nanoparticles on human health and environment.

CONSENT FOR PUBLICATION

Not applicable.

CONFLICT OF INTEREST

The authors declare no conflict of interest, financial or otherwise.

ACKNOWLEDGEMENT

Declared none.

REFERENCES

[1] Kochhar S. Economic Botany in the Tropics. 4th ed., Delhi: Macmillan Publishers India Ltd. 2011.

[2] Dickison WC. Integrative Plant Anatomy. Academic press 2000.

[3] Schwarze FWMR, Engels J, Mattheck C. Fungal strategies of wood decay in trees. Springer Science & Business Media 2000.
[http://dx.doi.org/10.1007/978-3-642-57302-6]

[4] Spear MJ, Curling SF, Dimitriou A, Ormondroyd GA. Review of Functional Treatments for Modified Wood. Coatings 2021; 11(3): 327.
[http://dx.doi.org/10.3390/coatings11030327]

[5] Evans P, Matsunaga H, Kiguchi M. Large-scale application of nanotechnology for wood protection. Nat Nanotechnol 2008; 3(10): 577.
[http://dx.doi.org/10.1038/nnano.2008.286] [PMID: 18838987]

[6] Taghiyari HR, Moradi-Malek B, Kookandeh MG. FarajpourBibalan OF. Effects of silver and copper nanoparticles in particleboard to control *Trametes versicolor* fungus. Int Biodeterior Biodegradation 2014; 94: 69-72.
[http://dx.doi.org/10.1016/j.ibiod.2014.05.029]

[7] Harandi D, Ahmadi H, Mohammadi Achachluei M. Comparison of TiO_2 and ZnO nanoparticles for the improvement of consolidated wood with polyvinyl butyral against white rot. Int Biodeterior Biodegradation 2016; 108: 142-8.
[http://dx.doi.org/10.1016/j.ibiod.2015.12.017]

[8] Fufa SM, Hovde PJ. Nano-based modifications of wood and their environmental impact: Review in: World Conference on Timber Engineering, Riva del Garda, Italy, 2010; 1-2.

[9] Hubbe MA, Rojas OJ, Lucia LA. Green modification of surface characteristics of cellulosic materials at the molecular or nano scale: A review. BioResources 2015; 10(3): 6095-206.

[http://dx.doi.org/10.15376/biores.10.3.Hubbe]

[10] Iavicoli I, Leso V, Beezhold DH, Shvedova AA. Nanotechnology in agriculture: Opportunities, toxicological implications, and occupational risks. Toxicol Appl Pharmacol. 2017. 15;329:96-111
[http://dx.doi.org/10.1016/j.taap.2017.05.025]

[11] Papadopoulos AN, Bikiaris DN, Mitropoulos AC, Kyzas GZ. Nanomaterials and Chemical Modifications for Enhanced Key Wood Properties: A Review. Nanomaterials (Basel) 2019; 9(4): 607.www.mdpi.com/journal/nanomate
[http://dx.doi.org/10.3390/nano9040607] [PMID: 31013808]

[12] Teng TJ, Mat Arip MN, Sudesh K, *et al.* Conventional technology and nanotechnology in wood preservation: A Review. BioResources 2018; 13(4): 9220-52.
[http://dx.doi.org/10.15376/biores.13.4.Teng]

[13] Winandy Wood Properties In JE. Arntzen, Charles J., ed. Encyclopedia of Agricultural Science. Orlando, FL: Academic Press: 1994; 4: pp. 549-61.

[14] Reijnders L. The environmental impact of a nanoparticle-based reduced need of cleaning product and the limitation thereof.Self-Cleaning Materials and Surfaces: A Nanotechnology Approach. John Wiley & Sons, Ltd 2013.
[http://dx.doi.org/10.1002/9781118652336.ch11]

[15] https://www.azonano.com/article.aspx?ArticleID=377

[16] Liu Y, Yan L, Heiden P, Laks P. Use of nanoparticles for controlled release of biocides in solid wood. J Appl Polym Sci 2001; 79(3): 458-65.
[http://dx.doi.org/10.1002/1097-4628(20010118)79:3<458::AID-APP80>3.0.CO;2-H]

[17] Wei J, Zhang G, Dong J, *et al.* Li Cet al. Facile, Scalable Spray-Coating of Stable Emulsion for Transparent Self-Cleaning Surface of Cellulose-Based Materials. ACS Sustain Chem& Eng 2018; 6(9): 11335-44.
[http://dx.doi.org/10.1021/acssuschemeng.8b00962]

[18] Yao Y, Gellerich A, Zauner M, Wang X, Zhang K. Differential anti- fungal effects from hydrophobic and superhydrophobic wood based on cellulose and glycerol stearoyl esters. Cellulose 2018; 25(2): 1329-38.
[http://dx.doi.org/10.1007/s10570-017-1626-x]

[19] Gao L, Lu W, Li J, Sun Q. Superhydrophobic conductive wood with oil repellency obtained by coating with silver nanoparticles modified by fluoroalkyl silane. Holzforschung 2015.

[20] Jasmani L, Rusli R, Khadiran T, Jalil R, Adnan S. Application of nanotechnology in wood-based products industry: a review. Nanoscale Res Lett 2020; 15(1): 207.
[http://dx.doi.org/10.1186/s11671-020-03438-2] [PMID: 33146807]

[21] Jin L, Walcheski P, Preston A. Laboratory studies on copper availability in wood treated with soluble amine copper and micronized copper systems - IRG/WP 08-30489 39th Annual Meeting of The International Research Group on Wood Protection. Guanacaste, Costa Rica. 2008; pp. 1-10.

[22] Moya R, Rodríguez-Zúñiga A, Vega-Baudrit J, Puente-Urbina A. Effects of adding TiO_2 nanoparticles to a water-based varnish for wood applied to nine tropical woods of Costa Rica exposed to natural and accelerated weathering. J Coat Technol Res 2017; 14(1): 141-52.
[http://dx.doi.org/10.1007/s11998-016-9848-7]

[23] Shiny KS, Sundararaj R, Mamatha N, Lingappa B. A new approach to wood protection: preliminary study of biologically synthesized copper oxide nanoparticle formulation as an environmental friendly wood protectant against decay fungi and termites. Maderas Cienc Tecnol 2019; 21(3): 347-56.
[http://dx.doi.org/10.4067/S0718-221X2019005000307]

[24] FilpoG D. Palermo AM, Rachiele F, Nicoletta F P.Preventing fungal growth in wood by titanium dioxide nanoparticles. International Biodeterioration & Biodegradation 2013; 85: 217-22.
[http://dx.doi.org/10.1016/j.ibiod.2013.07.007]

CHAPTER 14

Nano-Fungicides: Synthesis and Applications

Shweta Meshram[1,*], **Lham Dorjee**[1] and **Sunaina Bisht**[2]

[1] *Division of Plant Pathology, Indian Agricultural Research Institute, Pusa, New Delhi-110012, India*

[2] *Rani Lakshmi Bai Central Agricultural University, Jhansi, Uttar Pradesh-284 003, India*

Abstract: Demand for food, fibre and medicines has been boosted tremendously for the explosive population, which has certainly built pressure on the agriculture-based sector to meet the requirements by various means. Nano-fungicides are fungicidal formulation that contains fungicide particle size 10^{-9}. These nano-fungicides contain antimicrobial properties, which could be utilized against plant pathogens such as fungi and bacteria as a potent pesticide. Nanoparticles of fungicidal properties can be synthesized using different metals *via.* copper, silver, *etc.* Recent reports suggest that nanoparticles can also be synthesized using biological means such as fungi which pose effective fungicidal actions. Nanopesticides have their application in various areas such as agriculture, food, medical industries, storage packaging of food, *etc.* The present chapter will light upon the types and methods of nanoparticle synthesis and their applications. Categories of nano pesticides based on their nature application and source of synthesis will also be covered. Inventions in nano pesticides could lead us to less dependence upon conventional chemical pesticides which have adverse effects on climate, animal and human health.

Keywords: Agriculture industry, Disease management, Nanoparticles, Nano-fungicides, Synthesis.

INTRODUCTION

In recent times nanotechnology is one of the vital approaches to detect nano-substances/particles, which has found innumerable applications in various fields such as, agriculture, pharmaceutical and biomedical sciences. Changing the environment has led to several outbreaks of infectious diseases among plants and humans which are caused by different pathogenic bacteria, fungi and viruses. These pathogens have also developed resistance to antibiotics, fungicides and virucides. Therefore, chemical industries and researchers are searching for new effective chemicals.

[*] **Corresponding Author Shweta Meshram**: Division of Plant Pathology, Indian Agricultural Research Institute, Pusa, New Delhi-110012; Tel: +917065079135; Email: shwetavinay04@gmail.com

Savita, Anju Srivastava, Reena Jain & Pratap Kumar Pati (Eds.)

Various natural and artificial methods for the management and protection of plants from these diseases have been applied so far; among them, a method for plant disease management, use of pesticides is the most prevalent. Here, nano-particle based fungicides and antimicrobial compounds have the potential to be offered a promising path for sustainable plant disease management. Nanoparticles (NPs) of metals contain some suppressive properties against some foliar and soil-borne fungi called nano-fungicides. Nano fungicides are mainly formed by metals such as (Cu-NPs, CuO-NPs), silver (Ag-NPs) and zinc (ZnO-NPs). The average nano-particles size ranging from 1-100 nm are suitable to fall under nanoparticles [1]. In recent years, the environmental damage caused by excessive and indiscriminate use of pesticides has been widely discussed; therefore, scientists in the agricultural field are searching for alternative measures against pesticides, where the use of nano-fungicides has emerged as an eco-friendly, effective and less pollutant option. The current chapter describes recently identified and applied metal-based nano-pesticides and their synthesis.

COPPER-BASED NANO-FUNGICIDES

Copper-based chemicals are well-proven antimicrobials in agriculture since the discovery of synthetic fungicides. Copper is an essential trace element with well understood biological roles in all living organisms, including plants, microorganisms or animals, when present in trace amounts. Further, copper-based chemical formulations have been reported to be one of the first fungicides used for plant disease management [2]. In an *in vitro* study of copper nanoparticles (CuNPs) on plant pathogenic fungi, oomycete, bacteria, and beneficial microbes *Trichoderma harzianum* and *Rhizobium spp.*, along with wheat seeds, were conducted [3]. This study showed that CuNPs with non-nano copper-like copper oxychloride (CoC) at 50 mg/L concentration effectively inhibit 76% of *Phytophthora cinnamomi*. Copper nano fungicidal assessment on *Botrytis cinerea* suggested that maximum inhibition of mycelium is obtained in EC50 value range of 162 and 310 µg/mL. Copper nanoparticles were most effective in the form of $CuSO_4$ [4]. Copper nano-fungicides were also tested effectively against *Aspergillus flavus, Aspergillus niger and Candida albicans* [5]. Interestingly this study showed the non-biocidal effect of CuNPs on beneficial microbes, which suggests potential utilization as part of an integrated pest management strategy in the agri-ecosystem.

SILVER-BASED NANO-FUNGICIDES

As an alternative to chemically manufactured fungicides, the use of silver nanoparticles as antimicrobial or pesticide agents has become more common, as their technological advances make production more economical [6]. Silver nano-

fungicidal assessment on fungi showed inhibition of fungal mycelium exclusively against *B.Gray*mould fungi *B. cinerea* [4]. Another composition of silver nano-particle was suggested where nano-sized silica-silver consisted of nano-silver combined with silica molecules and water-soluble polymer, prepared by exposing a solution including silver salt, silicate and water-soluble polymer to radioactive rays [7]. This study demonstrated that mentioned composition is more effective than Ag-NPs alone as it shows that silver silica nanoparticles can inhibit phytopathogenic fungi at 3 ppm whereas Ag-NPs alone inhibition can range from 10 ppm to 100 ppm. This has even reported inhibition of powdery mildew of pumpkin at 0.3 ppm. Fungicidal activity of copper nanoparticles was assessed on PDA (Potato Dextrose Agar) media against various fungi above fungi including *Alternaria alternata (Alternaria* leaf blight*) A. brassicicola* (Blackspot) *Cladosporium*, Leaf spot fungi *Corynespora,* scab fungi *Clados porium cucumerinum,* Anthracnosepathogen *Glomerellacingulata,* root rot fungi *Pythium aphanidermatum* and *Fusarium* spp., these fungi were significantly inhibited at 100 ppm concentration [8]. Similarly, nanoparticles at 100 ppm were also reported effective for pepper anthracnose pathogen caused by *Colletotrichum* species [9].

ZINC BASED NANO-FUNGICIDES

Zinc nano fungicidal assessment shows that it exhibits fungicidal activity in form of ZnO-NPs and in EC50 range of 235 and 848 μg/mL. ZnO-NPs were more effective than $ZnSO_4$ against fungi such as *B. cinerea, Alternaria alternata and Moniliniafructicola* [4]. Another study demonstrated that ZnO NPs are most effective against post-harvest fungi *viz. Botrytis cinerea* and *Penicillium expansum.* It recorded maximum activity at 3 mmol l^{-1} and both fungi were significantly inhibited but *P. expansum* showed more sensitivity to ZnO NPs. ZnO deform the fungal hyphae and prevent the formation of conidiophore and conidia [10].

A study on lentil pathogens *Alternaria alternata, Fusarium oxysporum* f. sp. *lentis, Xanthomonas axonopodis* pv. *phaseoli, Pseudomonas syringae*pv. *Syringae* and *Meloidogyne incognita* conducted which demonstrates Spray of ZnO NPs reduces nematode multiplication, wilt, leaf spot and blight disease severity indices [11].

GOLD-BASED NANO-FUNGICIDES

The novel metal gold is also being studied for its potential use as nano-fungicides (AuNPs). A study on the green synthesis of gold nanoparticles using 1 mM gold chloride solution with leaf extract of *Annona muricata* reveal that the synthesized gold nanoparticles exhibited good antimicrobial activity depending on the concentration level [12]. Gold particles can also be synthesized using biological species such as *Bacillus subtilis* isolated from Hatti Gold Mine, which is assumed to be resistant against AuNPs toxicity is also reported [13] and with *Streptomyces griseoruber* with a size 5–50 nm is also reported [14].

SYNTHESIS OF NANOPARTICLES

A profound number of metallic nanoparticles have been successfully synthesized by using various plant extracts and microbes, including bacteria, fungi viruses and algae. Various methods for the synthesis of copper nanoparticles using chemical, physical and biological techniques are available. For copper nano-particle directly depend on their synthesis procedures. Among various techniques such as the Chemical reduction method, Microemulsion/colloidal method and sonochemical methods are most popular. Among them, chemical reduction methods are most suitable for the synthesis of copper nanoparticles [15] such as the Chemical method and non-chemical method (green synthesis).

CHEMICAL REDUCTION METHOD

This is the most convenient and cost-friendly method for synthesis for copper nanoparticles, chemical reduction of copper fairly maintains morphologies and sizes of nanoparticles [16, 17]. In this method, PVP (polyvinylpyrrolidone) and PEG polyethylene glycol are used as capping agents [15] and examples of reduction reactions are provided in Table **1**.

Table 1. Examples of the important component for reduction reaction for the synthesis of Cu nanoparticles.

Reducing Agent	Conducive Temperature	Solvent	Particle Size	References
Sodium borohydride	Ambient	Water, Toluene water	2-10 nm	[18]
Polyol	>120°C	glycol solvent	50 nm	[19]
Isopropyl alcohol	70°-100°C	Water	10–13 nm	[20]
Sodium phosphonate	85–95°C	Di-Ethylene glycol	45 nm	[21]
Hydrazine	<70°C	Water	20-100 nm	[22]
Ascorbic acid	~80°C	Deionized water	>10 nm	[23]
sodium formaldehyde sulfoxylate	~75-80°C	Di-Ethylene glycol	50 nm	[24]

MICRO-EMULSION/COLLOIDAL METHOD

It is a technique in which two immiscible fluids, such as water in oil or oil in water or water in supercritical carbon dioxide become a thermodynamically stable dispersion along with surfactant. The most common microemulsion is water, oil and a surfactant here surfactant act as a bridge [25, 26]. Microemulsions where water, oil, surfactant and alcohol- or amine-based co-surfactant produced clear and homogeneous solutions [27]. Micro-emulsion particles range from 1 nm to 100 nm, which aggregates to form an emulsion. Micelles are product which is oil in water and surfactant and aggregates to reduce free energy. Hydrophobic surfactants in nanoscale oil and micelles point toward the centre of aggregate, whereas the hydrophobic head groups toward the water, the bulk solvent. Completely metallic nanoparticles (Cu, Ag, Co, Al), metal sulphides (CdS, ZnS), oxides (TiO_2, SiO_2), and various other nanomaterials are prepared using this technique [28, 29]. Most commonly used solvents are water, Isooctane, cyclohexane, *etc.* and some popularly used stabilizers are hexametaphosphate (HMP), $Na(AOT)_2$, Quercetin, bis(Ethylhexyl) hydrogen phosphate (HDEHP), polyvinylpyrrolidone (PVP), Sodium dodecyl sulphate (SDS), *etc.* (Fig. **1**) [16, 30 - 32].

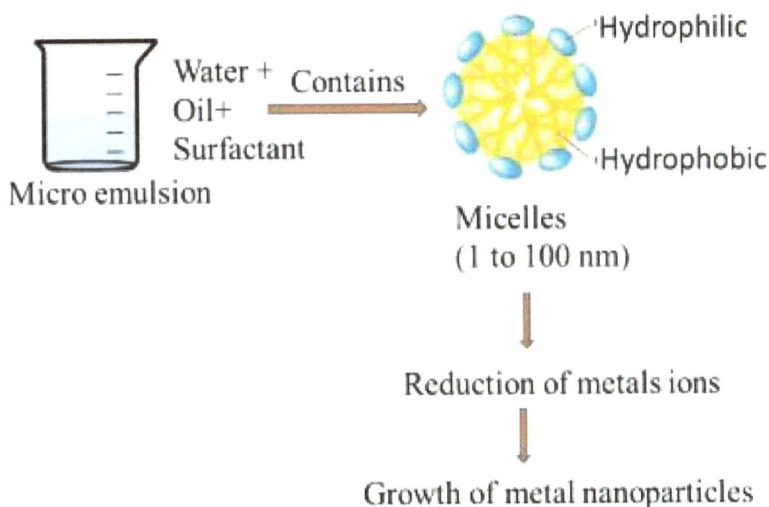

Fig. (1). Flow chart to illustrate synthesis of nano-particles mediated by surfactant head [15,33].

SONOCHEMICAL METHOD

Sonochemical is the process where powerful ultrasound radiations ranging from 20 kHz to 10 MHz are applied to molecules to intensify the chemical reaction [34]. The main physical phenomenon which is involved in the sonochemical reaction is Acoustic cavitation [35]. The mechanism of sonochemical reaction requires passing sound waves of constant frequency through a slurry or solution of carefully selected metal complex precursors. Sonoelectrochemical synthesis involves both electrolytes and ultrasonic pulses for the production of nanoparticles [36, 37]. Commonly used solutions for sonochemistry are $CuSO_4$ H_2SO_4, $CuSO_4$: CTAB, Copper hydrazine carboxylate (CHC), *etc.* Further, conducive ultrasound pulse time ranges 100-600 TUS/ms, whereas current pulse time (TON/ms) and rest time (TOFF/ms) range 250-900 and 150-300, respectively. Successful control of the formation of copper nanoparticles requires bath temperature, current density, current pulse time, ultrasound pulse time, ultrasound intensity and stabilizer are required.

BIOLOGICAL SYNTHESIS OF NANOPARTICLES OR GREEN SYNTHESIS

The development of easy methods for the preparation of nano-sized metal particles has gained significant attention because of their future appeal. Biosynthesis of nanoparticles mainly requires the three important parameters to be evaluated are (1) the choice of the solvent medium used, (2) the choice reducing agent and (3) the choice of a non-toxic material for the stabilization of the nanoparticles. The mechanism involved in the biosynthesis of metal compounds is usually reduced into their respective nanoparticles because of the plant phytochemicals or microbial enzymes with antioxidant or reducing properties (Fig. **2**) [15].

Ag-NP scan be synthesized using *Fusarium oxysporum* which generated stable nano-silver and gold particles in water. These nano-particles were reported to be effective against bacterial microbes such as *Staphylococcus aureus*. Microwave-assisted synthesis to achieve a stable Cu –Ag bimetallic system was developed using green approaches. In this, they used aqueous starch solution as a biopolymer matrix and ascorbic acid as a green reducing agent [38]. Microwaves (MW) have been utilized for the preparation of nanoparticles of metals like gold, silver, sulphide and oxides [39 - 42]. In another study, copper nanoparticles were synthesized using the modified polyol method by the reduction of copper acetate hydrate in the presence of Tween 80 by refluxing between 190° and 200 °C [5]. Another novel silver nano-particle was synthesised using bacterial species *Streptomyces sp* [43].

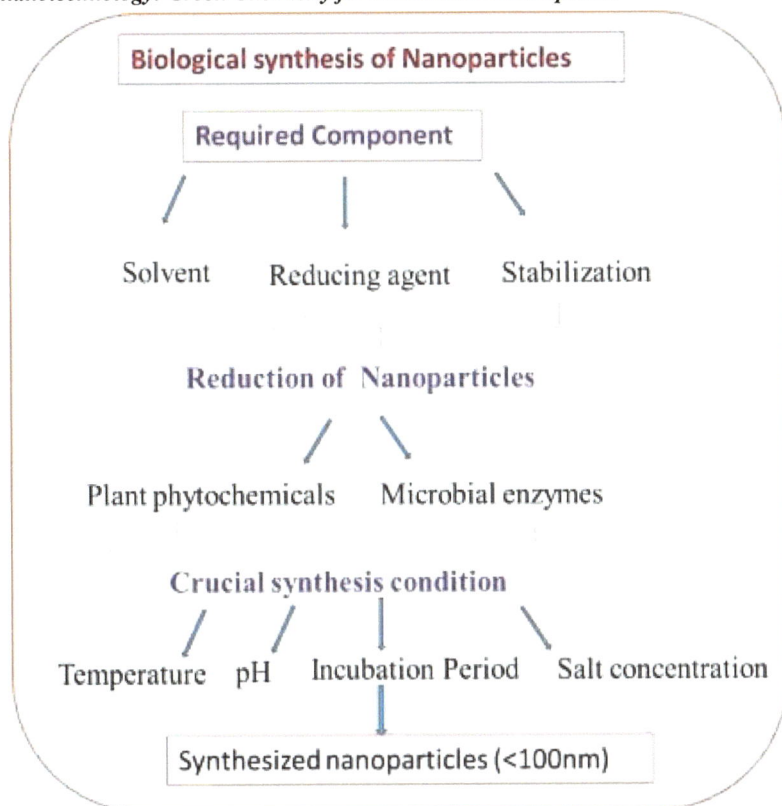

Fig. (2). Putative representation of biological synthesis/green of nano-particles.

ADVANTAGE OF VARIOUS SYNTHESIS METHODS

Chemical Reduction

- The simple method provides control over nanoparticle size and produces homogeneous nano-powders.
- High yield.
- Rapid reaction.
- NP size control is very good.

Disadvantages

- Toxicity and flammability of chemicals.
- Size distribution is narrow.

SONOCHEMICAL: The basic advantages of the sonochemical method are its simplicity, easy control and operating conditions of the size of nanoparticles by using precursors with variable concentrations in the solution.

DISADVANTAGES

•Long reaction time.

•Low yield.

•A very small concentration of prepared NP's, particle agglomeration.

•Size distribution is very narrow.

BIOLOGICAL: Selectivity and precision for nanoparticle formation, cost reactive, Eco-friendly and (chemical-free).

• A surface capping agent does not require.
• NP size control is good.

DISADVANTAGES

• Reaction time variable from Minutes to days.

• Low yield.

• Little knowledge about the mechanism.

• Narrow size particle distribution.

APPLICATIONS OF NANO FUNGICIDES AND NANOPARTICLES IN AGRICULTURE

Nanoparticles have a wide range of applications such as nano-fungicides, nano-bactericides, nano-fertilizers, food packaging, nutrient delivery, *etc*. Nanofungicides are capable of controlling fungal pathogen directly after spray. On the other hand, it has also the capability to suppress or inactivate mycotoxins produced by several fungi (Table **2**).

Nanoparticles as Carriers for Fungicides

Nanoparticles exclusively as fungicides have the potential to be part of the integrated disease management system as synthetic and green fungicides (Table **3**) A wide range of studies has been conducted on nanoparticles with conventional fungicides and biocides to assess their efficacy of nano-fungicides (Table **4**). Most popularly studied nanocarriers are polymer mixes, silica, and chitosan [53].

Table 2. Examples of management of mycotoxins using nano-fungicides.

Mycotoxins	Remarks	References
Aflatoxin	Potential antifungal activity was shown by NPs ZnO and Fe_2O_3 against isolated aflatoxigenic and non-aflatoxigenic *A. flavus* recovered from animal and poultry feeds. Effective minimum concentration 50 µg ml^{-1}) and effectively at 100–200 µg ml^{-1}	[44]
Citrinin	*Citrinin* was effectively inhibited using surface active maghemite nanoparticles (SAMNs) at the concentration of 1 g L^{-1} were able to remove 70% of citrinin.	[45]
Fumonisin	ZnNPs showed an inhibitory effect at 10 µg ml−1 zinc NPs treatment on *Fumonisin* synthesizing. Scanning Electron Microscope suggested rupture of cell wall	[46]
Ochratoxin	The antifungal potential was assessed using ZnO and Fe_2O_3 NPs for *Aspergillus ochraceus* and *Aspergillus niger* strains with 50 µg ml^{-1} and above concentration.	[47]
Patulin	*Penicillium expansum* and *F. oxysporum* fungi were successfully inhibited using ZnO NPs with the concentration of 3, 6 and 12 mmol L^{-1}	[48]

Table 3. Examples of some popularly used nano green nano pesticides [49 - 52].

Nano-biopesticides	Nanocarrier
Garlic essential oil	Poly-ethyleneglycol
Validamycin	Porous hollow, silicanano particles
Avermectin	Porous hollow, silicanano particles
Azadirachtin-A	Nanomicelles
Neem oil	poly-ε-caprolactone
Neem seed kernel	PCL poly-ε-caprolactone
Neem oil	Silica nanoparticles

Table 4. Examples of nanoparticles as carriers of fungicides along with target fungi.

Fungicides	Nanocarrier	Target Fungi	References
Tebuconazole, Chlorothalonil	PVP and PVP copolymer	Brown rot fungi *Gloeophyllum trabeum*	[54]
Validamycin	Porous hollow silica nanoparticles (PHSNs) Calcium carbonate	Powdery mildew of cucurbits *Sphaerotheca fuliginea* Soil born pathogen *Rhizoctonia solani*	[25]
Metalaxyl	Mesoporous silica Nano-particles (MSNs).	*Rhizoctonia solani* and oomycetes	[55]
Ferbam	Gold	*Aspergillus niger, F. oxysporum*	[56]

(Table 4) cont.....

Fungicides	Nanocarrier	Target Fungi	References
Carbendazim	Chitosan/Pectin	*F. oxysporum and A. parasiticus*	[57]
Bioactive compounds from *Chaetomium spp.* *	Polylactide (PLA)	Wide range of pathogens	[58]

METAL-BASED NANO-FUNGICIDES DIFFERENCES IN ACTION

As eco-friendly alternatives to current synthetic fungicides, nano-fungicides are predicted to play a major role in future plant disease management. Different metal-based nano-fungicides (Cu-NPs, CuO-NPs), silver (Ag-NPs) and zinc (ZnO-NPs) are available which have different levels of efficacy. One such study demonstrates that in almost all fungal species tested, Cu-NPs were more fungi toxic in suppression of mycelial growth than a protective fungicide containing Cu(OH)2, which was employed as a control. Except for *B. cinerea, A. alternata,* and *M. fructicola,* fungitoxicity studies with the NPs and bulk size reagents containing the corresponding metals demonstrated that ZnO-NPs were more harmful to all fungal species examined than ZnSO4, but Cu-NPs were more fungitoxic than CuSO4 in all the treatments. Although there was considerable variance between fungal species, colony formation bioassays demonstrated that all NPs were 10 to 100 times more fungitoxic to spores than hyphae and, in the majority of cases, more effective than $Cu(OH)_2$ [59]. Nanoparticles, such as nanozeolites, basic building blocks of silicate $[SiO_4]$ and aluminates $[AlO_4]$ tetrahedrons and hydrogels, composed of diverse polymers such as chitosan and alginate, which aid in soil quality improvement, as well as nanosensors, are preferably used in agriculture for monitoring plant and soil health [60]. Plants were shown to be harmless to silica nanoparticles [61], although some reports noticed a harmful effect due to changes in pH of the media after nanoparticles were added [62]. Titanium dioxide (TiO_2), silver, zinc oxide, cerium dioxide, copper, copper oxide, aluminium, nickel, and iron are the most extensively used metal and metal oxide nanoparticles in industry, hence their effects on plants are largely explored. Some non-metal nanoparticles, such as single-walled carbon nanotubes and fullerene, have been thoroughly investigated to learn more about their nanotoxicity mechanisms [63].

INTERACTION OF NANOPARTICLES WITH SOIL

The impact of discharged waste on the environment Metal-based NPs are crucial in establishing their mobility, reactivity, toxicity, and possible dangers in agricultural systems. Metal-based NPs may aggregate and subsequently precipitate onto the soil matrix, or they may be stabilised by dispersion in soil solution, or they may undergo physical or chemical reactions with other environmental components, modifying their properties. Metal-based NPs' environmental behaviour in agriculture is influenced by their properties as well as the features of both soil and soil pore water. Because of size-dependent 'quantum effects,' NPs have distinct physical and chemical properties [64, 65]. Surface characterizations of metal-based NPs regulate their surface charge and possible interactions with ambient components, both of which are important factors in NP aggregation. The stability of uncoated NPs is dependent on the surface element, element speciation, and crystal structure. Metal-based NPs commonly have a surface charge due to the presence of oxygen atoms or hydroxyl groups that can release or absorb protons when they are not coated. A charged surface layer and a diffused layer containing ions attracted from the soil solution can form an electrostatic double layer (EDL) surrounding these charged surfaces [65]. The stability of metal-based NPs declines as pH approaches zero-point charge, where the overall surface charge is close to zero [66]. DOM is one of the most reactive and mobile soil organic components, and it has a significant impact on metal-based NPs in the soil matrix [33]. DOM interacts with metal-based NPs in a variety of ways, causing them to change their environmental behaviour. Electrostatic interaction, van der Waals forces, and hydrophobic interaction between the DOM and NPs surface play a role in the interaction process. Higher electrostatic stability or increased hydrophobicity of DOM deposited on NPs surfaces can usually reduce NP aggregation [67]. The interaction between colloids and metal-based NPs is unavoidable due to the widespread presence of natural clay minerals in the soil matrix. Collisions can cause NPs to connect to inorganic colloids, resulting in heteroaggregation. The formation of primary heteroaggregates with natural clay colloids was favoured by opposite surface charges and/or high ionic strength *via* the attachment of nanoparticles to natural clay colloids; therefore, pH plays an important role in controlling heteroaggregation because the surface charge on NP and soil colloid surface are both pH-dependent [68, 69].

PLANT NANO-FUNGICIDE INTERACTION

Nanoparticles have recently been produced for application in agriculture as nanopesticides and nanofertilizers, which include the use of nanoparticles as nanocarriers for pesticides, fertilizers [70, 71]. Some non-metal nanoparticles, such as single-walled carbon nanotubes and fullerene, have been thoroughly investigated to determine their nanotoxicity processes. On the other hand, single-walled carbon nanohorns (SWCNHs) have been shown to improve growth and seed germination in several organs of corn, tomato, rice, and soybean [72]. The characteristics of NPs and environmental circumstances can influence the rate of NPs accumulation by plant roots. Partially by mechanical adhesion, CuO-NPs can be strongly adsorbed on the plant root surface; the adsorbed CuO-NPs cannot be desorbed by competing ions [73]. TiO2-NPs have been found to collect in wheat roots and subsequently spread throughout the plant without dissolving or changing the crystal phase. According to a study comparing the ability of four plant species to accumulate Au NPs in their roots, radish and ryegrass roots accumulated higher amounts of AuNPs (14–900 ng mg1) than rice and pumpkin roots (7–59 ng mg1) [74, 75]. Rice roots have also been reported to readily absorb Ag NPs [76, 77]. Positively charged Au NPs are more readily taken up by plant roots than negatively charged Au NPs, according to a study on Au NPs [75].The presence of microbial siderophores and root exudates is one of the possible mechanisms for the higher bioavailability of metal-based NPS in the rhizosphere. Plants and microorganisms are known to create organic ligands to solubilize minerals from inaccessible sources. Microbial siderophores are low molecular weight organic compounds produced by a wide range of species to chelate iron in iron-deficient environments. Other metals, such as Cu, Zn, and Ag, show a high affinity for siderophores, according to a recent study. As a result of the chelation between siderophores and metals, metal-based NPs should dissolve faster and have higher bioavailability. Exudates are also released by plant roots to aid in the uptake of nutrients from insoluble sources [78, 79]. Metal based NPs have a typical harmful effect on seed germination. NPs were found to reduce seed germination in ryegrass, barley, and flaxseed germination [80, 81]. However, NPs have been shown to have a favourable influence on seed germination. TiO_2 NPs at 2 and 10 mg L1 concentrations can improve wheat seed germination and seedling growth [82].

MERITS OF NANO-FUNGICIDES

- Eco-friendly and safer for the environment.
- These fungicides are required in smaller quantities and are sufficient to work over some time.
- The durability of the formulation after application and during storage, further, fewer doses of active ingredients.
- Better resist environmental elimination process better than conventional fungicides such as leaching, evaporation and photolysis, biodegradation and chemical hydrolysis.
- NPs largely have a uniform size which enables stability and fair distribution.
- Some NPs formulations are also reported synergistic effects with other NPs and pesticides.

DEMERITS OF NANO-FUNGICIDES

- Nanosize of the particles enables the high toxicity of nano-fungicides may pose a greater threat to humans, animals and beneficial organisms.
- In several cases, the synthesis and validation of nano-particles become a challenge.
- Knowledge of phytotoxicity is required in-depth.
- Higher cost of synthesis and formulation preparation initially.
- Large scale commercial production is required, most of the agriculture-based formulations are under initial trials.

CONCLUSION AND FUTURE PROSPECTS

Increasing food demand by exploding population and increasing pollution due to conventional pesticides needs urgent attention. Nano-fungicides are expected to play an important role in future plant disease management as eco-friendly alternatives to conventional synthetic fungicides. Nano fungicides can also be synthesized biologically using natural fungi or microorganisms, which makes it less risky and cost-friendly (Fig. **3**). There are differences among fungicidal effectiveness of different metal-based nano-fungicides. These differences can be utilized in exploring compound nano-fungicides. On the other hand, a deep investigation is required in plant systems to understand the behaviour of nano metals in plant physiology and morphology, such as penetration pattern of metal their transport inside plant vascular system. Another aspect is to study germination and growth patterns, photosynthesis and transpiration behaviour after nano fungicide spray. In addition, nanoparticles are subject to environmental circumstances and can vary their aggregation state, oxidation state, secondary phase precipitation, and so on, depending on the environment. The stability of nanoparticles is influenced by physical characteristics and chemicals present in various environments. As a result, nanoparticles may behave differently depending on the situation affecting their availability and reactivity in the ecosystem. This is important to understand the phytotoxicity nature of nano fungicides. Although there are several recent synthesis methods are available in the market, there is also a need to develop a simple, easy and high-yielding method of synthesis. Proper commercialization, field demonstration and marketing strategies are required as these nano-fungicides are new in the agriculture sector.

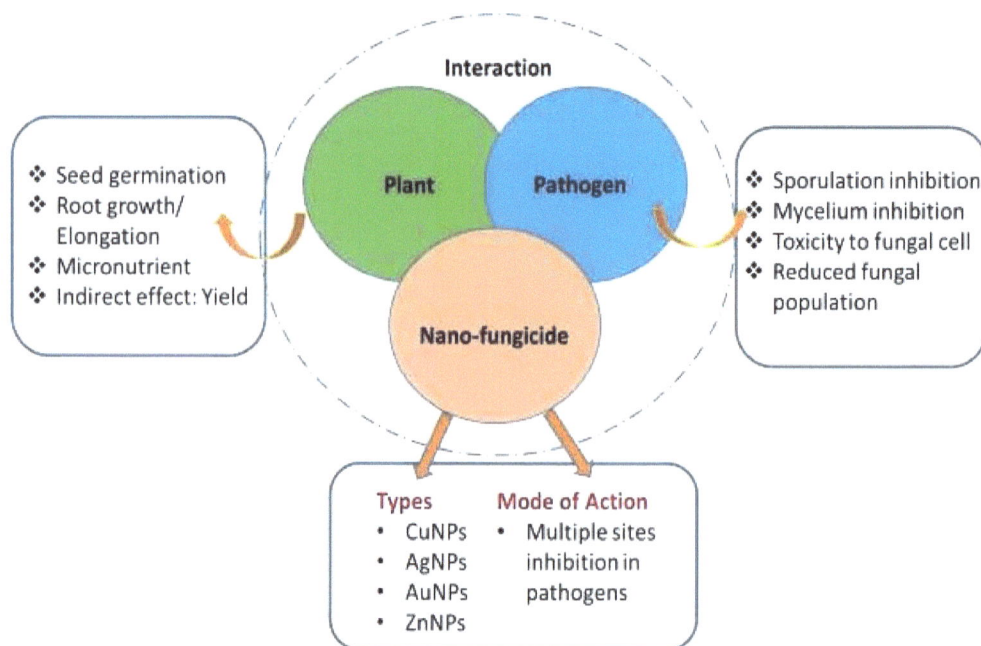

Fig. (3). Overview of Plant-Pathogen interaction in presence of Nano-fungicide and their possible effects.

CONSENT FOR PUBLICATION

Not applicable.

CONFLICT OF INTEREST

The authors declare no conflict of interest, financial or otherwise.

ACKNOWLEDGEMENTS

We are grateful to the division of Plant Pathology ICAR-Indian Agricultural Research Institute, Pusa, New Delhi and Rani Lakshmi Bai Central Agricultural University, Jhansi, Uttar Pradesh, for their continuous support.

REFERENCES

[1]   Ruparelia J P, Chatterjee A K, Duttagupta S P, Mukherji S. 2008.Strain specificity in antimicrobial activity of fl

[2]   Johnson GF. The early history of copper fungicides. Agric Hist 1935; 9: 67-79.

[3]   Banik S, Luque AP. In vitro effects of copper nanoparticles on plant pathogens, beneficial microbes and crop plants. Span J Agric Res 2017; 15(2): 23.
        [http://dx.doi.org/10.5424/sjar/2017152-10305]

[4]   Malandrakis A, Kavroulakis N, Chrysikopoulos C. Nano-fungicides against plant pathogens: Copper, silver and zinc NPs. In Geophysical Research Abstracts (Vol. 21). 2019.

[5]   Ramyadevi J, Jeyasubramanian K, Marikani A, Rajakumar G, Rahuman AA. Synthesis and antimicrobial activity of copper nanoparticles. Mater Lett 2012; 71: 114-6.
[http://dx.doi.org/10.1016/j.matlet.2011.12.055]

[6]   Jo YK, Kim BH, Jung G. Antifungal activity of silver ions and nanoparticles on phytopathogenic fungi. Plant Dis 2009; 93(10): 1037-43.
[http://dx.doi.org/10.1094/PDIS-93-10-1037] [PMID: 30754381]

[7]   Park HJ, Kim SH, Kim HJ, Choi SH. A new composition of nanosized silica-silver for control of various plant diseases. Plant Pathol J 2006; 22(3): 295-302.
[http://dx.doi.org/10.5423/PPJ.2006.22.3.295]

[8]   Kim SW, Jung JH, Lamsal K, Kim YS, Min JS, Lee YS. Antifungal effects of silver nanoparticles (AgNPs) against various plant pathogenic fungi. Mycobiology 2012; 40(1): 53-8.
[http://dx.doi.org/10.5941/MYCO.2012.40.1.053] [PMID: 22783135]

[9]   Lamsal K, Kim SW, Jung JH, Kim YS, Kim KS, Lee YS. Application of silver nanoparticles for the control of colletotrichum species in vitro and pepper anthracnose disease in field. Mycobiology 2011; 39(3): 194-9.
[http://dx.doi.org/10.5941/MYCO.2011.39.3.194] [PMID: 22783103]

[10]   He L, Liu Y, Mustapha A, Lin M. Antifungal activity of zinc oxide nanoparticles against Botrytis cinerea and Penicillium expansum. Microbiol Res 2011; 166(3): 207-15.
[http://dx.doi.org/10.1016/j.micres.2010.03.003] [PMID: 20630731]

[11]   Siddiqui ZA, Khan A, Khan MR, Abd-Allah EF. Effects of zinc oxide nanoparticles (ZnO NPs) and some plant pathogens on the growth and nodulation of lentil (Lens culinaris Medik.). Acta Phytopathol Entomol Hung 2018; 53(2): 195-211.
[http://dx.doi.org/10.1556/038.53.2018.012]

[12]   Folorunso A, Akintelu S, Oyebamiji AK, *et al.* Biosynthesis, characterization and antimicrobial activity of gold nanoparticles from leaf extracts of Annona muricata. J Nanostructure Chem 2019; 9(2): 111-7.
[http://dx.doi.org/10.1007/s40097-019-0301-1]

[13]   Ahmed E, Kalathil S, Shi L, Alharbi O, Wang P. Synthesis of ultra-small platinum, palladium and gold nanoparticles by Shewanellaloihica PV-4 electrochemically active biofilms and their enhanced catalytic activities. J Saudi Chem Soc 2018; 22(8): 919-29.
[http://dx.doi.org/10.1016/j.jscs.2018.02.002]

[14]   Ranjitha V R, Rai V R. Actinomycetes mediated synthesis of gold nanoparticles from the culture supernatant of Streptomyces griseoruber with special reference to catalytic activity. 3 Biotech 2017; 7(5): 299.

[15]   Umer A, Naveed S, Ramzan N, Rafique MS. Selection of a suitable method for the synthesis of copper nanoparticles. Nano 2012; 7(05): 1230005.
[http://dx.doi.org/10.1142/S1793292012300058]

[16]   Song X, Sun S, Zhang W, Yin Z. A method for the synthesis of spherical copper nanoparticles in the organic phase. J Colloid Interface Sci 2004; 273(2): 463-9.
[http://dx.doi.org/10.1016/j.jcis.2004.01.019] [PMID: 15082381]

[17]   Mott D, Galkowski J, Wang L, Luo J, Zhong CJ. Synthesis of size-controlled and shaped copper nanoparticles. Langmuir 2007; 23(10): 5740-5.
[http://dx.doi.org/10.1021/la0635092] [PMID: 17407333]

[18]   Lisiecki I, Billoudet F, Pileni MP. Control of the shape and the size of copper metallic particles. J Phys Chem 1996; 100(10): 4160-6.
[http://dx.doi.org/10.1021/jp9523837]

[19]   Cho SJ, Ouyang J. Attachment of platinum nanoparticles to substrates by coating and polyol reduction of a platinum precursor. J Phys Chem C 2011; 115(17): 8519-26.

[http://dx.doi.org/10.1021/jp2001699]

[20]   Goia DV, Matijević E. Preparation of monodispersed metal particles. New J Chem 1998; 22(11): 1203-15.
[http://dx.doi.org/10.1039/a709236i]

[21]   Zhu HT, Zhang CY, Yin YS. Rapid synthesis of copper nanoparticles by sodium hypophosphite reduction in ethylene glycol under microwave irradiation. J Cryst Growth 2004; 270(3-4): 722-8.
[http://dx.doi.org/10.1016/j.jcrysgro.2004.07.008]

[22]   Grouchko M, Kamyshny A, Ben-Ami K, Magdassi S. Synthesis of copper nanoparticles catalyzed by pre-formed silver nanoparticles. J Nanopart Res 2009; 11(3): 713-6.
[http://dx.doi.org/10.1007/s11051-007-9324-5]

[23]   Dang TMD, Le TTT, Fribourg-Blanc E, Dang MC. Synthesis and optical properties of copper nanoparticles prepared by a chemical reduction method. Advances in Natural Sciences: Nanoscience and Nanotechnology 2011; 2(1): 015009.
[http://dx.doi.org/10.1088/2043-6262/2/1/015009]

[24]   Khanna PK, More P, Jawalkar J, Patil Y, Rao NK. Synthesis of hydrophilic copper nanoparticles: effect of reaction temperature. J Nanopart Res 2009; 11(4): 793-9.
[http://dx.doi.org/10.1007/s11051-008-9441-9]

[25]   Liu F, Wen L-X, Li Z-Z, Yu W, Sun H-Y, Chen J-F. Porous hollow silica nanoparticles as controlled delivery system for water-soluble pesticide. Mater Res Bull 2006; 41: 2268-75.
[http://dx.doi.org/10.1016/j.materresbull.2006.04.014]

[26]   Kapoor S, Joshi R, Mukherjee T. Influence of I− anions on the formation and stabilization of copper nanoparticles. Chem Phys Lett 2002; 354(5-6): 443-8.
[http://dx.doi.org/10.1016/S0009-2614(02)00159-8]

[27]   Hirai T, Tsubaki Y, Sato H, Komasawa I. Mechanism of formation of lead sulfide ultrafine particles in reverse micellar systems. J Chem Eng of Jpn 1995; 28(4): 468-73.
[http://dx.doi.org/10.1252/jcej.28.468]

[28]   Bagwe RP, Khilar KC. Effects of intermicellar exchange rate on the formation of silver nanoparticles in reverse microemulsions of AOT. Langmuir 2000; 16(3): 905-10.
[http://dx.doi.org/10.1021/la980248q]

[29]   Cason JP, Miller ME, Thompson JB, Roberts CB. Solvent effects on copper nanoparticle growth behavior in AOT reverse micelle systems. J Phys Chem B 2001; 105(12): 2297-302.
[http://dx.doi.org/10.1021/jp002127g]

[30]   Lisiecki I, Pileni MP. Copper metallic particles synthesized" *in situ*" in reverse micelles: influence of various parameters on the size of the particles. J Phys Chem 1995; 99(14): 5077-82.
[http://dx.doi.org/10.1021/j100014a030]

[31]   Pileni MP. Nanocrystals: Fabrication, organization and collective properties. C R Chim 2003; 6(8-10): 965-78.
[http://dx.doi.org/10.1016/j.crci.2003.07.007]

[32]   Egorova EM, Revina AA. Synthesis of metallic nanoparticles in reverse micelles in the presence of quercetin. Colloids Surf A Physicochem Eng Asp 2000; 168(1): 87-96.
[http://dx.doi.org/10.1016/S0927-7757(99)00513-0]

[33]   Kitchens CL, Roberts CB. Copper nanoparticle synthesis in compressed liquid and supercritical fluid reverse micelle systems. Ind Eng Chem Res 2004; 43(19): 6070-81.
[http://dx.doi.org/10.1021/ie0497644]

[34]   Suslick K S, Choe S B, Cichowlas A A, Grinstaff M W. Sonochemical synthesis of amorphous iron. nature, 1991; 353(6343): 414-6.

[35]   Khalil H, Mahajan D, Rafailovich M, Gelfer M, Pandya K. Synthesis of zerovalent nanophase metal

particles stabilized with poly(ethylene glycol). Langmuir 2004; 20(16): 6896-903.
[http://dx.doi.org/10.1021/la0497402] [PMID: 15274601]

[36] Gedanken A. Using sonochemistry for the fabrication of nanomaterials. Ultrason Sonochem 2004; 11(2): 47-55.
[http://dx.doi.org/10.1016/j.ultsonch.2004.01.037] [PMID: 15030779]

[37] Ali MA, Ahmed T, Wu W, *et al.* Advancements in plant and microbe-based synthesis of metallic nanoparticles and their antimicrobial activity against plant pathogens. Nanomaterials (Basel) 2020; 10(6): 1146.
[http://dx.doi.org/10.3390/nano10061146] [PMID: 32545239]

[38] Durán N, Marcato PD, De Souza GI, Alves OL, Esposito E. Antibacterial effect of silver nanoparticles produced by fungal process on textile fabrics and their effluent treatment. J Biomed Nanotechnol 2007; 3(2): 203-8.
[http://dx.doi.org/10.1166/jbn.2007.022]

[39] Zhu J, Zhou M, Xu J, Liao X. Preparation of CdS and ZnS nanoparticles using microwave irradiation. Mater Lett 2001; 47(1-2): 25-9.
[http://dx.doi.org/10.1016/S0167-577X(00)00206-8]

[40] Yin H, Yamamoto T, Wada Y, Yanagida S. Large-scale and size-controlled synthesis of silver nanoparticles under microwave irradiation. Mater Chem Phys 2004; 83(1): 66-70.
[http://dx.doi.org/10.1016/j.matchemphys.2003.09.006]

[41] Pal A, Shah S, Devi S. Synthesis of Au, Ag and Au–Ag alloy nanoparticles in aqueous polymer solution. Colloids Surf A Physicochem Eng Asp 2007; 302(1-3): 51-7.
[http://dx.doi.org/10.1016/j.colsurfa.2007.01.054]

[42] Gu J, Fan W, Shimojima A, Okubo T. Microwave-induced synthesis of highly dispersed gold nanoparticles within the pore channels of mesoporous silica. J Solid State Chem 2008; 181(4): 957-63.
[http://dx.doi.org/10.1016/j.jssc.2008.01.039]

[43] Shirley AD, Dayanand A, Sreedhar B, Dastager SG. Antimicrobial activity of silver nanoparticles synthesized from novel Streptomyces species. Dig J Nanomater Biostruct 2010; 5(2): 447-51.

[44] Nabawy GA, Hassan AA, Sayed El-Ahl RH, Refai MK. Effect of metal nanoparticles in comparison with commercial antifungal feed additives on the growth of Aspergillus flavus and aflatoxin b1 production. J Glob Biosci 2014; 3: 954-71.

[45] Magro M, Moritz DE, Bonaiuto E, *et al.* Citrinin mycotoxin recognition and removal by naked magnetic nanoparticles. Food Chem 2016; 203: 505-12.
[http://dx.doi.org/10.1016/j.foodchem.2016.01.147] [PMID: 26948644]

[46] Hassan AA, Howayda ME, Mahmoud HH. Effect of zinc oxide nanoparticles on the growth of mycotoxigenic mould. SCPT 2013; 1(4): 66-74.

[47] Mouhamed AE, Hassan AA, Manal AH, El Hariri M, Refai M. Effect of metal nanoparticles on the growth of Ochratoxigenic moulds and Ochratoxin A production isolated from food and feed. Int J Res Stud Biosci 2015; 3(9): 1-14.

[48] Yehia RS, Ahmed OF. In vitro study of the antifungal efficacy of zinc oxide nanoparticles against *Fusarium oxysporum* and *Peniciliumexpansum*. Afr J Microbiol Res 2013; 7(19): 1917-23.
[http://dx.doi.org/10.5897/AJMR2013.5668]

[49] Li ZZ, Chen JF, Liu F, *et al.* Study of UV-shielding properties of novel porous hollow silica nanoparticle carriers for avermectin. Pest Manag Sci 2007; 63(3): 241-6.
[http://dx.doi.org/10.1002/ps.1301] [PMID: 17177171]

[50] Chen L, Zhang D, Chen J, Zhou H, Wan H. The use of CTAB to control the size of copper nanoparticles and the concentration of alkylthiols on their surfaces. Mater Sci Eng A 2006; 415(1-2): 156-61.
[http://dx.doi.org/10.1016/j.msea.2005.09.060]

[51] Yang F-L, Li X-G, Zhu F, Lei C-L. Structural characterization of nanoparticles loaded with garlic essential oil and their insecticidal activity against *Tribolium castaneum* (Herbst) (*Coleoptera: Tenebrionidae*). J Agric Food Chem 2009; 57(21): 10156-62.
[http://dx.doi.org/10.1021/jf9023118] [PMID: 19835357]

[52] El-Samahy MF, El-Ghobary AM, Khafagy IF. Using silica nanoparticles and neemoil extract as new approaches to control Tutaabsoluta (meyrick) in tomato under field conditions. Int J Plant Soil Sci 2014; 3: 1355-65.
[http://dx.doi.org/10.9734/IJPSS/2014/8435]

[53] Worrall EA, Hamid A, Mody KT, Mitter N, Pappu HR. Nanotechnology for plant disease management. Agronomy (Basel) 2018; 8(12): 285.
[http://dx.doi.org/10.3390/agronomy8120285]

[54] Liu Y, Yan L, Heiden P, Laks P. Use of nanoparticles for controlled release of biocides in solid wood. J Appl Polym Sci 2001; 79(3): 458-65.
[http://dx.doi.org/10.1002/1097-4628(20010118)79:3<458::AID-APP80>3.0.CO;2-H]

[55] Wanyika H. Sustained release of fungicide metalaxyl by mesoporous silica nanospheres. Nanotechnology for Sustainable Development. Cham: Springer 2013; pp. 321-9.
[http://dx.doi.org/10.1007/978-3-319-05041-6_25]

[56] Hou R, Zhang Z, Pang S, Yang T, Clark JM, He L. Alteration of the nonsystemicbehavior of the pesticide ferbam on tea leaves by engineered gold nanoparticles. Environ Sci Technol 2016; 50(12): 6216-23.
[http://dx.doi.org/10.1021/acs.est.6b01336] [PMID: 27254832]

[57] Sandhya , Kumar S, Kumar D, Dilbaghi N. Preparation, characterization, and bio-efficacy evaluation of controlled release carbendazim-loaded polymeric nanoparticles. Environ Sci Pollut Res Int 2017; 24(1): 926-37.
[http://dx.doi.org/10.1007/s11356-016-7774-y] [PMID: 27761863]

[58] Dar J, Soytong K. Construction and characterization of copolymer nanomaterials loaded with bioactive compounds from Chaetomium species. Agric Technol Thail 2014; 10(4): 823-31.

[59] Malandrakis AA, Kavroulakis N, Chrysikopoulos CV. Use of copper, silver and zinc nanoparticles against foliar and soil-borne plant pathogens. Sci Total Environ 2019; 670: 292-9.
[http://dx.doi.org/10.1016/j.scitotenv.2019.03.210] [PMID: 30903901]

[60] Fraceto LF, Grillo R, Gerson A. "Nanotechnology in agriculture: Which innovation potential does it have?" Frontiers in Environmental Science, 4 (March). Front Environ Sci 2016; 4.

[61] Slomberg DL, Schoenfisch MH. Silica nanoparticle phytotoxicity to Arabidopsis thaliana. Environ Sci Technol 2012; 46(18): 10247-54.
[http://dx.doi.org/10.1021/es300949f] [PMID: 22889047]

[62] Hairat S, Khurana P. Evaluation of Aegilops tauschii and Aegilops speltoides for acquired thermotolerance: Implications in wheat breeding programmes. Plant Physiol Biochem 2015; 95: 65-74.
[http://dx.doi.org/10.1016/j.plaphy.2015.07.009] [PMID: 26188500]

[63] Joner EJ, Hartnik T, Amundsen CE. Norwegian Pollution Control Authority Report no 2008. TA 2304/2007

[64] Schmid G, Ed. Nanoparticles: from theory to application. John Wiley & Sons 2011.

[65] Chen H. Metal based nanoparticles in agricultural system: behavior, transport, and interaction with plants. Chem Spec Bioavail 2018; 30(1): 123-34.
[http://dx.doi.org/10.1080/09542299.2018.1520050]

[66] Wilhelm C, Billotey C, Roger J, Pons JN, Bacri JC, Gazeau F. Intracellular uptake of anionic superparamagnetic nanoparticles as a function of their surface coating. Biomaterials 2003; 24(6): 1001-11.

[http://dx.doi.org/10.1016/S0142-9612(02)00440-4] [PMID: 12504522]

[67]   Deb SK, Shukla MK. A review of dissolved organic matter transport processes affecting soil and environmental quality. J Environ Anal Toxicol 2011; 1(106.10): 4172.

[68]   Praetorius A, Labille J, Scheringer M, Thill A, Hungerbühler K, Bottero JY. Heteroaggregation of titanium dioxide nanoparticles with model natural colloids under environmentally relevant conditions. Environ Sci Technol 2014; 48(18): 10690-8.
[http://dx.doi.org/10.1021/es501655v] [PMID: 25127331]

[69]   Labille J, Harns C, Bottero JY, Brant J. Heteroaggregation of titanium dioxide nanoparticles with natural clay colloids. Environ Sci Technol 2015; 49(11): 6608-16.
[http://dx.doi.org/10.1021/acs.est.5b00357] [PMID: 25913600]

[70]   Joner E J, Hartnik T, Amundsen C E. Norwegian Pollution Control Authority Report no. TA 2304/2007. Bioforsk; Environmental fate and ecotoxicity of engineered nanoparticles 2008; 1-6.

[71]   Lahiani MH, Chen J, Irin F, Puretzky AA, Green MJ, Khodakovskaya MV. Interaction of carbon nanohorns with plants: uptake and biological effects. Carbon 2015; 81: 607-19.
[http://dx.doi.org/10.1016/j.carbon.2014.09.095]

[72]   Li L, Li G, Smith RL, Inomata H. Microstructural evolution and magnetic properties of NiFe2O4 nanocrystals dispersed in amorphous silica. Chem Mater 2000; 12(12): 3705-14.
[http://dx.doi.org/10.1021/cm000481l]

[73]   Zhou D, Jin S, Li L, Wang Y, Weng N. Quantifying the adsorption and uptake of CuO nanoparticles by wheat root based on chemical extractions. J Environ Sci (China) 2011; 23(11): 1852-7.
[http://dx.doi.org/10.1016/S1001-0742(10)60646-8] [PMID: 22432310]

[74]   Larue C, Laurette J, Herlin-Boime N, *et al.* Accumulation, translocation and impact of TiO2 nanoparticles in wheat (Triticum aestivum spp.): influence of diameter and crystal phase. Sci Total Environ 2012; 431: 197-208.
[http://dx.doi.org/10.1016/j.scitotenv.2012.04.073] [PMID: 22684121]

[75]   Zhu ZJ, Wang H, Yan B, *et al.* Effect of surface charge on the uptake and distribution of gold nanoparticles in four plant species. Environ Sci Technol 2012; 46(22): 12391-8.
[http://dx.doi.org/10.1021/es301977w] [PMID: 23102049]

[76]   Thuesombat P, Hannongbua S, Akasit S, Chadchawan S. Effect of silver nanoparticles on rice (Oryza sativa L. cv. KDML 105) seed germination and seedling growth. Ecotoxicol Environ Saf 2014; 104: 302-9.
[http://dx.doi.org/10.1016/j.ecoenv.2014.03.022] [PMID: 24726943]

[77]   Doolette CL, McLaughlin MJ, Kirby JK, Navarro DA. Bioavailability of silver and silver sulfide nanoparticles to lettuce (Lactuca sativa): Effect of agricultural amendments on plant uptake. J Hazard Mater 2015; 300: 788-95.
[http://dx.doi.org/10.1016/j.jhazmat.2015.08.012] [PMID: 26322966]

[78]   Patel PR, Shaikh SS, Sayyed RZ. Modified chrome azurol S method for detection and estimation of siderophores having affinity for metal ions other than iron. Environmental Sustainability 2018; 1(1): 81-7.
[http://dx.doi.org/10.1007/s42398-018-0005-3]

[79]   Jones DL, Darrah PR. Role of root derived organic acids in the mobilization of nutrients from the rhizosphere. Plant Soil 1994; 166(2): 247-57.
[http://dx.doi.org/10.1007/BF00008338]

[80]   Lin D, Xing B. Phytotoxicity of nanoparticles: inhibition of seed germination and root growth. Environ Pollut 2007; 150(2): 243-50.
[http://dx.doi.org/10.1016/j.envpol.2007.01.016] [PMID: 17374428]

[81]   El-Temsah YS, Joner EJ. Impact of Fe and Ag nanoparticles on seed germination and differences in bioavailability during exposure in aqueous suspension and soil. Environ Toxicol 2012; 27(1): 42-9.
[http://dx.doi.org/10.1002/tox.20610] [PMID: 20549639]

[82]   Feizi H, Rezvani Moghaddam P, Shahtahmassebi N, Fotovat A. Impact of bulk and nanosized titanium dioxide (TiO2) on wheat seed germination and seedling growth. Biol Trace Elem Res 2012; 146(1): 101-6.
[http://dx.doi.org/10.1007/s12011-011-9222-7] [PMID: 21979242]

CHAPTER 15

Role of Fungi in Biofuel Production Chain

Renuka Agrawal[1,*] and **Ankur Maheshwari**[2]

[1] *Department of Botany, Miranda House, University of Delhi, Delhi, India*

[2] *Department of Zoology, Zakir Husain Dehli College, University of Delhi, Delhi, India*

Abstract: The demand of fuels as a source of energy for various operations is increasing daily. This has led to increased demand of fossil fuels, particularly by transportation and industrial sectors. There are multiple problems related to conventional fossil fuels like firstly, they are non-renewable resources with limited reserves. Secondly, fossil fuels pose serious environmental and health issues. Fossil fuels are one of the leading sources of emission of atmospheric greenhouse gases (GHG), resulting in global warming and thus climate change. These limitations and adverse effects of the use of fossil fuels have warranted scientists and policymakers to look for renewable and greener alternatives such as biofuels. Based on the type of feedstock used, biofuels are classified as first-generation, second-generation and third-generation. First-generation biofuels are based on edible resources which are already scanty. This has led to increased interest in second and third-generation biofuels. The agricultural waste and inedible crops constituting lignocellulosic materials are important second-generation biofuel feedstocks. The second-generation feedstocks can be a great alternative to conventional fossil fuels, but there are a few limitations, such as the cost and efficiency of production. Currently, scientists are looking at the role of fungi and utilization of various fungal enzymes in the hydrolysis of the lignocellulosic substrates for efficient and cost-effective production of biofuels. Nanomaterials have the ability for the better utilization of enzymes, biofuels, biodiesels and other microbial fuels. Therefore, nanotechnology can be utilized to address the challenges through various mechanisms and processes. This chapter is an attempt to focus on the role of fungi and fungal enzymes for better utilization of feedstock and sustainable production of renewable, cost-effective, environment-friendly biofuels.

Keywords: Biofuels, Fossil fuels, Fungi, Fungal enzymes, Lignocellulosic materials.

INTRODUCTION

There is increased demand of energy due to fast growing world's population along with the unprecedented speed of industrialization leading to energy crises. Guo *et al.* 2015, has predicted increased world energy demand to 8.6×10^{20} J in

* **Corresponding Author Renuka Agrawal**: Department of Botany, Miranda House, University of Delhi; Tel: +91-1--35617930; Email: renuka.agrawal@mirandahouse.ac.in

Savita, Anju Srivastava, Reena Jain & Pratap Kumar Pati (Eds.)
All rights reserved-© 2022 Bentham Science Publishers

2040 and was predicted 6.6×10^{20} J for the year 2020 [1]. This will lead to increased fossil fuel utilisation particularly by industry and transportation sector. There are multiple problems which are related to the utilization of conventional fossil fuels to meet this increased demand. First of all, they are non-renewable resources with limited reserves, which are depleting at a very fast rate. This has also led to increased crude oil prices globally. Secondly, utilization of fossil fuel is one of the leading factors contributing to global warming, environmental pollution and health issues [2]. The increase in the levels of greenhouse gases (GHG) as a consequence of burning of fossil fuels will lead to global warming and, therefore climate change [3, 4]. Consequently, there is an immediate need to shift the focus from fossil fuels to alternative renewable, environmentally friendly sources of energy. Over the past few decades, policy makers and scientists have been exploring ways and means to utilise agricultural crops and wastes as a source (feedstock) for fuel. Biofuels or agrofuel are mainly derived from biomass or biological waste such as agricultural waste. Nanomaterials/Nanoparticles are emerging technology for efficient and speeding up biofuel production and their role can't be ignored in biofuel production now and in future. Nanotechnology can help us in producing biofuel at a lesser cost than conventional methods. These biofuels have emerged as one of the potential renewable and environment friendly alternatives to conventional fossil fuels.

First-generation: these are based on edible plants, specifically starch or oilseeds crops such as corn, potato, wheat, rapeseed, soybean, sunflower [5].

Second-generation: these are based on inedible crops or agricultural wastes mainly consisting of lignocellulosic materials [5].

Third-generation: these are based on the use of different types microorganisms like microalgae for the production of different types of biofuels like bioethanol and biodiesel [6].

CLASSIFICATION OF BIOFUELS BASED ON FEEDSTOCK TYPE

Biofuels can be classified into three generations based on the type of feedstock starting material used (Fig. **1**).

First Generation Biofuels	Second Generation Biofules	Third Generation Biofuels
• based on edible plants specifically starch or oilseeds crops • e.g. corn, potato, wheat, rapeseed, soybean	• based on inedible crops or agricultural wastes mainly consisting of lignocellulosic materials • e.g. straw from cereal crops, bagasse, spent compost and leaf litter as feedstock	• based on use of different types microorganisms like microalgae for the production of different types of biofuels like bioethanol and biodiesel

Fig. (1). Classification of biofuels on the basis of type of feedstocks.

Biofuels can be further classified into three forms based on the physical state of biofuels (Fig. **2**) [1, 5]:

Solid Biofuels
• Woody Biomass
• Herbaceous Biomass
• Fruit Biomass
• Blends and Mixtures

Liquid Biofuels
• Bioethanol (produced by common yeast by fermentation)
• Biodiesel (Produced by Transesterification
• Biobutanol
• Drop in biofuel (Biomass derived liquid biofuels ready to use. such as syngas, liquefied biomass, and butanol complexes, sugar hydrocarbons

Gaseous Biofuels
• Biogas (generated through the process of anaerobic digestion)
• Syngas (Pyrolysis of lignocellulosic biomass or any other renewable material can be used to produce syngas. It's a mixture of carbon monoxide (30–60%), hydrogen (25–30%), and carbon dioxide (5–15%) along with CH4, H2S, COS, NH3, and water vapor)

Fig. (2). Classification of biofuels based on their physical state.

1. Solid biofuels: these include starting materials like firewood, wood charcoal, wood pellets and wood chips.

2. Liquid biofuels: these are mainly based on bioethanol, biodiesel, drop in biofuels, and pyrolysis bio-oil.

3. Gaseous biofuels: these mainly included biogas and syngas.

Solid Biofuels

Solid biofuels can be divided into four types, namely woody biomass, herbaceous biomass, fruit biomass, and blends and mixtures [1]. Some of these plants derived biomass like wood and wood chips derived from tree trunks, grasses, crop residues like straw and nutshells have been utilized for direct burning for thousands of years. Ancient people also used woody material to convert it into a carbon-enriched, porous, and black solid, charcoal by the pyrolysis [5].

Liquid Biofuels

Bioethanol, biodiesel, and biobutanol are the most commonly used liquid biofuels.

Bioethanol

For bioethanol production, most common yeast (*Saccharomyces cerevisiae*) is used for fermenting organic matter into ethanol. It's an eco-friendly fuel [7], but most commonly it is produced from starchy edible crops like corn, sugarcane potato, sorghum, wheat *etc* making it first-generation biofuel. This results in competition between the biofuel and food industry for the same set of resources like agricultural land, water, raw material and other resources [4, 8, 9]. The better alternative is to utilize nonedible plants and lignocellulosic biomass for the production of second-generation bioethanol.

Biodiesel

Biodiesel is produced by the transesterification" process [10]. Chemically biodiesel is mono-alkyl esters of vegetable oils, algal andmicrobial oils, and animal fats [11]. Diesel is utilized in transportation sector. Most commonly edible vegetable oils like soybean, canola, rapeseed, palm, safflower, almond, barley, coconut, cotton seed, *etc.* are used for the production of biodiesel [12, 13]. This again has a negative impact on global food security. Secondly the high cost of these edible oils as feedstock makes this option economically non-viable. Answer to this problem is to utilize waste animal fats, nonedible oils, and used vegetable oils for production of biodiesel. Among the front runners' non-edible plants for biodiesel production are physic nut (*Jatropha curcas*), castor beans (*Ricinus communis* L.), camelina (*Camelina sativa*), karanja (*Pongamia pinnata* L.), mahua (*Madhuca*spp.), neem (*Azadirachta indica A.*), simarouba (*Simarouba glauca* DC.), and cheura (*Diploknemabutyracea*).

Pyrolysis Bio-oil

When lignocellulosic biomass is heated at 300–900 °C in the absence of air, the process is called pyrolysis [5]. Three types of products are produced after pyrolysis namely, biochar, bio-oil and syngas [14]. Pyrolysis can be slow or fast. Fast pyrolysis results mainly in production of Bio-oil and other chemicals [15 - 17]. Fast pyrolysis of plant biomass results in about 10–30% biochar, 50–70% bio-oil, and 15–20% syngas [17]. This is a cost-effective method for bio-oil production from lignocellulosic materials. The lignocellulosic biomass has low nitrogen and ash which are known to be most efficient for bio-oil production. Crude pyrolysis bio-oil can be directly used in industrial scale combustion systems but also for making bio-oil as a petrol substitute [5].

Drop-in Biofuels

This category includes different biofuels, such as syngas, liquefied biomass, and butanol complexes, sugar hydrocarbons. These are basically biomass derived liquid biofuels having similar fuel specifications to petroleum-based fuels, these does not require any change in existing infrastructure and are ready to"drop in" to the existing fuel systems [1, 18]. Presently, this is not a cost-effective and easy to go option.

GASEOUS BIOFUELS

Biogas

Biogas is generated through the process of anaerobic digestion (AD) [19]. Biogas has many advantages as it is economical, environment friendly, it can also be efficient in waste management and also to mitigate water, soil, and air pollution sources. Different types of organic wastes like livestock manure, food processing waste, lignocellulosic materials (straw, bagasse, branches, *etc.*), municipal solid waste, sewage sludge, yard trimmings, *etc.* can be utilized for the production of biogas.

Syngas

Pyrolysis of lignocellulosic biomass or any other renewable material can be used to produce syngas. It is typically a mixture of carbon monoxide (30–60%), hydrogen (25–30%), and carbon dioxide (5–15%) along with CH_4, H_2S, COS,

NH_3, and water vapor. It can be directly burned for the generation of electricity. After purification, it can also be used for the production of transportation fuels, such as methanol, ethanol, methane, and dimethyl ether.

UTILIZATION OF LIGNOCELLULOSE AS FEEDSTOCK IN BIOFUEL PRODUCTION

The use of first-generation biofuel swhich are dependent on the availability of edible starch and oil seed crops like corn, wheat, potato, rapeseed, soybean, sunflower *etc.* has led to competition for available land, water resources, raw material and other resources [4, 8, 9]. This has resulted in the price hike for food items. Therefore, the focus has shifted from first generation biofuels to second generation biofuels which are mainly based on agricultural waste and renewable lignocellulosic materials. The second-generation biofuels can utilize lignocellulosic materials such as straw from cereal crops, bagasse, spent compost and leaf litter as feedstock, thus reducing pressure on food crops for generation of biofuels [20]. Lignocellulose is composed mainly of three different polymers that are cellulose, hemicellulose and lignin [21 - 23].

Cellulose is the one of the major components of lignocellulosic biomass and it is composed of a linear β-1,4- linked polymer of D-glucose units [24, 25]. Cellulose is a polysaccharide which through the process of hydrolysis, can be converted into simple sugars. The resulting sugars can easily use by yeasts for bioethanol production.

Hemicellulose is the second most abundant component of lignocellulose. It is a heterogeneous polysaccharide composed of two pentoses (D-xylose and D-arabinose), three hexoses (D-mannose, D-glucose and D- galactose) along with sugar acids [22, 26, 27]. Hemi-cellulose consists of both a linear β-1,4-linked backbone polymer and branched polymers [26]. Hemicellulose is another polysaccharide that through the process of hydrolysis can be converted into the simple sugars and sugars in turn can readily uses by yeasts for bioethanol production.

Lignin is the third most abundant polymer found in lignocellulose. It generally exists as a heterogeneous polymer consisting of dimethoxylated, mono-methoxylated and non-methoxylated phenylpropanoid units [22].

Apart from three major components, lignocellulosic biomass also contains varying amounts of other materials such as proteins, ash and pectic materials. Pectin is a third major structural component of the plant cell wall after cellulose and hemicellulose [22].

Biorefining

The lignocellulose can be converted to biofuels through four major steps (pre-treatment, hydrolysis, fermentation and separation (Fig. **3**).

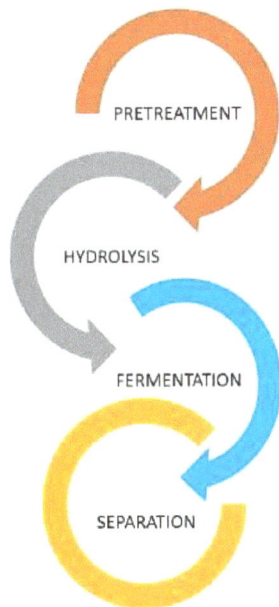

Fig. (3). Biorefining: The process of Lignocellulose conversion into biofuels.

Pre-Treatment

The first step of biorefining *i.e.,* pre-treatment is most crucial step in the biorefining process. The pre-treatment enhances the hydrolysis leading to higher yields of sugar which in turn is used for the fermentation process. The hydrolysis is improved because of better cellulose accessibility and increased pore sizes as pre-treatment step removes lignin and hemicellulose. The pre-treatment involves various treatment methods which can be classified as physical, chemical or biological [22, 25, 28]. The physical pre-treatment methods include processes such as milling, irradiation and steam explosion. Chemical pre-treatment methods involve the use of ammonium fibre explosion (AFEX), organosolv treatment and the addition of either acid or alkali [22, 28, 29]. Biological pre-treatment methods utilize microorganisms to remove lignin from the lignocellulose material [22]. The removal is carried out by the enzymes produced by these microorganisms. The enzymes specifically disrupt the fibril and lignin structures of the plant cell wall. The enzyme use is advantageous over physical and chemical methods as it requires lower amount of energy, produces lesser amount of waste and therefore minimal effect on the environment [22, 29].

The choice of method for pre-treatment depends on the type of lignocellulosic material and the type of hydrolysis to be done later. The acid pre-treatment is the choice of method if the hydrolysis utilizes the fungal enzymes, which requires lower pH of around 4–5 [22].

Hydrolysis

Hydrolysis is the process by which the lignocellulose polymers are saccharified to produce simple sugars like hexoses and pentoses, which can be readily fermented [30]. There can be two types of methods that can be employed for the process of hydrolysis; acid hydrolysis and enzymatic hydrolysis [22, 28]. Acid hydrolysis is the older method and are less common because of lower yields and environmental concerns [22, 31]. On the other hand, in enzymatic hydrolysis, the specific enzymes produced by fungi and bacteria are utilized for the breakdown of the lignocellulose into the simple monomeric sugars [22, 28]. The by-products are almost negligible in case of enzymatic hydrolysis, but this method is more complex, expensive and time consuming as compared to acid hydrolysis [28].

Fermentation

Fermentation is the third step of bioconversion process. During fermentation, hydrolysates such as glucose, xylose, arabinose and mannose are converted to bioethanol by the action of microorganisms such as Saccharomyces cerevisiae [22, 28]. Generally, the hydrolysates are first detoxified and then the fermentation is carried out because many inhibitory compounds such as phenolic and furan derivatives are produced during the first two steps (Pre-treatment and Hydrolysis) [22, 28]. Some strains of Saccharomyces cerevisiae, such as TMB 3,400, can efficiently convert glucose, arabinose and xylose to bioethanol [22]. Thus, making Saccharomyces cerevisiae as the most widely used microorganism for the utilization of large amounts of hydrolysates for the production of bioethanol with higher percentage yields.

Combinations

In order to reduce the production costs and time, and to achieve higher end-product yields, investigators are trying to use a different combination of first three steps of biorefinery process. There are different strategies that can be adopted and in some cases the cost is higher but gives better yield, whereas in others, the cost can be cut but then the yield is also compromised. For example, separate hydrolysis and fermentation (SHF) optimises each process separately but it requires the use of large amounts of enzymes like β-glucosidase thus making this a costly process [22]. Another strategy is simultaneous saccharification and fermentation (SSF), where both steps are combined together into one reaction, which allows the direct fermentation of hydrolysates into bioethanol. This strategy reduces the enzyme costs but end- products yields are also lower [22, 28]. Another strategy called consolidated bioprocessing (CBP) combine all the three steps of biorefining through the use of one or more microorganisms [22]. This strategy can be effective in reducing the cost of bioethanol, but this requires thorough research on the microorganisms, enzymes and pH, and temperature optima.

ROLE OF FUNGI IN BIOREFINING

Many different microorganisms, including fungi have shown a tremendous potential in biorefining processes especially in biofuel production. In biofuel production, fungi are important in the industrial hydrolysis and pre-treatment steps. There are many important criteria on the basis of which fungi are selected, such as their ability to simultaneously or selectively degrade the lignocellulosic biomass, their thermostability, their engineering capabilities and on the basis of their enzymes [20]. Fungal members especially those belonging to Basidiomycota and Ascomycota are responsible for degradation and modification of lignocellulosic material in nature such as forest litter, wood, compost and soil into different end products depending on the fungal genus involved [23]. Different fungal species have been proposed for the production of different biofuels such as biohydrogen, biodiesel and cellulosic ethanol. Different procedures have been developed to convert lignocellulosic material into biofuels using filamentous fungi [6]. The possible use of *Trichoderma*, *Aspergillus*, and *Fusarium* species through the consolidated bioprospecting(CBP) for biofuel production has also been explained in the review [4].

White-rot basidiomycetes (*Trametes*, *Pleurotusostreatus*, and *Phanerochaet-echrysosporium*), Brown-rot basidiomycetes along with some ascomycetes (*Trichodermareesei)* play an important role in degradation of the lignocellulose material [32]. These fungal members secrete extracellular enzymes capable of degrading lignocellulosic materials. White-rot basidiomycetes can produce an array of different enzymes capable of degrading all components of the lignocellulosic plant wall whereas brown- rot and ascomycetes produces some of these enzymes.

The enzymes which are capable of degrading cellulose such as exo-β-1-4-glucanases and endo-β-1,4-glucanases, are commonly produced by different strains of *T. reesei* in large concentrations. At least 10 different lignocellulose-hydrolyzing enzymes under the control of more than 200 glycoside hydrolase genes have been isolated and characterised in *T. reesei*. Many other fungal species such as *Myceliophthora thermophile*, *Talaromycesemersonii*, *Tribulus terrestris*,

Chaetomium thermophilum, and *Thermoascus aurantiacus* have been identified which can produce thermostable cellulases having high cellulose-degrading activity at high temperatures [1].

The filamentous fungi are most commonly utilized in biofuel production during the processes of pre-treatment and enzymatic hydrolysis of lignocellulosic biomass to simple sugar monomers.

Some filamentous fungi are capable of producing different types of cellulases in higher quantities (about 100 g/ L). *Trichoderma* and *Chrysosporium lucknowense* are the fungal members which are commercially exploited for cellulases production in large quantities which are higher than what is obtained from bacterial counterparts [5]. Scientist have increased the level of β-glucosidase production to improve the overall cellulase activity on biomass residues through the expression of *Periconia* sp. β-glucosidase gene in *T. reesei* using genetic engineering technologies [33].

Fungi can also be used for production of biodiesel. Fungi can be utilized in two different ways as fungi have the capacity to generate high concentrations lipids which can be directly used for the production of biodiesel. *S. cerevisiae* have beengenetically modified to produce biodiesel [34]. Secondly, fungi can be utilized as biocatalyst for the transesterification process. As compared to chemical acidic or basic catalysts, utilization of fungal lipases as catalyst is more desirable as it is environment friendly and does not pose any health risks. Also, fungal lipases do not require any cofactors in the process, they have wide range of substrate specificity and are highly stable in different types of organic solvents.

Fungal Enzymes

The fungal enzymes produce by different fungi function as systems and they degrade each lignocellulosic component separately. For example, lignin-modifying enzymes (LMEs) disrupt the recalcitrant lignin components, whereas cellulase systems work on the cellulosic microfibrils to convert polymers into single glucose units. Similarly a different group of hemicellulases breaks down different side chains and backbone polymers of hemicelluloses.

Lignin-Modifying Enzymes

Lignin-modifying enzymes (LMEs) are a group of different fungal enzymes involved in lignin degradation. It included laccase, lignin peroxidase, manganese peroxidase and versatile peroxidase. Phenol oxidase enzymes are capable of oxidising various phenolic compounds; these include the LMEs as well as tyrosinase, catechol oxidase and catalase-phenoloxidase [35].

Laccase

Laccases mainly include extracellular glycoproteins that belongs to multi-copper oxidase (MCO) superfamily. These are mainly secreted by most of the white-rot basidiomycetes and some ascomycetes members, and also a number of brown-rot basidiomycetes produce laccases in liquid culture [23, 36]. Typically, fungal laccase has molecular mass ranging between 60-80 kDa with an isoelectric point between 3 to 6 [29, 36 - 38]. In fungi, multiple genes are involved in encoding the laccase glycoproteins resulting in secretion of several isoforms [29, 36]. Fungal laccases have wide applications in biofuel production and they largely function by lignin depolymerisation and biodegradation *via* oxidation of phenolic substrates [23, 29, 36, 38 - 40]. They also function along with peroxidases in the polymerisation process of monolignols into lignin polymers, during the detoxification step of hydrolysates by the removal of inhibitors before the fermentation process and they also play an important role in the treatment of waste waters [26, 36, 41, 42]. Laccases are considered green biocatalysts because they utilize molecular O_2 instead of H_2O_2, although they possess low redox potential in contrast to the peroxidases involved in lignin degradation [43].

Lignin Peroxidase

Peroxidases and laccases function in a synergistic way in the process of lignification of plants as well in the degradation and depolymerisation process of the lignin [26, 38, 42, 44]. These peroxidases (LiP, MnP and VP) are extracellular, heme-containing hydrogen peroxide-dependent enzymes which oxidises the high redox potential components of the lignin [40, 44]. These peroxidases are non-specific and are commonly secreted by white-rot basidiomycetes for degradation of lignin [45].

Lignin peroxidase (LiP) was first discovered from the white-rot *Phanerochaetechrysosporium*in 1983 by Tien and Kirk [46, 47]. Lignin peroxidases are monomeric proteins with molecular mass of approximately 40 kDa. They have similarities with the classical peroxidases in having iron coordinated to four tetrapyrrole rings and to a histidine residue [29]. LiPs can oxidise a number of different phenolic compounds such as syringic acid, guaiacol and vanillyl alcohol, and it can also oxidise nonphenolic aromatic compounds [40, 47, 48]. Both LiPs and laccases interact directly with the lignin structure while MnPs act upon the lignin indirectly. Hydrogen peroxide plays a key role in the degradation of lignin by peroxidases.

Manganese Peroxidase

Manganese peroxidases (MnP) are also H_2O_2-dependent heme-containing peroxidase enzymes which are capable of degrading lignin substrates. This was first discovered in *P. chrysosporium* (1983–1984) by Kuwahara *et al.* [49]. MnP have a molecular mass of approximately 40–50 kDa. Manganese peroxidase requires manganese (Mn^{2+}) as a cofactor for its oxidative function [29, 46, 47]. As compared to LiPs, these are more diverse enzymes. The production of MnP and LiP by a fungus regulated by the presence of Mn^{2+} molecules. If Mn^{2+} is present in low concentrations, predominately LiP is produced, whereas high concentrations of Mn^{2+} will result in increased production of MnP [44]. MnPs requires the presence of cofactors and chelators for oxidising the lignin substructures.

Versatile Peroxidase

Versatile peroxidases (VP) were first described from the fungus *Pleurotuseryngii*, which possesses both the functionalities of LiPs and MnPs. Versatile Peroxidases can oxidise both phenolic and nonphenolic aromatic compounds [5].

Hydrogen Peroxide-Producing Enzymes

White-rot basidiomycetes and some other fungal members require H_2O_2 for the functioning of the extracellularperoxidase enzymes involved in the lignin degradation. Two oxidases produced by the fungus: glyoxal oxidase and aryl alcohol oxidase provide the H_2O_2 and they function through reducing molecular O_2 to H_2O_2 along with the oxidation of a co-substrate [22, 29].

Phenol Oxidases

These enzymes include a range of copper enzymes without glycosyl hydrolase and peptidase activity. In the presence of molecular O2, they can oxidise phenolic compounds. The phenol oxidases include the LMEs as well as tyrosinases, catechol oxidases and catalase phenoloxidases.

Cellulase

Cellulases are the hydrolytic enzymes involved in the degradation of the cellulosic component of lignocellulosic material. Cellulase systems involve three different enzymes; endoglucanases, cellobiohydrolase (CBHs) and β-glucosidases [50, 51]. The overall extent of biomass saccharification is dependent on the properties of biomass, such as accessibility of the cellulose chains to these cellulases, lignin composition, degree of cellulose crystallinity and degree of polymerisation [52]. The hydrolysis of cellulose is hindered by the lignin composition through adsorption of cellulase enzymes [53]. It has been recently demonstrated that the cellulases work by penetrating the porous regions of the cellulose fibrils (at amorphous regions) and then the depolymerisation of the chains occurs [54]. This stage of cellulose saccharification is known as morphogenesis and includes the swelling and fragmentation of the cellulose fibres [55]. Cellulases can be grouped into 11 glycoside hydrolase (GH) families and are classified as the modular proteins composed of at least two distinct modules; a) the catalytic module and b) the carbohydrate binding module (CBM) [22, 55, 56].

Fungal Endoglucanases (EG) generally monomeric proteins have little or no glycosylation and are characterised by an open binding cleft [22, 51]. EGs works by cleaving the internal β-1, 4-glycosidic bonds of the cellulose chains [50, 53]. Different fungal species produce a number of different isozymes of EGs. The ascomycete *T. reesei* produces at least five different isozymes of EG and white rot basidiomycete *P. chrysosporium* produces three isozymes of EGs [22].

Cellobiohydrolase (CBH) are monomeric, exo-acting enzymes, with little or no glycosylation. CBH acts by attacking the cellulose chains at their terminal ends [51]. Different isozymes of CBH can act from either the reducing or non-reducing end of the cellulose chain. *T. reesei* can produce two CBH isozymes, one acting from the reducing end and whereas other acting from the non-reducing end of the cellulose chains, thus allowing a more efficient hydrolytic process. The β-glucosidases are important for commercial hydrolysis as they can increase the hydrolysis process by quickly converting cellobiose to glucose, and thus reducing the potential of end-point inhibition [55]. β-glucosidases can be classified into intracellular, extracellular and cell wall-associated groups. These can be monomeric (approximately 35 kDa), dimeric or even trimeric proteins which are greater than 146 kDa [22].

All the three cellulases act in a synergistic manner. Cellulases are the most important industrial enzymes comprising around 75% of the total industrial demand of enzymes [57].

Hemicellulases

In comparison to lignin and cellulose, hemicellulose components of lignocellulosic materials are more accessible to extracellular fungal enzymes because they do not form crystalline structures or microfibrils. The hemicellulose within the plant cell wall is heterogeneous, therefore requiring a variety of enzymes with different functionalities for complete hydrolysis of hemicellulose component of lignocellulosic material.

Xylanases are group of enzymes that act on the xylan heteropolymers. It comprises endo-1, 4-β-xylanases and β-xylosidases. Endo-1, 4-β-xylanases act by cleaving the glycosidic bonds in the polymer backbone and releasing xylo-oligosaccharides whereas β-xylosidases hydrolyse the xylo-oligosaccharides and xylose are released [27]. If substituted xylan units are present on the polymer backbone, then most of the xylanases are unable to cleave. This requires other accessory enzymes to remove these substituted residues. Some fungal members like *Penicillium capsulatum* and *Talaromyces emersonii* have the complete sets of xylan-degrading enzymes. Therefore, these fungal members can be exploited for use in the biorefinery system [27].

Mannanases are enzymes that cleave the galactomannan or glucomannan polymers present in the plant cell wall. Mannanases also work with the help of accessory enzymes which can remove the substituted residues. It consists of β-1, 4-mannanase and β-mannosidases. The β-1, 4-mannanases act on the backbone, whereas β-mannosidases act on the mannan substitutions. Both enzymes, xylanases and mannanases act in a synergistic manner along with accessory enzymes for a complete breakdown during hydrolysis. The accessory enzymes acting on the hemicellulose hydrolysis include α-L-arabinofuranosidase, α-glucuronidase, acetylxylan esterase and ferulic acid esterase [27]. The microorganisms can ferment a number of different hexose and pentose sugars produced by hemicellulose into ethanol [26, 43].

NANOTECHNOLOGY AND BIOFUELS

Nanomaterials have special characteristics like greater surface areas, crystalline structure, stability, catalytic activity, better storage ability and adsorption capacity which makes them suitable candidate in biofuel production directly/indirectly. Nanoparticles such as nanofibers, nanotubes, inorganic nanoparticles enhance the metabolic reactions of bioprocessing involved in biofuel production.

Nanomaterials such as Gold, Silver, Silica nanoparticles are found to enhance biohydrogen production when used in optimal concentration. Gold Nanoparticles enhance the activity of enzymes for substrate utilization by 56%, which induced 46% increase in biohydrogen production. Silver nanoparticles do the same by enhancing the substrate utilization up to 62%. Silica nanoparticles with Iron oxide are more promising nanocomposites with more stability and catalytic activity [58 - 62]. Nanoparticles play roles not only in biohydrogen but also in biogas production, nanoparticles play a redox role and therefore enhances the activity of various enzymes in anaerobic condition [62, 63]

In biodiesel production, during the transesterification reaction, oil may react with an alcohol in the presence of an appropriate nanocatalyst or catalyst [64]. As the enzyme's biocatalytic activity gets reduced in methanol presence. A method has been reported by researchers that how to reuse the lipase by adsorbing it on various carriers [65]. A new potential is created by the Magnetic nanoparticles due to their stability and low-cost catalyst ability and their easy recovery from the reaction mixtures during biodiesel production [66].

Various nanomaterials, including Manganese oxide nanoparticles have been reported to increase bioethanol production from sugarcane leaves. Multiple reports mentioned that immobilizing *Saccharomyces cerevisiae* with magnetic nanoparticles/alginates enhances bioethanol production [62, 67]. Conclusively, nanomaterials play a major role in biofuel production with low cost and high efficiency making them available for developing countries too.

CONCLUSION

The extensive, indiscriminate utilisation of fossil fuels, rise in crude oil prices and increase in GHG emissions leading to global warming have necessitated the requirement for alternative, renewable and environment friendly options. Biofuels are one such potential renewable and environment-friendly option. Biofuels can be classified as first, second and third generation on the basis of feedstock used. The second-generation biofuels based on renewable lignocellulosic materials and wastes such as leaf litter spent compost, bagasse and the straw left after harvesting cereal crops are the most promising and sustainable option. The lignocellulosic materials are mainly composed of three types of polymers; lignin, cellulose and hemicellulose. The biorefinery process required for production of biofuel based on lignocellulosic materials involves four stages; pre-treatment, hydrolysis, fermentation and separation. The pre-treatment and hydrolysis of lignocellulosic material can be carried out using fungal enzymes in more environmentally friendly manner. But the current technologies pertaining to biofuel production are costly and not very efficient. The use of fungal enzymes such as lignin-modifying enzymes, hemicelluloses and cellulases have the potential for efficient hydrolysis. Many fungal groups such as white-rot basidiomycetes, brown-rot basidiomycetes and a number of ascomycetes members have the capability of degrading lignocellulosic biomass efficiently. The fungal enzymes are key to efficient biorefinery systems with higher yields and cost-effective production of biofuels. There are three important fungal enzyme systems for efficient utilization of lignocellulosic biomass for biofuel production. These include: lignin-modifying enzymes (LMEs) that disrupt the recalcitrant lignin components, whereas cellulase systems work on the cellulosic microfibrils to convert polymers into single glucose units and hemicellulases breaks down different side chains and backbone polymers of hemicelluloses.

CONSENT FOR PUBLICATION

Not applicable.

CONFLICT OF INTEREST

The authors declare no conflict of interest, financial or otherwise.

ACKNOWLEDGEMENT

Declared none.

REFERENCES

[1]　Guo M, Song W, Buhain J. Bioenergy and biofuels: history, status, and perspective. Renew Sustain Energy Rev 2015; 42: 712-25.
[http://dx.doi.org/10.1016/j.rser.2014.10.013]

[2]　Hosseinzadeh-Bandbafha H, Tabatabaei M, Aghbashlo M, *et al.* A comprehensive review on the environmental impacts of diesel/biodiesel additives. Energy Convers Manage 2018; 174: 579-614.
[http://dx.doi.org/10.1016/j.enconman.2018.08.050]

[3]　Strobel GA. Bioprospecting--fuels from fungi. Biotechnol Lett 2015; 37(5): 973-82.
[http://dx.doi.org/10.1007/s10529-015-1773-9] [PMID: 25650344]

[4]　Jouzani GS, Taherzadeh MJ. Advances in consolidated bioprocessing systems for bioethanol and butanol production from biomass: a comprehensive review. Biofuel Res J 2015; 2: 152-95.
[http://dx.doi.org/10.18331/BRJ2015.2.1.4]

[5]　Jouzani GS, Aghbashlo M, Tabatabaei M. Biofuels: Types, Promises, Challenges, and Role of Fungi.Fungi in Fuel Biotechnology. 1st ed. Springer Nature Switzerland AG 2020; pp. 1-14.
[http://dx.doi.org/10.1007/978-3-030-44488-4_1]

[6]　Raven S, Francis A, Srivastava C, *et al.* Fungal biofuels: innovative approaches.Recent Advancement in White Biotechnology Through Fungi. Dordrecht: Springer 2019; pp. 385-405.
[http://dx.doi.org/10.1007/978-3-030-14846-1_13]

[7]　Zabed H, Sahu JN, Suely A, *et al.* Bioethanol production from renewable sources: current perspectives and technological progress. Renew Sustain Energy Rev 2017; 71: 475-501.
[http://dx.doi.org/10.1016/j.rser.2016.12.076]

[8]　Manochio C, Andrade BR, Rodriguez RP, Moraes BS. Ethanol from biomass: a comparative overview. Renew Sustain Energy Rev 2017; 80: 743-55.
[http://dx.doi.org/10.1016/j.rser.2017.05.063]

[9]　Dutta K, Daverey A, Lin J-G. Evolution retrospective for alternative fuels: first to fourth generation. Renew Energy 2014; 69: 114-22.
[http://dx.doi.org/10.1016/j.renene.2014.02.044]

[10]　Aghbashlo M, Hosseinpour S, Tabatabaei M, Dadak A. Fuzzy modeling and optimization of the synthesis of biodiesel from waste cooking oil (WCO) by a low power, high frequency piezo-ultrasonic reactor. Energy 2017; 132: 65-78.
[http://dx.doi.org/10.1016/j.energy.2017.05.041]

[11]　Tabatabaei M, Aghbashlo M, Dehhaghi M, *et al.* Reactor technologies for biodiesel production and processing: a review. Pror Energy Combust Sci 2019; 74: 239-303.
[http://dx.doi.org/10.1016/j.pecs.2019.06.001]

[12]　Taher H, Al-Zuhair S. The use of alternative solvents in enzymatic biodiesel production: a review. Biofuels Bioprod Biorefin 2017; 11: 168-94.
[http://dx.doi.org/10.1002/bbb.1727]

[13]　Hegde K, Chandra N, Sarma SJ, Brar SK, Veeranki VD. Genetic engineering strategies for enhanced biodiesel production. Mol Biotechnol 2015; 57(7): 606-24.

[http://dx.doi.org/10.1007/s12033-015-9869-y] [PMID: 25902752]

[14] Vamvuka D. Bio-oil, solid and gaseous biofuels from biomass pyrolysis processes—an overview. Int J Energy Res 2011; 35: 835-62.
[http://dx.doi.org/10.1002/er.1804]

[15] Kabir G, Hameed BH. Recent progress on catalytic pyrolysis of lignocellulosic biomass to high-grade bio-oil and bio-chemicals. Renew Sustain Energy Rev 2017; 70: 945-67.
[http://dx.doi.org/10.1016/j.rser.2016.12.001]

[16] Carpenter D, Westover TL, Czernik S, Jablonski W. Biomass feedstocks for renewable fuel production: a review of the impacts of feedstock and pretreatment on the yield and product distribution of fast pyrolysis bio-oils and vapors. Green Chem 2014; 16: 384-406.
[http://dx.doi.org/10.1039/C3GC41631C]

[17] Laird DA, Brown RC, Amonette JE, Lehmann J. Review of the pyrolysis platform for coproducing bio-oil and biochar. Biofuels Bioprod Biorefin 2009; 3: 547-62.
[http://dx.doi.org/10.1002/bbb.169]

[18] Karatzos S, McMillan JD, Saddler JN. The potential and challenges of drop-in biofuels. 2014.

[19] Aghbashlo M, Tabatabaei M, Soltanian S, Ghanavati H. Biopower and biofertilizer production from organic municipal solid waste: an exergoenvironmental analysis. Renew Energy 2019; 143: 64-76.
[http://dx.doi.org/10.1016/j.renene.2019.04.109]

[20] Coyne JM, Gupta VK, O'Donovan A, Tuohy MG. The role of fungal enzymes in global biofuel production technologies. 2013.
[http://dx.doi.org/10.1007/978-3-642-34519-7_5]

[21] Aghbashlo M, Tabatabaei M, Nadian MH, et al. Prognostication of lignocellulosic biomass pyrolysis behavior using ANFIS model tuned by PSO algorithm. Fuel 2019; 253: 189-98.
[http://dx.doi.org/10.1016/j.fuel.2019.04.169]

[22] Dashtban M, Schraft H, Qin W. Fungal bioconversion of lignocellulosic residues; opportunities & perspectives. Int J Biol Sci 2009; 5(6): 578-95.
[http://dx.doi.org/10.7150/ijbs.5.578] [PMID: 19774110]

[23] Martínez AT, Speranza M, Ruiz-Dueñas FJ, et al. Biodegradation of lignocellulosics: microbial, chemical, and enzymatic aspects of the fungal attack of lignin. Int Microbiol 2005; 8(3): 195-204.
[PMID: 16200498]

[24] Turner P, Mamo G, Karlsson EN. Potential and utilization of thermophiles and thermostable enzymes in biorefining. Microb Cell Fact 2007; 6: 9.
[http://dx.doi.org/10.1186/1475-2859-6-9] [PMID: 17359551]

[25] Howard R, Abotsi E, Jansen van Rensburg E, Howard S. Lignocellulose biotechnology: issues of bioconversion and enzyme production. Afr J Biotechnol 2003; 2: 602-19.
[http://dx.doi.org/10.5897/AJB2003.000-1115]

[26] Chandel AK, Chandrasekhar G, Radhika K, Ravinder R, Ravindra P. Bioconversion of pentose sugars into ethanol: a review and future directions Biotechnol Mol Biol Rev 2011; 6: 008-20.

[27] Saha BC. Hemicellulose bioconversion. J Ind Microbiol Biotechnol 2003; 30(5): 279-91.
[http://dx.doi.org/10.1007/s10295-003-0049-x] [PMID: 12698321]

[28] Ong LK. Conversion of lignocellulosic biomass to fuel ethanol: a brief review. Planter 2004; 80: 517-24.

[29] Isroi MR, Syamsiah S, Niklasson C, Nur Cahyanto M, Lundquist K, Taherzadeh M. Biological pre-treatment's of lignocelluloses with white-rot fungi and its applications: a review. BioResources 2011; 6: 5224-59.
[http://dx.doi.org/10.15376/biores.6.4.Isroi]

[30] Harris D, DeBolt S. Synthesis, regulation and utilization of lignocellulosic biomass. Plant Biotechnol J

2010; 8(3): 244-62.
[http://dx.doi.org/10.1111/j.1467-7652.2009.00481.x] [PMID: 20070874]

[31]   Hernon AT, O'Donovan A, Shier MC. Applications of fungal enzymes in bioconversion.Lambert Academic Publishing. Germany 2010.

[32]   Chandel AK, Gonçalves BCM, Strap JL, da Silva SS. Biodelignification of lignocellulose substrates: An intrinsic and sustainable pretreatment strategy for clean energy production. Crit Rev Biotechnol 2015; 35(3): 281-93.
[http://dx.doi.org/10.3109/07388551.2013.841638] [PMID: 24156399]

[33]   Dashtban M, Qin W. Overexpression of an exotic thermotolerant β-glucosidase in trichoderma reesei and its significant increase in cellulolytic activity and saccharification of barley straw. Microb Cell Fact 2012; 11: 63.
[http://dx.doi.org/10.1186/1475-2859-11-63] [PMID: 22607229]

[34]   Sanchez S, Demain AL. Bioactive products from fungi. Food bioactives. Cham: Springer 2017; pp. 59-87.
[http://dx.doi.org/10.1007/978-3-319-51639-4_3]

[35]   Bakir U, Phillips SEV, McPherson MJ, Ogel ZB. Purification, characterization, and identification of a novel bifunctional catalase-phenol oxidase from scytalidiumthermophilum. Appl Microbiol Biotechnol 2008; 79: 407-15.
[http://dx.doi.org/10.1007/s00253-008-1437-y] [PMID: 18369615]

[36]   Lundell TK, Mäkelä MR, Hildén K. Lignin-modifying enzymes in filamentous basidiomycetes--ecological, functional and phylogenetic review. J Basic Microbiol 2010; 50(1): 5-20.
[http://dx.doi.org/10.1002/jobm.200900338] [PMID: 20175122]

[37]   Widiastuti H. Activity of ligninolytic enzymes during growth and fruiting body development of white rot fungi *Omphalinasp* and *Pleurotusostreatus*. Hayati J Biosci 2008; 15: 140-4.
[http://dx.doi.org/10.4308/hjb.15.4.140]

[38]   Bonnen AM, Anton LH, Orth AB. Lignin-degrading enzymes of the commercial button mushroom, agaricusbisporus. Appl Environ Microbiol 1994; 60(3): 960-5.
[http://dx.doi.org/10.1128/aem.60.3.960-965.1994] [PMID: 16349223]

[39]   Baldrian P. Fungal laccases - occurrence and properties. FEMS Microbiol Rev 2006; 30(2): 215-42.
[http://dx.doi.org/10.1111/j.1574-4976.2005.00010.x] [PMID: 16472305]

[40]   D'Souza TM, Merritt CS, Reddy CA. Lignin-modifying enzymes of the white rot basidiomycete Ganoderma lucidum. Appl Environ Microbiol 1999; 65(12): 5307-13.
[http://dx.doi.org/10.1128/AEM.65.12.5307-5313.1999] [PMID: 10583981]

[41]   Vanholme R, Demedts B, Morreel K, Ralph J, Boerjan W. Lignin biosynthesis and structure. Plant Physiol 2010; 153(3): 895-905.
[http://dx.doi.org/10.1104/pp.110.155119] [PMID: 20472751]

[42]   Campbell MM, Sederoff RR. Variation in lignin content and composition (mechanisms of control and implications for the genetic improvement of plants). Plant Physiol 1996; 110(1): 3-13.
[http://dx.doi.org/10.1104/pp.110.1.3] [PMID: 12226169]

[43]   Cañas AI, Camarero S. Laccases and their natural mediators: biotechnological tools for sustainable eco-friendly processes. Biotechnol Adv 2010; 28(6): 694-705.
[http://dx.doi.org/10.1016/j.biotechadv.2010.05.002] [PMID: 20471466]

[44]   Blanchette RA. Delignification by wood-decay fungi. Annu Rev Phytopathol 1991; 29: 381-98.
[http://dx.doi.org/10.1146/annurev.py.29.090191.002121]

[45]   Brambl R. Fungal physiology and the origins of molecular biology. Microbiology 2009; 155(Pt 12): 3799-809.
[http://dx.doi.org/10.1099/mic.0.035238-0] [PMID: 19850621]

[46]   Chen M, Zeng G, Tan Z, *et al.* Understanding lignin-degrading reactions of ligninolytic enzymes: binding affinity and interactional profile. PLoS One 2011; 6(9): e25647-7.
[http://dx.doi.org/10.1371/journal.pone.0025647] [PMID: 21980516]

[47]   Dashtban M, Schraft H, Syed TA, Qin W. Fungal biodegradation and enzymatic modification of lignin. Int J Biochem Mol Biol 2010; 1(1): 36-50.
[PMID: 21968746]

[48]   Piontek K, Antorini M, Choinowski T. Crystal structure of a laccase from the fungus *Trametes versicolor* at 1.90-A resolution containing a full complement of coppers. J Biol Chem 2002; 277(40): 37663-9.
[http://dx.doi.org/10.1074/jbc.M204571200] [PMID: 12163489]

[49]   Kuwahara M, Glenn JK, Morgan MA, Gold MH. Separation and characterization of two extracellular H2O2-dependent oxidases from ligninolytic cultures of *Phanerochaetechrysosporium.* FEBS Lett 1984; 169: 247-50.
[http://dx.doi.org/10.1016/0014-5793(84)80327-0]

[50]   Mtui GYS. Recent advances in pre-treatment of lignocellulosic wastes and production of value added products. Afr J Biotechnol 2009; 8: 1398-415.

[51]   Grassick A, Murray PG, Thompson R, *et al.* Three-dimensional structure of a thermostable native cellobiohydrolase, CBH IB, and molecular characterization of the cel7 gene from the filamentous fungus, *Talaromyces emersonii.* Eur J Biochem 2004; 271(22): 4495-506.
[http://dx.doi.org/10.1111/j.1432-1033.2004.04409.x] [PMID: 15560790]

[52]   Agbor VB, Cicek N, Sparling R, Berlin A, Levin DB. Biomass pretreatment: fundamentals toward application. Biotechnol Adv 2011; 29(6): 675-85.
[http://dx.doi.org/10.1016/j.biotechadv.2011.05.005] [PMID: 21624451]

[53]   Alvira P, Tomás-Pejó E, Ballesteros M, Negro MJ. Pretreatment technologies for an efficient bioethanol production process based on enzymatic hydrolysis: A review. Bioresour Technol 2010; 101(13): 4851-61.
[http://dx.doi.org/10.1016/j.biortech.2009.11.093] [PMID: 20042329]

[54]   Thygesen LG, Hidayat BJ, Johansen KS, Felby C. Role of supramolecular cellulose structures in enzymatic hydrolysis of plant cell walls. J Ind Microbiol Biotechnol 2011; 38(8): 975-83.
[http://dx.doi.org/10.1007/s10295-010-0870-y] [PMID: 20852928]

[55]   Arantes V, Saddler JN. Access to cellulose limits the efficiency of enzymatic hydrolysis: the role of amorphogenesis. Biotechnol Biofuels 2010; 3: 4.
[http://dx.doi.org/10.1186/1754-6834-3-4] [PMID: 20178562]

[56]   Brás JL, Cartmell A, Carvalho AL, *et al.* Structural insights into a unique cellulase fold and mechanism of cellulose hydrolysis. Proc Natl Acad Sci USA 2011; 108(13): 5237-42.
[http://dx.doi.org/10.1073/pnas.1015006108] [PMID: 21393568]

[57]   Chandel AK, Chandrasekhar G, Silva MB, Silvério da Silva S. The realm of cellulases in biorefinery development. Crit Rev Biotechnol 2012; 32(3): 187-202.
[http://dx.doi.org/10.3109/07388551.2011.595385] [PMID: 21929293]

[58]   Zhang Y, Shen J. Enhancement effect of gold nanoparticles on biohydrogen production from artificial waste water Int J Hydrog energy 2007; 32: 17-23.

[59]   Zhao W, Zhang Y, Du B, Wei D, Wei Q, Zhao Y. Enhancement effect of silver nanoparticles on fermentative biohydrogen production using mixed bacteria. Bioresour Technol 2013; 142: 240-5.
[http://dx.doi.org/10.1016/j.biortech.2013.05.042] [PMID: 23743428]

[60]   Abbas M, *et al.* Highly stable silica encapsulating magnetic nanoparticles (Fe_3O_4/SiO_2) synthesized using single surfactanting-polyol process. Ceram Int 2014; 40: 1379-85.
[http://dx.doi.org/10.1016/j.ceramint.2013.07.019]

[61]   Yang G, Wang J. Improving mechanisms of biohydrogen production from grass using zero-valent iron nanoparticles. Bioresour Technol 2018; 266: 413-20.
[http://dx.doi.org/10.1016/j.biortech.2018.07.004] [PMID: 29982065]

[62]   Singh V, Yadav VK, Mishra V. Nanotechnology: An application in biofuel production.Nanomaterials and Biofuels Research. Clean Energ Prod Technol 2020; pp. 143-60.
[http://dx.doi.org/10.1007/978-981-13-9333-4_6]

[63]   Romero-Guiza M, *et al.* The role of additives on anaerobic digestion: A review Renew Sust energy Rev 2016; 58: 1486-1499.
[http://dx.doi.org/10.1016/j.rser.2015.12.094]

[64]   Seffati K, *et al.* Enhanced biodiesel production from chicken fat using CaO/CuFe2O4 nanocatalyst and its combination with diesel to improve fuel properties. Fuel 2019; 235: 1238-44.
[http://dx.doi.org/10.1016/j.fuel.2018.08.118]

[65]   Zhang XL, *et al.* Biodiesel production from heterotrophic microalgae through transesterification and nanotechnology application in the production. Renew Sustain Energy Rev 2013; 26: 216-23.
[http://dx.doi.org/10.1016/j.rser.2013.05.061]

[66]   Verma ML, Barrow CJ, Puri M. Nanobiotechnology as a novel paradigm for enzyme immobilisation and stabilisation with potential applications in biodiesel production. Appl Microbiol Biotechnol 2013; 97(1): 23-39.
[http://dx.doi.org/10.1007/s00253-012-4535-9] [PMID: 23132346]

[67]   Cherian E, Dharmendirakumar M, Bhaskar G. Immobilization of cellulose onto MnO_2 nanoparticles for bioethanol production by enhanced hydrolysis of agricultural waste. Chin J Catal 2015; 36: 1223-9.
[http://dx.doi.org/10.1016/S1872-2067(15)60906-8]

SUBJECT INDEX

A

Absorption 69, 270
  band, resonance 69
  spectra, broad 270
Acenetobacter baumanii 8
Acetylxylan esterase 365
Acid(s) 15, 16, 71, 72, 80, 83, 86, 87, 145,
    162, 163, 171, 174, 190, 191, 195, 294,
    298, 269, 303, 305, 311, 312, 335, 337,
    358, 359, 362
  Acetic 72
  Ascorbic 72, 80, 83, 86, 335, 337
  citric 269
  deoxyribonucleic 162
  gallic 71
  humic 171
  jasmonic 195
  nucleic 15, 16, 145, 294, 298, 303, 305, 311,
    312
  onic 87
  organic 190, 191
  salicylic 163
  Selenious 72
  syringic 362
Activity 16, 31, 33, 47, 49, 50, 51, 52, 73, 74,
    75, 76, 77, 80, 81, 82, 83, 84, 85, 86, 89,
    90, 117, 118, 158, 163, 164, 174, 208,
    209, 215, 232, 252, 334, 361, 363, 365
  agricultural 83
  antagonistic 52
  anthropogenic 16, 158, 163, 174
  anti-bacterial 215
  antibiotic 16
  anti-leukaemia 89
  antiproliferative 51
  bactericidal 118
  broad-spectrum 209, 232
  cellulase 361
  enzyme 33, 164, 365
  enzyme's biocatalytic 365
  free radical scavenging 86

  fungicidal 334
  metabolic 47
  mycoparasitic 49
  mycotoxic 50
  natural microbial 31
  nephroprotective 89
  oxidoreductase 84
  peptidase 363
  photocatalytic 90
  photosynthetic 164
  phytopathogenic 51
  prophylactic 89
  tyrokinase inhibitory 252
Aedes aegypti 12, 13, 140
Aeromonas hydrophilla 118
Agaricus bisporus 16, 167, 168
Agents 2, 32, 37, 38, 52, 65, 85, 88, 106, 107,
    131, 145, 187, 192, 250, 254, 247
  anti-infective 65
  antiproliferative 254
  cancer prevention 250
  chemotherapeutic 88, 145
  microbial 37, 38, 52
  reducing agent and stabilizing 106, 131
Ag nanoparticles 45
Agricultural 160, 165, 298, 353
  crops 353
  production systems 160
  productivity 298
  sustainable development 165
Agrochemicals wastes 170
Agrostis tenuifolia 167
Air 119, 163
  disinfection 119
  pollutants 163
Alcohol, vanillyl 362
Allium sativum 80
Aloe vera leaf extract 87
Alpha NADPH-dependent nitrate reductase 40
Alternaria alternata 12, 47, 49, 51, 76, 113,
    118, 134, 139, 141, 146, 147, 195
Amanita submembranacea 174

R

Radiolysis 66, 105
Reaction 44, 67, 71, 92, 106, 119, 134, 145,
    149, 164, 239, 261, 267, 293, 295, 337,
    359, 365
  antibody-antigen 267
  electron transfer chain 164
  immunological 239
  polymerase chain 261, 293, 295
  sonochemical 337
  transesterification 365
Reactive oxygen species (ROS) 14, 16, 140,
    141, 162, 163, 209, 210, 217, 220, 225,
    227, 231
Real-time diagnostic techniques 271
Receptors145, 299
  biological recognition 299
  produced DNA probe 145
Reductases 10, 30, 40, 44 77, 78
  bacterial cell membrane 77, 78
  extracellular enzyme-nitrate 44
  nitrate-dependent 44
  suggested NADPH-dependent 10
Reduction 9, 37, 40, 44, 66, 68, 103, 105, 136,
    141, 142, 143, 144, 148, 335
  enzymatic 9
  properties 66
  reactions 335
Reduction process 129, 141, 142
  three-electron 142
Respiratory chain, mitochondrial 16
Reverse osmosis (RO) 32
Rhizoctonia solani 47, 48, 49, 51, 118, 140,
    195, 197, 212, 219, 222
*Rhizopus 12, 43, 45, 50, 135, 144, 169, 173,
    174, 274*
  oryzae 12, 144, 169
  stolonifer 43, 45, 50, 135, 173, 174, 274
Rhododendron ponticum 297
RNA 14, 141, 193, 266, 300, 305, 308, 309
  damage 14, 141
  double-stranded 193
  viruses 309

RNAi molecules 198
Rust disease 302

S

Sabourand dextrose agar (SDA) 214, 224, 226
Saccharomyces 77, 85, 86, 87, 88, 89,90, 169,
    355, 359
  boulardii 89
  cerevisiae 77, 85, 86, 87, 88, 90, 169, 355,
    359
Salmonella typhi 51
Salt stress 82
Scanning electron microscopy 48, 67, 114,
    146, 148, 196
Schizosaccharomyces pombe 8
SDS-PAGE analysis 9
Selenite 64, 65, 72, 75, 78
  detoxification 78
  pentahydrate 72
  reduction of 72, 78
Selenium 62, 64, 65, 69, 72, 74, 84, 89, 91,
    159, 190
  chloride 74
  dioxide 72
  metal 83
  monoclinic 69
Selenium nanoparticles
  biogenic 89, 91
  green synthesised 89
SELEX process 306
Sensitivity 264, 265, 266, 269, 271, 272, 276,
    277, 301, 302, 306, 307, 309
  analytical 269
  hypomolecular 276
  transverse 276
Sensors 33, 254, 255, 265, 266, 270, 271, 272,
    273, 274, 275, 277, 302, 303, 305, 306
  antibody-biotin-streptavidin-HRP 272
  nanotechnology-enabled detection 277
Silver 9, 12, 14, 15, 29, 37, 47,104, 109, 110,
    130, 131, 136, 149, 190, 208, 209, 247,
    341
  alloy 190